3rd International Conference on Precision Machinery and Manufacturing Technology (ICPMMT 2018)

MATEC Web of Conferences Volume 185 (2018)

Auckland, New Zealand
4 - 8 February 2018

Editor:

Gow-Yi Tzou

ISBN: 978-1-5108-6859-5

Printed from e-media with permission by:

Curran Associates, Inc.
57 Morehouse Lane
Red Hook, NY 12571

CURRAN ASSOCIATES INC. proceedings.com

Some format issues inherent in the e-media version may also appear in this print version.

Printed by Curran Associates, Inc. (2018)

For additional information, please contact EDP Sciences – Web of Conferences
at the address below.

EDP Sciences – Web of Conferences
17, Avenue du Hoggar
Parc d'Activité de Courtabœuf
BP 112
F-91944 Les Ulis Cedex A
France

Phone: +33 (0) 1 69 18 75 75
Fax: +33 (0) 1 69 28 84 91

contact-edps@webofconferences.org

Additional copies of this publication are available from:

Curran Associates, Inc.
57 Morehouse Lane
Red Hook, NY 12571 USA
Phone: 845-758-0400
Fax: 845-758-2633
Email: curran@proceedings.com
Web: www.proceedings.com

TABLE OF CONTENTS

Preface

The 1st International Conference on Precision Machinery and Manufacturing Technology 2015 (ICPMMT 2015) has been held at Howard Plaza Resort, Kenting, Taiwan on May 22-24, 2015. The papers submitted to ICPMMT 2015 were recommended to KEM (EI Compendex) through the peer reviewing process.

The 2nd ICPMMT 2017 has been taken place at Howard Plaza Resort, Kenting, Taiwan on May 19-21, 2017. After conference, all papers submitted to ICPMMT 2017 were be recommended to MATEC Web of Conferences, Open-access Journal in Materials science, Engineering and Chemistry (Google Scholar, CAS, EI Compendex, Inspec, DOAJ, Web of Science, Scopus). All papers through the peer reviewing process after the conference were published in MATEC Web of Conferences.

First time the 3rd ICPMMT 2018 has been held at Crowne Plaza in Auckland, New Zealand on 4-8 February, 2018. The purpose of ICPMMT 2018 was to provide opportunities for institutions and industries to share ideas and present the works of scientists, engineers, educators and students from all over the world. The organizing committee of the conference was delighted to invite prospective authors to submit their original manuscripts to ICPMMT 2018. The participants came from Taiwan, Australia, New Zealand, China, Slovenia, Russia, India, Malaysia, Korea etc.; initially over 120 paper abstracts were submitted to ICPMMT 2018, finally 80 full papers have been accepted in ICPMMT 2018. All papers were reviewed by at least two reviewers. The papers accepted for publication have been considered to be sufficiently original and to contain new developed contributions or technologies deserved to be published in MATEC Web of Conferences with EI index. After the peer reviewing process, just only 40 papers could be published.

The 4th ICPMMT 2019 will be back to at Howard Plaza Resort, Kenting, Pingtung, Taiwan on 24-26 May, 2019. After conference, all papers submitted to ICPMMT 2019 through the peer reviewing process will be also recommended to publish in the publications with EI Compendex. Please refer to the website **www.icpmmt.org**. **Welcome all of you to join ICPMMT 2019.**

Chair Prof. Gow-Yi Tzou

Dean of College Engineering, Chung-Chou University, Taiwan, ROC

Honor President of Fasteners Industry Technology Development Association, Taiwan, ROC

July, 2018

Committees

Conference Chair:

Chair Prof. Gow-Yi Tzou (Chung-Chou University of Science and Technology, Taiwan

Steering Committee:

Prof. Gow-Yi Tzou (Chung-Chou University of Science and Technology, Taiwan)

Prof. Thomas Neitzert (Auckland University of Technology, New Zealand)

Prof. Jeong-Whan Yoon (Deakin University, Australia; KAIST Korea)

Scientific Committee:

Gow-Yi Tzou (Chung-Chou University of Science and Technology, Taiwan)

Ho-Chi Cheng (Chung-Chou University of Science and Technology, Taiwan)

Thomas Neitzert (Auckland University of Technology, New Zealand)

Timotius Pasang (Auckland University of Technology, New Zealand)

Guy Littlefair (Auckland University of Technology, New Zealand)

Maziar Ramezani (Auckland University of Technology, New Zealand)

Richard Lin (The University of Auckland, New Zealand)

Jeong-Whan Yoon (Deakin University, Australia; KAIST Korea)

Cheng Lu (Wollongong University, Australia)

Zengyi Jiang (Wollongong University, Australia)

Jing-Jing Li (Penn State University, USA)

Scott Miller (Hawaii University, USA)

Peter Groche (TU Darmstadt, Germany)

Fuh-Kuo Chen (Taiwan University, Taiwan)

Jen-Yuan Chang (Tsing Hua University, Taiwan)

De-Shin Liu (Chung-Cheng University, Taiwan)

Jong-Ning Aoh (Chung-Cheng University, Taiwan)

Yung-Chou Kao (Chung-Cheng University, Taiwan)

You-Min Huang (Taiwan University of Science & Technology, Taiwan)

Yeong-Maw Hwang (Sun Yat-Sen University, Taiwan)

Cheng-Tang Pan (Sun Yat-sen University, Taiwan)

Jinn-Jong Sheu (Kaohsiung University of Applies Science, Taiwan)

Chao-Cheng Chang (Kaohsiung University of Applies Science, Taiwan)

Cho-Pei Jiang (Formosa University, Taiwan)

Dyi-Cheng Chen (Changhua University of Education, Taiwan)

Yeong-Lin Lai (Changhua University of Education, Taiwan)

Shao-Yi Hsia (Kao-Yuan University of Science and Technology, Taiwan, Taiwan)

Chih-Liang Chu (Southern Taiwan University of Science and Technology, Taiwan)
Cheng-Chien Wang (Southern Taiwan University of Science and Technology, Taiwan)
Yen-Ming Tseng (Fuzhou Polytechnic, China)

edp sciences

Web of Conferences

Statement of Peer review

In submitting conference proceedings to *Web of Conferences*, the editors of the proceedings certify to the Publisher that

1. They adhere to its **Policy on Publishing Integrity** in order to safeguard good scientific practice in publishing.
2. All articles have been subjected to peer review administered by the proceedings editors.
3. Reviews have been conducted by expert referees, who have been requested to provide unbiased and constructive comments aimed, whenever possible, at improving the work.
4. Proceedings editors have taken all reasonable steps to ensure the quality of the materials they publish and their decision to accept or reject a paper for publication has been based only on the merits of the work and the relevance to the journal.

2018 The 3rd International Conference on Precision Machinery and Manufacturing Technology, 4 Feb-8 Feb 2018, Auckland, New Zealand

Proceedings editor(s): Gow-Yi Tzou

Date and editor's signature 16.7.2018

17, avenue du Hoggar - PA de Courtabœuf – BP 112 - 91944 Les Ulis Cedex A (France)
Tél. : 33 (0)1 69 18 75 75 - Fax : 33(0)1 69 07 45 17 – www.edpsciences.org

MATEC Web of Conferences **185**, 00001 (2018) https://doi.org/10.1051/matecconf/201818500001
ICPMMT 2018

Design of intelligent surfaces for energy intensive processing industry

Jeremy R. Smith[1,2], *Jérôme* Leveneur[1], and *John V.* Kennedy[1,*]

[1]National Isotope Centre, GNS Science, P O Box 31312, Lower Hutt, 5010, New Zealand
[2]Massey Institute of Food Science & Technology, Massey University, Palmerston North, New Zealand

Abstract. There are three different factors that can affect adhesion: the process fluid, the processing conditions and the surface of the processing equipment. Of these three factors, the surface properties of the processing equipment are the factor that offers the greatest opportunity for manipulation. The two key surface properties that have been identified to reduce adhesion are the surface energy and the surface topography. The surface energy of a material determines its degree of wettability and, a surface's affinity for water. In previous studies the surface energy of materials have been leveraged in order to create a surface with reduced levels of fouling through surface modification or the addition of polymer coatings with varying degrees of hydrophobicity. In addition, the topography of surfaces has been modified to reduce the level of particle adhesion. These modifications involve creating either a structured or random porous microstructure on the surface. Additional methods identified to reduce fouling include the application of liquid infused porous surfaces at low shear conditions and the use of non-contact heating through techniques such as microwave processing.

1 Introduction

Stainless steel is one of the most commonly used materials in processing industries due to its strength, corrosive resistance, cleanability and durability [1]. However, stainless steel is prone to fouling due to its relatively high surface energy and hydrophilic nature [2]. In order to combat the effects of fouling, a number of different techniques have been explored to reduce/eliminate adhesion of compounds to a surface. These technologies range from the application of a coating through physical or vapour deposition, doping or electroplating [3]. Two main approaches can be taken to have a positive effect on fouling and cleaning: modification of the energy specific surface properties and the modification of the topographical surface properties [4].
One possibility that presents itself if a durable, corrosive resistant surface modification where stainless steel may be replaced with a lower cost material such as carbon steel [5]. This would have the additional benefit of being able to utilise material with better heat transfer coefficients that may mitigate the effects of any surface coating. However, for a

* Corresponding author: J.Kennedy@gns.cri.nz

surface modification to be applied to an industrial setting, such as the food industry, it needs to be able to withstand high temperature for short time treatments as well as the intensity of clean in place (CIP) techniques [6]. There is also significant difficulty in comparing reports of different anti-fouling coatings/systems as the experimental conditions vary widely [7].

2 Surface structure

The roughness of a surface is generally accepted as one of the main factors affecting adhesion forces between particles and a surface due to the modification of the true area of contact [8]. Having a high level of surface roughness has been identified as enhancing mineral deposition, possibly due to the increase in the number of sites available for nucleation to occur [9]. This suggests that by increasing the smoothness of a surface the level of adhesion and subsequent fouling will be reduced. However, here are conflicting reports regarding whether roughness diminishes or enhances adhesion to a surface. This is likely due to the scale of the roughness, with larger scale roughness promoting adhesion whereas micro-scale roughness can act to reduce adhesion. This is demonstrated in a large number of publications that show that micro/nano scale roughness can result in reductions in adhesion compared to a smooth surface [10].

2.1 Structured topography

The majority of surface topographies are controlled at the macro-scale, but not often controlled at the micro and nano scale [11]. As food components exist over a range of length scales, the structure of the solid surface affects the likelihood and extent of fouling. For example, bacteria range in the micrometre size and their surface appendages in the nanometre range, the control of surface topography at these scales could strongly affect bacterial attachment [11]. There are a number of different approaches to create a nano structured surface, from 0D nanoparticles, 1D wires and individual fibres, 2D thin films and 3D fibre assembly [12].

Engineered surface topographies can be achieved using a number of different methods including: nanoimprint lithography, photolithography, [11, 13, 14]. A number of these methods, although very effective at creating structured topographies, are not practical for large scale development of surfaces. Etching has be used to create superhydrophobic stainless steel surfaces [15]. This allows the creation of a surface with the same anti-corrosive properties as native stainless steel but with much higher water contact angles (in the 150° range). The key advantage of being able to create roughness using etching directly on a surface, opposed to adding roughness through deposition techniques, is the inherent stability of the structures formed [15]. As different compounds within a system can vary over a wide range of different sizes, surface topography that contains only one length scale may not be effective as an antifouling material against all of the components in the system [11]. By creating hierarchical topographies, by superimposing small anti-algal topography onto a larger anti-barnacle topography a 97% reduction in microorganism attachment was identified in a marine setting [16].

Dynamic interactions between proteins and nanoscale surface topographies are complex due to the combination of both attractive and repulsive forces which are a function of the changes in local surface properties [17]. In addition, the flow of fluid over a textured surface may enhance the removal of deposits in comparison with a smooth surface, enhancing the benefits of a structured surface [11]. Graham et al. [18] identified that patterned holes were more effective at reducing attachment of E.coli than linear features under both static and flow conditions. In addition, all structured topographies were found to reduce the attachment of E.coli in comparison with a smooth surface, with 1 μm patterned holes being found to be the most effective [18].

2.2 Biomimetic

Biomimetics is the inspiration of ideas or designs adapted from biology [19]. It involves the understanding of biological functions, structures, design of various materials and principles of objects found in nature [20]. The most discussed biological source for inspiration in the area of antifouling has been the lotus leaf [21, 22]. Other natural sources looked at in regards to antifouling include the pitcher plant, cicada wings, butterfly wings, snail shells, fish scales and shark skin [23].

The lotus leaf has attracted a lot of attention due to its strong water repelling behaviour. This is due to the leaf being a superhydrophobic surface, producing a water contact angle of above 150° (producing the water beading effect) [24]. The superhydrophobic nature of the leaf is due to its hierarchical surface roughness and the low surface energy epicuticula wax coating the leaf [25], as shown in Figure 1. When water droplets come into contact with the leaf they sit on the apex of the nanostructures as air fills the structure under the droplet [24]. They are considered to be 'self-cleaning' surfaces due to the low contact angle hysteresis of water droplets causing water to roll off the surface whilst taking other particles from the surface [26]. Due to this phenomenon, the term 'lotus effect' is used to refer to surfaces that display superhydrophobic properties. Inspired by the lotus effect, many researchers have sort to mimic the behaviour by creating artificial lotus-like surfaces [27]. However, as these surfaces rely on the trapped air to repel liquids, they tend to fail when the air escapes under conditions involving high temperatures, pressure, humidity and when exposed to low surface tension liquids [28].

Another natured inspired micro-topographical design is the Sharklet AFTM based on the skin of a shark [10]. A sharks skin is made up of tiny scales that are themselves composed of small 'riblets' which are aligned in the direction of fluid flow [11]. The use of the Sharklet design by Schumacher et al. [16] was found to reduce microorganism attachment by 97%, when their hierarchical topography was applied to a marine environment. In their work they also identified a correlation between increasing aspect ratio and a reduction in attachment of bacteria.

As the surface fouling microorganisms in the food and biomedical industry are considerably smaller in size to those in marine applications (typically 1-2 μm vs 5-10 μm) research has been conducted using the Sharklet design with these types of microorganisms [11]. The design was applied in the biomedical field to investigate its effect on the biofilm formation of Staphylococcus aureus [30]. Chung et al. [30] found that although the Sharklet design did disrupt the formation of a S.aureus biofilm in comparison with a smooth surface, there were still bacterial cells in the recesses between the design features. Additional work in the biomedical area to investigate the effect of the Sharklet design to reduce microbial contamination has been with E.coli [31]. In this work, Reddy et al. [31] tried a number of modified Sharklet designs to reduce E.coli adhesion and got an average reduction of 47% in colony forming units (CFUs). However, the question still remains as to whether the Sharklet design is scalable to produce significant antifouling efficiency against the smaller microorganisms present in both the food and biomedical industry [11]. It has been found that superhydrophobic surfaces with hierarchical structures combining the distinct & complementary roles of microstructure (pressure dissipation) and nanostructure (high capillary pressure) can further decrease contact angle hysteresis and increase pressure tolerance [28]. However, hierarchical topographical features can be difficult to generate, especially at a scale that can be applied to commercial equipment.

3 Surface free energy

The surface free energy of a solid surface is a measure of its intermolecular or interfacial attractive forces [32]. It is the available energy resulting from surface groups, molecules or atoms that are able to interact with other groups, molecules or atoms approaching the surface [33]. The work of adhesion (Wsl) is the work required to separate the liquid from the solid and is equal to the sum of the surface free energy of the solid (ɣs) and the surface tension (ɣl) minus the interfacial tension between the solid and the liquid (ɣsl) [33]:

$$W_{sl} = \gamma_s + \gamma_l - \gamma_{sl} \tag{1}$$

Thus, with lower surface free energy of the solid (ɣs), the weaker the adhesion of a particle to the surface [34]. Numerous studies have been conducted investigating the effect of surface energy and adhesion, in particular bacterial adhesion [35]. These studies identified that low energy surfaces have less bacterial adhesion and are easier to clean due to weaker binding at the interface [36]. An optimal range of surface free energy was identified by Baier [37] where adsorption was at a minimum between 22-28 mN/m [38], as shown in Figure 2.

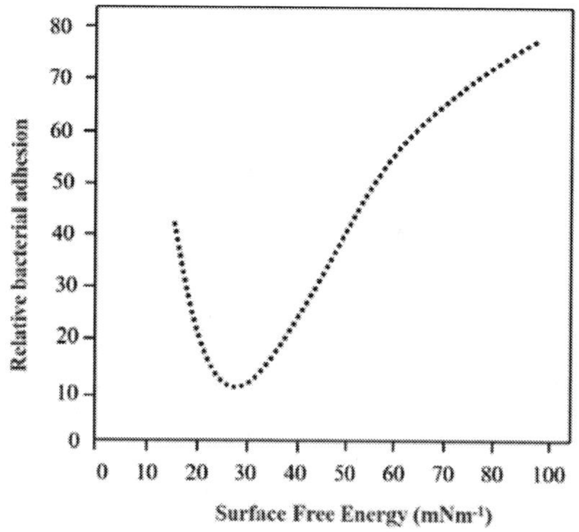

Fig. 1. Effect of surface energy on bacterial attachment.

In this range the adhesive force between the deposits and the surface are at a minimum resulting in either reduced fouling or increased ease of cleaning. This has additionally been demonstrated with mineral deposits [32], backed on tomato deposits [39], cell attachment to a surface [18]. The surface free energy of a material is inversely proportional to its water contact angle [18]. Thus the contact angle can be used as a means to measure the surface free energy of a material [33]. This can be calculated using the equation below [40]:

$$\gamma_s = \gamma_{sl} + \gamma_l \cos\theta \tag{2}$$

Low energy surfaces tend to be hydrophobic, as surface free energy can be used as measure to the degree to which water can be adsorbed to a surface [33]. Metals have high surface energies while polymers tend to have a low surface energy [41]. This is why particles tend to adhere to metallic surfaces more than polymer based surfaces. High surface energy materials are more prone to nucleation and thus have more nucleation sites than lower surface energy materials [42]. It is these nucleation sites that are the starting point for mineral adhesion. Changing the surface energy modifies the adhesive forces between deposits and a surface; not the cohesive forces within the deposits [38]. This means that it will have an effect on the interaction between the surface and a potential fouling component, but not on the interaction between the components. As a result, changes to the surface energy can alter the initial onset of adhesion, however, following the initial formation of a fouling layer the subsequent addition to the layer will progress at the same rate as if the fouling rate was on an untreated surface. However, there have also been a number of reports finding both no correlation between cell attachment and surface free energy or finding adhesion is reduced at higher free energies [43]. Therefore to identify the underlying component related to adhesion of material to a surface, the surface energy of a solid was broken down into component parts [44]. This considers the surface energy of a solid is comprised of an apolar Lifshitz-van der Waal component (γ_2^{LW}) and a polar Lewis acid-base component (γ_2^{AB}) [43]. The acid-base component consists of an electron donor component (γ_2^-) and an electron acceptor component (γ_2^+):

$$\gamma_2^{AB} = 2\sqrt{\gamma_2^+ \gamma_2^-} \tag{3}$$

The electron acceptor component of the surface energy is close to 0 for most solid materials. This suggests that the lower the electron donor component of surface energy, the lower the levels of adhesion. Liu & Zhao [45] proposed a simple expression to determine the adhesion strength of bacteria that they coined the CQ ratio. The CQ ratio is the apolar Lifshitz-van der Waal component of the solid's surface energy divided by the electron donor component (γ_2^{LW}/γ_2^-) [45]. A strong correlation was found between the total interaction energy of various bacteria in a biofilm and the CQ ratio. In addition, a strong correlation was identified between the CQ ratio and the number of adhering bacterial cells as well as the removal of bacterial cells [45]. In an additional study published in the same year, Liu & Zhao [43] once again found strong correlations between bacterial adhesion and the CQ ratio, this time on a Ni-P-TiO_2-PTFE composite material. However, following the two studies conducted in 2011, no further published work related to the CQ value could be found beyond referencing the original experiments.

4 Wettability

The wettability of a solid surface is dependent on the interfacial tension at the phase boundary between the surface and what it is in contact with [46]. Wetting is determined by the properties of both the solid surface and the liquid in contact with it [47]. The resulting interaction between the surface and the liquid is determined by the balance of intermolecular forces. Wettability has been identified as a factor that influences fouling, including as a contributing factor in the adhesion of bacteria to a surface [18]. The wettability of a solid surface is related to the surface energy of the surface [41]. There are two possible wetting regimes or states that occur on rough surfaces: the homogeneous Wenzel state and the composite Cassie-Baxter state [20]. The Wenzel model states that a droplet maintains contact with the surface and penetrates the asperities, and the surface contact area is increased [48]. In the Cassie-Baxter model, a droplet is suspended on the asperities thus residing on a composite phase made of both a solid-liquid and liquid-vapour interface [49]. An image depicting each of the models is displayed in Figure 3.

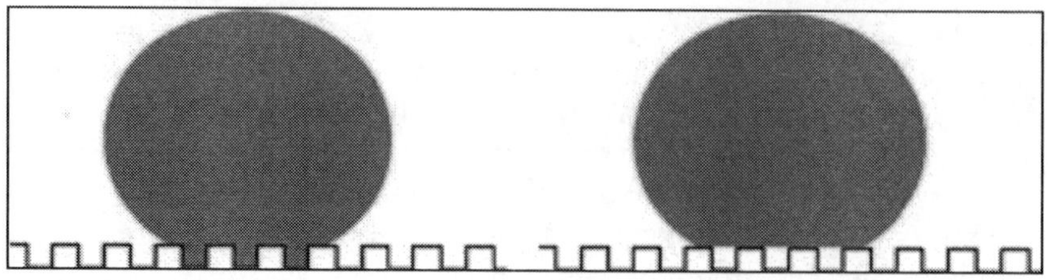

Fig. 2. Wenzel model (left) and Cassie-Baxter model (right) [48].

For complete wetting, the Wenzel model takes into account that the droplet follows the roughened surface and therefore has an equation for the resulting contact angle that takes into account a roughness factor [48]. This finds the contact angle in the Wenzel state to be equal to the contact angle on a smooth surface multiplied by the roughness factor [20]:

$$\cos\theta = r\cos\theta_0 \tag{4}$$

The roughness factor is the actual surface area divided by the planar area [21, 48]. The model predicts that roughness enhances hydrophobicity if θ0 is greater than 90°, while increasing roughness for surfaces with θ0 less than 90° will result in a decrease in the contact angle [20]. In the Cassie-Baxter regime, the contact angle can be calculated using the equation below:

$$\cos\theta = \varphi_s(\cos\theta_0 + 1) - 1 \tag{5}$$

where Φs is the solid fraction in contact with the droplet [50]. Surfaces that are wetted in the Wenzel regime are usually sticky, while those wetted in the Cassie-Baxter regime tend to be slippery [48].

4.1 Super hydrophobic surfaces

Hydrophobic surfaces can be enhanced to become super hydrophobic by modifying the surface by means of the addition of roughness or more accurately by adding a certain type of morphology [48]. A super hydrophobic surface is a solid surface that when in contact with water produces a contact angle of greater than 150° [51]. They can work to limit/prevent adhesion as water is neither adsorbed or does not spread over the surface, so the water dynamics of the flowing system keep the underlying surface clean by sweeping away solid material [52]. This ability of fluid flow to sweep away particles from the surface is referred to as self-cleaning [53]. The two main requirements for a super-hydrophobic surface are that the surface should be hydrophobic, and hence have a low surface energy, and they should also be rough [54]. There are two conventional ways to produce a super-hydrophobic surface: creating a rough structure on a hydrophobic surface and the other is by modifying rough surfaces with low surface free energy materials [55]. The same techniques utilised in the fabrication of micro and nanostructures are used in the production of super-hydrophobic surfaces, namely lithography, etching, deposition and self-assembly [54].

A number of factors affect the ability of these super hydrophobic surfaces to prevent adhesion from occurring. These factors are related to the conditions to which the surfaces are exposed to. This is due to these surfaces relying on air as a lubricating fluid which can fail when the air escapes under conditions involving high temperatures, pressure, humidity and when exposed to low surface tension liquids [28]. In addition, the fluid flow conditions also play an important role, as a surface that is super-hydrophobic that can reduce bacteria adhesion when there are flow conditions; however, in stagnant conditions they may promote surface adhesion [56]. As discussed earlier in the report, the Lotus leaf was one of the initial inspirations for researchers to look at wettability to reduce particle adhesion due to its super-hydrophobic properties [24]. From this inspiration, a wide range of super-hydrophobic surfaces have been created to investigate their antifouling properties. Examples include clusters of carbon nanotubes (CNTs) which have been shown to produce good super-hydrophobic surfaces as they provide substantial surface roughness, strength, thermal conductivity and flexibility [52]. Additionally, by creating roughness on a hydrophobic polysiloxane surface the water contact angle was found to increase from 75° to 169°, making the surface a super-hydrophobic surface [57].

4.2 Super hydrophilic surfaces

A super hydrophilic surface produces a contact angle of less than 5° when water is brought into contact with it [52]. The super hydrophilic surface works by forming a hydration layer on the surface [52]. The hydration layer also known as the "lubricating liquid film" or "water shield" lessens the attractive electrostatic interactions between foulants [58]. The hydration layer acts to provide a barrier between the potential fouling compounds and the surface. Examples of this includes water attracting nano-composites surfaces with a TiO_2 coating [52]. In the work of Yoon & co-workers [52] they found that highly uniform coating was produced with the creation of tiny structures around groups of agglomerated TiO_2 allowing water to instantly spread onto the surface. This is caused by capillary effects and the water surface tension, resulting in a tightly bound layer of water near the super hdrophilic surface [59].

In an experiment conducted using E.coli as an adhesive substrate, it was found that in stagnant conditions a super hydrophilic TiO_2 coating was superior to a superhydrophobic carbon nanotube/PTFE coated system in limiting adsorption of the bacteria [52]. However, the opposite trend was exhibited in flowing conditions, where the shear exerted by the flowing fluid caused the super hydrophobic surface to have a superior performance. This is likely to be down to the effect of shear at the interface of the solid/liquid. As the shear increases there will be a greater force acting on any solids on the hydrophobic surface, pushing them along and conforming to its 'self-cleaning' properties under shear. However, the super hydrophilic, which relies on an aqueous layer between the surface and the rest of the process fluid, in the higher shear environment will reduce the thickness of this layer and hence reduce its effectiveness at preventing adhesion. This indicates the importance of the flow conditions on the adhesion of particles to a surface. The results caused the authors to suggest that a hybrid system could be considered in order to be implemented throughout an entire processing system where there is both stagnant and flow conditions.

4.3 Amphiphilic material

An amphiphilic material is one that contains both hydrophilic and hydrophobic regions. The development of surfaces that contain both hydrophobic and hydrophilic components so that they exhibit both anti-fouling and fouling release properties[7]. The intrinsic incompatibility between the hydrophobic and hydrophilic components results in phase segregation which leads to compositional, morphological and topographical surface heterogeneity, which makes the interaction between biomolecules and a surface energetically unfavourable [60]. Amphiphilic materials have been found to be effective at preventing the adsorption of a range of proteins [60, 61] as well as exhibiting anti-biofouling characteristics [62]. These surfaces have been particularly applied to investigations surrounding marine related fouling. Examples of amphiphilic materials included: branched fluro-polymers with polyethylene glycol (PEG) chains [62, 63]; perfluroethylene and PEG [64]; polysolfone-graft- poly (ethylene glycol) methyl ether methacrylate [65]. One issue is that many of these materials contain fluro-polymers that are present in the form of long perfluroalkyl chains, which can degrade to form toxic perfluroalkyl acids [7]. Therefore alternative compounds would need to be utilised if amphiphilic surfaces are to be applied to the food industry.

5 Coating systems

The basic idea behind coating systems is to provide a barrier layer between the surface of the equipment and the processing environment. In addition, the coating can provide a 'new' surface with a lower surface energy or enhanced smoothness that can act to reduce the adhesion of fouling compounds. One issue with the application of a coating layer to heat transfer equipment, such as heat exchanges, is that the coatings tend to have a lower thermal conductivity and thus reduces the overall heat transfer coefficient. However, as fouling layers tend to have a similar effect in reducing the heat transfer coefficient, the use of a coating layer generally provides a higher overall heat transfer coefficient over the duration of processing compared to the progressive reductions as a result of fouling.

5.1 Polymer coatings

Polymers have been used to coat processing equipment to mitigate fouling by reducing the adhesion of compounds to surfaces. Polymers tend to have a low surface energy [41] and be hydrophobic in nature. The low surface energy of these polymers is likely to be the reason for their ability to reduce the level of particle adhesion on a surface [34]. A number of different polymers have been utilised as a coating material including, Perfluoroalkoxy (PFA) [66], Polyetheretherketone (PEEK) [2], polydimethylsiloxane (PDMS) [67] and polyurethane [68]. These polymer coatings have been applied to a range of different surfaces and equipment to investigate the fouling of a number of different food related compounds. Low energy polymer coatings were found to have a reduced level of starch adhesion than stainless steel [2]. Polymer coatings have also been identified as being able to limit mineral adhesion on heat exchangers [69]. In addition, polymer coatings have been found to be effective at reducing bacterial adhesion to a processing surface [70, 71]. Many of these polymer coatings have also been identified as having an enhanced cleanability in contrast with stainless steel [2]. However, the most common polymer for antifouling coatings is PTFE.

5.2 PTFE & composites

Polytetrafloroethylene or PTFE is a widely used solid lubricant [72]. A possible reason for the low friction of PTFE is the long chains, $-(CF_2\text{-}\ CF_2)\text{-}$, which orient on the transfer surface during sliding, resulting in a low shear-strength interface with the bulk surface on which the PTFE is coated [72]. Due to the molecular arrangement of PTFE, with its 13/1 or 15/1 helical structure, the fluorine atoms are close together and form a smooth cylindrical surface which molecules can easily slide against [73]. Kananeh et al.[46] investigated the effect of various coatings on whey protein adsorption to plate heat exchangers. They identified that PTFE resulted in a greater degree of adhesion but had a significant reduction in CIP time, whereas polyurethane coatings reduced adhesion and reduced cleaning (CIP) time, although not to the same degree as PTFE. One key drawback with the use of PTFE as a coating system is that it has poor wear resistance [72, 74]. In order to improve PTFE's ability to withstand wear, filler particles are often added to the PTFE matrix [74]. A number of different PTFE based composites are outlined in the sections below.

5.2.1 Ni-P-PTFE

A promising technique to reduce fouling is the use of electroless nickel (EN) deposition with embedded PTFE particles [4, 75]. The application of a Ni-P-PTFE coating is achieved through an autocatolytic plating process [76]. This involves the incorporation of PTFE nanoparticles into the Ni-P matrix acquiring advantages of properties of both materials [32]. As this coating is metal based with only a small percentage of PTFE, the thermal conductivity is only slightly lower than stainless steel making it suitable for applications involving heat transfer [45].
Ni-P-PTFE has been found to be an effective coating in the reduction of fouling during milk processing. Barish & Goddard [4] found a 96% reduction in the accumulation of fouling on plate heat exchangers used to process raw milk at 85°C over an 8 hour period, in comparison with stainless steel. One of the most significant alterations to the surface is in regards to the surface energy, with stainless steel having an energy of 41.44 mN/m compared with the Ni-P-PTFE coating of 24.69 mN/m [4]. Over the 8 hours of processing, the Ni-P-PTFE coated heat exchanger was identified as having a reduction in the heat transfer coefficient of less than 10% while the standard stainless steel heat exchanger had a reduction in the heat transfer coefficient of close to 70%. The harsh conditions of the CIP process were found to influence the surface composition of EN-PTFE, in particular the acid used was found to cause a decrease in the fluorine concentration after completing 6 cycles [6]. However, the alkali sanitizer, although found to reduce the fluorine over extended periods of time, did not lose its hydrophobicity, surface energy and roughness, hence retaining its anti-fouling character [4]. These authors stated that acid may not be needed in the cleaning of modified surfaces as its main job is to remove any inorganic fouling which may not accumulate on the modified surface.

5.2.2 Ni-Cu-P-PTFE

The addition of copper to Ni-P-PTFE aids in improving the corrosive resistance of the material [77]. As with other PTFE composites, the increasing content of PTFE in the material was found to decrease the surface free energy of the composite [32]. In the work of Zhao et al. [32] where a Ni-Cu-P-PTFE coating was utilised to investigate adhesion of both bacteria and calcium sulphate, the adhesion of both substances showed a direct link with surface free energy following the Baier curve. One interesting observation that was made in this work was that the structure of the calcium sulphate differed significantly on the stainless steel and the low surface energy material. Instead of the usual densely packed structure, the low surface energy Ni-Cu-P-PTFE had a loose porous deposit of calcium sulphate.

5.2.3 CNT-PTFE

The use of carbon nanotubes as a coating material is somewhat limited due to their high degree of wear [46]. However, the addition of CNT's to a matrix such as PTFE can create a nano composite material that has improved properties. The use of both ultrasonication and annealing assist in the uniform distribution of CNT nano particles into a PTFE matrix, which produces a composite that has a high level of wear resistance and is also non-wettable [78]. Increasing the percentage of CNT in the matrix was found to decrease the friction coefficient, suggesting the CNTs can be self-lubricating [79].
A CNT-PTFE nanocomposite surface was found to have weaker attachment for protein nucleation than both PTFE and uncoated stainless steel [46]. The weaker attachment allowed the shear exerted by the flowing milk to sweep away protein molecules from the

surface, resulting in a reduction in the nucleation rate and reduced level of fouling. Kananeh et al. [46] found that over 5 hours of pasteurisation, the CNT-PTFE coating reduced the energy consumption by 10.2% due to the reduced level of protein accumulation on the surface of the heat exchanger. In addition, due to the reduction in accumulation, it is possible that the production runs could be extended as a result so that there is a reduction of down time during CIP. CNT-PTFE has been identified as an effective anti-fouling coating in regards to bacterial adhesion [80]. This was in the form of reducing bacterial (E.coli) adhesion in flowing conditions, as it produced a superhydrophobic surface that was in effect self-cleaning [52]. However, when the same surface coating was used in stagnant conditions the reduction in bacterial adhesion in comparison to stainless steel was much lower than in flow conditions (25% reduction compared to ~80%) [80].

The incorporation of PTFE nanoparticles into a silver coating has been done to incorporate the antibacterial properties of silver with the non-stick properties of PTFE [81]. It was found in this work that by increasing the content of PTFE nanoparticles in the silver coating reduced the surface energy and as a result reduced the level of E.coli cells adhering to the surface. In addition, Zhao et al. [81] found that the Ag-PTFE coatings had a greater level of corrosion resistance than stainless steel 316L.

5.3 Hydrophilic polymers

Hydrophilic polymers have been used as coatings to reduce the adhesion of fouling material [7]. As discussed earlier, the common properties of these polymers include their hydrophilic nature, electrical neutrality and being hydrogen bond acceptors (not donors) [82, 83]. The polymers can be applied as a simple coating to surfaces or grafted onto a surface in the form of polymer brushes [7].

The key factor that results in the reduction of fouling on hydrophilic polymers is the strong association with water molecules that can result in a hydration layer. The association with water can provide a barrier for fouling material and lessens the attractive electrostatic interactions between foulants [58]. This, in combination with the application of the polymer as a coating or in brush form, can act to reduce the formation of a fouling layer. As proteins tend to preferentially adsorb onto hydrophobic surfaces such as stainless steel, the creation of surfaces with a hydrophilic nature can act to reduce protein related fouling [7]. The reduction in protein fouling has been the primary application of hydrophilic polymers in experimental work conducted. One of the most common hydrophilic polymers that is used to resist protein adsorption is polyethylene glycol, or PEG as it is often referred to. Immobilizing PEG is a common method in producing a surface that resists the adhesion of proteins [67]. The mechanism preventing protein adsorption to PEG is still not understood in detail, however, the main reason is due to the strong association, through hydrogen bonding, between PEG and water leading to an excluded volume effect making protein adsorption unfavourable [84]. In addition, PEG brushes have been found to provide a steric barrier preventing compounds coming in contact with the surface, preventing adhesion [85]. A issue with the use of PEG is that it is susceptible to oxidation which leads to a reduction in its adsorption resistance over time [86].

5.4 Zwitterionic polymers

Although many hydrophilic surfaces can prevent/reduce non-specific protein adsorption, they are often not capable of preventing the adhesion of other compounds such as bacteria [87]. Zwitterionic polymers are defined as ampholytic polymers where ionic groups of opposite charges form a part of the same pendant groups [88]. Zwitterionic polymers were inspired by the anti-biofouling properties of blood cell membranes [70]. Despite having charged groups, zwitterionic compounds have an overall neutral charge [89]. In studies comparing zwitterionic polymers with hydrophilic polymers, such as PEG , comparable results have been found related to protein adsorption [7, 90]. Zwitterionic materials that have been used as anti-fouling coatings include phosphobetaine, sulfobetaine and carboxybetaine [89].

The hydration of zwitterionic polymers is critical to their ability to repel proteins as the hydration layer generates the repulsive force [7]. Unlike PEG, zwitterionic polymers bind to water electrostatically as a result of ionic interactions [91]. The electrostatic binding of water to the surface is stronger than that which occurs on the hydrophilic polymer coating [91]. This high water-binding capacity allows for the formation of a highly hydrated surface which is key to zwitterionic material having anti-fouling properties [89].

As with the hydrophilic polymers the surface grafting density and chain length determine the zwitterionic polymers resistance to protein adhesion [7].

5.5 Electroless plating

In addition to its use with PTFE nanoparticles (as discussed above), electroless plating can be applied in other anti-fouling coatings. Electroless plating is a chemical reduction of a metallic ion in a solution containing a reducing agent, and the ensuing deposition of the metal without electrical energy [92]. The incorporation of TiO_2 nanoparticles into the electroless plated Ni-P matrix has been identified as a means of increasing the hardness, corrosion resistance [93] and wear resistance of the coating [94]. The production of an electroless plating of Ni-P-TiO_2 on stainless steel was found to have anti-adhesion properties when tested with three different strains of bacteria, after exposure to UV light [95]. In this work, Zhao et al. [95] incorporated titanium dioxide nanoparticles at a number of different levels into the Ni-P matrix, identifying that adhesion reduced as the percentage of TiO_2 in the composite increased. This decrease in adhesion was found to correlate with an increase in electron donor component of surface energy. The use of UV light on the TiO_2 prior to bacterial adhesion testing was due to the microbiocidal effect due to a photocatalytic reaction produced when TiO_2 is exposed to wavelengths of light below 385 nm [96]. This work stems from that done on bacterial adhesion on TiO_2, which found that pre-exposure to UV light caused TiO_2 to temporarily become a superhydrophilic substance (dropping the water contact angle from 59 to 0°) and reduced bacterial adhesion to the surface [97]. When this was applied to the Ni-P-TiO_2 composite, the use of UV was also found to reduce the water contact angle from 52 to 40°, and as a result making the surface more hydrophilic [95].

An additional surface coating trialled using electroless Ni-P is Ni-P-BN, which was used to investigate its ability to restrict mineral deposit formation [98]. In this work, Al-Janabi et al. [98] found that Ni-P-BN reduced the rate and amount of calcium sulphate deposition in comparison to a number of different coatings and stainless steel. The authors evaluated the fouling rate with a number of different components of the interaction energy, finding that the Ni-P-BN to have the lowest van der Waal component but the highest electron donor component. The Ni-P-BN coating was found to exhibit a degree of corrosion following the cleaning of the surface [98].

5.6 Antimicrobial coatings

Antimicrobial surfaces can prevent the formation of biofilms by either acting to repel microbes or kill those in the vicinity of the surface [99]. The latter, active antimicrobial materials incorporate physically entrapped or covalently immobilised antimicrobial agents [100]. These materials utilise the diffusion of the antimicrobial agent out of the matrix as a dispersion mechanism. A number of different antimicrobials have been incorporated into the matrix of these antimicrobial materials including silver [101], antibiotics [102], covalently coupled quaternary ammonium [103] and nitric oxide [104]. Although these materials often perform relatively well, they are often hindered by their limited lifetime resulting from the mechanism responsible for their antimicrobial action [100]. Another approach, avoiding the limited life of antimicrobial coatings, is to implant the silver ions into the metal surface.

Another approach is to develop antimicrobial surfaces that kill bacteria on contact; not releasing biocides and thus not exhausting the antimicrobial properties [99]. These contact active antimicrobial surfaces can be created using antimicrobial polymers attached to the processing surface [105]. There are two approaches of achieving this: polymeric biocides, biocidal groups attached to a polymer backbone, and biocidal polymers [99]. Polymeric biocides are small repeating units of biocidal groups attached to a polymer. Biocidal polymers such as polycationic antimicrobial polymers work by disrupting the outer membrane of the cell, causing cell lysis to occur resulting in cell death [99]. Examples of antimicrobial polycations that are used include quaternary ammonium and phosphonium.

6 Ion implantation

Ion implantation is the introduction of atoms into surface of a solid material via the bombardment with ions in the keV to MeV energy range [106]. As the ions only penetrate the uppermost layers of the surface, the surface properties can be modified whilst the bulk properties remain the same [107]. As a surface alloy is created, there are no issues regarding adhesion such as in chemical vapour deposition (CVD) [76]. The local environment of a particle on the surface of a solid differs from that of particles within the solid [108]. This is due to interior particles experiencing a uniform force from all sides, while for the particles at the surface an unsaturated force field exists. As a result, there is a tendency for the surface particles to be pulled into the solids interior in order to reduce the surface area with a consequence of producing surface tension.

The Electric Double Layer theory states that if an element with weak metal properties (e.g. F, H, Si, C…) is implanted into a surface; both the free electrons on the surface and the surface energy will be reduced [109]. In addition, the dose and ion implantation energy both have a significant effect on the surface energy and adhesion of the modified surface [108]. Unlike a number of different surface modification techniques, ion implantation does not have a significant effect on the overall heat transfer coefficient [110]. This is due to the ions being embedded in the surface of the material and only a few atoms thick. The implantation of N^+, O^+ and SiF_3^+were evaluated in regards to their effect on the adhesion of two different strains of bacteria [111]. All of the implanted surfaces outperformed native stainless steel. They were found to have a lower surface energy, increased water contact angle and had a decreased level of roughness. In this work, Zhao et al. [111] found lower bacterial adhesion with stainless steel surfaces that had SiF_3^+ implanted in comparison to those with N^+ and O^+. Over a time frame of 24 hours, the adhesion to the SiF_3^+ implanted surface remained statistically significantly lower than native stainless steel for both bacterial strains. Although a significant difference was identified for the adhesion of one bacterial strain to the N^+, O^+ implanted surfaces; no statistically significant difference was identified for the other strain compared to native stainless steel.

The implantation of SiF_4 into stainless steel was found to result in a significant reduction in $CaSO_4$ deposits forming in pool boiling experiments [108]. Additional work done investigating ion implantation and its effect on mineral fouling mitigation includes the work of Nejim et al. [110]. In their work they investigated the effect of Si, F & H ions implanted into stainless steel to reduce the adhesion of $CaSO_4$. It was found that all three surfaces

reduced the amount of mineral deposits significantly compared to untreated stainless steel. Out of the three modified surfaces the F implanted surface had superior performance to the others [110]. This benefit in the reduction in $CaSO_4$ deposits manifested in a final heat transfer coefficient of up to 120% greater than the control surface, following 50 hours of pool boiling.
Buef et al. [112] in their comparison of various surface modification techniques found that the implantation of SiF^+ was the best performer in reducing adhesion. However, the reduction from native stainless steel was only 18% and little difference in adhesion was identified between stainless steel and the SiF^+ following cleaning. In the same experiment MoS_2 was implanted into a surface by Beuf et al. [112], and although there was a slight reduction in adhesion compared to stainless steel, it was a smaller reduction than the SiF^+ sample. The use of antimicrobial compounds, such as silver ions, have been used to try and prevent biofilm formation [113]. Dhillon [107] investigating the effect of Ag ion implantation on bacterial adhesion and biofilm formation. This work identified that although the Ag ions had an antimicrobial effect; bacteria could still bind to a conditioning film of protein or were protected by EPS, reducing the effect of the surface modification. The key drawback in regards to the use of ion implantation is the high capital and operating costs associated with the technique [76].

7 Physical vapour deposits

Physical vapour deposition (PVD) is a form of vacuum deposition involving physical processes to deposit thin coatings via the condensation of a vaporised form of a solid precursor [114]. There are a number of varieties of PVD, including: cathodic arc deposition, electron beam, thermal evaporation deposition, pulsed laser deposition and sputtering deposition [114]. The key drawback associated with PVD methods are the low deposition rate, however, this was circumvented with the development of magnetron sputtering [114]. This involves the addition of magnets to the cathode resulting in the trapping of a greater quantity of electrons just above it, producing a higher ionization yield.

7.1 Sputtering

Sputtering coating is a physical vapour deposition (PVD) process occurring at sub-atmospheric pressures [76]. The high voltage that is applied between the surface and the coating material ionises the gas in the chamber until a plasma is ignited [76]. Sputtering deposition uses a glow discharge plasma to deposit conductive coatings on a surface via the ion bombardment of a metal [114]. The advantages of using sputtering coatings is the development of hard and very adhesive films [76].
TiN was used as a surface coating to investigate its effect at reducing adhesion of calcium phosphate before and after cleaning [115, 116]. In the experiment, Rosmaninho et al. [116] used sputtering to apply a coating on a stainless steel surface, with different proportions of Ti and N giving the surfaces a different electron donor component of surface free energy (ɣ-). In their testing conditions they found that surfaces with lower ɣ-, deposition started at a slower rate and reached a lower amount than the higher ɣ- surfaces. Rosmaninho et al. [116] found that all of their TiN surfaces retained lower quantities of mineral deposits after cleaning than stainless steel and those with lower ɣ- values had a lower mass of deposits remaining. The titanium content was found to influence the ɣ- value, with the samples with the lowest quantity of titanium having the lower values of ɣ-. In addition, Rosmaninho et al. [115] found an exponential growth curve associated between the particle adhesion coefficient and ɣ- values, with the lower ɣ- values having the lowest particle adhesion coefficient.

7.2 Aerosol assisted CVD

Aerosol assisted CVD is a variation of the conventional CVD process whereby aerosol droplets are used to transport the precursor, with the aid of inert or reactive carrier gases [117]. This has advantages over traditional CVD processes as the precursor does not need to be volatile, just soluble in any solvent that can be made into an aerosol [118]. It combines the advantages of both conventional CVD and spray pyrolysis processes [119].
In work conducted using a silicon based elastomer, AACVD was found to produce a highly developed microstructure in comparison to the flat surface created from dip coating, resulting in a contact angle of 165° compared to 95° [120]. In this work by Crick et al.[120] the AACVD sample was found to limit the attachment of bacteria following an hour long submersion in both E. coli and S. aureus compared to both the dip coated surface and an uncoated glass surface.

7.3 Plasma deposition

Plasma enhanced CVD is a chemical vapour deposition process whereby the chemical reaction is not initiated thermally, but through the formation of plasma through the use of RF or microwaves [76]. Differing from sputtering techniques in which the coating material is introduced as a solid, CVD use gaseous precursors [76]. It uses the energy of a plasma discharge to deposit a coating on a surface [114]. It is a uniform, reproducible an tunable means of treating a wide range of substrates and geometries [84]. It can be used for the synthesis of organic coatings in which monomers are activated in the plasma phase which allows the growth of materials close to the chemical composition of the monomer [114]. The synthesised materials are commonly referred to as plasma polymers (pp) and have been found to be useful in a number of different applications due to their good adherence to a surface as well as their high level of crosslinking and branching [84]. Plasma CVD has the ability to implement additional atoms into a coating, for example the addition of silicon containing precursors and oxygen to the working gas can be used to create a DLC-Si-O coating [76]. Disadvantages of the use of plasma CVD coatings is their reduced density and weaker adherence to the surface [76].

7.4 Atmospheric plasma deposition

As plasma is defined as a fully or partially ionised gas, atmospheric plasma refers to such a media that is developed at atmospheric pressure [114]. Atmospheric plasma deposition methods have advantages over vacuum techniques as they avoid the use of expensive pumping systems and having to build complex transfer chambers from air to vacuum with a high pressure gap [114].
It was found that an AP Liquid Deposition PEG coated surface significantly reduced the adsorption of bovine serum albumin (BSA) [a protein] by 90% compared to a bare Si surface, independent of plasma power [84]. Protein (BSA) adsorption reduction on an AP CVD PEG coated surface was found to increase with decreasing plasma power, from a 50 to 85% reduction in protein adsorption [84].

8 Diamond like carbon (DLC)

Hydrogenated amorphous carbon, also called diamond like carbon (DLC) has been investigated as a possible anti-fouling coating material [121]. DLC has attracted attention due to good thermal conductivity, corrosion resistance, very smooth surface and hardness

[122]. In addition, the use of DLC coatings has been approved for use in the food industry by the FDA due to its atoxicity and corrosion resistance [122]. Carbon-carbon bonds can be either the near-planar trigonal (sp^2) found in graphite or the tetragonal (sp^3) form present in diamond [123]. DLC has a combination of both types of bonding, with its characteristics determined by the sp^3:sp^2 ratio. DLC also contains hydrogen in percentages ranging from just a few percent to in the region of 50% [124]. The hydrogen in DLC is important in providing a number of key properties including stabilising the network and preventing it collapsing into a graphite phase.

There are a number of techniques that can be used to deposit DLC on a surface including: Direct ion beam deposition, pulsed laser ablation, filtered cathode arc deposition, magnetron sputtering, RF plasma-activated chemical vapour deposition, plasma source ion implantation and deposition [123]. The deposition of the DLC coating onto a surface will influence the behaviour that the coating exhibits [124]. Diamond like carbon has been used in the research or anti-fouling materials for a number of different industries, including medical implants [125] and food processing equipment [122]. It has been found to be able to reduce the adhesion of both whey protein and calcium phosphate adhesion in batch processes [75]. However, in the work done by Beuf [112] DLC was found not to have any significant reduction in fouling and also did not clean as well as native stainless steel.

8.1 Modified DLC

The amorphous nature of DLC allows the addition of additional elements, such as Si, F, N, O, into the coating whilst maintaining the amorphous structure [121]. It has been found that the introduction of these additional elements into the DLC structure can be an effective method to reduce fouling [126]. Modified DLC coatings are usually produced using magnetron sputtering techniques or by plasma enhanced chemical vapour deposition (PECVD) [126]. The addition of both N and Si to DLC coatings has been performed to evaluate the effect that the coatings had on bacterial adhesion [127]. Liu et al. [127] evaluated two levels of both N & Si in the DLC surface on the adhesion in a shaken container over an hour long period. They found that the modified DLC coatings performed better than pure DLC in regards to limiting bacterial adhesion to the surface. The N doped DLC coating was found to out-perform the Si based coatings, with a reduction of adhesion of 89.7% found in the 2% N sample compared to the best performing Si doped sample's 38.5%. A good correlation was found between the electron donor component of surface energy and the level of cellular attachment to the surface. This identified a trend of decreased attachment with increasing γ-, which contrasts a number of other studies showing the opposite trend. However, the different coatings also produced a different level of roughness on the surface – with a strong correlation found between the number of adhering cells and the surface roughness.

Fluorinated DLC has also been found to have antibiological properties in regards to decreasing surface adhesion as well as increasing the removability of adhered cells [128, 129]. This became more pronounced as the proportion of F content was increased in the coating, as it was found to reduce the surface energy of the surface [128]. The C-F bond is believed to decrease the dielectric constant and the cross-linked C-C bonds contribute to the materials thermal stability [129]. Su [126] found that F-doped DLC reduced both bacterial adhesion and increase removal, improved with increasing F content (decreased SE) with the 39% w.t. sample performing the best (67% reduced adhesion & 32% increased removal compared to DLC). In addition to the decrease in surface energy with increasing fluorine content [128], structure of the film coatings also became more like a polymer structure than diamond like [129].

In addition to experiments looking at fluorinated DLC, Su's [126] PhD thesis reviewed a number of different types of modified DLC coatings. The experiments carried out for this work were conducted using the various DLC coatings and evaluating the degree of bacterial adhesion. The experimental work was conducted in stagnate conditions over a range of contact times from 1 to 18 hours. The various DLC coatings tested during the PhD [126], included Si-N, F, B, Ti and Si-O_x, were all identified as being able to reduce the adhesion of bacteria to a surface. The boron doped DLC was found to be most effective at the highest dose rate trialled (9.5%) with the reduced total bacterial adhesion of 70% of the pure DLC coating. In addition, the B-DLC coating was found to increase the removal rate of bacteria by 164% compared to the pure DLC coating. The results exhibited by the boron doped DLC was linked to the surface energy, with increasing boron content leading to a lower surface energy and decrease in bacterial adhesion [126]. A similar trend was exhibited with the Ti-doped DLC coating trialled as part of the work of Su [126], with an increasing content of Ti resulting with a decrease in the level of bacterial adhesion (the best performing 66% better than pure DLC). However, unlike the B-DLC coating, the surface energy of the Ti-DLC coating remained relatively unchanged with the alteration to the level of titanium in the coating. A trend was identified in the electron donor component of the surface energy, with an increase in the γ- found to correlate with the increase in Ti content, and subsequent decrease in cell adhesion.

The Fraunhofer Institute in Germany has been investigating a number of different antifouling techniques for a number of years. Among the different coatings developed are SICAN (C:H:Si) and SICON® (C:H:Si:O), which are DLC based coatings comprising Silicon. These coatings have been used in a number of different experiments to identify their resistance to fouling. This includes an investigation carried out to identify whether the coatings were able to reduce the accumulation of both protein and minerals in a simulated milk ultrafiltrate with whey protein isolate added [130]. In this work, Boxler et al. [130] used a pilot scale facility to pass their milk based fluid through a plate heat exchanger over a 2 hour period at 85°C. In their work they found that the formation of the initial fouling layer, subsequent build-up and composition of the deposit were affected by the surface energetic properties, in particular the electron-donor component (γ-) which showed a polynomial function with protein adhesion (curve depicting best γ- as being mid-range for the experiments conducted). The SICAN coating was found to have the highest level of fouling mitigation of the various PECVD DLC coatings trialled in the study.

However, work has been done looking at doped-DLC coatings identifying no benefit to their use over stainless steel [1]. One constraint of interpreting this research is that the compounds used to dope the DLC are not disclosed in the publication. The results that Patel et al. [1] were initially promising at the laboratory scale where they found that for whey protein isolate (WPI) fouling was reduced by 35-40% compared to stainless steel over a 120 minutes trial period. However, when this period was extended to 480 minutes the reduction in fouling on the doped-DLC surface reduced to less than 15%. At a full pilot scale even the slight benefit seen in the laboratory scale was lost, with the doped-DLC sample not performing as well as the stainless steel control surfaces. In addition to WPI, Patel et al. [1] performed testing on raw milk where they found only a slight difference at a pilot scale and no significant difference between the doped-DLC samples and stainless steel at the pilot scale. As well as looking at initial adhesion, they also investigated the removal of foulants during the cleaning process where they found no significant difference between the control and the DLC surfaces. These results lead to Patel et al. [1] stating that they found no commercial benefit from using surface modification to reduce fouling in processing conditions. The testing conditions used in these trials were more robust than a number of others in literature as they used full scale processing equipment, processing conditions and used times closer to commercial run times.

9 Liquid infused surfaces

Liquid infused structures steam from yet another biomimetic related idea, the pitcher plant [131]. The pitcher plant has a water film that is locked into its surface under humid conditions [132]. The water film causes insects to aquaplane when they step on the wetted pitcher rim and results in the insects falling into the pitcher plant [133] as the water repels the oil on the insects feet [134]. This action has attracted attention, in particular in the area of antifouling coatings. It has advantages over the lotus-inspired surfaces, which rely on air as a lubricating fluid and tend to fail when the air escapes under conditions involving high temperatures, pressure, humidity and when exposed to low surface tension liquids [28], unlike the liquid infused surfaces.

Slippery liquid-infused porous surfaces, or SLIPS, consist of a film of lubricating liquid that is locked within the micro/nano-porous substrate. The lubricating liquid provides an ultra-smooth, continuous and chemically homogeneous interface that presents a slippery surface for a range of different fluids [28]. The lubricating fluid is stabilised by the capillary forces arising from the micro-structured texture [135]. There are three key criteria in designing a stable SLIPS: i) the chemical affinity between the lubricating fluid and the solid should be higher than that between the solid and the bulk fluid; ii) the solid should be roughened to increase the surface area for adhesion and immobilization of the lubricating fluid; iii) the lubricating fluid and bulk fluid should be largely immiscible [56].

The molecularly smooth and chemically homogenous lubricating liquid provides a low contact angle hysteresis (typically <2°) and a low sliding angle [136]. The liquid plays a similar role to air pockets created in the lotus effect [134]. However, it has advantages over the air pockets in the fact that it can withstand pressures and is effective against destabilisation resulting from organic solvents [131]. Wong et al. [131], while investigating the properties of SLIPS, found that the surfaces' ability to repel both water and liquid hydrocarbons was stable when subjected to pressures of 676 atm. By choosing lubricating fluids that are immiscible with both oils and water, SLIPS can exhibit omniphobic properties [134]. One of the biggest weaknesses of the liquid infused porous surfaces is their durability [134]. Over time, the lubricating fluid may be lost due to evaporation or due to a high shear environment, compromising the capabilities of the surface [131]. Thus, choosing an appropriate lubricating fluid with a minimal evaporation rate or enhanced viscosity to minimise losses is advantageous, or alternatively a fluid reservoir can be used to enable self-replenishment of the lubricating fluid. The surfaces are designed to be self-healing, with the lubricating fluid flowing to any damaged area via surface energy driven capillary action [137]. It has been found that uniform nanostructured surfaces provide a superior performance in high simulated shear conditions compared to hierarchical or micro-structured surfaces [28]. This is due to the lubricating fluid creating a stable smooth overcoating until the thickness of the lubricating layer becomes comparable to the nanoscale features. So despite the nanotextured surface in the study of Kim et al. [28] losing more lubricant than the hierarchical structure studied, the quantity of liquid needed to maintain the SLIPS capabilities is lower.

The loss of lubricating fluid in higher shear environments does not necessarily rule out these surfaces, with the potential use of the lubricant as a sacrificial fluid. As a sacrificial liquid, the lubricant could be used to facilitate the processing of a fluid by reducing fouling and product losses. In this situation, the lubricant would have to be a liquid that does not negatively affect the quality of the product as the lost lubricant would end up in the product. One method to assist with the replenishment of a lubricating fluid is the inclusion of a vascular-like system [138]. In the work of Howell and others [138] they found that a vascular system built into a liquid infused surface was able to continuously replenish the depleted lubricating liquid, maintaining the antifouling properties of the surface over time.

This type of system would utilise a reservoir to replenish the vascular system. Depending on the surface tension of the lubricant, the porous solid can be either a high or low energy surface [139]. This influences the activation barrier for nucleation to occur on the surface, with an impregnating lubricant with a low surface tension resulting in a high activation barrier and vice versa.
Micro-textured surfaces have a critical contact angle below which the lubricating liquid can impregnate the texture and remain held by the capillary forces [140]. It can be calculated using the Equation 2 below:

$$\boldsymbol{\theta_c = \cos^{-1}[(1-\emptyset)/(r-\emptyset)]} \tag{6}$$

The equation calculates the critical contact angle using the fraction of projected area of the textured surface that is occupied by the solid (Φ) and the ratio of total surface area of the textured surface to its projected area (r) [140]. Liquid infused surfaces have been created on a wide range of different materials including: polymers [133], stainless steel [140] and glass. These liquid infused surfaces have been applied to a number of different process fluids to test their repelling qualities. These include simple fluids through to complex fluids including blood [140], SLIPS has been used to investigate bacterial adhesion to the liquid infused surface [141]. In this work, Epstein et al. [141] identified that the modified surface was capable of preventing between 96-99% of common bacterial biofilms from attaching over a 7 days period in low flow conditions. The authors explored a number of different biofilm forming bacteria species including E.coli and S.aureus. In addition, the liquid infused surface was stable when submerged under extreme pH, salinity and UV exposure. The surface that was used by Epstein et al. [141] was composed of a porous PTFE substrate or a fluro-silanised Si wafer, which was then infused with the lubricating fluid. Essentially the SLIPS surface prevents bacterial adhesion due to the lubricating liquid preventing the bacteria coming in contact with the solid below [142]. Additional studies have looked at liquid infused surfaces and bacterial adhesion, with a reduction in adhesion of both S.aureus and E.coli found to be at least 3 orders of magnitude compared to non-SLIPS surfaces [138]. Liquid-impregnated surfaces have been used to investigate their ability to resist deposits of calcium sulphate forming [139]. In this work, Subramanyam et al. [139] used ion etching to texture a silicon surface and coating it with octadecyltrichlorosilane, to lower its surface energy, and then impregnated the surface with different lubricating fluids. They identified that the surface tension and spreading coefficient of the lubricant had a profound effect on the surface properties and subsequent adhesion. All of their impregnated surfaces were identified as having a lower nucleation rate and thus a lowering the mass of calcium sulphate deposits, with a 10-fold reduction found in adhesion for one of the surfaces trialled. In addition to the silicon based surface, Subramanyam et al. [139] also developed a liquid-impregnated surface using stainless steel as the substrate, textured using simple sandblasting. They found that there was a reduction in deposits formed and that those formed could be easily washed away with low flow water.
SIPS has also been applied to investigate ice adhesion to surfaces [143, 144]. In their work, Kim et al. [143] developed a liquid infused porous surface using aluminium with the electrodeposition of highly textured polypyrrole, followed by fluorination and then infiltration of the lubricant. The found that the SLIPS suppressed the formation of ice and was also found to be effective at removing condensing moisture. The degree of the ice reduction and adhesion strength on the surface was found to be an order of magnitude greater than other state-of-the-art materials that were tested and two orders of magnitude lower than bare aluminium. The application of LiquiGlide to containers to reduce adhesion of products is one of the company's leading applications. This is done by applying an aerosol coating of wax and ethanol to the surface of the container. The ethanol evaporates off, leaving a porous wax coating on the surface of the container. An oil is then added to the

container and is infused into the porous wax matrix. Residue oil is then removed from the surface leaving a liquid infused porous surface that is capable of preventing the adhesion of various materials that are filled into the container.

10 Ultrasonics

Ultrasound is the use of soundwaves with a frequency in excess of human hearing, 20kHz [145]. The application of ultrasound in the food industry can be divided into two: low energy, using frequencies above 100 kHz at intensities below 1 Wcm^{-2}, and high energy ultrasound, using frequencies between 20 and 500 kHz as well as intensities higher than 1 Wcm^{-2} [145]. Whereas low energy ultra sound is used for non-invasive monitoring, high energy ultrasound is used to influence the physical, mechanical and biochemical properties of food.The propagation of ultrasound through a biological media causes compression and decompression through the material which imparts a high amount of energy [146]. High energy ultrasound has been used as an alternative processing tool for a range of operations ranging from controlling microstructure, emulsification, defoaming, accelerating enzyme activity, freezing, thawing and microbial inactivation [145].
The chemical effect of ultrasound does not occur as a direct result of interactions with molecular species, rather from acoustic cavitation, which involves the formation, growth and collapse of cavities (bubbles) in liquids [147]. At a surface, the collapse of the bubbles results in asymmetric microjets that can cleanse a surface of contaminants [145]. In addition, it can prevent adhesion from occurring as it causes the surface to be constantly moving, making it difficult for fouling to occur.
Sonication is thought to increase nucleation and crystal growth rate in a saturated or supercooled material [148]. This can be used in processing to assist in the development of ice crystals. When ultrasound is used to enhance crystallisation there is the benefit of the prevention of encrustation of crystals on the cooling element [148]. This means that ultrasound can be used in freezing processes to reduce the adhesion of ice and the subsequent product losses that can occur due to ice adhesion to cooling elements. High energy ultrasound has also been used to disrupt/inactivate bacteria as well as deagglomerate bacterial flocs (clusters) through the effect of cavitation [149]. It has therefore been utilised as an alternative processing tool in the inactivation of microorganisms as it also has the advantage of not degrading the processing fluid, which can occur in thermal processes [150]. However, it is in combination with different techniques such as thermal processing that ultrasound can be used to reduce processing time due to accelerated food sterilisation [145].

11 Microwave processing

In contrast to traditional thermal processing techniques whereby energy is transferred by convection, conduction or radiation, microwave energy is transferred directly to a material through molecular interaction with the electromagnetic field [151]. One key attraction of microwave heating, in the context of this report, is that the process produces minimal fouling deposits as a result of eliminating heat transfer surfaces [152].
Microwaves are a part of the electromagnetic spectrum, with wavelengths from 1 mm to 1 m [151] and corresponding frequencies between 0.3 and 300 GHz [153]. Microwaves can be reflected, absorbed and/or transmitted by a material, with reflection and absorption requiring the interaction between the electromagnetic field and the material [153]. The fundamental electrical property that describes the interaction between electromagnetic energy and material is the relative complex permittivity [154]. The relative complex

permittivity, ε*, is the permittivity of the material, ε, divided by the permittivity of free space, ε0 [155]. It can be broken down into two components by the following equation:

$$\varepsilon^* = \varepsilon' - j\varepsilon'' \tag{7}$$

This interaction with the electromagnetic field can be broken down to how the material stores energy, the dielectric constant (ε'), and how it dissipates energy as heat, the dielectric loss factor (ε") [156]. These properties quantify the capacitive and conductive components of the dielectric response respectively [151]. The dielectric constant is dominated by polarisation (in particular dipolar rotation) while the loss factor is dominated by ionic conduction within the sample being examined [156]. In foods, water is the primary component responsible for dipolar rotation while salts are responsible for the ionic interactions [157]. The polarisation involves the short-range displacement of charge through both the formation and rotation of electric (or magnetic) dipoles, while conduction requires the (relatively) long range transport of charge [153]. Ionic loss is typically predominant at frequencies below 1 GHz [158]. Whereas dipolar polarisation is typically more significant above 1 GHz, although it still influences the dielectric response of a material at lower frequencies as well [159]. The frequencies used are also of particular interest in regards to gaining an understanding of microwave heating [154]. This is due to specific frequencies, including 915 MHz & 2.45 GHz, being set aside for use in microwave ovens & industrial microwave heating [160].

When water molecules are exposed to microwave radiation, the molecular dipoles try to align in the direction of the applied electromagnetic field [161]. As the applied field oscillates, the dipoles attempt to follow the oscillations, and the electromagnetic energy it transferred into thermal energy as a result of molecular friction and dielectric loss. The alternating electric field causes the charged ions to oscillate back and forth, resulting in heating due to the agitation caused by the ions bumping into other molecules [161]. The dielectric properties of a food material are dependent on its composition, especially that of water and minerals. In addition, the degree of freedom of those elements in the food material impacts the dielectric properties [154]. Ionic conduction at relatively low frequencies has a much greater ability to generate heat than dipolar rotation, with ionic solutions heating faster than pure water [161].

Microwave heating is also affected by the relative freedom of components in a system, such as whether ions are bound [162] and the degree of freedom of the water molecules present. The dielectric properties of a material can change during processing due to physical and structural transformations [154]. As a result, the ability for microwaves to generate heat can vary during processing [151]. The penetration depth (dp) of microwave processing is commonly defined as the depth into a sample where the power has dropped to 1/e or 36.8% of its transmitted value [163]. The penetration depth is inversely proportional to the microwave frequency. Therefore microwave processing using a frequency of 915 MHz have a higher penetration depth into a product than processing using a frequency of 2.45 GHz [164].

A number of advantages exist in the use of microwave heating including: it does not rely on the diffusion of heat from the surface so can apply rapid & uniform heating of large volumes [151], selective heating of components possible [153], minimal come up time (CUT) required [152], energy saving due to the high degree of efficiency in microwave heating [165], One key disadvantage with most common microwave heating processes is the non-uniform heat distribution that can be generated that results in hot and cold spots[166]. This non-uniform heating is more common in solid food processing rather than in the heating of fluids [165].

Microwave drying offers the means to shorten the drying time of a product as well as improving the product quality compared to conventional drying processes [167]. In drying applications microwave processing has been found to reduce processing time due to the build-up of internal vapour pressure that drives out moisture of the product [161]. The improvement in product quality steams from the short processing time required, resulting in a decrease in the loss of heat sensitive compounds such as colours, nutrients and flavour [165].

Limitations of microwave drying processes include the possible development of scorching at the edges of a product due to the development of excessive heat at those points during the final stages of drying [168]. This is the result of the difficulty in controlling the final temperature of a product during microwave drying; unlike air drying where the product cannot exceed the temperature of the hot air. One method used to assist in overcoming the limitations of microwave drying or the limitation of other conventional drying techniques is the coupling of microwave drying with another drying method [167]. This includes the use of microwave (MW) assisted air drying, MW assisted vacuum drying, MW assisted freeze drying and MW assisted spouted bed drying. This microwave assisted drying can improve the drying efficiency of conventional drying processes and produce product with improved quality [168].

Microwave drying is considered to be very useful during the falling rate period. This is the period of the drying process when the diffusion is rate limiting, resulting in shrinkage and a reduced surface moisture content [168]. In microwave drying, due to the volumetric heating, the vapours are generated inside and an internal pressure gradient is created that forces the water out without any shrinkage occurring. However, this rapid mass transport can result in a phenomenon known as 'puffing', which may or may not be deemed desirable in a final product [167].

Microwave or microwave assisted processing has been applied to a number of different fluid and solid food systems [165]. This includes: the drying of honey [169], minimal processing of apple puree [170], avocado puree [171], and the pasteurisation of orange juice [172]. In these types of applications, microwave processing has been found to be capable of providing logarithmic reductions in the level of bacteria present in a product [173], the rapid inactivation of enzymes [174], and retaining antioxidant activity [175]. Industrial microwave processing systems are available including microwave dryers. These techniques have been adopted predominantly in Europe and Japan, with other countries such as the USA being more hesitant [152].

12 Summary

There are three key parameters that can affect adhesion in processing applications: the process fluid, the processing conditions and the surface that the fluid interacts with. The surface has the greatest scope for manipulation to try and control the undesirable adhesion of matter that results in fouling. Manipulation of the surface energy and surface topography of a material, both independently and collectively, have been proven to be effective methods for altering the rate of adhesion to a surface. Although this is an area of research gaining increasing attention, no silver bullet has been found that produces a surface that is durable and reduces adhesion universally for all species and applications. The key to developing a surface that is capable of controlling adhesion requires ensuring that it is suitable for processing applications, and thus can withstand the rigours of the shear stresses and cleaning regimes that are commonplace in processing industries.

References

1. M. Jimenez, H. Hamze, A. Allion, G. Ronse, G. Delaplace, M. Traisnel, Antifouling stainless steel surface: competition between roughness and surface energy, in: Materials Science Forum, Trans Tech Publ, 2012, pp. 2523-2528
2. J.A. Barish, J.M. Goddard, Anti-fouling surface modified stainless steel for food processing, Food and Bioproducts Processing, **91** (2013) 352-361
3. J.S. Patel, B. Bansal, M.I. Jones, M. Hyland, Fouling behaviour of milk and whey protein isolate solution on doped diamond-like carbon modified surfaces, Journal of Food Engineering, **116** (2013) 413-421
4. M. Mauermann, U. Eschenhagen, T. Bley, J.-P. Majschak, Surface modifications–application potential for the reduction of cleaning costs in the food processing industry, Trends in Food Science & Technology, **20** (2009) S9-S15
5. L. Gomes da Cruz, E.M. Ishiyama, C. Boxler, W. Augustin, S. Scholl, D.I. Wilson, Value pricing of surface coatings for mitigating heat exchanger fouling, Food and Bioproducts Processing, (2014)
6. J.A. Barish, J.M. Goddard, Stability of nonfouling stainless steel heat exchanger plates against commercial cleaning agents, Journal of Food Engineering, **124** (2014) 143-151
7. T. Mérian, J.M. Goddard, Advances in Nonfouling Materials: Perspectives for the Food Industry, Journal of agricultural and food chemistry, **60** (2012) 2943-2957
8. F. Podczeck, Investigations into the reduction of powder adhesion to stainless steel surfaces by surface modification to aid capsule filling, International Journal of Pharmaceutics, **178** (1999) 93-100
9. D. Gunn, Effect of surface roughness on the nucleation and growth of calcium sulphate on metal surfaces, Journal of Crystal Growth, **50** (1980) 533-537
10. J.F. Schumacher, M.L. Carman, T.G. Estes, A.W. Feinberg, L.H. Wilson, M.E. Callow, J.A. Callow, J.A. Finlay, A.B. Brennan, Engineered antifouling microtopographies–effect of feature size, geometry, and roughness on settlement of zoospores of the green alga Ulva, Biofouling, **23** (2007) 55-62
11. M.V. Graham, N.C. Cady, Nano and Microscale Topographies for the Prevention of Bacterial Surface Fouling, Coatings, **4** (2014) 37-59
12. Y. Engel, J.D. Schiffman, J.M. Goddard, V.M. Rotello, Nanomanufacturing of biomaterials, Materials Today, **15** (2012) 478-485
13. Biswas, I.S. Bayer, A.S. Biris, T. Wang, E. Dervishi, F. Faupel, Advances in top–down and bottom–up surface nanofabrication: Techniques, applications & future prospects, Advances in colloid and interface science, **170** (2012) 2-27
14. A.K. Epstein, J. Aizenberg, Biomimetic nanostructured surfaces with designer mechanics and geometry for broad applications, in: MRS Proceedings, Cambridge Univ Press, 2009, pp. 1236-SS1209-1207
15. L. Li, V. Breedveld, D.W. Hess, Creation of superhydrophobic stainless steel surfaces by acid treatments and hydrophobic film deposition, ACS applied materials & interfaces, **4** (2012) 4549-4556
16. J.F. Schumacher, N. Aldred, M.E. Callow, J.A. Finlay, J.A. Callow, A.S. Clare, A.B. Brennan, Species-specific engineered antifouling topographies: correlations between the settlement of algal zoospores and barnacle cyprids, Biofouling, **23** (2007) 307-317
17. P. Roach, D. Eglin, K. Rohde, C.C. Perry, Modern biomaterials: a review—bulk properties and implications of surface modifications, Journal of Materials Science: Materials in Medicine, **18** (2007) 1263-1277

18. M.V. Graham, A.P. Mosier, T.R. Kiehl, A.E. Kaloyeros, N.C. Cady, Development of antifouling surfaces to reduce bacterial attachment, Soft Matter, **9** (2013) 6235-6244
19. M. Sarikaya, C. Tamerler, A.K.-Y. Jen, K. Schulten, F. Baneyx, Molecular biomimetics: nanotechnology through biology, Nature materials, **2** (2003) 577-585
20. B. Bhushan, E.K. Her, Fabrication of superhydrophobic surfaces with high and low adhesion inspired from rose petal, Langmuir, **26** (2010) 8207-8217
21. F. Guittard, T. Darmanin, Recent Advances in the Potential Applications of Bioinspired Superhydrophobic Materials, Journal of Materials Chemistry A, (2014)
22. O.-U. Nimittrakoolchai, S. Supothina, Deposition of organic-based superhydrophobic films for anti-adhesion and self-cleaning applications, Journal of the European Ceramic Society, **28** (2008) 947-952
23. S. Nishimoto, B. Bhushan, Bioinspired self-cleaning surfaces with superhydrophobicity, superoleophobicity, and superhydrophilicity, Rsc Advances, **3** (2013) 671-690
24. M. Sun, C. Luo, L. Xu, H. Ji, Q. Ouyang, D. Yu, Y. Chen, Artificial lotus leaf by nanocasting, Langmuir, **21** (2005) 8978-8981
25. M. Qu, G. Zhao, X. Cao, J. Zhang, Biomimetic fabrication of lotus-leaf-like structured polyaniline film with stable superhydrophobic and conductive properties, Langmuir, **24** (2008) 4185-4189
26. G.D. Bixler, B. Bhushan, Bioinspired rice leaf and butterfly wing surface structures combining shark skin and lotus effects, Soft Matter, **8** (2012) 11271-11284
27. K. Liu, X. Yao, L. Jiang, Recent developments in bio-inspired special wettability, Chemical Society Reviews, **39** (2010) 3240-3255
28. P. Kim, M.J. Kreder, J. Alvarenga, J. Aizenberg, Hierarchical or not? Effect of the length scale and hierarchy of the surface roughness on omniphobicity of lubricant-infused substrates, Nano letters, **13** (2013) 1793-1799
29. C. Mao, C. Liang, W. Luo, J. Bao, J. Shen, X. Hou, W. Zhao, Preparation of lotus-leaf-like polystyrene micro- and nanostructure films and its blood compatibility, Journal of Materials Chemistry, **19** (2009) 9025-9029
30. K.K. Chung, J.F. Schumacher, E.M. Sampson, R.A. Burne, P.J. Antonelli, A.B. Brennan, Impact of engineered surface microtopography on biofilm formation of Staphylococcus aureus, Biointerphases, **2** (2007) 89-94
31. S.T. Reddy, K.K. Chung, C.J. McDaniel, R.O. Darouiche, J. Landman, A.B. Brennan, Micropatterned surfaces for reducing the risk of catheter-associated urinary tract infection: an in vitro study on the effect of sharklet micropatterned surfaces to inhibit bacterial colonization and migration of uropathogenic Escherichia coli, Journal of Endourology, **25** (2011) 1547-1552
32. Q. Zhao, Y. Liu, C. Wang, S. Wang, H. Müller-Steinhagen, Effect of surface free energy on the adhesion of biofouling and crystalline fouling, Chemical Engineering Science, **60** (2005) 4858-4865
33. M.E. Callow, R.L. Fletcher, The influence of low surface energy materials on bioadhesion—a review, International biodeterioration & biodegradation, **34** (1994) 333-348
34. E. Lindner, A low surface free energy approach in the control of marine biofouling, Biofouling, **6** (1992) 193-205
35. Q. Zhao, C. Liu, Y. Liu, S. Wang, Bacterial and protein adhesion on Ni-P-PTFE coated surfaces, (2007)
36. Hamza, V.A. Pham, T. Matsuura, J.P. Santerre, Development of membranes with low surface energy to reduce the fouling in ultrafiltration applications, Journal of Membrane Science, **131** (1997) 217-227

37. R. Baier, Substrata influences on adhesion of microorganisms and their resultant new surface properties, in, Wiley-Interscience, New York, 1980
38. P.J. Fryer, P.T. Robbins, I. Asteriadou, Current Knowledge in Hygienic Design: Can We Minimise Fouling and Speed Cleaning?, in: Advances in Food Process Engineering Research and Applications, Springer, 2013, pp. 209-227
39. W. Liu, P. Fryer, Z. Zhang, Q. Zhao, Y. Liu, Identification of cohesive and adhesive effects in the cleaning of food fouling deposits, Innovative Food Science & Emerging Technologies, **7** (2006) 263-269
40. M.C. Michalski, S. Desobry, J. Hardy, Food materials adhesion: a review, Critical reviews in food science and nutrition, **37** (1997) 591-619
41. B. Bhandari, T. Howes, Relating the stickiness property of foods undergoing drying and dried products to their surface energetics, Drying Technology, **23** (2005) 781-797
42. R. Rosmaninho, L.F. Melo, Calcium phosphate deposition from simulated milk ultrafiltrate on different stainless steel-based surfaces, International Dairy Journal, **16** (2006) 81-87
43. C. Liu, Q. Zhao, Influence of surface-energy components of Ni–P–TiO2–PTFE nanocomposite coatings on bacterial adhesion, Langmuir, **27** (2011) 9512-9519
44. C.J. Van Oss, Interfacial forces in aqueous media, CRC press, 2006
45. C. Liu, Q. Zhao, The CQ ratio of surface energy components influences adhesion and removal of fouling bacteria, Biofouling, **27** (2011) 275-285
46. A.B. Kananeh, E. Scharnbeck, U. Kück, N. Räbiger, Reduction of milk fouling inside gasketed plate heat exchanger using nano-coatings, Food and Bioproducts Processing, **88** (2010) 349-356
47. K.R. Goode, K. Asteriadou, P.T. Robbins, P.J. Fryer, Fouling and cleaning studies in the food and beverage industry classified by cleaning type, Comprehensive Reviews in Food Science and Food Safety, **12** (2013) 121-143
48. E. Celia, T. Darmanin, E. Taffin de Givenchy, S. Amigoni, F. Guittard, Recent advances in designing superhydrophobic surfaces, Journal of colloid and interface science, **402** (2013) 1-18
49. A.B.D. Cassie, S. Baxter, Wettability of porous surfaces, Transactions of the Faraday Society, **40** (1944) 546-551
50. B.R. Bhandari, N. Datta, T. Howes, Problems associated with spray drying of sugar-rich foods, Drying Technology, **15** (1997) 671-684
51. M. Miwa, A. Nakajima, A. Fujishima, K. Hashimoto, T. Watanabe, Effects of the surface roughness on sliding angles of water droplets on superhydrophobic surfaces, Langmuir, **16** (2000) 5754-5760
52. S.H. Yoon, N. Rungraeng, W. Song, S. Jun, Superhydrophobic and superhydrophilic nanocomposite coatings for preventing Escherichia coli K-12 adhesion on food contact surface, Journal of Food Engineering, **131** (2014) 135-141
53. M. Nosonovsky, B. Bhushan, Superhydrophobic surfaces and emerging applications: non-adhesion, energy, green engineering, Current Opinion in Colloid & Interface Science, **14** (2009) 270-280
54. B. Bhushan, Fabrication Techniques Used for Structures with Superhydrophobicity, Self-Cleaning, Low Adhesion/Low Drag with Antifouling Properties, in: Biomimetics, Springer, 2012, pp. 67-78
55. L. Feng, S. Li, Y. Li, H. Li, L. Zhang, J. Zhai, Y. Song, B. Liu, L. Jiang, D. Zhu, Super-hydrophobic surfaces: from natural to artificial, Advanced materials, **14** (2002) 1857-1860
56. A.K. Epstein, T.-S. Wong, R.A. Belisle, E.M. Boggs, J. Aizenberg, Liquid-infused structured surfaces with exceptional anti-biofouling performance, Proceedings of the National Academy of Sciences, **109** (2012) 13182-13187

57. H. Zhang, R. Lamb, J. Lewis, Engineering nanoscale roughness on hydrophobic surface—preliminary assessment of fouling behaviour, Science and Technology of Advanced Materials, **6** (2005) 236-239
58. Tiraferri, Y. Kang, E.P. Giannelis, M. Elimelech, Superhydrophilic thin-film composite forward osmosis membranes for organic fouling control: fouling behavior and antifouling mechanisms, Environmental science & technology, **46** (2012) 11135-11144
59. S. Song, L. Jing, S. Li, H. Fu, Y. Luan, Superhydrophilic anatase TiO2 film with the micro- and nanometer-scale hierarchical surface structure, Materials Letters, **62** (2008) 3503-3505
60. V. Kumar, J. Pulpytel, G. Giudetti, H. Rauscher, F. Rossi, F. Arefi-Khonsari, Amphiphilic Copolymer Coatings via Plasma Polymerisation Process: Switching and Anti-Biofouling Characteristics, Plasma Processes and Polymers, **8** (2011) 373-385
61. J. Peng, Y. Su, Q. Shi, W. Chen, Z. Jiang, Protein fouling resistant membrane prepared by amphiphilic pegylated polyethersulfone, Bioresource Technology, **102** (2011) 2289-2295
62. C.S. Gudipati, J.A. Finlay, J.A. Callow, M.E. Callow, K.L. Wooley, The antifouling and fouling-release perfomance of hyperbranched fluoropolymer (HBFP)-poly (ethylene glycol)(PEG) composite coatings evaluated by adsorption of biomacromolecules and the green fouling alga ulva, Langmuir, **21** (2005) 3044-3053
63. S. Arifuzzaman, A.E. Özçam, K. Efimenko, D.A. Fischer, J. Genzer, Formation of surface-grafted polymeric amphiphilic coatings comprising ethylene glycol and fluorinated groups and their response to protein adsorption, Biointerphases, **4** (2009) FA33-FA44
64. Y. Wang, D.E. Betts, J.A. Finlay, L. Brewer, M.E. Callow, J.A. Callow, D.E. Wendt, J.M. DeSimone, Photocurable amphiphilic perfluoropolyether/poly (ethylene glycol) networks for fouling-release coatings, Macromolecules, **44** (2011) 878-885
65. Z. Yi, L.-P. Zhu, Y.-Y. Xu, Y.-F. Zhao, X.-T. Ma, B.-K. Zhu, Polysulfone-based amphiphilic polymer for hydrophilicity and fouling-resistant modification of polyethersulfone membranes, Journal of Membrane Science, **365** (2010) 25-33
66. D. Tsuchida, C. Kang, M. Okada, K. Matsumoto, T. Kawagoe, Ice formation process by cooling water–oil emulsion with stirring in a vessel, International journal of refrigeration, **25** (2002) 250-258
67. Banerjee, R.C. Pangule, R.S. Kane, Antifouling coatings: recent developments in the design of surfaces that prevent fouling by proteins, bacteria, and marine organisms, Advanced materials, **23** (2011) 690-718
68. A.B. Kananeh, E. Scharnbeck, D. Hartmann, Application of antifouling surfaces in plate heat exchanger for food production, in: International conference on Heat Exchanger Fouling and Cleaning VIII, Schladming, Austria, 2009
69. E. Sadeghinezhad, S.N. Kazi, A. Badarudin, M.N.M. Zubair, B.L. Dehkordi, C.S. Oon, A review of milk fouling on heat exchanger surfaces, Reviews in Chemical Engineering, **29** (2013) 169-188
70. S. Krishnan, C.J. Weinman, C.K. Ober, Advances in polymers for anti-biofouling surfaces, Journal of Materials Chemistry, **18** (2008) 3405-3413
71. J.W. Costerton, P.S. Stewart, E.P. Greenberg, Bacterial Biofilms: A Common Cause of Persistent Infections, science, **284** (1999) 1318-1322
72. W.G. Sawyer, K.D. Freudenberg, P. Bhimaraj, L.S. Schadler, A study on the friction and wear behavior of PTFE filled with alumina nanoparticles, Wear, **254** (2003) 573-580
73. F.W. Billmeyer Jr, Textbook of polymer science in: New York, 1984

74. B.A. Krick, J.J. Ewin, G.S. Blackman, C.P. Junk, W. Gregory Sawyer, Environmental dependence of ultra-low wear behavior of polytetrafluoroethylene (PTFE) and alumina composites suggests tribochemical mechanisms, Tribology International, **51** (2012) 42-46
75. R. Rosmaninho, O. Santos, T. Nylander, M. Paulsson, M. Beuf, T. Benezech, S. Yiantsios, N. Andritsos, A. Karabelas, G. Rizzo, Modified stainless steel surfaces targeted to reduce fouling–evaluation of fouling by milk components, Journal of Food Engineering, **80** (2007) 1176-1187
76. Santos, T. Nylander, R. Rosmaninho, G. Rizzo, S. Yiantsios, N. Andritsos, A. Karabelas, H. Müller-Steinhagen, L. Melo, L. Boulangé-Petermann, C. Gabet, A. Braem, C. Trägårdh, M. Paulsson, Modified stainless steel surfaces targeted to reduce fouling—surface characterization, Journal of Food Engineering, **64** (2004) 63-79
77. S. Armyanov, J. Georgieva, D. Tachev, E. Valova, N. Nyagolova, S. Mehta, D. Leibman, A. Ruffini, Electroless Deposition of Ni-Cu-P Alloys in Acidic Solutions, Electrochemical and solid-state letters, **2** (1999) 323-325
78. N. Rungraeng, Y.-C. Cho, S.H. Yoon, S. Jun, Carbon nanotube-polytetrafluoroethylene nanocomposite coating for milk fouling reduction in plate heat exchanger, Journal of Food Engineering, **111** (2012) 218-224
79. W. Chen, F. Li, G. Han, J. Xia, L. Wang, J. Tu, Z. Xu, Tribological behavior of carbon-nanotube-filled PTFE composites, Tribology Letters, **15** (2003) 275-278
80. N. Rungraeng, S. Jun, Carbon Nanotubes–Polytetrafluoroethylene Nanocomposite Coatings, Polymer Nanocomposite Coatings, (2013) 61
81. Q. Zhao, Y. Liu, C. Wang, Development and evaluation of electroless Ag-PTFE composite coatings with anti-microbial and anti-corrosion properties, Applied Surface Science, **252** (2005) 1620-1627
82. E. Ostuni, R.G. Chapman, R.E. Holmlin, S. Takayama, G.M. Whitesides, A survey of structure-property relationships of surfaces that resist the adsorption of protein, Langmuir, **17** (2001) 5605-5620
83. R.G. Chapman, E. Ostuni, S. Takayama, R.E. Holmlin, L. Yan, G.M. Whitesides, Surveying for surfaces that resist the adsorption of proteins, Journal of the American Chemical Society, **122** (2000) 8303-8304
84. B. Nisol, G. Oldenhove, N. Preyat, D. Monteyne, M. Moser, D. Perez-Morga, F. Reniers, Atmospheric plasma synthesized PEG coatings: non-fouling biomaterials showing protein and cell repulsion, Surface and Coatings Technology, **252** (2014) 126-133
85. D. Perera-Costa, J.M. Bruque, M.a.L. González-Martín, A.C. Gómez-García, V. Vadillo-Rodriguez, Studying the Influence of Surface Topography on Bacterial Adhesion using Spatially Organized Microtopographic Surface Patterns, Langmuir, **30** (2014) 4633-4641
86. J. Cui, Y. Ju, K. Liang, H. Ejima, S. Lörcher, K.T. Gause, J.J. Richardson, F. Caruso, Nanoscale engineering of low-fouling surfaces through polydopamine immobilisation of zwitterionic peptides, Soft Matter, **10** (2014) 2656-2663
87. S. Jiang, Z. Cao, Ultralow-Fouling, Functionalizable, and Hydrolyzable Zwitterionic Materials and Their Derivatives for Biological Applications, Advanced materials, **22** (2010) 920-932
88. M. Hess, R. Jones, J. Kahovec, T. Kitayama, P. Kratochvíl, P. Kubisa, W. Mormann, R. Stepto, D. Tabak, J. Vohlídal, Terminology of polymers containing ionizable or ionic groups and of polymers containing ions (IUPAC Recommendations 2006), Pure and Applied Chemistry, **78** (2006) 2067-2074

89. H.-W. Chien, C.-C. Tsai, W.-B. Tsai, M.-J. Wang, W.-H. Kuo, T.-C. Wei, S.-T. Huang, Surface conjugation of zwitterionic polymers to inhibit cell adhesion and protein adsorption, Colloids and Surfaces B: Biointerfaces, **107** (2013) 152-159
90. W. Feng, S. Zhu, K. Ishihara, J.L. Brash, Protein resistant surfaces: Comparison of acrylate graft polymers bearing oligo-ethylene oxide and phosphorylcholine side chains, Biointerphases, **1** (2006) 50-60
91. Y. He, J. Hower, S. Chen, M.T. Bernards, Y. Chang, S. Jiang, Molecular simulation studies of protein interactions with zwitterionic phosphorylcholine self-assembled monolayers in the presence of water, Langmuir, **24** (2008) 10358-10364
92. H. Zhang, J. Zou, N. Lin, B. Tang, Review On Electroless Plating Ni–P Coatings For Improving Surface Performance Of Steel, Surface Review and Letters, (2014)
93. J. Novakovic, P. Vassiliou, K. Samara, T. Argyropoulos, Electroless NiP–TiO< sub> 2</sub> composite coatings: Their production and properties, Surface and Coatings Technology, **201** (2006) 895-901
94. W. Chen, W. Gao, Y. He, A novel electroless plating of Ni–P–TiO< sub> 2</sub> nano-composite coatings, Surface and Coatings Technology, **204** (2010) 2493-2498.
95. Q. Zhao, C. Liu, X. Su, S. Zhang, W. Song, S. Wang, G. Ning, J. Ye, Y. Lin, W. Gong, Antibacterial characteristics of electroless plating Ni–P–TiO2 coatings, Applied Surface Science, **274** (2013) 101-104
96. X.-P. Wang, Y. Yu, X.-F. Hu, L. Gao, Hydrophilicity of TiO< sub> 2</sub> films prepared by liquid phase deposition, Thin Solid Films, **371** (2000) 148-152.
97. B. Li, B.E. Logan, The impact of ultraviolet light on bacterial adhesion to glass and metal oxide-coated surface, Colloids and Surfaces B: Biointerfaces, **41** (2005) 153-161
98. Al-Janabi, M. Malayeri, H. Müller-Steinhagen, Minimization of CaSO4 deposition through surface modification, Heat transfer engineering, **32** (2011) 291-299
99. F. Siedenbiedel, J.C. Tiller, Antimicrobial polymers in solution and on surfaces: overview and functional principles, Polymers, **4** (2012) 46-71
100. P.N. Coneski, P.A. Fulmer, J.H. Wynne, Enhancing the fouling resistance of biocidal urethane coatings via surface chemistry modulation, Langmuir, **28** (2012) 7039-7048
101. R. Kumar, H. Münstedt, Silver ion release from antimicrobial polyamide/silver composites, Biomaterials, **26** (2005) 2081-2088
102. A.W. Smith, Biofilms and antibiotic therapy: is there a role for combating bacterial resistance by the use of novel drug delivery systems?, Advanced drug delivery reviews, **57** (2005) 1539-1550
103. M. Simões, L.C. Simões, M.J. Vieira, A review of current and emergent biofilm control strategies, LWT - Food Science and Technology, **43** (2010) 573-583
104. P.N. Coneski, M.H. Schoenfisch, Synthesis of nitric oxide-releasing polyurethanes with S-nitrosothiol-containing hard and soft segments, Polymer chemistry, **2** (2011) 906-913
105. J.C. Tiller, C.-J. Liao, K. Lewis, A.M. Klibanov, Designing surfaces that kill bacteria on contact, Proceedings of the National Academy of Sciences, **98** (2001) 5981-5985
106. J.W. Mayer, L. Eriksson, J.A. Davies, Ion implantation in semiconductors: silicon and germanium, (1970)
107. M. Dhillon, The effect of silver ion-implantation of stainless steel on bacterial adhesion and biofilm formation, in: Institute of Food, Nutrition and Human Health, Massey University, Palmerston North, 2012
108. H. Müller-Steinhagen, Q. Zhao, Investigation of low fouling surface alloys made by ion implantation technology, Chemical Engineering Science, **52** (1997) 3321-3332
109. Q. Zhao, B. Burnside, Dropwise condensation of steam on ion implanted condenser surfaces, Heat Recovery Systems and CHP, **14** (1994) 525-534

110. Nejim, C. Jeynes, Q. Zhao, H. Muller-Steinhagen, Ion implantation of stainless steel heater alloys for anti-fouling applications, in: Ion Implantation Technology Proceedings, 1998 International Conference on, IEEE, 1999, pp. 869-872
111. Q. Zhao, Y. Liu, C. Wang, S. Wang, N. Peng, C. Jeynes, Reduction of bacterial adhesion on ion-implanted stainless steel surfaces, Medical engineering & physics, **30** (2008) 341-349
112. M. Beuf, G. Rizzo, J. Leuliet, H. Müller-Steinhagen, S. Yiantsios, A. Karabelas, T. Benezech, Fouling and cleaning of modified stainless steel plate heat exchangers processing milk products, Heat Exchanger Fouling and Cleaning: Fundamentals and Applications, **14** (2003)
113. D. Quain, E. Storgårds, The extraordinary world of biofilms, PloS Biology, **5** (2009) 2458-2461
114. D. Merche, N. Vandencasteele, F. Reniers, Atmospheric plasmas for thin film deposition: A critical review, Thin Solid Films, **520** (2012) 4219-4236
115. R. Rosmaninho, F. Rocha, G. Rizzo, H. Müller-Steinhagen, L. Melo, Calcium phosphate fouling on TiN-coated stainless steel surfaces: Role of ions and particles, Chemical Engineering Science, **62** (2007) 3821-3831
116. R. Rosmaninho, G. Rizzo, H. Müller-Steinhagen, L. Melo, Deposition from a milk mineral solution on novel heat transfer surfaces under turbulent flow conditions, Journal of Food Engineering, **85** (2008) 29-41
117. X. Hou, K.L. Choy, Processing and Applications of Aerosol-Assisted Chemical Vapor Deposition, Chemical vapor deposition, **12** (2006) 583-596
118. R.G. Palgrave, I.P. Parkin, Aerosol assisted chemical vapour deposition of photochromic tungsten oxide and doped tungsten oxide thin films, Journal of Materials Chemistry, **14** (2004) 2864-2867
119. K. Choy, Chemical vapour deposition of coatings, Progress in materials science, **48** (2003) 57-170
120. C.R. Crick, S. Ismail, J. Pratten, I.P. Parkin, An investigation into bacterial attachment to an elastomeric superhydrophobic surface prepared< i> via</i> aerosol assisted deposition, Thin Solid Films, **519** (2011) 3722-3727
121. R. Hauert, A review of modified DLC coatings for biological applications, Diamond and Related Materials, **12** (2003) 583-589
122. C. Boxler, W. Augustin, S. Scholl, Fouling of milk components on DLC coated surfaces at pasteurization and UHT temperatures, Food and Bioproducts Processing, **91** (2013) 336-347
123. G. Dearnaley, J.H. Arps, Biomedical applications of diamond-like carbon (DLC) coatings: A review, Surface and Coatings Technology, **200** (2005) 2518-2524
124. C. Donnet, Recent progress on the tribology of doped diamond-like and carbon alloy coatings: a review, Surface and Coatings Technology, **100** (1998) 180-186
125. Soininen, E. Kaivosoja, T. Sillat, S. Virtanen, Y.T. Konttinen, V.M. Tiainen, Osteogenic differentiation on DLC-PDMS-h surface, Journal of Biomedical Materials Research Part B: Applied Biomaterials, (2014)
126. X. Su, Development and evaluation of anti-biofouling nano-composite coatings, in, University of Dundee, 2013
127. C. Liu, Q. Zhao, Y. Liu, S. Wang, E. Abel, Reduction of bacterial adhesion on modified DLC coatings, Colloids and Surfaces B: Biointerfaces, **61** (2008) 182-187
128. X. Su, Q. Zhao, S. Wang, A. Bendavid, Modification of diamond-like carbon coatings with fluorine to reduce biofouling adhesion, Surface and Coatings Technology, **204** (2010) 2454-2458

129. M. Ishihara, T. Kosaka, T. Nakamura, K. Tsugawa, M. Hasegawa, F. Kokai, Y. Koga, Antibacterial activity of fluorine incorporated DLC films, Diamond and Related Materials, **15** (2006) 1011-1014
130. C. Boxler, W. Augustin, S. Scholl, Influence of surface modification on the composition of a calcium phosphate-rich whey protein deposit in a plate heat exchanger, Dairy Science & Technology, **94** (2014) 17-31
131. T.-S. Wong, S.H. Kang, S.K. Tang, E.J. Smythe, B.D. Hatton, A. Grinthal, J. Aizenberg, Bioinspired self-repairing slippery surfaces with pressure-stable omniphobicity, Nature, **477** (2011) 443-447
132. H.F. Bohn, W. Federle, Insect aquaplaning: Nepenthes pitcher plants capture prey with the peristome, a fully wettable water-lubricated anisotropic surface, Proceedings of the National Academy of Sciences of the United States of America, **101** (2004) 14138-14143
133. L. Xiao, J. Li, S. Mieszkin, A. Di Fino, A.S. Clare, M.E. Callow, J.A. Callow, M. Grunze, A. Rosenhahn, P.A. Levkin, Slippery Liquid-Infused Porous Surfaces Showing Marine Antibiofouling Properties, ACS applied materials & interfaces, **5** (2013) 10074-10080
134. M. Nosonovsky, Materials science: Slippery when wetted, Nature, **477** (2011) 412-413
135. J.D. Smith, R. Dhiman, S. Anand, E. Reza-Garduno, R.E. Cohen, G.H. McKinley, K.K. Varanasi, Droplet mobility on lubricant-impregnated surfaces, Soft Matter, **9** (2013) 1772-1780
136. P.W. Wilson, W. Lu, H. Xu, P. Kim, M.J. Kreder, J. Alvarenga, J. Aizenberg, Inhibition of ice nucleation by slippery liquid-infused porous surfaces (SLIPS), Physical Chemistry Chemical Physics, **15** (2013) 581-585
137. C. Ishino, M. Reyssat, E. Reyssat, K. Okumura, D. Quere, Wicking within forests of micropillars, EPL (Europhysics Letters), **79** (2007) 56005
138. C. Howell, T.L. Vu, J.J. Lin, S. Kolle, N. Juthani, E. Watson, J.C. Weaver, J. Alvarenga, J. Aizenberg, Self-replenishing vascularized fouling-release surfaces, ACS applied materials & interfaces, (2014)
139. S.B. Subramanyam, G. Azimi, K.K. Varanasi, Designing Lubricant-Impregnated Textured Surfaces to Resist Scale Formation, Advanced Materials Interfaces, **1** (2014)
140. K.S. Khalil, S.R. Mahmoudi, N. Abu-dheir, K.K. Varanasi, Active surfaces: Ferrofluid-impregnated surfaces for active manipulation of droplets, Applied Physics Letters, **105** (2014) 041604
141. A.K. Epstein, D. Hong, P. Kim, J. Aizenberg, Biofilm attachment reduction on bioinspired, dynamic, micro-wrinkling surfaces, New Journal of Physics, **15** (2013) 095018
142. S.Y.-E. Tan, S.C. Chew, S.Y.-Y. Tan, M. Givskov, L. Yang, Emerging frontiers in detection and control of bacterial biofilms, Current opinion in biotechnology, **26** (2014) 1-6
143. P. Kim, T.-S. Wong, J. Alvarenga, M.J. Kreder, W.E. Adorno-Martinez, J. Aizenberg, Liquid-infused nanostructured surfaces with extreme anti-ice and anti-frost performance, ACS Nano, **6** (2012) 6569-6577
144. S.B. Subramanyam, K. Rykaczewski, K.K. Varanasi, Ice adhesion on lubricant-impregnated textured surfaces, Langmuir, **29** (2013) 13414-13418
145. T. Awad, H. Moharram, O. Shaltout, D. Asker, M. Youssef, Applications of ultrasound in analysis, processing and quality control of food: A review, Food Research International, **48** (2012) 410-427
146. J. Mason, F. Chemat, M. Vinatoru, The extraction of natural products using ultrasound or microwaves, Current Organic Chemistry, **15** (2011) 237-247

147. K.S. Suslick, The chemical effects of ultrasound, Scientific American, **260** (1989) 80-86
148. F. Chemat, H. Zill e, M.K. Khan, Applications of ultrasound in food technology: Processing, preservation and extraction, Ultrasonics Sonochemistry, **18** (2011) 813-835
149. E. Joyce, S. Phull, J. Lorimer, T. Mason, The development and evaluation of ultrasound for the treatment of bacterial suspensions. A study of frequency, power and sonication time on cultured< i> Bacillus</i> species, Ultrasonics Sonochemistry, **10** (2003) 315-318
150. P. Piyasena, E. Mohareb, R. McKellar, Inactivation of microbes using ultrasound: a review, International journal of food microbiology, **87** (2003) 207-216
151. E.T. Thostenson, T.W. Chou, Microwave processing: fundamentals and applications, Composites Part A: Applied Science and Manufacturing, **30** (1999) 1055-1071
152. J. Ahmed, H.S. Ramaswamy, Microwave pasteurization and sterilization of foods, Food Science And Technology-New York-Marcel Dekker-, **167** (2004) 691
153. D.E. Clark, W.H. Sutton, Microwave processing of materials, Annual Review of Materials Science, **26** (1996) 299-331
154. J. Smith, The assessment of structure and component mobility of Mozzarella cheese, in: Department of Food Science & Technology, Massey University, Palmerston North, 2013
155. S.O. Nelson, Dielectric properties of agricultural products: measurements and applications, IEEE Transactions on Electrical Insulation, **26** (1991) 845-869
156. J. Smith, A. Carr, M. Golding, D. Reid, L. Zhang, Assessing the use of dielectric spectroscopy to analyse calcium induced compositional and structural changes in a model cheese, Procedia Food Science, **1** (2011) 1833-1840
157. M.A. Rao, S.S. Rizvi, A.K. Datta, J. Ahmed, Engineering properties of foods, CRC Press, 2014
158. E. Nyfors, P. Vainikainen, Industrial microwave sensors, Artech House Norwood (MA, USA), 1989
159. S. Ryynänen, The electromagnetic properties of food materials: a review of the basic principles, Journal of Food Engineering, **26** (1995) 409-429
160. Herve, J. Tang, L. Luedecke, H. Feng, Dielectric properties of cottage cheese and surface treatment using microwaves, Journal of Food Engineering, **37** (1998) 389-410
161. H. Feng, Y. Yin, J. Tang, Microwave drying of food and agricultural materials: basics and heat and mass transfer modeling, Food Engineering Reviews, **4** (2012) 89-106
162. R.V. Decareau, R.A. Peterson, Microwave processing and engineering, VCH, 1986
163. M. Venkatesh, G. Raghavan, An overview of dielectric properties measuring techniques, Canadian biosystems engineering, **47** (2005) 15-30
164. C. Tong, R. Lentz, J. Rossen, Dielectric properties of pea puree at 915 MHz and 2450 MHz as a function of temperature, Journal of food science, **59** (1994) 121-122
165. C. Salazar-González, M.F. San Martín-González, A. López-Malo, M.E. Sosa-Morales, Recent studies related to microwave processing of fluid foods, Food and bioprocess technology, **5** (2012) 31-46
166. R. Vadivambal, D. Jayas, Non-uniform temperature distribution during microwave heating of food materials—A review, Food and bioprocess technology, **3** (2010) 161-171
167. M. Zhang, J. Tang, A. Mujumdar, S. Wang, Trends in microwave-related drying of fruits and vegetables, Trends in Food Science & Technology, **17** (2006) 524-534
168. S. Chandrasekaran, S. Ramanathan, T. Basak, Microwave food processing—A review, Food Research International, **52** (2013) 243-261

169. Z.-W. Cui, L.-J. Sun, W. Chen, D.-W. Sun, Preparation of dry honey by microwave–vacuum drying, Journal of Food Engineering, **84** (2008) 582-590
170. P.A. Picouet, A. Landl, M. Abadias, M. Castellari, I. Viñas, Minimal processing of a Granny Smith apple purée by microwave heating, Innovative Food Science & Emerging Technologies, **10** (2009) 545-550
171. R.I. Guzmán-Gerónimo, M.G. López, L. Dorantes-Alvarez, Microwave processing of avocado: Volatile flavor profiling and olfactometry, Innovative Food Science & Emerging Technologies, **9** (2008) 501-506
172. Fratianni, L. Cinquanta, G. Panfili, Degradation of carotenoids in orange juice during microwave heating, LWT-Food Science and Technology, **43** (2010) 867-871
173. S. Tajchakavit, H. Ramaswamy, P. Fustier, Enhanced destruction of spoilage microorganisms in apple juice during continuous flow microwave heating, Food Research International, **31** (1998) 713-722
174. K.N. Matsui, J.A.W. Gut, P.V. De Oliveira, C.C. Tadini, Inactivation kinetics of polyphenol oxidase and peroxidase in green coconut water by microwave processing, Journal of Food Engineering, **88** (2008) 169-176
175. M. Igual, E. García-Martínez, M. Camacho, N. Martínez-Navarrete, Effect of thermal treatment and storage on the stability of organic acids and the functional value of grapefruit juice, Food Chemistry, **118** (2010) 291-299

Generalized optimizations of two-stage forging of micro/meso copper fastener

Shih-Hsien Lin[1], *Un-Chin* Chai, *Gow-Yi* Tzou[2,*], and *Dyi-Cheng* Chen[1]
[1]Department of Industrial Education and Technology, National Changhua University of Education, Changhua, 500, Taiwan
[2]Department of Mechanical and Automation Engineering, Chung-Chou University of Science and Technology, Yuanlin, Changhua, 510, Taiwan

Abstract. Three are generalized simulation optimizations considering the forging force, the die stress, and the dual-goals in two-stage forging of micro/meso copper fastener. Constant shear friction between the dies and workpiece is assumed to perform multi-stage cold forging forming simulation analysis, and the Taguchi method with the finite element simulation has been used for mold-and-dies parameters design simulation optimizations considering the forging force, die stress, and dual-goals. The die stress optimization is used to explore the effects on effective stress, effective strain, velocity field, die stress, forging force, and shape of product. The influence rank to forging process of micro/meso copper fastener for three optimizations can be determined, and the optimal parameters assembly consider die stress can be obtained in this study. It is noted that the punch design innovation can reduce the forging force and die stress.

1 Introduction

Shah et al. [1] used FEM to do cold forging simulation analysis of thread head, besides comparing simulation results with experiment to verify the acceptance of this model. Vickers et al. [2] used plasticine, aluminum, 6061-T6 aluminum alloy as the simulated materials, using the experimental to analyze cold forging of thread head. MacCormack et al. [3] used FEM simulation to analyze the die of hexagonal bolt in the multi-stage cold forging, to increase die's life and successfully reduce die stress 17.7%. Asnafi [4] used FEM to analyze the dies of cold forging, the reasons for failure of dies can be found through the experimental method. Sun et al. [5] aimed at stainless automotive battery fastener to do the processing animation simulation and FEM analysis of multi-stage cold forging; propose the pass schedule plan to carry out the experiments comparing with FEM simulation results the good agreement has been done. Shih et al. [6] proposed FEM simulation and experimental verification on multi-stage forming of flange sleeve; using Deform 3D commercial software to design the pass schedule and the dies dimensions, and realistic experiments have been performed to verify the acceptance of FEM simulation.

[*]Corresponding author: gowyitzou@gm.ccut.edu.tw

Engel et al. [7] firstly used basic research to realize what is micro-forming. Wang et al. [8] explore size effects of the cavity dimension on the micro-forming ability during coining process. Gau et al. [9] utilized an experimental study to investigate into the size effects on flow stress and formability of aluminum and brass for micro-forming; the size effect reduces the flow stress. Engel et al. [10] continued to explore tribology in micro-forming; the tribology concept for micro-forming is different from traditional tribology. Chen [11] proposed the robust design on equal channel angle extrusion of Ti-6Al-4V using Taguchi method; combine FEM simulation with Taguchi method to obtain the optimal parameters assembly to reach robust design. G.Y. Tzou et al.[12] first proposed a study to change the traditional punch design of micro/meso copper fastener forging, and to conduct the forging force optimization to explore the effects on effective stress, effective strain, velocity field, die stress, and shape of product. The optimal parameters assembly and the control factors influence could be obtained. This present research then proposes three generalized optimizations and comparisons including the forging force, die stress, and dual-goals.

2 FEM analyses

2.1 Material properties and product specifications

Figure 1 is a diagram of micro/meso copper fastener for each pass. The copper fastener in the second stage is the specification of final product. The workpiece is a copper, C2600; the flow stress is $\sigma = 553.547\varepsilon^{0.26}$, which is an elastic-plastic material. This study uses Deform-2D to perform FEM simulation; the simulation conditions are summarized in table 1. The friction is assumed to be constant shear friction. Figure 2 shows schematic diagram before and after forming for the first pass. Figure 3 shows a schematic diagram of punch and die for the first pass. In Figure 3, the major four parameters are shown as, punch angle (α), die angle (β), die bottom height (H), and die filet (R); Taguchi method can be used to carry out the force optimization with these four control factors. The orthogonal table, $L_9(3^4)$, can be used to do experimental plan including four control factors and three levels. The signal noise ratio can be used to obtain the optimal parameters assembly of reducing forging force.

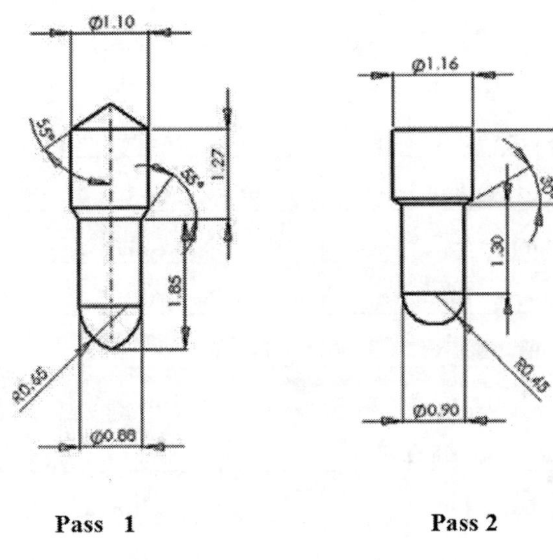

Fig. 1. Diagram of micro/meso copper fastener for each pass.

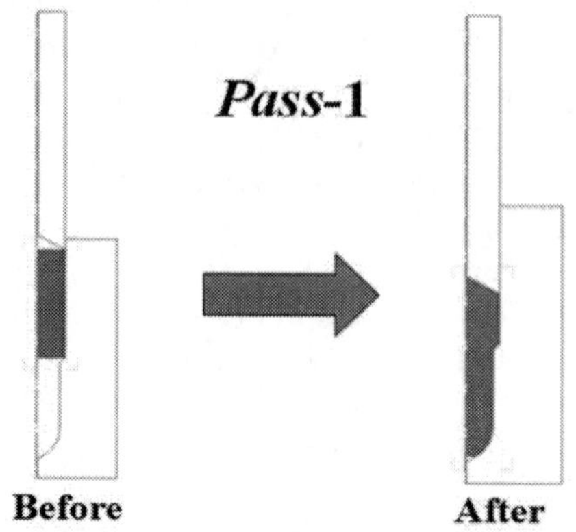

Fig. 2. Schematic diagram before and after forming for the first pass.

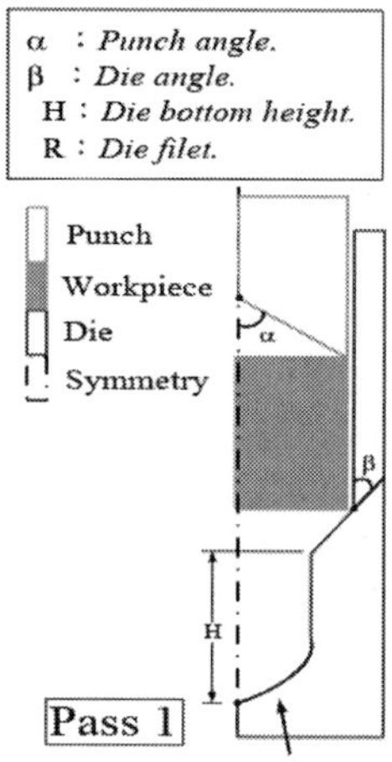

Fig. 3. Schematic diagram of punch and diefor the first pass.

Table 1. FEM simulation conditions.

Workpiece	**C2600**
Young's Modulus (*E*)	110GPa
Poisson's Ratio (υ)	0.28
Power law	$\sigma = 553.547\varepsilon^{0.26}$
Object type	Elasto-Plastic
Number of Elements	8800
Front die	**Rigid body**
Velocity (V_o)	0.1mm/s
Shear Friction (*m*)	0.2
Rear die	**Rigid body**
Velocity (V_o)	0 mm/s
Shear Friction (*m*)	0.2

3 Optimization and results

3.1 Optimization of forging force

The first part is doing the optimization of forging force shown in Figure 4. Figure 4 shows control factors and levels for forging force optimization.

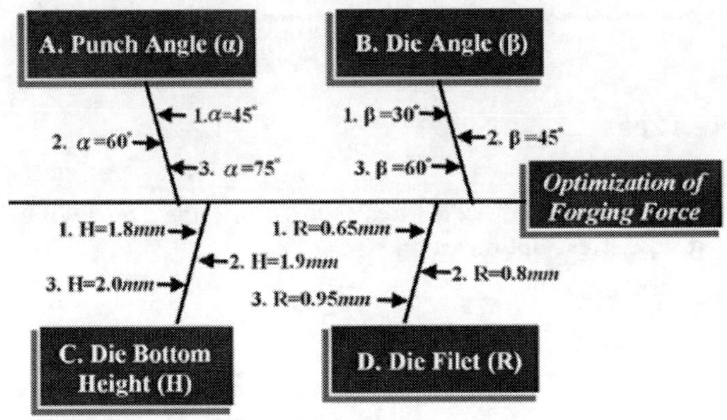

Fig. 4. Control factors and levels for forging force optimization.

According to the layout in Figure 4, L_9 (3^4) can be used to obtain S/N of forging force, the analysis results are summarized in Table 2. In Table 2, the fourth simulation experiment (IV) has the maximum forging force and the minimum S/N; the ninth simulation experiment (IX) has the minimum forging force and the maximum S/N, where the best parameters assembly in orthogonal table is $A_3B_3C_2D_1$(i.e. α=75°, β=60°, H=1.9 mm, R=0.65mm). Table 3 shows S/N response characteristics, form this table, the punch angle (α) influences to forging force very much, then next is die angle (β). The optimal parameters assembly is $A_1B_3C_2D_1$ (i.e. α=45°, β=60°, H=1.9 mm, R=0.65mm). Using this assembly to run Deform-2D, the forging force is 1.53kN. The optimal force, 1.53 kN, is lower than that obtained from the orthogonal table, 1.72 kN. Let this optimal force compare to the worst force in Table L_9 (3^4), the forging force is improved 47.67%. If let this optimal force compare to the best force in Table 2, L_9 (3^4), the forging force is improved 12.4%.

Table 2. $L_9(3^4)$ simulation results for forging force.

EXP	A (α)	B (β)	C (H)	D (R)	Force (kN)	S/N
I	1	1	1	1	2.05	-6.235
II	1	2	2	2	1.90	-5.575
III	1	3	3	3	1.90	-5.575
IV	**2**	**1**	**2**	**3**	**2.54**	**-8.097**
V	2	2	3	1	2.20	-6.848
VI	2	3	1	2	2.29	-7.197
VII	3	1	3	2	2.32	-7.310
VIII	3	2	1	3	1.88	-5.483
IX	3	3	2	1	1.72	-4.711

Table 3. S/N response characteristics for forging force.

S/N	A (α)	B (β)	C (H)	D (R)
Level 1	-5.7951	-7.2138	-6.3050	-5.9314
Level 2	-7.3806	-5.9689	-6.1274	-6.6938
Level 3	-5.8345	-5.8275	-6.5778	-6.3850
Effect	1.5855	1.3864	0.4503	0.7625
Rank	1	2	4	3

3.2 Optimization of die stress

The second part is doing the optimization of die stress shown in Figure 5. Figure 5 shows control factors and levels for die stress optimization.

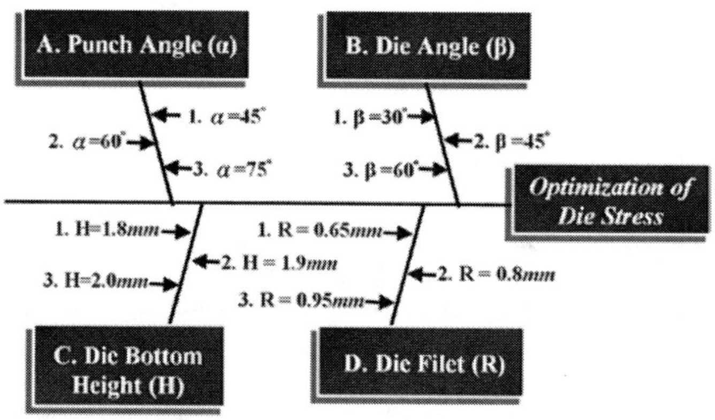

Fig. 5. Control factors and levels for die stress optimization.

Table 4. $L_9(3^4)$ simulation results for die stress.

EXP	A (α)	B (β)	C (H)	D (R)	Stress (MPa)	S/N
I	1	1	1	1	3480	-70.832
II	**1**	**2**	**2**	**2**	**3570**	**-71.053**
III	1	3	3	3	3550	-71.005
IV	2	1	2	3	3430	-70.706
V	2	2	3	1	3240	-70.211
VI	2	3	1	2	3370	-70.553
VII	3	1	3	2	3290	-70.344
VIII	3	2	1	3	3510	-70.906
IX	3	3	2	1	3390	-70.604

According to the layout in Figure 5, L_9 (3^4) can be used to obtain S/N of die stress, the analysis results are summarized in Table 4. In Table 4, the second simulation experiment (II) has the maximum die stress and the minimum S/N; the fifth simulation experiment (V) has the minimum die stress and the maximum S/N, where the best parameters assembly in orthogonal table is $A_2B_2C_3D_1$(i.e. α=60°, β=45°, H=2.0 mm, R=0.65mm). Table 5 shows

S/N response characteristics, form this table, the punch angle (α) influences to die stress very much, then next is die filet (R). The optimal parameters assembly is $A_2B_1C_3D_1$ (i.e. α=60°, β=60°, H=2.0 mm, R=0.65mm). Using this assembly to run Deform-2D, the die stress is 2500MPa. The optimal dies stress, 2500MPa, is lower than that obtained from the orthogonal table, 3240MPa. Let this optimal die stress compare to the worst force in Table L_9 (3^4), the die stress is improved 42.8%. If let this optimal force compare to the best force in Table 5, and the die stress is improved 29.6%.

Table 5. S/N response characteristics for die stress.

S/N	A (α)	B (β)	C (H)	D (R)
Level 1	-70.9632	-70.6271	-70.7634	-70.5488
Level 2	-70.4898	-70.7235	-70.7877	-70.6500
Level 3	-70.6180	-70.7204	-70.5198	-70.8722
Effect	0.4734	0.0963	0.2680	0.3234
Rank	1	4	3	2

3.3 Optimization of dual-goals

In this study, the minimization of die stress is set as the primary goal and the minimization of forging force is set as the secondary goal. The optimum combination of parameters is shown in Table 6. Table 6 shows the comparison of the optimum combinations of parameters and the influence ranks of forging parameters between the die stress optimization and forging force optimization. As shown in Table 6, the optimum combination of parameters for the die stress optimization is $A_2B_1C_3D_1$ (α=60°, β=30°, H=2.0mm, R=0.65mm). Punch angle (α) has the most effect on the die stress, following by die filet (R), die bottom height (H), and the die angle (β). In the forging force optimization, the optimum combination of parameters is $A_2B_3C_2D_1$ (α=60°, β=30°, H=1.9mm, R=0.65mm). Punch angle (α) has the largest influence on the forging force, following by die angle (β), die fillet (R), and the die bottom height (H).

As mentioned before, the minimization of die stress is the primary goal in this part and the minimization of forging force is the secondary goal. Therefore, the first two forging parameters of the die stress optimization are chosen first. They are punch angle (A2) and die filet (D1). Then, the first two forging parameters of the die stress optimization are chosen. However, the first (A2) and the fourth (D1) forging parameters of the forging force optimization overlap with the forging parameters of the die stress optimization (A2 and D1). Therefore, the second (B3) and the third (C2) forging parameters of the forging force optimization are chosen. The optimum combination of parameters for the dual-goals optimization is formed as $A_2B_3C_2D_1$ (α=60°, β=60°, H=1.9mm, R=0.65mm).

Table 6. S/N response characteristics for the dual-goals optimization.

Primary goal: Die stress	Optimum combination of parameters	A_2 (α)	B_1 (β)	C_3 (H)	D_1 (R)
	Influence rank	1	4	3	2
Secondary goal: Forging force	Optimum combination of parameters	A_1 (α)	B_3 (β)	C_2 (H)	D_1 (R)
	Influence rank	1	2	4	3

3.4 Analysis of generalized optimization

The optimum combinations of parameters of three generalized optimizations are summarized as follows:

Optimum combination of parameters for dual-goals：$A_2B_3C_2D_1$, i.e. α=60°, β=60°, H=1.9 mm, R=0.65mm.

Optimum combination of parameters for die stress：$A_2B_1C_3D_1$, i.e. α=60°, β=30°, H=2mm, R=0.65mm.

Optimum combination of parameters for forging force：$A_1B_3C_2D_1$, i.e. α=45°, β=60°, H=1.9 mm, R=0.65mm.

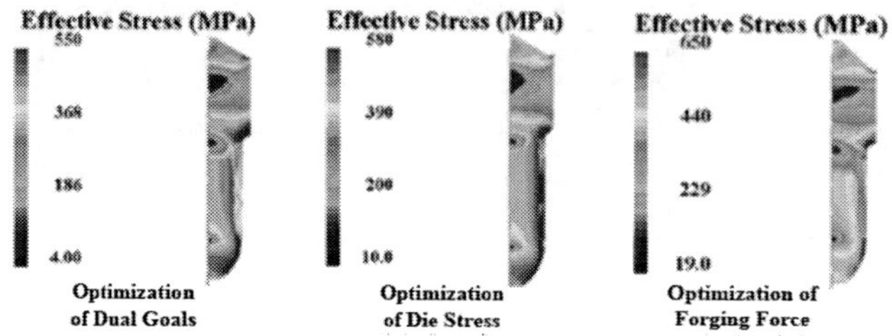

Fig. 6. Effective stresses for three generalized optimization at pass 1.

Figure 6 is effective stresses for three generalized optimization at pass 1. The maximun effective stress is 550MPa for dual-goals optimization, 580MPa for dies stress optimization, and 650MPa for forging force optimization. The smaller effective stress is induced in the dual-goals optimization.

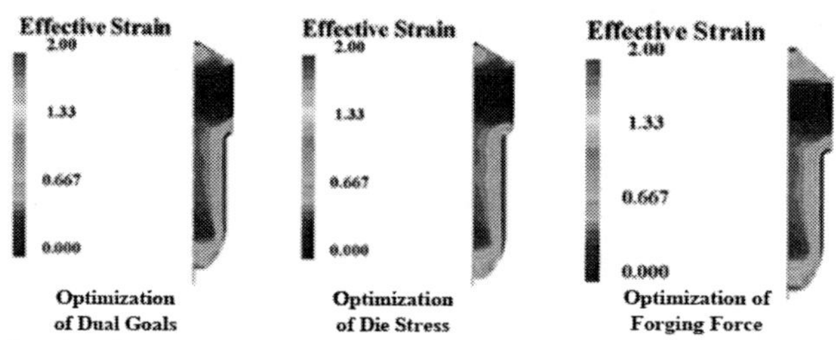

Fig. 7. Effective strains for three generalized optimization at pass 1.

Figure 7 is effective strains for three generalized optimization at pass 1. The maximun effective strain is 2 mm/mm for generalized optimization at pass 1.

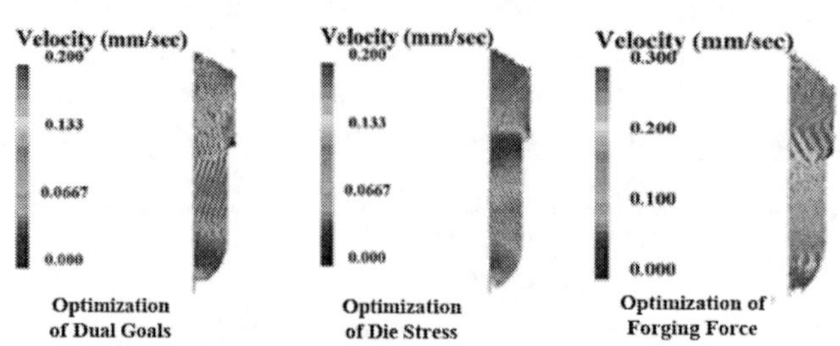

Fig. 8. Velocity fields for three generalized optimization at pass 1.

Figure 8 is velocity fields for three generalized optimization at pass 1. The maximun velocity is 0.3mm/sec for forging force optimization at pass 1.

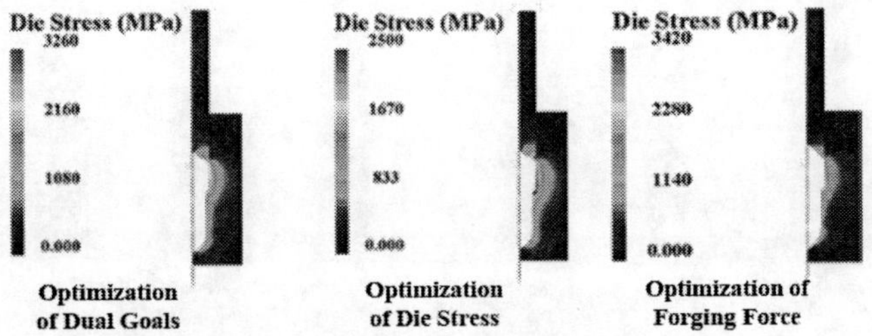

Fig. 9. Die stresses for three generalized optimization at pass 1.

Figure 9 is dies stresses for three generalized optimization at pass 1. The maximun die stress is 3420 MPa for forging force optimization, 3260MPa for dual-goals optimization, and 2500MPa for die stress optimization .

The forging parts in the first pass can be transferred to the second pass to form the final product shown in Figure 10.

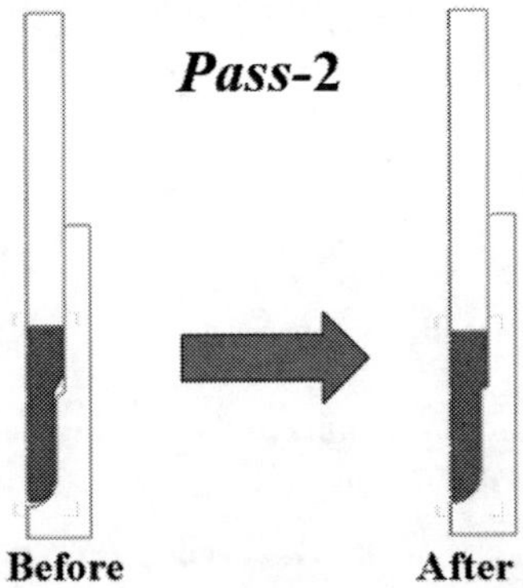

Fig. 10. Schematic diagram before and after forming for the second pass.

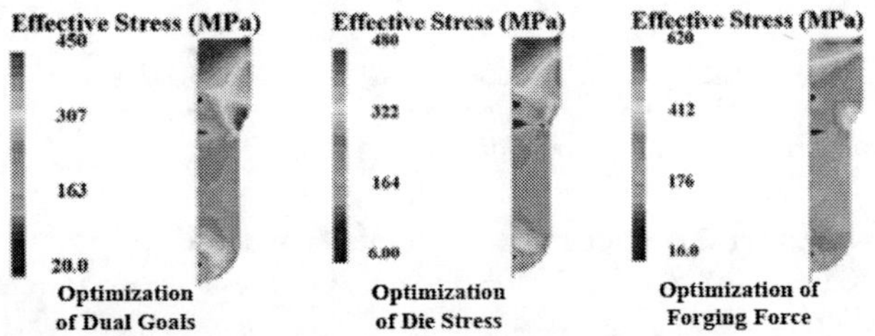

Fig. 11. Effective stresses for three generalized optimization at pass 2.

Figure 11 is effective stresses for three generalized optimization at pass 2. The maximun effective stress is 450MPa for dual-goals optimization, 480MPa for dies stress optimization, and 620MPa for forging force optimization. The smaller effective stress is induced in the dual-goals optimization.

 https://doi.org/10.1051/matecconf/201818500002

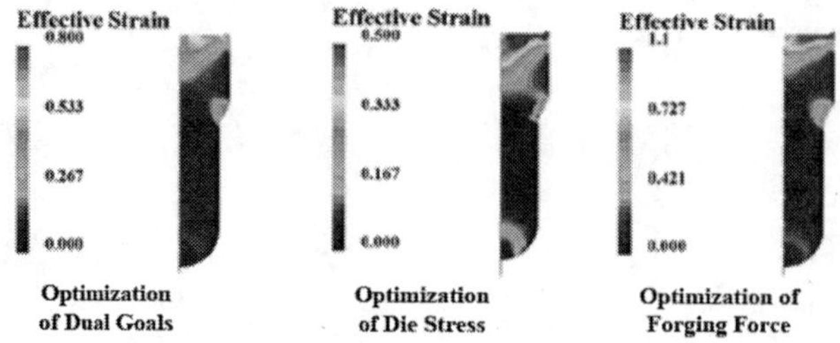

Fig. 12. Effective strains for three generalized optimization at pass 2.

Figure 12 is effective strains for three generalized optimization at pass 2. The maximun effective strain is 0.8 mm/mm for dual-goals optimization, 0.55 mm/mm for dies stress optimization, and 1.1 mm/mm for forging force optimization. The smaller effective strain is induced in the die stress optimization.

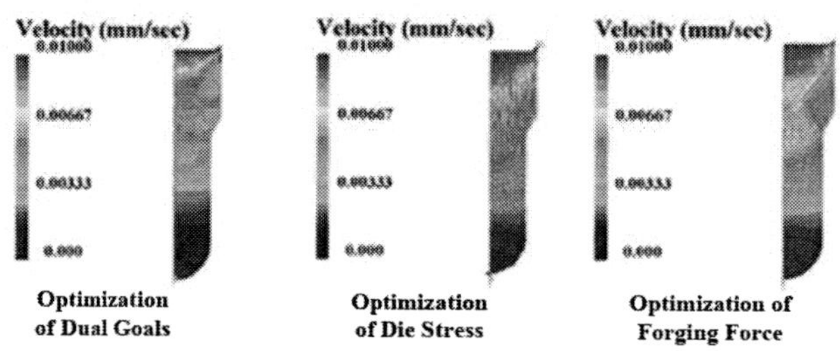

Fig. 13. Velocity fields for three generalized optimization at pass 2.

Figure 13 is velocity fields for three generalized optimization at pass 2. The maximun velocity is 0.1mm/sec for forging force optimization at pass 2.

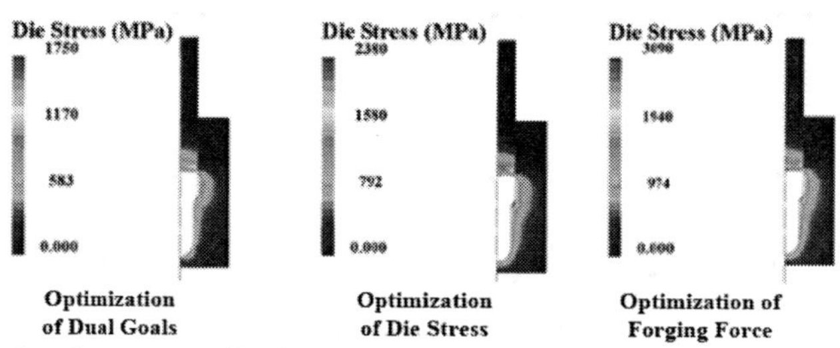

Fig. 14. Die stresses for three generalized optimization at pass 2.

Figure 14 is dies stresses for three generalized optimization at pass 2. The maximun die stress is 1750 MPa for dual-goals optimization, 2380MPa for die stress optimization, and 3090MPa for forging force optimization .

4 Conclusions

This study proposes the three generalized optimizations combining FEM simulations and Taguchi method to obtain the optimal parameters assembly, the effective stress, the effective strain, the velocity field, and the die stress. Moreover the influence rank to the forging process can be determined for three generalized optimizations in this study. The optimum combinations of parameters of three generalized optimizations are summarized as follows:

1. Optimum combination of parameters for dual-goals：$A_2B_3C_2D_1$, i.e. α=60°, β=60°, H=1.9 mm, R=0.65mm. The influence rank to the dual-goals is punch angle ＞die filet＞die angle＞die bottom height.
2. Optimum combination of parameters for die stress：$A_2B_1C_3D_1$, i.e. α=60°, β=30°, H=2mm, R=0.65mm. The influence rank to the die stress is punch angle ＞die filet＞die bottom height＞die angle.
3. Optimum combination of parameters for forging force：$A_1B_3C_2D_1$, i.e. α=45°, β=60°, H=1.9 mm, R=0.65mm. The influence rank to the forging force is punch angle ＞die angle＞die filet ＞die bottom height.
4. At the die angle location, the maximum effective stress, the maximum effective strain, and the die stress are occurred. Especially for the die stress, the die stress is still under the allowable stress of WC mold.
5. The dimensions simulated can compare with the specifications, the error is under 2 % shown in Figure 14. It indicates the FEM simulations can be accepted.

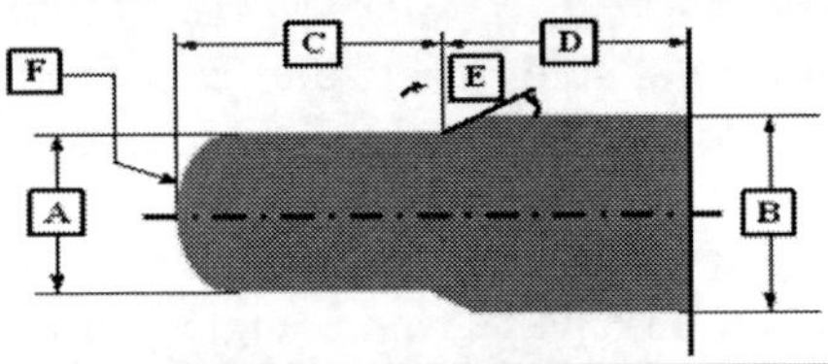

Pass_2			
	Specification (S)	Deform-2D (F)	Error(%)
A	Φ0.9±0.01mm	Φ0.9	0
B	Φ1.16±0.01mm	Φ1.16	0
C	1.75±0.02mm	1.746	0.22%
D	1.1±0.02mm	1.085	1.3%
E	30°±2°	30°	0
F	0.45R±0.1R	0.45R	0

Fig. 15 . Diensions comparisons of simulations and specifications.

Acknowledgments

Thanks for the project's supports of Ministry of Economic Affairs (MOEA), and Ministry of Education (MOE) to complete this research and the financial supports.

References

1. S. N. Shah, S. Kobayashi, Rigid-plastic analysis in cold heading by the matrix method, Proc. 15th International Conference on Machine Tool Design Research, 603-608 (1974)
2. G.W. Vickers, A. Plumtree, R. Sowerby, J.L. Duncan, Simulation of the heading process, J. Engineering Materials and Technology, 126-135 (1975)

3. C. MacCormack, J. Monaghan, A finite element analysis of cold-forging dies using two-and three-dimensional models, J. of Materials Processing Technology, **118**, 286-292 (2001)
4. N. Asnafi, On tool stresses in cold heading of fasteners, Engineering Failure Analysis. **6**, 321-335 (1999)
5. M.C. Sun, G.Y Tzou., L.A. Zheng, Processing animation simulation and FEM analysis of multi-stage cold forging of stainless automotive battery fastener, Indian Journal of Engineering & Materials Sciences, **20**, 219-224 (2013).
6. C.W. Shih, G.Y. Tzou, K.H. Chang, FEM simulation and experimental verification on multi-stage forming of flange sleeve, Journal of Chinese Society Mechanical Engineers, **35**, 243-250 (2014)
7. U. Engel and R. Eckstein, Microforming—from basic research to its realization, J. of Materials Processing Technology, **125-126**, 35-44 (2002)
8. C.J. Wang, D.B. Shan, J. Zhou, B. Guo, L.N. Sun, Size effects of the cavity dimension on the microforming ability during coining process, J. of Materials Processing Technology, **187-188**, 256-259 (2007)
9. J. T. Gau, C. Ptincipe, J. Wang, An experimental study on size effects on flow stress and formability of aluminum and brass for microforming, Journal of Materials Processing Technology, **184** , 42-46 (2007)
10. U. Engel, Tribology in micro forming, Wear, **260**, 265-273 (2006)
11. D.C. Chen, G.Y. Tzou, Y.J. Li, Robust design on equal channel angle extrusion of Ti-6Al-4V using Taguchi method, International Journal of Material Science, **5**, 1 (2015), doi:10.12783/ijmsci.2015.0501
12. G.Y. Tzou, U.C. Chai, S.H. Lin, D.C. Chen, Optimal parameters design and analysis of multi-stage forging of a micro/meso copper fastener, Advances in Energy Science and Equipment Engineering, **II**, 1307-1310 (2016)

Microstructure and mechanical properties of pure copper subjected to skin pass asymmetric rolling

Dongzhi Luo[1], *Cheng* Lu[1,*], and *Lihong* Su[1]

[1]School of Mechanical, Materials, Mechatronic and Biomedical Engineering, Northfields Avenue, Wollongong, NSW 2522, Australia

Abstract. Pure copper after 300°C annealing has been processed by skin pass asymmetric rolling up to small and medium total reductions in this work to study the evolution of microstructure, mechanical properties and texture. Subsequent EBSD characterization shows that after 300 °C annealing, the grains equiaxed but the texture is still typical rolling texture. At small rolling reductions (5.0% and 11.8%), the microstructure did not change much, the stress increased, and the ductility were still high (>20%), and the texture remained the same. At higher rolling reduction (32.0%), the grains were refined, the stress increased but the ductility decreased at the same time, and the texture remained the same. At medium rolling reduction (63.6%), the grains were further refined, the stress further increased, and the ductility further decreased, and the texture changed from typical rolling texture to shear texture at the edge and rolling texture at the centre.

1 Introduction

Ultrafine grained (UFG) or nanograined (NG) materials have been paid much attention in recent decades for their extremely high strength and high corrosive resistance. Severe plastic deformation is the most applausive method to produce bulk UFG materials, including accumulative roll bonding (ARB) [1] and equal channel angular pressing (ECAP) [2] and so on. The newly introduced asymmetric rolling [3] has also been widely used for manufacture bulk UFG or even NG materials [4,5].

In asymmetric rolling, the working pieces are rolled between rolls of different diameters or same diameter rolls rotating at different velocities. Asymmetric rolling is considered an innovative process for the strain state imposed on the sheet is an integration of plane strain and shear though it can be achieved in the conventional rolling system [6]. Several studies [7-10] have shown that asymmetric rolling can generate grain rotation and grain subdivision which result in grain refinement and texture transformation and improve the properties of the materials. Lee's work [7] indicated that strong <111> //ND orientation induced by the shear deformation during asymmetric rolling might improve the deep drawability and the refined grains could improve the mechanical property. In asymmetric

*Corresponding author: chenglu@uow.edu.au

rolling, rolling reduction per pass also plays a significant role in the final properties. Ma [9] studied the influence of thickness reduction per pass (TRPPs) in the range of 15-75% on the properties of 7075 aluminium alloy and found that smaller TRPPs led to fine sub-grains generation with an average size less than 0.5 μm which improved the mechanical properties.

Most studies to date focused on the properties and structures after asymmetric rolling up to medium and large reduction, or large rolling reduction per pass. In this regard, this work applied a so-called skin pass rolling method and combined with asymmetric rolling to obtain refined grains and rather high ductility. Subsequent study concentrates on the microstructure and mechanical properties of pure copper after skin pass asymmetric rolling up to small to medium reduction.

2 Experiments

The materials used in this work were pure copper with a purity of 99.95%. The as received copper was in sheet form with a thickness of 2.97 mm. Prior to rolling, the copper sheet was heat treated at 300 °C for two hours, and then air cooled to room temperature. After heat treatment, the working pieces were subjected to skin pass rolling using small rolling reduction per pass for 10, 20, 50 and 100 rolling passes, which gave total reductions of 5.0%, 11.8%, 32.0% and 63.6%, respectively. The reduction per pass is nearly 0.6%. A multi-function rolling mill with the roll diameter of 120 mm was adopted. The rolls were independently driven by two motors. The rolling speed ratio was kept at 1:1.4 during the skin pass asymmetric rolling.

Dog-bone shaped tensile samples were prepared from the skin pass rolled sheets according to ASTM E8 and the gauge length of the specimen was 25mm. The tensile tests were carried out on an Instron tensile machine at an initial strain rate of 5×10-4s-1. To ensure the accuracy, digital image correlation (DIC) was used to measure the strain of the samples.

Electron backscatter diffraction (EBSD) technique was adopted to characterize the microstructures and textures of each skin pass rolled sheet. The specimens for EBSD were Electron polished using Struers LectroPol-5 after grinding and mechanical polishing. An electrolytes of 500 ml distilled water with 250 ml phosphoric acid (85%) and 250 ml ethanol (85%) were used. Electron polishing was conducted at 22 V for 7-10 seconds with a flow rate of 13 at 25 °C. EBSD was carried out using JSM 7001F (JEOL) field emission scanning electron microscope (FESEM) on the RD-ND plane at both the edge and the centre of the sheets.

3 Results and discussions

Figure 1 shows the engineering stress-strain curves of the 300 °C annealed and skin pass rolled samples. As it is shown in the figure that the yield stress (YS) $\sigma_{0.2}$ and ultimate tensile strength (UTS) σ_{UTS} of the annealed sample is 184 MPa (using the 0.2% offset method) and 248 MPa, respectively. The uniform elongation (UE) ε_u of the anneal sample is 30.1%. After 10 passes asymmetric rolling, $\sigma_{0.2}$ increases by 36% to 250 MPa compared to the annealed state. As the rolling passes continue to increase, the increment rate of strength gradually decreases. After 20 passes rolling, $\sigma_{0.2}$ increases by 50% from the annealed state while the total reduction reaches 11.8%. The maximum strength σ_{UTS} reaches 312 MPa after 100 passes rolling. Although the total reduction from 50 passes to 100 passes increases by 100%, the strength does not increase dramatically. The detailed statistics of the tensile tests results are listed in Table 1.

The shape of the stress-strain curves also changes with the number of rolling passes. The annealed sample has a long work hardening stage and large uniform elongation. The work hardening stage drastically decreases with rolling passes which results in dramatic decrement of uniform elongation. The samples up to 20 passes maintain a rather stable deformation stage which keeps the total elongation larger than 25%. At larger reduction, such as the specimens of 50 and 100 passes, the stress-strain curves exhibit typical characteristics of high strength with lower ductility materials as it is referred in Ref [4,11]. When the total reduction reaches 63.6%, the strength does not increase much as compared to that of the 11.8% reduction. The uniform elongation, on the other hand, dramatically drops from 13.6% to 1.4%.

Compared to the 35% reduction specimen in Ref. [12], in which the yield stress is near 350 MPa while uniform elongation goes to peak immediately after yield, the 50 passes (32% reduction) sample in this study shows a similar strength with the 35% reduction specimen but have better ductility.

Figure 2 presents the EBSD images of pure copper at edge and centre after asymmetric rolling up to various strains. The upper images (Figure 2 a to e) shows the orientation maps at the edge while lower images (Figure 2 f to j) are from the centre of the sheets. According to Figure 2, the microstructure of the annealed, 10 and 20 passes rolled samples exhibits equiaxed grains at both edge and centre, but the grains gradually become smaller with increasing rolling deformation. After 50 passes, the grains start to elongate along the rolling direction. After 100 passes, the grains become pancake shaped at both edge and centre and the grain orientation also changes dramatically: from 85μm at 300 °C anneal sample and 48μm at 10 passes to about 4μm at 100 passes.

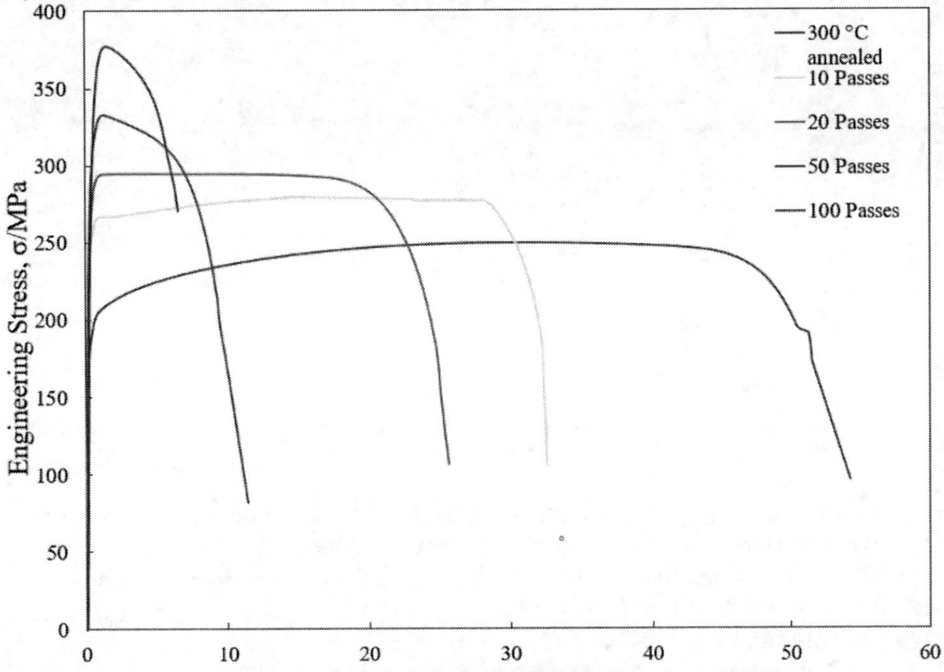

Fig. 1. The engineering stress-strain curves of annealed and skin pass rolled samples.

Table 1. The tensile properties of annealed and skin pass rolled pure copper samples.

Specimens	Total Reduction/ %	$\sigma_{0.2}$/MPa	ε_u/%	σ_{UTS}/MPa
300°C annealed	0	184	30%	248
10 passes rolled	5.0	250	13.6%	279
20 passes rolled	11.8	276	4.3%	294
50 passes rolled	32.0	303	1.2%	333
100 passes rolled	63.6	312	1.4%	375

Combining with the inverse pole figure, it can be seen that the initial orientation of the sample was mainly dominated by [111] direction. With the development of deformation, the grain orientation turns to [101] direction which is the result of grain rotation during the deformation.

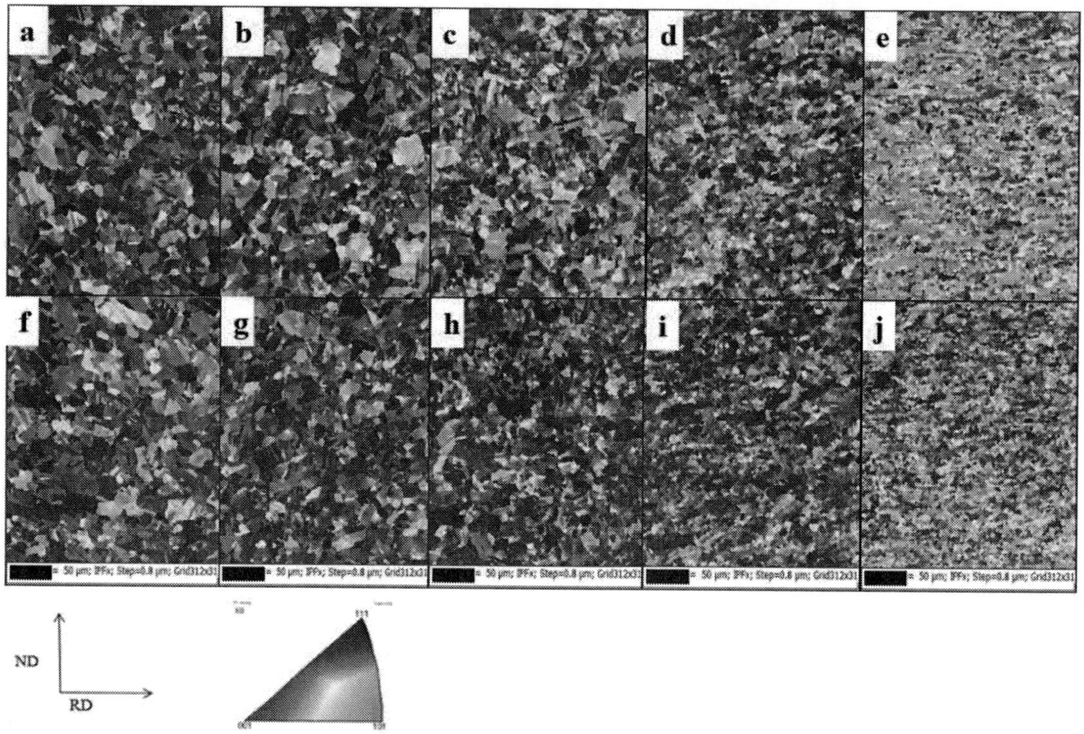

Fig. 2. The orientation maps at edge and centre of the examined specimens, a-e shows the edge and f-j is the centre (a, f: 300 °C annealed, b, g: 10 passes, c, h: 20 passes, d, i: 50 passes, e, j: 100 passes).

Figure 3 gives the distribution of misorientation angle at each rolling pass. As is shown in the figure, the misorientation angle in the 300 °C annealed sample presents two peaks at near 2° and 60° which means there exist large amount of both low and high angle grain boundaries. When rolled to 10 and 20 passes, the low angle grain boundaries begin to increase dramatically, and the peak around 2° comes to the maximum value at pass 20. After 50 passes, both peaks decrease and further decrease after 100 passes. The volume fraction of low angle grain boundaries (LAGBs), high angle grain boundaries (HAGBs) and twinning boundaries (TBs) of the annealed and asymmetrically rolled pure copper sheets is depicted in Figure 4. It can be seen from Figure 4 that both the edge and centre have the same trend. Before 20 passes, the volume fraction of LAGBs increases at both edge and centre. From 50 passes on, it begins to decrease. The trend of HAGBs is exactly the opposite to LAGBs: before 20 passes, the volume fraction decreases with the number of passes, then increases after 50 passes. The fraction of TBs is gradually decreasing before 50 passes but raises slightly at 100 passes.

Compared with Figure 3 and Figure 4, the rise of LAGBs indicates an increment of subgrain boundaries at the early stage of rolling as the result of dislocation kinking and tangling. With the development of rolling passes the newly formed subgrain boundaries evolve towards HAGBs which was well interpreted in Ref. [13]. Ma [11] has proposed that using lower TRPPs can help to store higher energy which increases the potential of recrystallization and the discontinuous recrystallization process (DDRX) will increase volume fraction of HAGBs. The phenomenon depicted in Figure 4 of both edge and centre can confirm that the rolling deformation consists of some DDRX process.

Figure 6 depicts further the evolution and transformation of texture via three major fibres at both edge and centre. Prior to rolling (red line), the texture type is Brass at the edge and Copper at the centre. With the development of deformation, the intensity along α fibre decrease. Along τ fibre, the highest intensity stays at the position of Copper component for the specimens rolled up to 50 passes at both edge and centre. The highest intensity shifts to Rotated Cube at the edge of the 100 passes sample, while at the centre, it stays close to the Copper component, but relative intensity is much lower than the 50 passes and the less reduction samples. At the same time, the intensity of Rotated Cube is the highest compared to the samples with less rolling passes. The intensity distribution along β fibre also confirms the mentioned trend.

In α fibre line, the Goss component is relatively intense in low strain and initial state, and then the intensity decreases at larger strain, because the Goss orientation is metastable and can be transformed into other components easily. S component is not strong as Copper but still maintains relatively high intensity in the rolling passes up to 50. Compared to the edge and centre part of specimen, the edge one shows more Rotated Cube texture than the centre part which indicates that at topmost of the sample, the shear stress dominates.

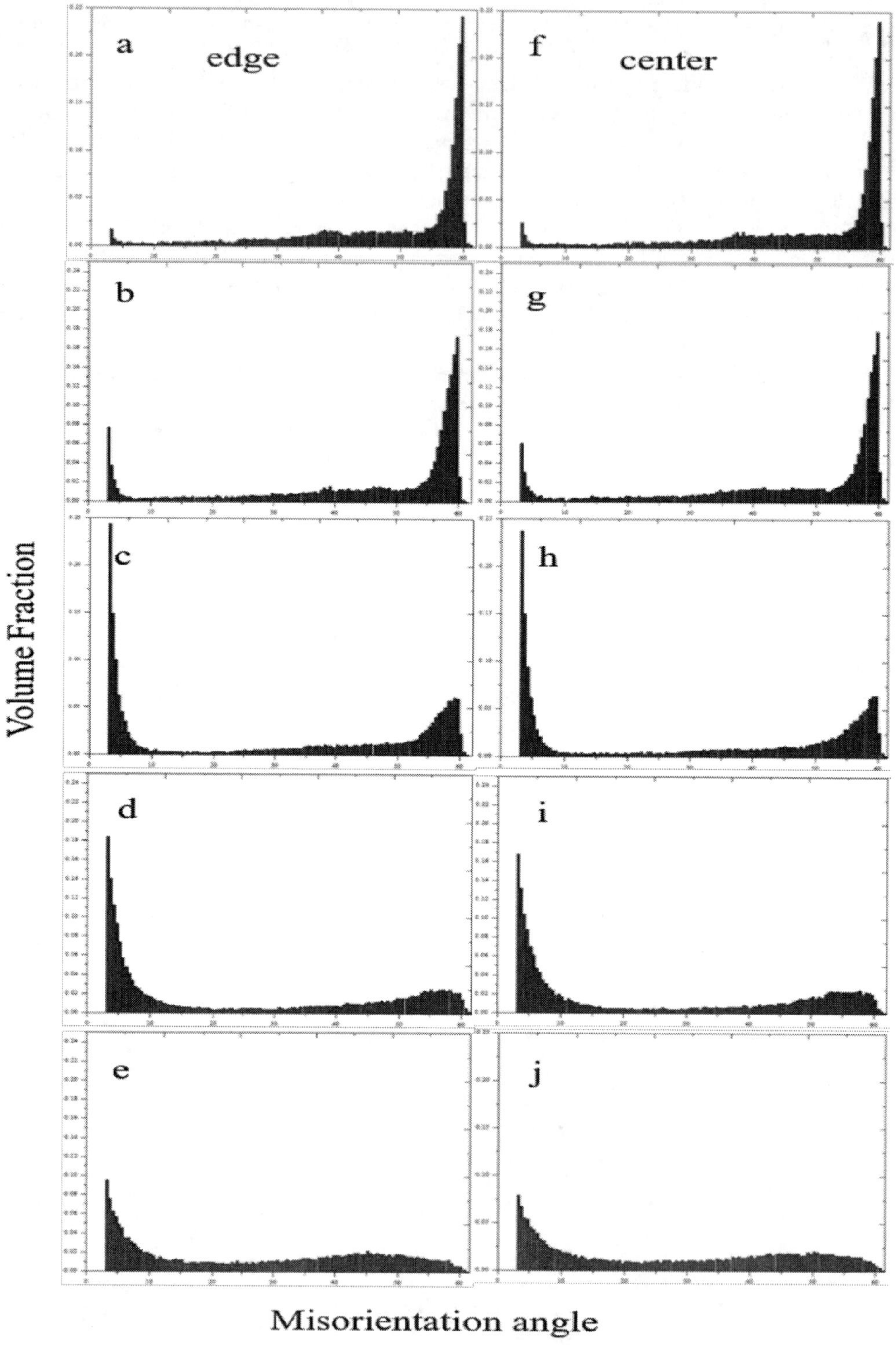

Fig. 3. Misorientation angle distribution of the annealed and skin pass rolled pure copper: (a, f: 300°C 2h, b, g: 10 passes, c, h: 20 passes, d, i: 50 passes, e, j: 100 passes). The horizontal axis range: 0~60°, the vertical axis ranges from 0 to 0.25 in intensity.

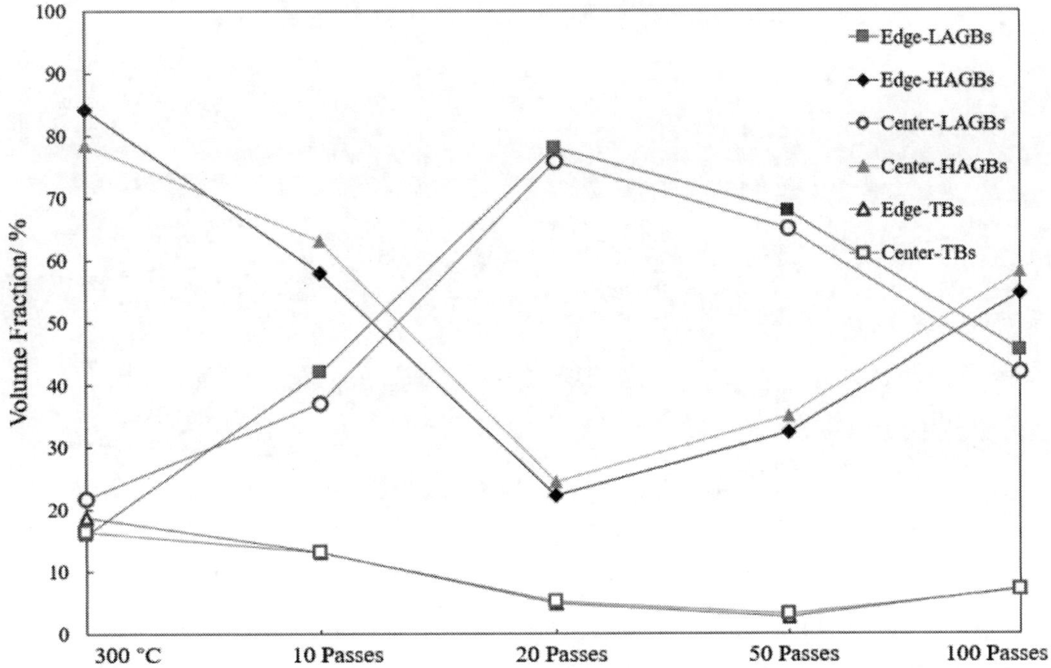

Fig. 4. Volume fraction distribution of LAGBs, HAGBs and TBs of annealed and skin pass rolled specimens.

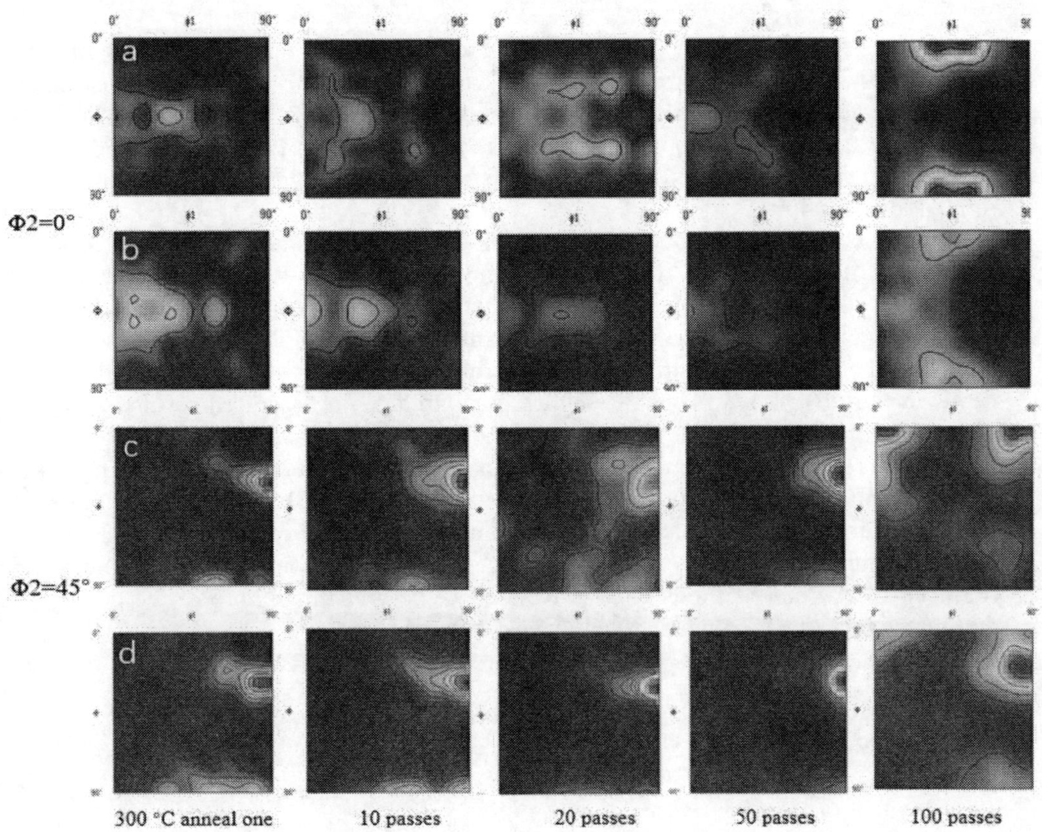

Fig. 5. ODF section of annealed and skin pass rolled specimens: a edge part at φ2=0°, b centre part at φ2=0°, c edge part at φ2=45°, d centre part at φ2=45°.

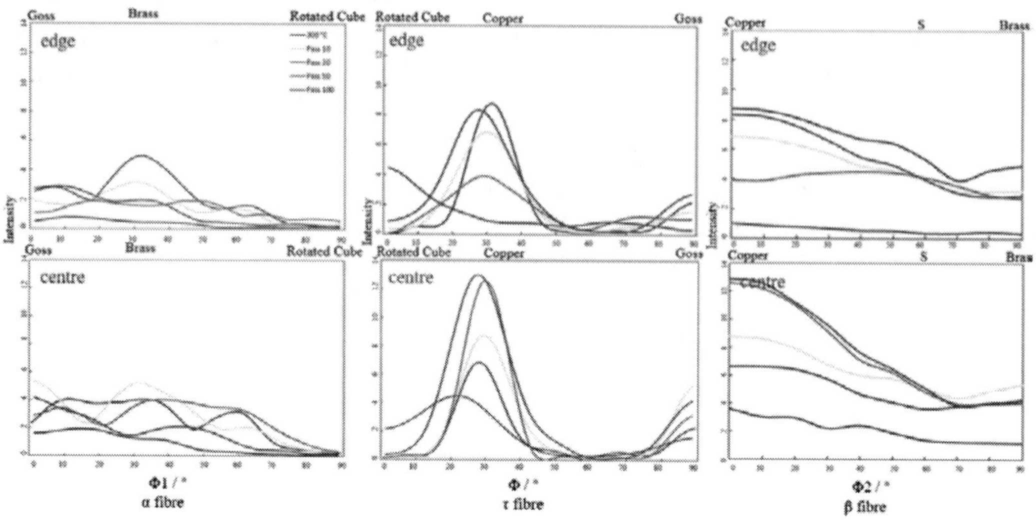

Fig. 6. Intensity distribution along α-fibre (Φ=45°, φ2=0°), τ-fibre (φ1=90°, φ2=45°) and β-fibre for the annealed and skin pass rolled specimens. The horizontal axis ranges from 0 to 90°, and vertical axis ranges from 0 to 14 in intensity of texture fibre.

4 Conclusions

Pure copper after 300°C annealing was processed by skin pass asymmetric rolling up to small and medium total reductions in this work to study the evolution of microstructure, mechanical properties and texture. The conclusions are:

1. The 300°C annealed samples still exhibited rolling/deformation microstructure and the initial grain size is about 85μm, after 100 passes skin pass rolling, the grains greatly refined to about 4μm.
2. Tensile testing results showed that at small rolling reductions (5.0% and 11.8%), the ductility of the sheets was still relatively high (more than 20 %) while the strength increased by 35.8%. At medium rolling reductions (32.0% and 63.6%), the strength of the sheets increased dramatically but at the same time, the ductility decreased drastically. Although the increment at strength with decrement of ductility is still existing, and the decrement of 10 passes is much smaller compared to the reported results of similar asymmetrically rolled samples.
3. At small rolling reductions, LAGBs dominated. With the increment of rolling passes and accumulated rolling reduction, LAGBs evolved into HAGBs.
4. The texture of the annealed and low rolling reductions samples was dominated by Copper component at both edge and centre. After rolling up to medium reduction (63.6%), the texture at the edge was Rotated Cube and the texture at the centre was dominated by Copper component with weak Rotated Cube.

References

1. L. H. Su, C. Lu, *J. Alloys Compd.*, **594**, 12-22 (2014)
2. J. K. Kim, H. G. Jeong, S. I. Hong, *Scripta Mater.* **45**, 8, 901-907 (2001)
3. L.L.Chang, J. H. Cho, and S. B. Kang, *J. Mater. Process. Technol.* **211,** 9, 1527-1533 (2011)
4. J. H. Jiang, Y. Ding, *Scripta Mater.* **60**, 10, 905-908 (2009)
5. F. Q. Zuo, J. H. Jiang, A.D. Shan, *Tans. Nonferrous Met. Soc. China*, **18**, 4, 774-777 (2008)
6. Y. H. Ji,J.J.Park, *Mater. Sci. Eng. A*, **499**, 1, 14-17 (2009)
7. J. K. Lee, D.N. Lee, *Int. J. Mech. Sci.* **50**, 5, 869-887 (2008)
8. F. J. Simões, R. J. de Sousa, *Int. J. Mater*, **2**, 1, 891-894 (2009)
9. C. Ma, L. Hou, *Mater. Sci. Eng. A* , **650**, 454-468 (2016)
10. A. Uniwersał., M. Wroński, *Acta Mater.*, **139**, 30-38 (2017)
11. Y. Wang, M. Chen, *Nat.* **419**, 6910, 912-915 (2002)
12. W. J. Kim, K.E. Lee, *Mater. Sci. Eng. A*, **506**, 1, 71-79 (2009)
13. D. A. Hughes., Q. Liu, *Acta Mater*, **45**, 1, 105-112(1997)
14. J. Y. Chang, G.H. Kim, *Scripta Mater.*, **44**, 2, 331-336 (2001)

Characteristic research on lithium iron phosphate battery of power type

Yen-Ming Tseng[1], *Hsi-Shan* Huang[1], *Li-Shan* Chen[2,*], and *Jsung-Ta* Tsai[1]

[1]College of Intelligence Robot, Fuzhou Polytechnic, No.8 Lianrong Road, Fuzhou University Town, 350108, Fuzhou City, Fujian Province, China
[2]School of Management, Fujian University of Technology, No.3 Xueyuan Road, Fuzhou University Town, 350108, Fuzhou City, Fujian Province, China

Abstract. In this paper, it is the research topic focus on the electrical characteristics analysis of lithium phosphate iron (LiFePO4) batteries pack of power type. LiFePO4 battery of power type has performance advantages such as high capacity, lower toxicity and pollution, operation at high temperature environment and many cycling times in charging and discharge and so on. The charging and discharging characteristics for LiFePO4 batteries of power type pack have been verified and discussed by the actual experiment. Base on the 12V10AH LiFePO4 battery was proceeding on charging and discharging test with over high current value and which investigate the parameters such as the internal resistance, the related charge and discharge characteristics of LiFePO4 battery pack , the actual value of internal voltage and internal resistance of the battery pack and by polynomial mathmatic model to approach the accury of inner resistance on discharging mode.

1 Introduction

The battery is storage and energy conversion components which can be stored in the original physical energy, chemical energy or other type energy and that can be converted into electricity and released to attach circuit for application [1, 2]. In accordance with the energy storage and release methods that can be divided into physical batteries and chemical batteries. The former are convert from the light, heat and other renewable energy to electricity and to storage in batteries or release in transmission line to loadings and consumption the power which the description type are such as solar cell [3], wind power[4], atomic force batteries and so on. Chemical batteries are chemical substances through the material of the redo reaction which by the active substances reactive effect convert from chemical energy to electricity that base on the charging and discharging apply recycle can divided into primary battery and secondary battery. Primary battery just can only discharge once and till to not drive the electrical loading so far. At the end of time in discharging mode that the battery does not chemical substances all played a chemical effect can no longer be able to provide electricity and it cannot be provided by the external power supply at after the full discharge make this battery useless. That is because its electrochemical

* Corresponding author: Sun56@ms8.hinet.net

reaction is irreversible by the reasons of electrolyte cannot be restored by recharging step. The secondary battery is chemically converted into electrical energy and can be recharged into a battery by recharging the battery that effect is re-converted into chemical energy and it can be reused. The number of times uses will vary with the material and the application. In general, that exists in the market such as the lead-acid batteries, nickel-cadmium batteries, nickel-metal hydride batteries and lithium family batteries [5]. Different chemical type batteries rely on its working voltage, capacitance and safety of the relationship that made different applications. A lithium battery mainly refers to lithium-ion batteries. At present in the market can be found for the lithium cobalt material, lithium manganese material but the both in the use of security on the doubt there will be cause high temperature resulting in explosion or the decline in capacity of the problem. To improve the lithium batteries to lithium phosphate iron (LiFePO4) batteries [6, 7, 8] for these problems, can eliminate the user's security concerns. In this paper, the charging and discharging characteristics of power type LiFePO4 batteries pack will be by the actual experiment to verify and discussion. The study steps are following :

1) Explore and compare the dynamic characteristics of different secondary batteries of power type.
2) Discuss and decide the lithium iron phosphate (LiFePO4) battery of power type pack rated to voltage and capacity as the subject of research.
3) Discussion on the parameters and characteristics of power type LiFePO4 pack in charging mode and discharging mode.
4) To investigate the power type LiFePO4 pack by charging and discharging test with heavy duty.
5) Analyze the charging and discharging curves of LiFePO4 of power type with the polynomial method.
6) To find the power type LiFePO4 pack the inner resistance in charging and discharging test with heavy duty to reach the relative errors.

2 Comparison with the LiFePO4 and the other different secondary battery

Currently on the market mainly using the most extensive secondary battery are lead-acid batteries pack, and lithium batteries such as lithium-cadmium batteries, nickel-metal hydride battery, lithium cobalt battery and LiFePO4 battery packs. Lead-acid battery because of the widely operating temperature, simple structure, technology is mature and low price characteristics to form the higher usage rate but the lower cycle life and discharge coefficient (or called *Crate*), higher internal resistance and high toxicity caused by high pollution shortcomings to make the replace effect by other chemical Battery packs. in this paper, it is discussed the LiFePO4 battery packs which have the advantages of high capacitance, low toxicity and no pollution, high temperature environment and good circulation performance under heavy duty charge and discharge mode, and wide sources of raw materials. Compared with other lithium family batteries packs which LiFePO4 battery packs have high efficiency energy conversion up to 95% and posses the more life cycle up to 2000 times than the other lithium family batteries packs life cycle about from 400 to 500 times. LiFePO4 battery packs is also very suitable with power supply for electric motors and for power management such as electricity scooter, pure electricity scooter and hybrid cars applications and so on, and in the future will become the mainstream of electric vehicles.

3 Internal resistance varying characteristics in charging and discharging mode of LiFePO4 battery pack

In Figure 1 which *Vb* is inside voltage of battery pack and Rin is inner resistance of battery pack. Generally, battery equivalent circuit will not show *Rcov* and *C*. The resistance Rcov and capacitor C in parallel configuration and in series with the former circuit that simulation the in over-voltage states will produce the phenomenon. So that made the analogy charging process of this battery equivalent circuit diagram is more realistic and therefore improves the shortcomings of the linear model and is therefore more accurate than the linear model. And *Req* is the equivalent battery resistance seen by side of the voltage source, since the capacitance *C* is the open state so that equivalent over-voltage will produce the phenomenon, and the general battery equivalent resistance *Req* is equal as equation (1).

$$Req=Rin+Rcov \tag{1}$$

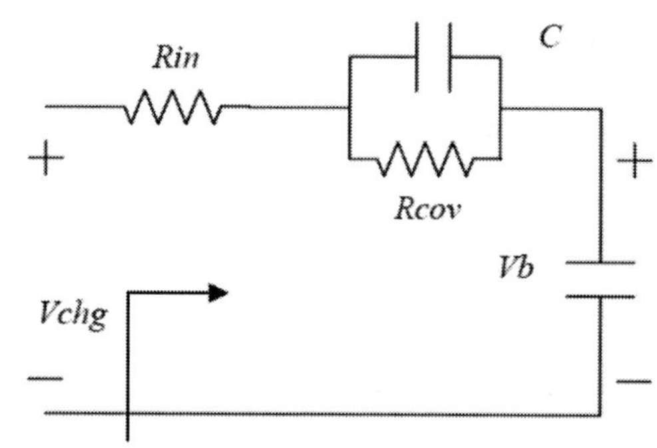

$$Req = Rin + Rcov$$

Fig. 1. Battery pack equivalent circuit diagram.

Form view of the battery pack capacity in fixed that the capacity is full or not with its battery pack inner resistance is closely related in charging mode or proceeding. In general, each battery pack capacity of the full percentage greater relative inner resistance will be made greater. In contrast, the smaller the degree of filling of each battery packs company with the smaller its internal resistance. The relationship between the external voltage and the magnitude of the battery pack resistance is by the actual circuit to test and explain.

3.1 The relationship between with the charging voltages and currents

Based on the specification of A123 26650 LiFePO4 battery cell as shown in the Table 1.

Table 1. LiFePO4 battery cell specification [9, 10].

Type	A123
Rate voltage	3.3V
Rate capacity	2300mAh
Weight	73g
Charging time	Conventional charging about 1C-2C or 3-4A and fast charging about 5C or 10A.

A constant voltage charging circuit is designed for a 12V 10Ah LiFePO4 battery pack to keep the charging voltage constant and allow the charging current to be less than 3C which let charging current between in conventional charging and fast charging areas. The charging data is shown in Table 2. When the time is 0 for start, the initial value of the voltage is 11.48V and charging current is 30A. When the time passes 5 seconds, the voltage rose to 11.66V and then the charging current dropped to 25A. When the time after 50 seconds which the inside battery pack voltage rose to 12.22V and then the charging current is14A. According to the design of charging voltage for the battery pack, when the battery pack inside voltage reach the charging voltage will let the charging current become about 0A, that is called full charging. When the time passes 100 seconds, the voltage rose to 12.61V and then the charging current is 11.7A.When the time passes 230 seconds, the voltage rose to 13.00V and then the charging current dropped to 8.5A. And the time after 240 seconds, voltage rose to 13.00 V and then the charging current dropped to 3.2A we can see that when the battery slowly filling the voltage will stabilize, the current will slowly decline. Base on the data we can drown the charging curve shown in Figure 2.

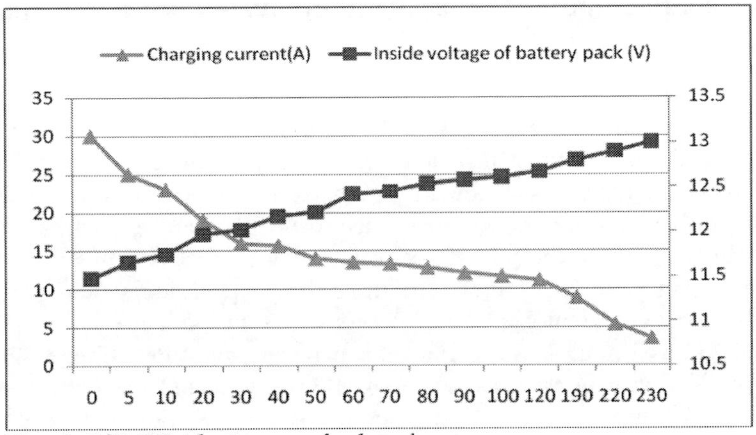

Fig. 2. LiFePO4 battery pack charging.

Table 2. LiFePO4 battery pack charging trace constant with voltage method.

Time(second)	Inner voltage of battery pack (V)	Charging current(A)
0	11.48	30
5	11.66	25
10	11.75	23
20	11.27	19
30	12.02	16
40	12.17	15.7
50	12.22	14
60	12.42	13.5
70	12.45	13.3

80	12.54	12.8
90	12.58	12.1
100	12.61	11.7
120	12.67	11.2
190	12.8	8.9
220	12.9	5.4
230	13	3.6
240	13	3.2

3.2 Calculation the dynamic internal voltage and equivalent inner resistance of the battery pack

Figure 3 is a 12V10AHLiFePO4 battery pack for the power source supply and by series with a DC motor which internal resistance R_M is 0.07 ohm to form a closed loop for measure the battery pack dynamic inner resistance in discharge mode under different voltage equalization. The equivalent resistance *Req* seen from the power supply side is shown in equation (2) and the dynamic internal resistance *Rin* of the battery pack is shown in equation (3). According to the discharge battery pack voltage level is divided into 13 measurements points sown in table 3. At first time the battery pack side voltage 13.2V, current is 2.8A, supply side equivalent resistance *Req* is 4.71 ohm and the battery pack internal resistance Rin is 4.64 ohm. At sixth time the battery pack side voltage 12.8V, current is 3.3A, supply side equivalent resistance *Req* is 3.88 ohm and the battery pack internal resistance *Rin* is 3.81 ohm. And at the last time the battery pack side voltage 10.7V, current is 3.3A, supply side equivalent resistance *Req* is 3.24 ohm and the battery pack internal resistance *Rin* is 3.17 ohm. From the table 3, it is made a result which is the higher the supply voltage with the smaller the discharging current will reach the resistance larger by loading with unchanged or fixed resistance.

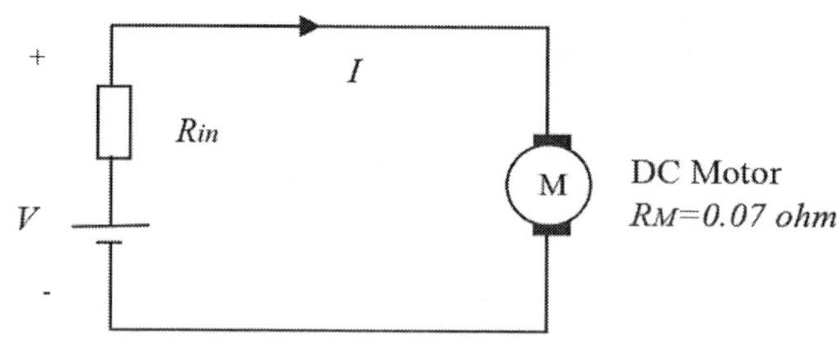

Fig. 3. LiFePO4 battery pack charging.

$$Req = V/I = Rin + R_M \tag{2}$$

$$Rin = Req - R_M \tag{3}$$

With the relationship between the battery internal resistance *Rin* and the voltage *V* in Table 3 that by using the third-order polynomial mathematical model of voltage *V* is used to approach the internal dynamic intermal resistance Rin and predictive internal resistance value of the battery, as shown in equation (4). Using the relative error *REv%* as shown in equation (5) to confirm the accuracy of the mathematical model and the Table 4 is shown in the table for each voltage class approaching the battery block internal resistance and relative error.

$$Rin = a_0 + a_1 x V + a_2 x V^2 + a_3 x V^3 \tag{4}$$

where

Rin: internal resistance of battery pack
V: discharging voltage

$$RE_V\% = \frac{|\,Rin,actual - Rin,fit\,|}{Rin,actual} \times 100\% \tag{5}$$

where

Rin, actual: internal resistance of battery pack which equal to Rin
Rin, fit: internal resistance approach by polynomial mathematic model.

Table 3. LiFePO4 battery pack charging trace constant voltage method.

Parameters / order	Voltage(V)	Current(A)	*Req*(ohm)	*Rin*(Ohm)
1	13.2	2.8	4.71	4.54
2	13.0	2.7	4.81	4.74
3	13.0	2.9	4.48	4.41
4	12.9	3.1	4.16	4.09
5	12.9	3.3	3.91	3.84
6	12.8	3.3	3.88	3.81
7	12.8	3.5	3.66	3.59
8	12.8	3.6	3.56	3.49
9	12.4	3.5	3.54	3.47
10	11.7	3.4	3.44	3.37
11	11.5	3.4	3.38	3.31
12	11.1	3.3	3.36	3.29
13	10.7	3.3	3.24	3.17

It can be seen from Table 4 that the results of the polynomial calculated by the relative error formula (5) and the measured internal resistance of the battery are within 2.21% which the maximum difference value is 0.07ohm and *REv%* is 2.21%, and the minimum difference value is only 0.01ohm and *REv%* is 0.26% and the total average *REv%* is 1.41% , so it can be resulted, the calculation result of formula (3) is very close to the actual value of internal resistance of battery pack. Figure 4 is relative error chart by polynomial mathematic model.

Table 4. The relative error by polynomial mathematic model.

Paramaters / order	Voltage(V)	*Rin,actual* (ohm)	*Rin,fit* (ohm)	*Rev%*
1	13.2	4.64	4.59	1.08%
2	13.0	4.26	4.31	1.17%
3	12.9	4.09	4.02	1.71%
4	12.8	3.81	3.80	0.26%
5	12.4	3.47	3.44	0.86%
6	11.7	3.37	3.32	1.48%
7	11.5	3.31	3.24	2.11%
8	11.1	3.29	3.23	1.82%
9	10.7	3.17	3.10	2.21%
average				1.41%

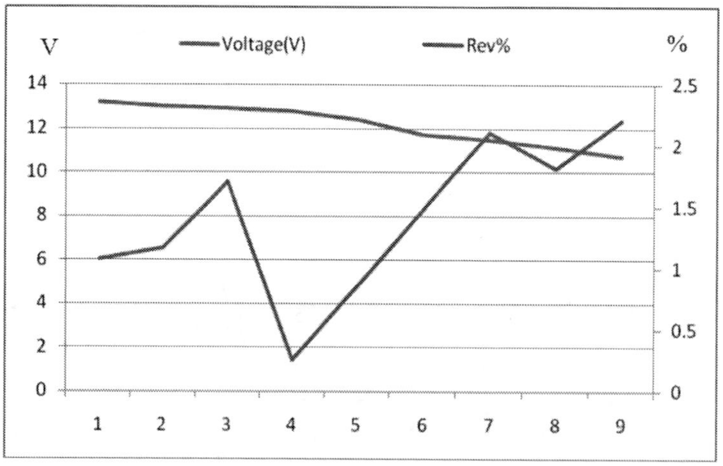

Fig. 4. The relative error chart by polynomial mathematic model.

4 Conclusions

For the actual circuit design to do charge and discharge test obtained the battery pack voltage, resistance and other parameters and to determine which each other dependency. And deduce the battery internal dynamic resistance and in order to other applications before the lead operations. In the charging mode, when the charging voltage is fixed that result of the battery voltage, current and internal resistance are closely related to each other.

Similarly, in the discharge mode is also the same. In the discharge mode, the battery internal resistance and the battery potential can be approximated by the mathematical model of the third-order polynomial to the internal resistance of the battery pack and obtain its value. Through the every voltage level to statistics the total relative error is 1.41%. In this article, through the relevant simple circuit can be accurate and quickly know the dynamic resistance of the battery pack and the relevant internal resistance and electricity residue of battery pack.

References

1. Le An, Dylan Dah-Chuan Lu, IEEE Transactions on Industrial Electronics, **62(2)**, (2015)
2. R. Rajitha, C. Balachandra Reddy, International Journal of Advanced Scientific Technologies in Engineering and Management Sciences (IJASTEMS-ISSN: 2454-356X), **1(7)**, (2015)
3. Alejandro Pena-Bello, Meinrad Burer Martin, K. PatelDavidParra, Journal of Energy Storage,**13**, (2017)
4. Sercan Teleke,, Mesut E. Baran, Alex Q. Huang, Subhashish Bhattacharya and Loren Anderson, IEEE Transactions on Energy Conversion, **24(3)**, (2009)
5. Vinodkumar Etacheri, Rotem Marom,'Ran Elazari, Gregory Salitra and Doron Aurbach, Journal of Energy & Environmental Science,**1(9)**, (2011)
6. Zhian Zhang, Zhiyong Zhang, Wei Chen, Guanchao Wang, Jie Li and Yanqing Lai,New Journal of Chemistry, **1(5)**, (2015)
7. Hu L, La Mantia F, Wu H , , Adv. Energy Mater, **1**, (2011)
8. Jiajun Wang and Xueliang Sun, Energy & Environmental Science, **8**,(2015)
9. A123 system LLC , *26650 Cylindrical Cell datasheet*, (2016)
10. A123 system LLC , *18650 Cylindrical Cell datasheet*, (2016)

Study on micro nosing process assisted by ultrasonic vibration

Chao-Cheng Chang* and *Sin-Siang* Huang

Department of Mold and Die Engineering, National Kaohsiung University of Science and Technology, No. 415, Jiangong Rd., Sanmin Dist., Kaohsiung City 80778, Taiwan

Abstract. This study used experiments to investigate the effect of ultrasonic vibration on the micro nosing process of copper cup. The billets with 1 mm diameter and length were backwards extruded to form the cups with 0.1 mm wall thickness and 1.8 mm height. The cups were then used in micro nosing processes. A forming system with the ultrasonic vibration has been developed and installed on a precision press to conduct the experiments. An optical fibre sensor and a precession load cell were equipped on the ram of the press in order to accurately measure the load and stroke of the process. Two forming conditions with and without ultrasonic vibrations of 20 kHz were considered in this study. The results show that the ultrasonic vibration can improve the materials flow which leads to an even diameter along the shrunk region of the nosed cup. Moreover, the ultrasonic vibration greatly reduces the load in the micro nosing process of the copper cup.

1 Introduction

Nosing process reducing the dimension of the open end of a container has been used for producing a wide variety of metal components with specific shapes and profiles. The process has also been developed to fabricate axisymmetric seamless reservoirs by a single-stage nosing process of thin-walled tubes [1]. As the demands of micro metal components constantly increase in the fields of portable products, precision instruments and medical devices, researchers are interested in the development of new nosing processes at micro scale. However, size effects may cause some problems such as nonhomogeneous materials flow and relatively higher friction in micro metal forming [2-5] that could also exist in the micro nosing process. By increasing the forming temperature and reducing the grain size of the billet, it is possible to improve materials flow and result in uniform deformations [6, 7]. Moreover, Yao et. al.[8] found that high-frequency vibration on the compression test of pure aluminium can reduce forming stress, surface roughness and friction. The high-frequency vibration leading to acoustic softening was considered to be a much more efficient method to activate plastic deformation than pure heating. In the study on the compression testing of martensitic stainless steel with superimposed ultrasonic vibration, Michalski et. al. [9] also reported that the force reduction and specimen heating can be

*Corresponding author: ccchang@nkust.edu.tw

achieved. Furthermore, Hu et. al. [10] investigated the effect of superimposed ultrasonic vibration on the compression tests of commercially pure aluminium, and showed that the ultrasonic dynamic impact effect was the dominating mechanism for stress reduction in compression condition. Bunget and Ngaile [11] studied the effect of ultrasonic vibration on the micro extrusion process and concluded that the forming load greatly decreased and the surface of the micro-extruded parts was significantly improved when the ultrasonic vibrations were imposed. The reviewed studies show that the ultrasonic vibration has high potential for improving the quality of the formed parts and reducing forming load in the metal forming process at micro scale.

The micro nosing process can be used to shrink the open end of a micro metal component or assemble micro metal parts. However, the process at micro scale may encounter high friction that could cause high forming load and defects. To reduce the friction effects, applying ultrasonic vibration on the micro nosing process could be one of methods. This study used experiments to investigate the effect of ultrasonic vibration on the micro nosing process of copper cup, and the deformation of the workpiece and forming load are discussed.

2 Experimental setup

2.1 Micro nosing process

In this study, the micro nosing process is to reduce the diameter of the open end of the cup which is fabricated by a micro backward extrusion process. As-received pure copper wires with 1 mm diameter were cut and ground into lengths of 1 mm as the billets for the backward extrusion processes to produce the cups with 1 mm outer diameter, 0.1 mm wall thickness and 1.8 mm height. The extruded cups were then used in micro nosing processes to shrink the diameter of the open end to 0.7 mm. Figures 1 and 2 present the schematic diagrams of the micro backward extrusion and micro nosing process, respectively, and the key dimensions. The conical die with a semi-die angle of 30°, as displayed in Figure 2(a), was used in the process.

2.2 Ultrasonic vibration forming system

To investigate the effect of the ultrasonic vibration on the micro nosing process, a forming system capable to superimpose ultrasonic vibration was established for the experiments. The system consists of an ultrasonic vibration signal generator, a piezoelectric transducer, a horn, and a die, as shown in Figure 3. The horn was made in titanium alloy and the die in tool steel was assembled with a tungsten carbide insert. Figure 4 shows the key dimensions and photos of the die for the micro nosing process. The established system was installed onto a precision press with a load capacity of 50 kN and a positioning accuracy of 10 μm as presented in Figure 5. Moreover, an optical fibre sensor with an accuracy of 1 μm and a load cell with an accuracy of 0.5 N, as shown in Figure 5(b), were installed on the ram of the press to measure the stroke and forming load of the process, respectively.

2.3 Procedure

The pure copper billets with 1 mm diameter and length were lubricated with oil and backwards extruded in a constant punch speed of 0.1 mm per second to form the cups for the use in the experiments of the micro nosing process. To investigate the influence of the ultrasonic vibration, the die was assisted by vibrations with a frequency of 20 kHz and an amplitude of 5 μm. The sampling rate for data acquisition was 7.2 kHz for the experiments with the ultrasonic vibration. Table 1 lists the conditions for the experiments of the micro backward extrusion and the nosing processes with and without ultrasonic vibration.

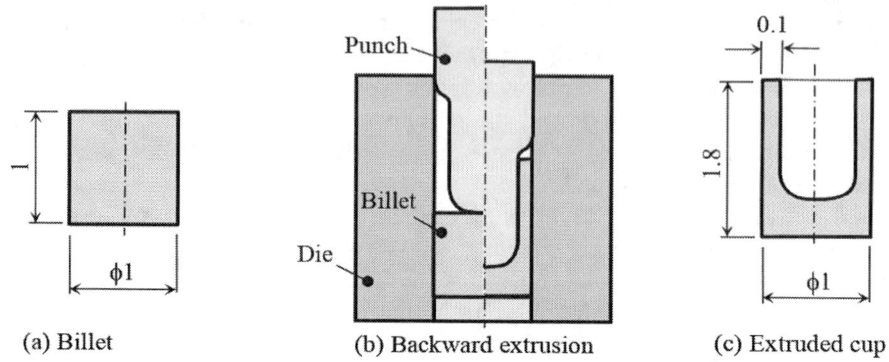

Fig. 1. Micro backward extrusion.

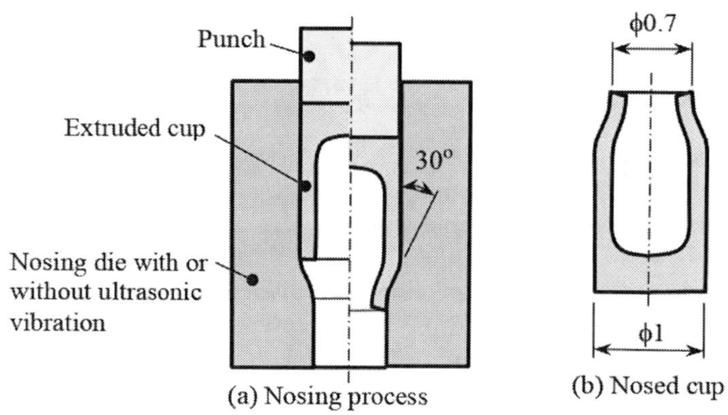

Fig. 2 Micro nosing process.

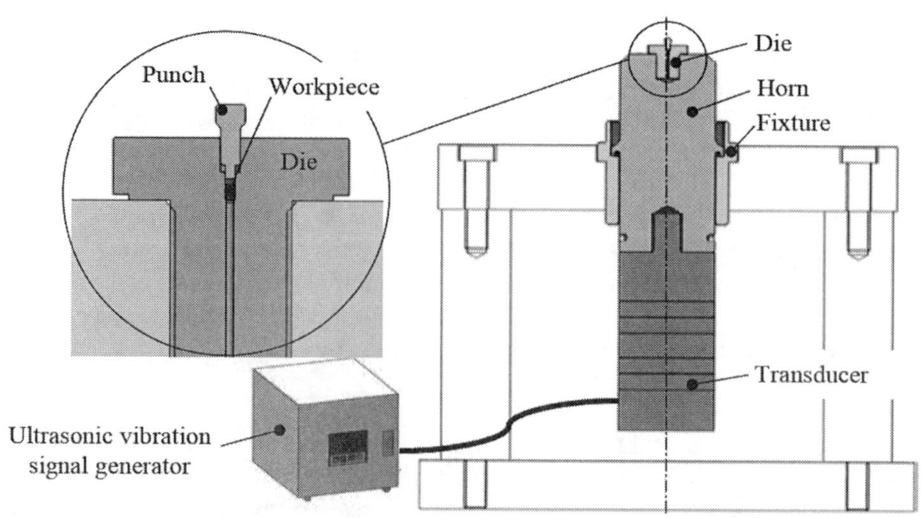

Fig. 3. Schematic diagram of the micro nosing system assisted by ultrasonic vibration.

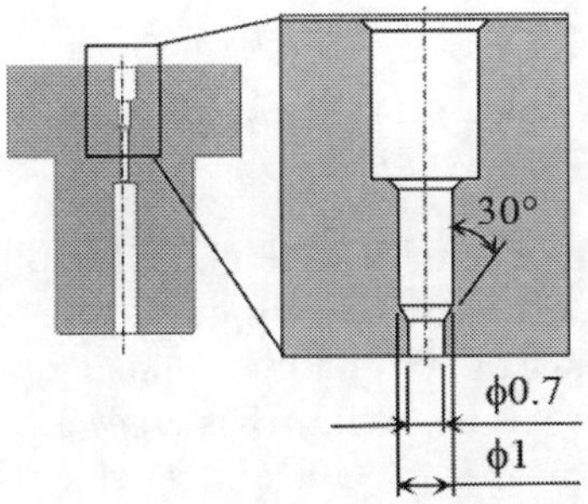

(a) Key dimensions of nosing die

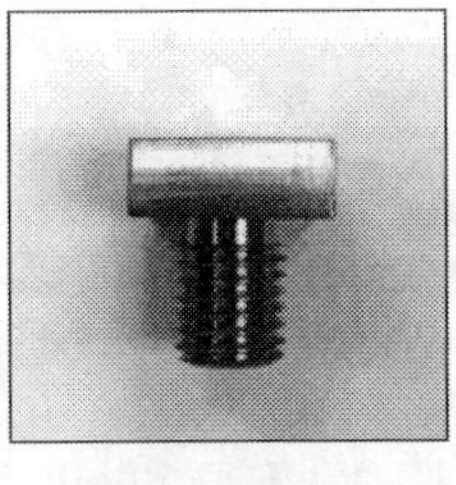

(b) Front view

(c) Top

Fig. 4. Key dimensions and photos of the nosing die.

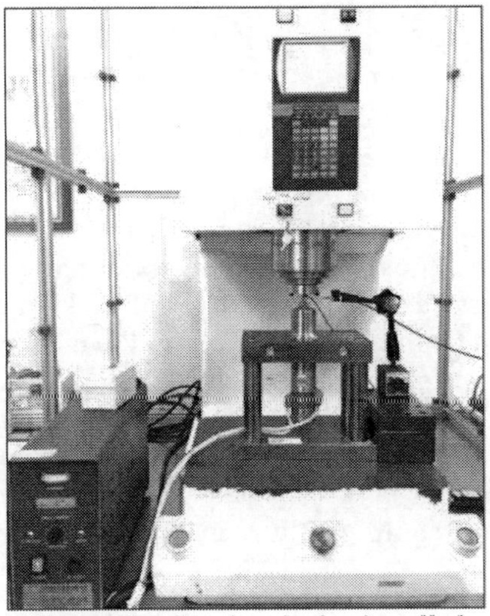

(a) Ultrasonic vibration device installed on the press

(b) Load cell installed on the ram of the press

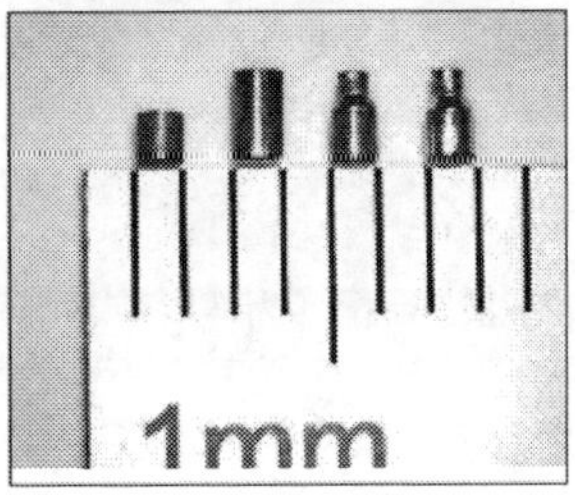

(c) Billet, extruded cup, nosed cup and nosed cup from the process with ultrasonic vibration (from left to right)

Fig. 5. Experimental setup.

Table 1. Conditions for experiments of micro backward extrusion and micro nosing process.

Conditions		Micro backward extrusion and micro nosing process without ultrasonic vibration	Micro nosing process with ultrasonic vibration
Punch speed, mm/s		0.1	
Stroke, mm		0.64	
Lubricant		Oil	
Sampling rate for data acquisition, Hz		120	7200
Ultrasonic vibration on the die	Frequency, kHz	None	20
	Amplitude, μm	None	5

3 Results and discussions

3.1 Deformation

The extruded cup, presented in Figure 6(a), with uniform rim and wall thickness implies the materials flowed homogeneously in the micro backward extrusion process. Figure 6(b) presents the nosed cup in which the open end has been shrunk by the micro nosing process without the ultrasonic vibration. It clearly shows the shrunk region has a neck with a smaller diameter. The phenomenon might be caused by the friction effect which led to difficulties in materials flow along the shrunk region as the nosing process progressed. With the die assisted by the ultrasonic vibration, an even outer diameter in the nosed cup can be produced, as shown in Figure 6(c). The above results show that the ultrasonic vibration can improve the materials flows and thus result in better quality of the product in the micro nosing process.

3.2 Load

The load from the nosing process without ultrasonic vibration gradually increased with the stroke to the maximum value of about 120 N as displayed in Figure 7. However, the load curve varied with different peak values in the nosing process with the ultrasonic vibration. The high frequency variation of the load curve is a typical phenomenon in an ultrasonic vibration assisted forming process. Moreover, the curve showed two ranges of peak values. In the early stage of the nosing process, the extruded cup might be sliding in the die cavity that resulted in an average load of about 5 N and the most peak values within a range of about 20 N. In the later stage after the stroke of about 0.2 mm, the average load clearly increased to the value of about 30 N, then gradually increased to the maximum value of about 45 N. The most peak values were in a range of 50 N. The comparison in the load curves clearly shows that the ultrasonic vibration can significantly reduce the load in the nosing process of the micro copper cup.

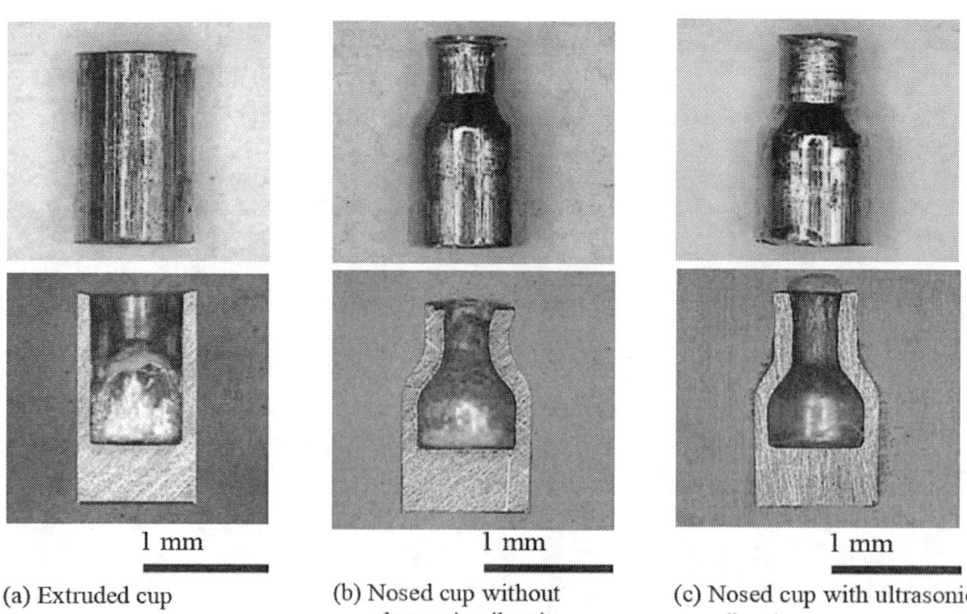

(a) Extruded cup (b) Nosed cup without ultrasonic vibration (c) Nosed cup with ultrasonic vibration

Fig. 6. Fabricated parts.

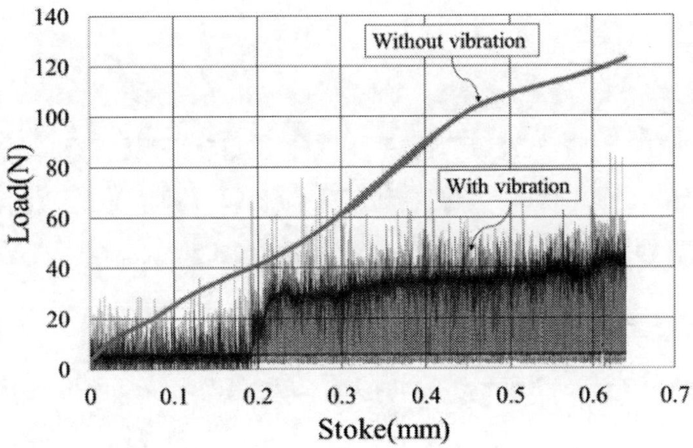

Fig. 7. Load curves of nosing processes.

4 Conclusions

The study investigated the effect of ultrasonic vibration on the deformation and forming load in the micro nosing process of copper cup. Two forming conditions with and without the ultrasonic vibrations were considered in the study. The results clearly show that the ultrasonic vibration can improve the materials flow which leads to an even diameter along the shrunk region of the nosed cup. Moreover, the ultrasonic vibration greatly reduces the load in the micro nosing process of the copper cup.

The authors gratefully acknowledge the financial support from the Ministry of Science and Technology, Taiwan (Grant No. MOST 105-2221-E-151-018).

References

1. L. Alves, T. Pardal, P. Martins, Journal of Cleaner Production **18**(16) 1740 (2010)
2. M. Geiger, M. Kleiner, R. Eckstein, N. Tiesler, U. Engel, CIRP Ann. **50**(2) 445 (2001)
3. F. Vollertsen, H. Schulze Niehoff, Z. Hu, Int. J. Mach. Tools. Manuf. **46**(11) 1172 (2006)
4. C. Wang, B. Guo, D. Shan, Manufacturing Review **1** 23 (2014)
5. W. Chan, M. Fu, B. Yang, Materials & Design **32**(7) 3772 (2011)
6. C.C. Chang, J.C. Lin, Proc. Inst. Mech. Eng. Part B-J. Eng. Manuf. **226**(B1) 183 (2012)
7. C.C. Chang, T.C. Wang, Advanced Materials Research **83-86** 1092 (2010)
8. Z. Yao, G.Y. Kim, L. Faidley, Q. Zou, D. Mei, Z. Chen, J. Mater. Process. Technol. **212**(3) 640 (2012)
9. M. Michalski, F. Piott, M. Merklein, Procedia Engineering **207** 1970 (2017)
10. J. Hu, T. Shimizu, M. Yang, Procedia Engineering **207** 1063 (2017)
11. C. Bunget, G. Ngaile, Ultrasonics **51**(5) 606 (2011)

Effects of substrate temperature on the properties of the indium tin oxide thin films deposited by sputtering method

Chia-Ching Wu[1,*] and Chien-Chen Diao[2]

[1]Department of Applied Science, National Taitung University, Taitung, Taiwan, R.O.C.
[2]Department of Electronic Engineering, Kao Yuan University, Kaohsiung, Taiwan, R.O.C.

Abstract. High-quality transparent conductive indium tin oxide (ITO) thin films were deposited on glass substrates using radio frequency sputtering method. The structure and electrical and optical properties of the ITO thin films were mainly investigated. The ITO thin films showed strong diffraction peak having a preferred orientation along the [111] direction as the deposition temperature at 120 to 160 °C. In the transmission spectra, the optical transmittance increased in the visible range as the deposition temperature increased from RT to 160 °C. A minimum resistivity of 3.06×10^{-3} Ω-cm was obtained for the ITO thin film deposition at 160 °C.

1 Introduction

Transparent conducting oxide (TCO) thin films, such as tin oxide (SnO_2), indium tin oxide (ITO), zinc oxide (ZnO), indium zinc oxide (IZO), gallium zinc oxide (GZO) and indium gallium zinc oxide (IGZO), are used for a wide range of important applications, such as transparent electrodes in light-emitting diode (LED) [1], organic light-emitting devices (OLED) [2], liquid crystal display (LCD) [3] and solar cell [4] applications. Among the TCO thin films, indium oxide doped with tin (ITO) is widely used due to its unique electrical and optical properties such as good conductivity (about 10^4 $\Omega^{-1}cm^{-1}$) and high transmittance (~85%) in the visible region due to its large bandgap of about 3.70 eV [5].

ITO thin film has been prepared by various techniques, such as pulsed laser deposition [6], radio-frequency (RF) magnetron sputtering [7], the sol–gel method [8], and electron beam evaporation [9] and spray pyrolysis [10]. Various growth conditions have the effect of increment in the carrier concentrations contributing to the enlargement of the bandgap, which is known as the Burstein– Moss shift [11]. RF magnetron sputtering is widely used because it is considered to be one of the best methods for preparing ITO thin film with high quality. In this study, ITO thin films of a fixed composition (In_2O_3=90 wt% and SnO_2=10 wt%) were deposited on the glass substrate with different substrate temperature by RF magnetron sputtering. The structural, optical, and electrical properties of the ITO thin films were characterized by scanning electron microscopy (SEM), X-ray diffraction (XRD) pattern, UV-visible spectroscopy, and Hall measurement.

*Corresponding author: ccwu@nttu.edu.tw

2 Experimental procedures

Indium tin oxide thin films were deposited on glass substrates with a In_2O_3 : SnO_2 (90 : 10 wt%) ceramic target (Summit-Tech. Co.) of 99.99% purity by using radio frequency (R.F.) magnetron sputtering system, as shown in Fig. 1. The glass substrate was cut into a square shape 20mmx20mm in size. The prepared glass substrate was cleaned ultrasonically in acetone, ethanol and IPA for 20 minutes and finally rinsed in distilled water. The working distance between the substrate and target was fixed at 12 cm. A heater was used for substrate heating from 120 to 160 °C. The temperature was obtained by a thermocouple on the substrate surface and was controllable with 1% measurement accuracy. The working pressure was set to 5×10^{-3} torr and the R.F. power varied from 40 W to 120 W. The surface morphology and thickness of the ITO thin films were measured by using the field emission scanning electron microscope (FE-SEM). The crystalline structural characterization of the ITO thin films was carried out by an x-ray diffractometer (XRD) using Cu $K\alpha$ radiation ($K\alpha$= 0.15418 nm). The optical transmission was taken in the wavelength range of 200–1100 nm using a Hitachi U-2900 model UV/Vis spectrophotometer. The resistivity (ρ), carrier concentration (n_e) and carrier mobility (μ) were obtained from Hall-effect measurements using the Van der Pauw method.

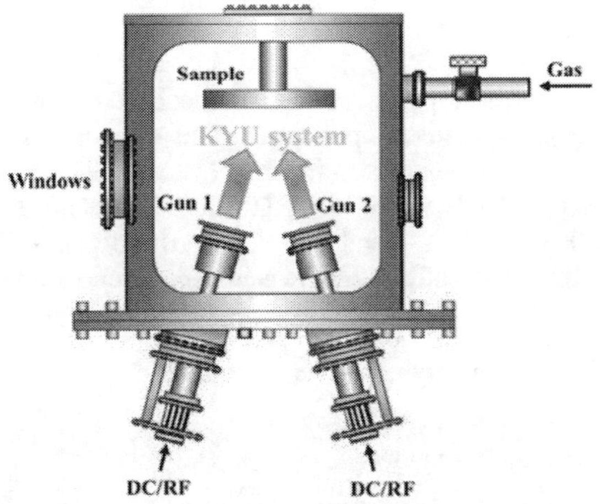

Fig. 1. The diagram of the radio frequency magnetron sputtering system.

3 Results and discussion

XRD patterns of the ITO thin films with different deposition powers are shown in Fig. 2. As the ITO thin film deposition at room temperature, it shows an amorphous structure and the ITO thin films are polycrystalline structure as the substrate temperature at 120 °C to 160 °C. As the substrate temperature increases to 120 °C to 160 °C, all patterns exhibited the (211), (222), (400), (400) and (620) peaks of the ITO thin films at diffraction angles (2θ) near 21.2°, 30.3°, 34.8°, 50.3° and 59.9°, respectively. The obtained diffraction peaks were in accordance with the standard JCPDS data (Card no. 65-3170). And no any second phase was obtained in the patterns. In addition, the intensity of the (222) diffraction peak increased as the substrate temperature increased from 120 °C to 160 °C. The fact that the spectra showed a diffraction peak of (222) means that the prepared films had a tendency to crystallize along the preferred orientation of [111] direction. From above results, the crystallization of the ITO thin films is enhanced as higher deposition temperature is used, for that the resistivity of the ITO thin films will closely depend on the deposition temperaature.

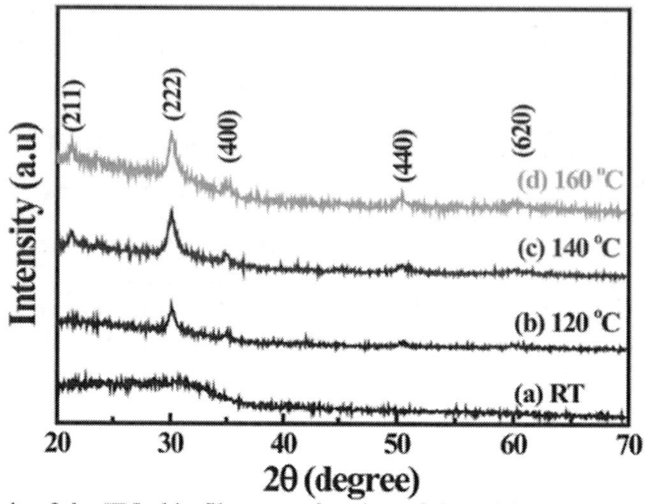

Fig. 2. XRD analysis of the ITO thin films as a function of deposition temperature. (a) RT, (b)120, (c)140 and (e)160 °C, respectively.

Figures 3 and 4 shows the top-view and cross-sectional SEM image of the ITO thin films prepared at various substrate temperatures. The film prepared at room temperature exhibits non-crystalline structure as shown in Fig. 3(a), and he surfaces of the ITO thin film are smooth and compact. Compared with the ITO thin film deposition at RT, the top-view micrographs of the 120 °C-, 140 °C-, and 160 °C-deposited ITO thin films had the similar morphologies, but the grain sizes slightly increased as the deposition temperature increased, as shown in Figs. 2(b)-2(d).

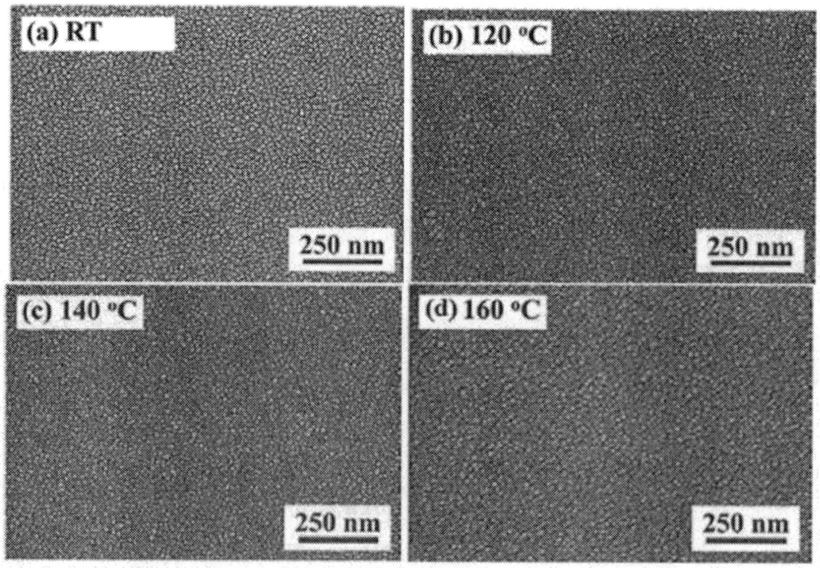

Fig. 3. Top-view SEM observations of the ITO thin films as a function of deposition temperature. (a) RT, (b)120, (c)140 and (e)160 °C, respectively.

Figure 4 shows the thickness of the ITO thin film as a function of deposition temperature. The thickness of ITO thin films increased with deposition temperature is considerable. It found that all the ITO thin films had a tendency to crystallize along the preferred orientation of [111] direction. The thickness increased from 125 to 288 nm as the deposition temperature increased from RT to 160 °C, as shown in Fig. 5.

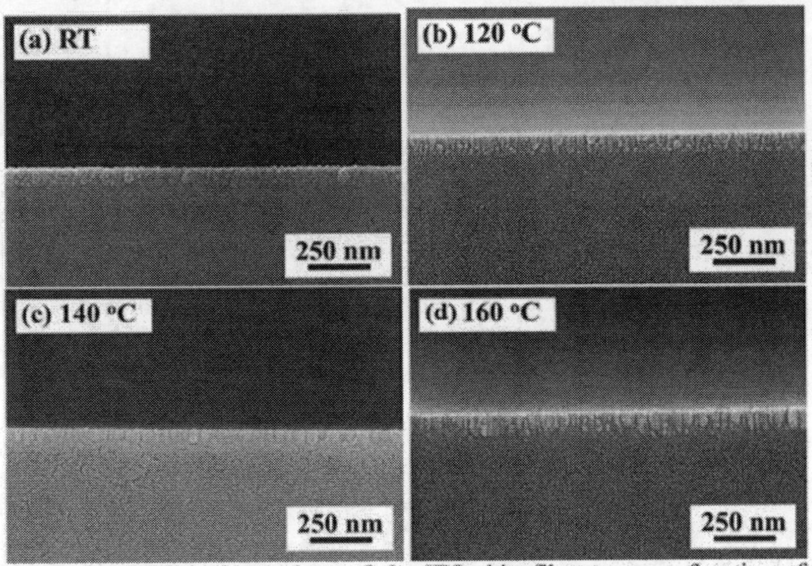

Fig. 4. Cross-sectional SEM observations of the ITO thin films as as a function of deposition temperature. (a) RT, (b)120, (c)140 and (e)160 °C, respectively.

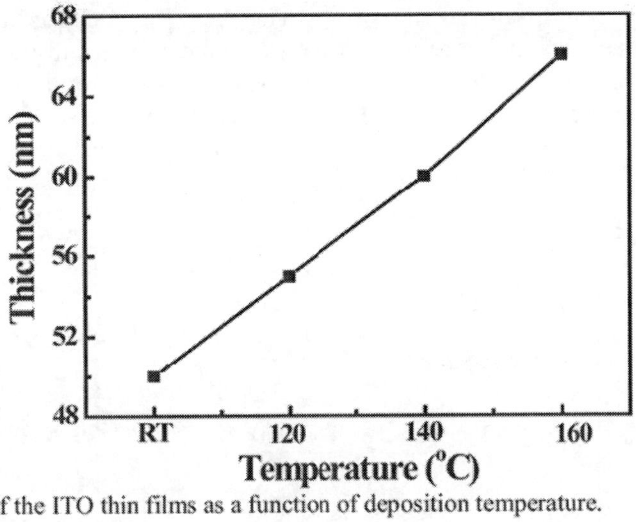

Fig. 5. Thickness of the ITO thin films as a function of deposition temperature.

Figure 6 shows the UV/Vis optical transmittance spectra of the ITO thin films recorded within the wavelength range of the 250-1100nm. It was found that the transmittance in the visible range increased with increasing substrate temperature, and that the absorption edge was at about 320 nm. The average transmittance of the ITO thin film is about 89.9% in the 400 to 700 nm range and 89.3% at 550 nm as deposition at room temperature. The average transmittance (400 to 700 nm) of the ITO thin films slightly increased from 90.13% to 91.71% as deposition temperature increased from 120 to 160 °C. This result caused by the transmittance is closely related to a scattering effect. It is considered that an increase in substrate temperature accelerated the degree of crystallization. Accordingly, the microstructure with the crystallization had a higher value of the transmittance by reducing the light scattering in the film [12].

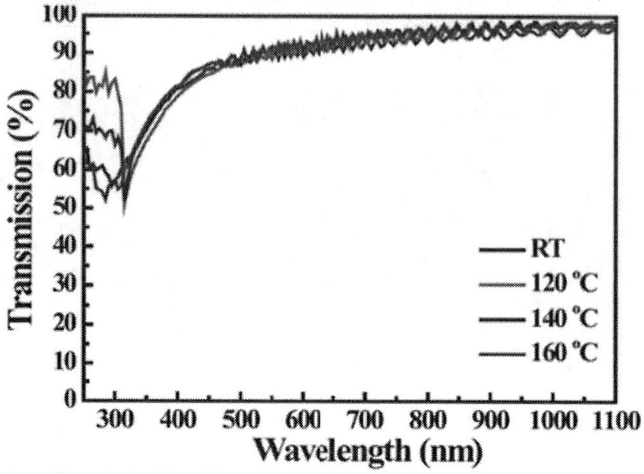

Fig. 6. UV-Vis spectra of the ITO thin films as a function of deposition temperature.

Figure 7 shows the resistivity, mobility and carrier concentration of the ITO thin films as a function of deposition temperature. As the deposition temperaature increased from RT to 160 °C, the mobility decreased from 22.3 to 14.5 cm^2/V-s. From the XRD and SEM results, the ITO thin film deposition at room temperature shows an amorphous structure and the less grain boundary exist in the film. However, the number of grain boundary of the ITO thin films increaseed and leading to an decrease in carrier mobility as the deposition temperaature increased from 120 to 160 °C. The resistivity decreased with increase in substrate temperature. The decrease in resistivity with increase in substrate temperature can also be explained by the fact that the grain size increases significantly with the increase in deposition temperature and increasing conductivity. The carrier concentration of ITO thin films slightly increased from 2.06×10^{19} to 1.99×10^{20} cm^{-3} as the deposition temperature increased from RT to 160 °C.

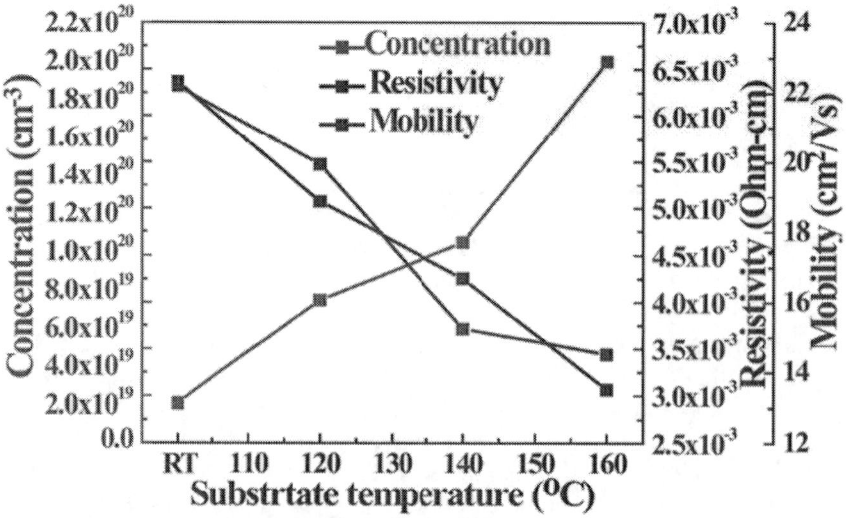

Fig. 7. Resistivity, Hall mobility, and carrier concentration of the ITO thin film as a function of deposition temperature.

4 Conclusions

In this study, crystalline the ITO thin films were deposited on glass substrates using R.F. magnetron sputtering technique as the deposition temperature at 120 to 160 °C. The resistivity of ITO thin films decreased from 6.36×10^{-3} to 3.06×10^{-3} Ω-cm with increasing in the deposition temperature, which could be explained by a grain size increases significantly with the increase in deposition temperature. The transmittance (400-700 nm) of the ITO thin films increased with an increase in the substrate temperature. Based on the above results obtained in this study, the ITO thin film was found to be important for future work on electron devices.

Acknowledgment

The authors acknowledge the financial support of the Ministry of Science and Technology of the Republic of China (MOST 106-2221-E-143-008- and 106-2622-E-143-002-CC3).

References

1. J. Gu, Z. Xiao, M. Xu, G. Wang, Z. Lu, S. Huang, J. Phys. Chem. Solids, **59**, 27 (1998)
2. Y.S. Tsai, F.S. Juang, T.H. Yang, M.C. Yokoyama, L.W. Ji, Y.K. Su, J. Phys. Chem. Solids, **69**, 764 (2008)
3. M. Gaillet, L. Yan, E. Teboul, Thin Solid Films, **516**, 170 (2007)
4. S. Marikkannu, M. Kashif, N. Sethupathy, V.S. Vidhya, Shakkthivel Piraman, A. Ayeshamariam, M. Bououdina, Naser M.Ahmed, M. Jayachandran, Mater. Sci. Semi. Process., **27,** 562 (2014)
5. W. Wohlmuth, I. Adesida, Thin Solid Films, **479,** 223 (2005)
6. F. O. Adurodija, L. Semple, R. Br€uning, J. Mater. Sci., **41**, 7096 (2006)
7. K. J. Patel, M. S. Desai, C. J. Panchal, J. Mater. Sci. Mater. Electron., **22**, 959 (2011).
8. S. Kundu, P. K. Biswas, Chem. Phys. Lett., **414**, 107 (2005)
9. Z. Xu, P. Chen, Z. Wu, F. Xu, G. Yang, B. Liu, C. Tan, Z. Xu, L. Zhang, R. Zhang, Y. Zheng, J. Mater. Sci. Mater. Electron., **25**, 2287 (2014)
10. Z.B. Zhou, R.Q. Cui, Q.J. Pang, Y.D. Wang, F.Y. Meng, T.T. Sun, Z.M. Ding, X.B. Yu, Appl. Surf. Sci., **172**, 245, (2001)
11. V. S. Reddy, K. Das, A. Dhar, S. K. Ray, Semicond. Sci. Technol. **21,** 1747, 2006
12. B. R. Cho, Solid State Phenomena, 124, 195 (2007)

Evaluation of the sinking processes for high-pressure-gas cylinders

Heng-Sheng Lin[1,*], *Chien-Yu* Lee[1], *Chia-Jung* Lin[2], and *Ho-Chung* Fu[1]

[1]Department of Mold and Die Engineering, National Kaohsiung University of Applied Sciences, Kaohsiung 807, Taiwan
[2]Mosa Industrial Corporation, Huwei, Yunlin 632, Taiwan

Abstract. High-pressure-gas cylinders are used in broad applications. Cracks on the open end would occur during the riveting stage. Such forming defects are caused by excessive hardening, although the open end has been annealed with induction heating prior to the sinking operation. Therefore, a proper design for the sinking dies is essential to the forming production of the HPG cylinders. In this paper, two die-design concepts were examined which included the conventional design for six-stage sinking with fixed die radius, and the economic design for five-stage sinking with incremental die radii. Finite element software DEFORM 2D was used to investigate the two sinking schemes. The effect of the sinking schemes on the sinking load, strain distribution, and lip thickness were analysed. The results show that the economic five-stage sinking with a large increment of die radii can provide less strain hardening as compared to other sinking schemes. Although the forming load level is acceptable and the change of lip thickness is insignificant, the production cost of the five-stage scheme is still high. A more economic measure by sinking with one-stage rotary swaging can provide an alternative scheme with advantages of simple die design and saving the lead for annealing.

1 Introduction

High-pressure-gas (HPG) cylinders are used in broad applications such as food dispensers, air pistols, fire extinguishers and so on. The manufacturing process of the 8-gram HPG cylinder, as shown in Fig. 1, comprises blanking from steel coil and followed by deep-drawn into a slender tube in six-stage transfer stamping. The workpiece is then sunk and annealed with induction heating on its open end. The preform is further sunk with another six-stage transfer stamping, as shown in Fig. 2. The tube is subsequently milled on its open end and filled with high pressure gas and riveted to its completion. Some detailed manufacturing processes can be found in literature [1] for other examples of HPG cylinders.

Cracks on the open end would occur during the riveting stage. Such forming defects are caused by excessive hardening, although the open end has been annealed with induction heating prior to the sinking operation. Therefore, a proper design for the sinking dies is essential to the forming production of the HPG cylinders. Amirhosseini et al. [2] analyze

*Corresponding author: hslin@kuas.edu.tw

the nosing process of empty and foam-filled circular metal tubes on semispherical die, and Lu [3] studies the preform and loading rate in the tube nosing process by spherical die. Sheu and Su [4] investigate the cold nosing process for the aluminum conical milk can. Kwan [5] investigates the cold eccentric nosing process of metal tubes with an eccentric conical die from circular tube billets. Oba et al. [6] use two-step forming for improvement of forming limit in rotary nosing with relieved die for reduction of tube tip.

In this paper, two die-design concepts were examined which included the conventional design for six-stage sinking with fixed die radius, and the economic design for five-stage sinking with incremental die radii. Finite element software DEFORM 2D was used to investigate the two sinking schemes. The effect of the sinking schemes on the sinking load, strain distribution, and lip thickness were analysed. Finally the multi-stage sinking schemes were compared with the one-stage rotary swaging process by experimental verification.

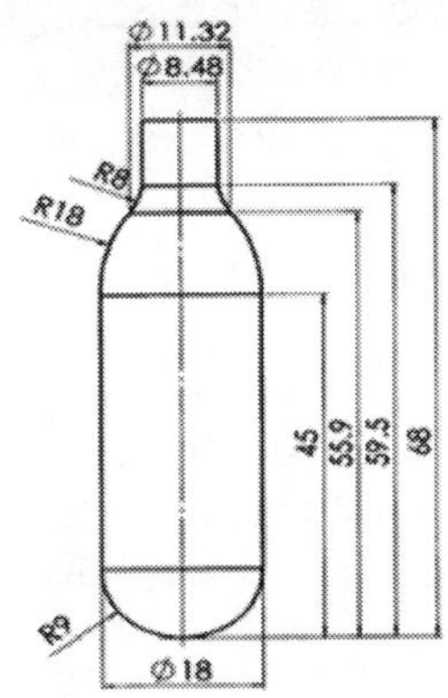

Fig. 1. Dimensions of HPG cylinder.

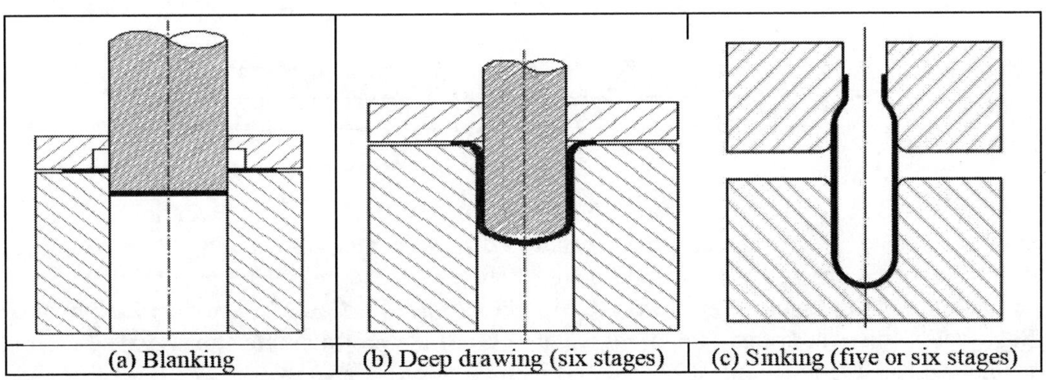

Fig. 2. Forming stages of HPG cylinders.

2 Methods

2.1 Sheet material

The sheet metal used was cold rolled steel SPCE of 0.8 mm thick. The flow stress of the sheet metal is shown in Fig. 3. The average yield strength σ_y was 160 MPa.

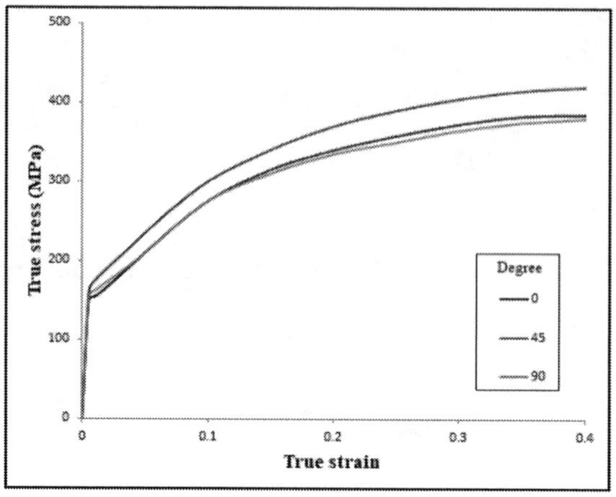

Fig. 3. Flow stress of SPCE steel sheet.

2.2 Sinking schemes

After the completion of deep drawing, one stage sinking and annealing, the workpiece's open end was to be sunk from 14.0 to 8.5 mm in its outer diameter. For the conventional six-stage sinking scheme, a sinking ratio of 0.92 was chosen, while for the economic five-stage sinking scheme, a sinking ratio of 0.90 was used. The target outer diameters used in both six-stage and five-stage sinking schemes are shown in Tables 1 and 2, respectively.

Table 1. Target outer diameters used in six-stage sinking scheme, sinking ratio 0.92.

	Stage 1	Stage 2	Stage 3	Stage 4	Stage 5	Stage 6
Outer diameter, mm	12.9	11.8	10.9	10.0	9.2	8.5

Table 2. Target outer diameters used in five-stage sinking scheme, sinking ratio 0.90.

	Stage 1	Stage 2	Stage 3	Stage 4	Stage 5
Outer diameter, mm	12.6	11.3	10.2	9.2	8.5

Die radii remain unchanged throughout each sinking stage in the conventional six-stage sinking scheme. However, die radii are reduced incrementally as sinking proceeds in the economic five-stage sinking schemes. Table 3 shows the die radii used in each stage of the respective sinking schemes. Increments from 0 to 9 mm were tested from the FE simulation. Fig. 4 shows the variations of die radius with outer tube half-diameter for a typical case of forming scheme "Incremental 1".

Table 3. Die radii used in each stage of forming schemes, unit: mm.

Forming schemes	Stage 1	Stage 2	Stage 3	Stage 4	Stage 5	Stage 6
Conventional 6	7	7	7	7	7	7
Incremental 0	7	7	7	7	7	
Incremental 1	11	10	9	8	7	
Incremental 3	19	16	13	10	7	
Incremental 5	27	22	17	12	7	
Incremental 7	35	28	21	14	7	
Incremental 9	43	34	25	16	7	

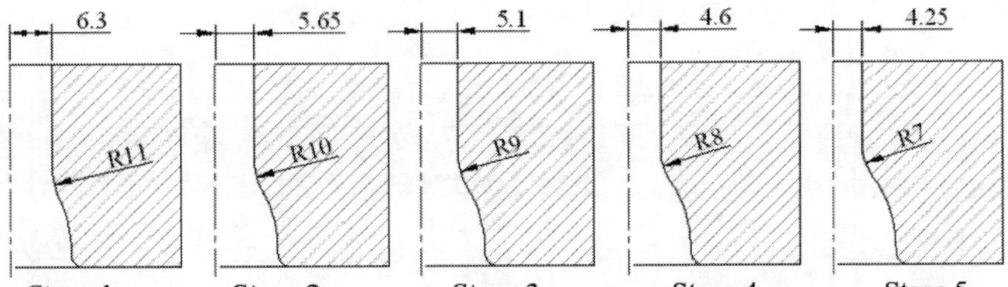

Fig. 4. Variations of die radius with outer tube half-diameter for forming scheme "Incremental 1".

2.3 FE simulation

Finite element software DEFORM was used in simulating all the sinking schemes. The 2D axis-symmetric module was chosen. The planar anisotropy of the sheet material was neglected. There were 5,000 elements used in meshing the workpiece, and all the forming tools were assumed to be rigid. Fig. 5 shows the schematic of the setup for the sinking operation. The stamping speed was fixed to 300 mm/s. Because the tribological behavior of the process was not the main trust of this work, friction factors at the interfaces between the tool and workpiece were fixed to 0.05.

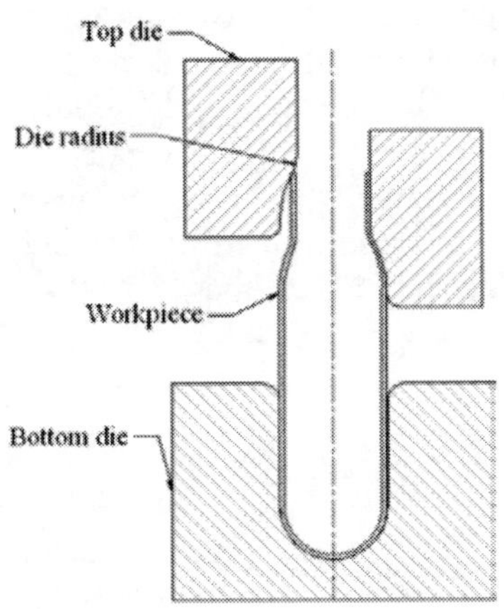

Fig. 5. Setup for the sinking operation, left: before sinking; right: at the lower dead point of sinking.

3 Results

3.1 Sinking load/stress

The sinking load occurring on the open end of tube should not exceed to cause distortion on the tube body which nests inside the bottom die. To ensure that the tube body is not distorted during sinking, the axial stress on the tube body should not exceed the yield stress of workpiece. Therefore, the sinking load obtained from the simulation is first divided by the cross-sectional area of tube body to obtain the axial sinking stress σ_s. The sinking stress is further non-dimensionalized by dividing by the initial yield stress σ_y of the sheet metal. When the value σ_s / σ_y is much greater than unity, there is possibility of distortion on the tube body.

The sinking stresses for both the conventional six-stage and the economic five-stage schemes are shown in Fig. 6. The sinking stress increases as the forming stage proceeds,

 https://doi.org/10.1051/matecconf/201818500007

attributed to the increase of strain hardening in the sinking zone. The stress level is higher with the economic five-stage scheme because of greater amount of deformation was used for the respective sinking stages.

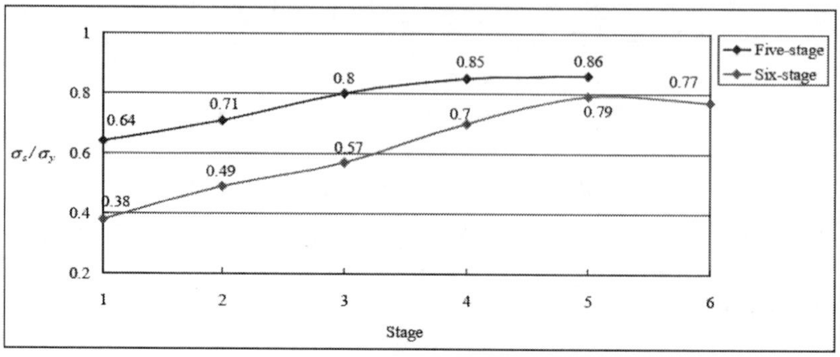

Fig. 6. Sinking stresses for both the conventional six-stage and the economic five-stage schemes.

Fig. 7 shows the sinking stresses for the five-stage schemes of various increments of die radius. The sinking stress increases as the increment of die radius increases. This is caused by the increasing frictional stress exerting on a larger contact interface between the die face and workpiece when the die radius becomes larger. Some values are greater than unity for the forming schemes with increments of die radius larger than 3 mm. Overall, the sinking stress shown in Fig. 7 are acceptable without causing distortion on the tube body. The tube body has experienced certain amount of strain hardening during the prior deep drawing stages. Therefore the yield stress on the tube body should be much higher than the initial yield stress 160 MPa of the sheet metal used in the calculation. The forming stress of Stage 5 is slightly less than that of Stage 4. This is because the sinking amount decreases substantially at the final stage as compared to that of Stage 4, although the sinking ratio remains constant throughout the sinking schemes.

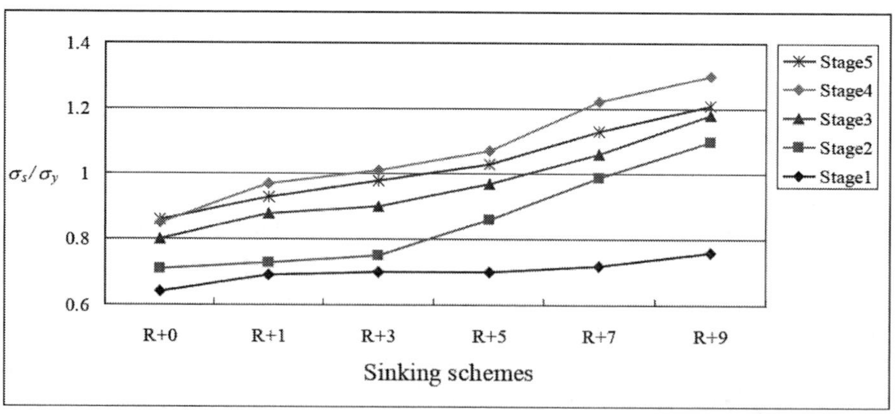

Fig. 7. Sinking stresses for the five-stage schemes of various increments of die radius.

3.2 Distribution of effective strain

Fig. 8 shows the distributions of effective strain at the final stage of selective sinking schemes. The maximum effective strain occurs on the tube inner surface and the value decreases as the die radius increases. This indicates that the strain inhomogeneity decreases with a small die inclusion angle in sinking with a large die radius [7]. High strain level is unfavourable, according to the field experience, because the sinking zone would become susceptible to crack formation in the subsequent riveting operation.

 https://doi.org/10.1051/matecconf/201818500007

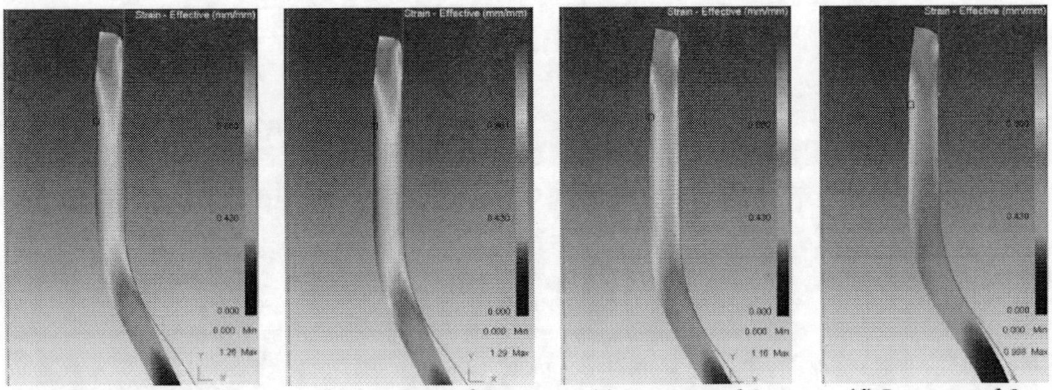

(a) Conventional 6 (b) Incremental 0 (c) Incremental 3 (d) Incremental 9

Fig. 8. Distributions of effective strain at the final stage of various sinking schemes.

Fig. 9 shows the distributions of effective strain along the radial direction at Stage 5 for the economic sinking schemes. The radial position is normalized by the outer radius of the sunk tube. The distributions show a monotonic decrease of effective strain as the increment of die radius increases. The phenomenon observes that the strain level, i.e. deformation inhomogeneity [7] decreases, as the die inclusion angle increases. This trend indicates that using a large increment of die radius can obtain a less hardened sunk tube.

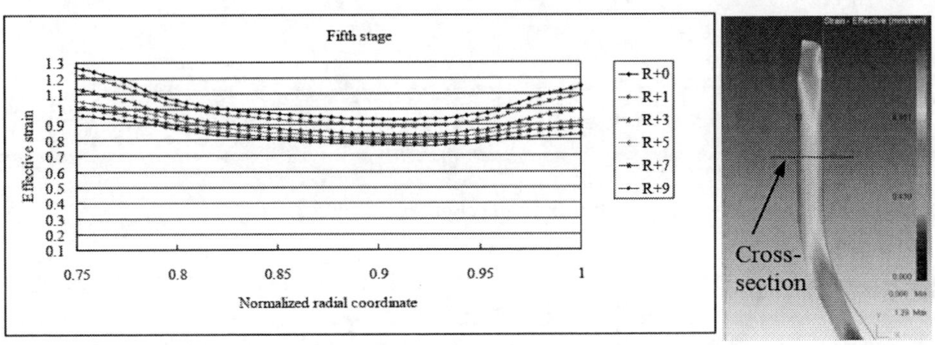

Fig. 9. Distributions of effective strain at Stage 5 of economic sinking schemes.

3.3 Lip thickness

Lip thickness increases during the sinking process. It is the primary straining direction as compared to the axial straining. Fig. 10 shows the increase of lip thickness of respective forming schemes, and the amount decreases sharply at the final stage, attributed to the accumulated strain hardening from the sinking operation. The six-stage sinking scheme exhibits a mild increase of lip thickness because of a large sinking ratio 0.92 was used for the respective stages. Fig. 11 shows the final lip thickness of respective sinking schemes. The lip thickness is the largest in sinking with an increment of die radius of 9 mm, followed by 7 mm, and so on. This is due to less strain level was generated in sinking with a large die radius, i.e. a small die inclusion angle. The analyses for the lip thickness all indicates that the variations with the change of die design are not significant. This guarantees that the amount of surface machining in the original process design remains valid.

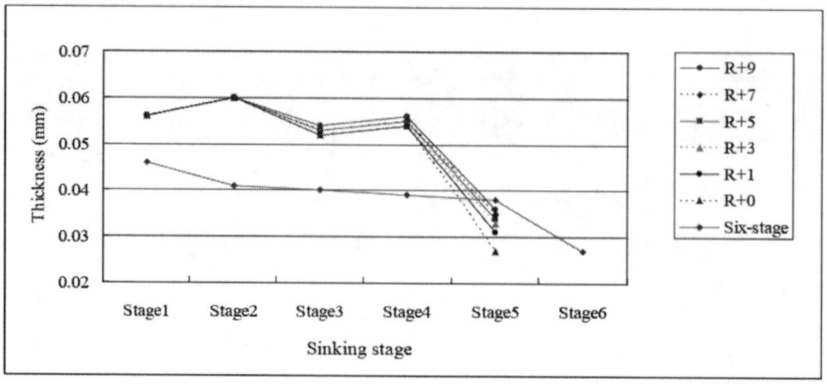

Fig. 10. Increase of lip thickness of respective sinking schemes.

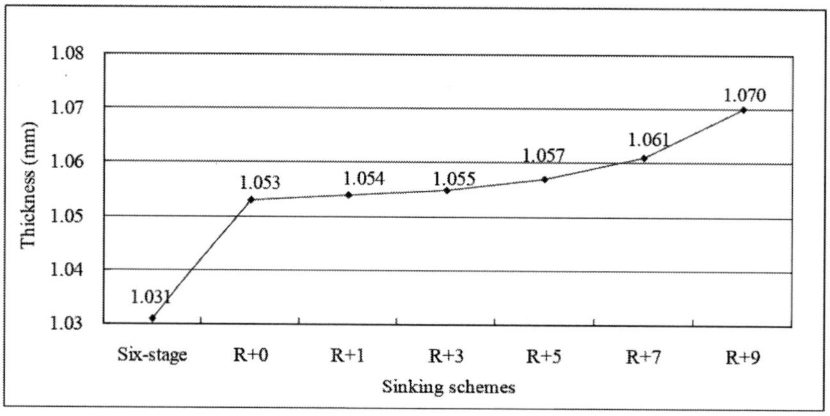

Fig. 11. Final lip thickness of respective sinking schemes.

4 Discussions

In Section 3, two die-design concepts are examined which include the conventional design for six-stage sinking with fixed die radius, and the economic design for five-stage sinking with incremental die radii. The results of the FE analysis show that economic five-stage with a large increment of die radii can provide less strain hardening as compared to other sinking schemes. Although the forming load level is acceptable and the change of lip thickness is insignificant, the production cost of the "economic" five-stage scheme is still high. The energy used in conduction heating, the effort in pickling the oxidation, and the maintenance of the multi-stage dies, still render the sinking scheme by transfer stamping not a lean and green manufacturing method.

A more economic measure by sinking with one-stage rotary swaging has been attempted by Lin et al. [8]. Figure 12 shows a typical hardness distribution along the axial direction for sinking with an un-annealed tube (preform). There is no significant difference between the un-annealed preform and the swaged workpiece, which indicates the feasibility of sinking with un-annealed preform. The distribution also shows that the variation of hardness number is less on the outer surface than that of inner surface. A similar trend is also observed in the FE analysis of the distribution of effective strain that the hardening level is higher on tube inner surface than that of outer surface, as shown in Fig. 9.

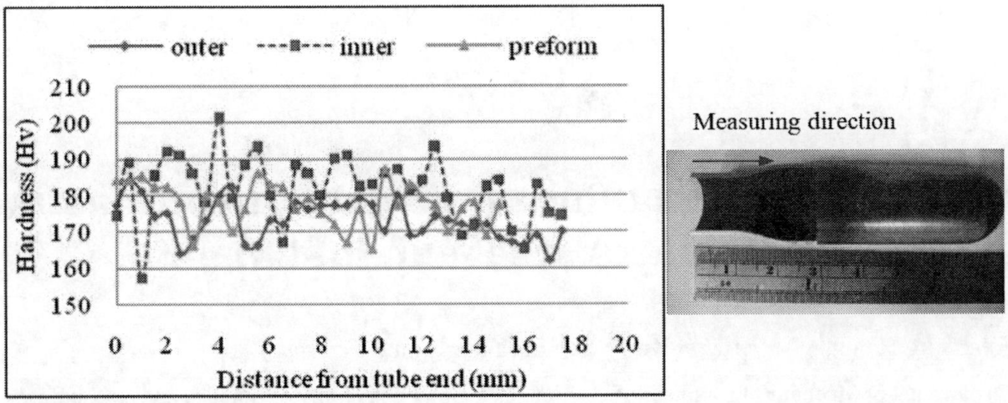

Fig. 12. Hardness distribution of sinking of one-stage rotary swaging with un-annealed preform. [8]

5 Conclusions

Two die-design concepts were examined in this work which included the conventional design for six-stage sinking with fixed die radius, and the economic design for five-stage sinking with incremental die radii. Finite element software DEFORM 2D was used to investigate the two sinking schemes. The results show that the economic five-stage sinking with a large increment of die radii can provide less strain hardening as compared to other sinking schemes. Although the forming load level is acceptable and the change of lip thickness is insignificant, the production cost of the five-stage scheme is still high. A more economic measure by sinking with one-stage rotary swaging can provide an alternative scheme with advantages of simple die design and saving the lead for annealing.

Acknowledgements

The authors acknowledge Mosa Industrial Corporation in providing the samples.

References

1. *ASM Metals Handbook,* Volume 14, Forming and Forging (1989)
2. S.G. Amirhosseini, A. Niknejad, N. Setoudeh, Trans. Nonferrous Met. Soc. China **27**, 1976 (2017)
3. Y.H. Lu, Comput. Methods Appl. Mech. Engrg. **194**, 2839 (2005)
4. J.J. Sheu, H.H. Su, Proceedings of the 2009 IEEE IEEM, 2319
5. C.T. Kwan, J. Mater. Process. Technol. **140**, 530 (2003)
6. W. Obaa, Y. Imaizumia, S. Kajikawaa, T. Kubokia, Procedia Eng. **207**, 1743 (2017)
7. W.F. Hosford, R.M. Caddell, *Meral Forming, Mechanics and Metallurgy*, 3rd edn. Cambridge University Press, UK (2007)
8. H.S. Lin, C.J. Lin, Y.C. Hsu, J.H. Lee, B.Y. Lee, Y.C. Chen, Steel Res. Int. **81**, 576 (2010)

Numerical simulation and design of casting system for stainless steel exhaust manifold

Pei-Hsing Huang[1,*], *Jenn-Kun* Kuo[2], *Te-Hua* Fang[3], and *Wei-ren* Wu[1]

[1]Department of Mechanical Engineering, National Pingtung University of Science and Technology, Pingtung 912, Taiwan

[2]Department of Greenergy, National University of Tainan, Tainan 70005, Taiwan

[3]Department of Mechanical Engineering, National Kaohsiung University of Applied Sciences, Kaohsiung 807,Taiwan

Abstract. During operation, exhaust manifolds must bear the corrosion of high temperatures and repeated stress, which can easily lead to cavitation, corrosion, and creeping damage in the casting structure and affect product safety. To improve the structure of exhaust manifolds and increase their service life, we employed AnyCasting mold flow analysis to SUS304 stainless steel exhaust manifolds. We examined the influence of casting system design and process parameters such as ceramic shell temperature, casting temperature, and pouring speed on the filling and solidification processes of the liquid metal. Finally, we used the Niyama criterion to predict the probability and distribution of shrinkage porosity defects in the exhaust manifold and made improvements to enhance the quality of exhaust manifold castings. Keywords: exhaust manifold, mold flow analysis, casting system.

1 Introduction

The casting industry has thousands of years of history, but developments in casting technology are still being made. Casting products must undergo strict quality inspections to meet relevant specifications. Improving various casting defects is the goal of the casting industry, and to enhance casting quality and yield while reducing labor costs, many researchers have developed molten metal flow analysis methods for casting defect prediction using computer-aided engineering (CAE) [1-10]. At present, the majority of casting manufacturers in Taiwan use trial and error to derive optimal casting schemes. During this process, manufacturers will produce castings with defects, which they then repair using welding, as shown in Fig. 1. Zhao et al. [11] used ProCAST to predict where shrinkage defects will appear in solidifying steel plates and compared the results with those of actual castings. Huang et al. [12] established that directly using open risers or blind risers cannot effectively prevent shrinkage cavity and porosity defects in stainless steel impeller castings, but coordinating the risers with cold iron can. Ma and Chuan [13] examined the

[*] Corresponding author: phhuang1970@gmail.com

differences in the internal mass of products before and after optimization and found that structural optimization can significantly enhance the performance of rear cover housing. Using the finite element method, Yang and Chi [14] simulated the temperature field of the casting solidification process and analyzed shrinkage cavity and porosity defects. Li and Peng [15] employed CAE software to simulate the solidification processes of a casting scheme and its improved scheme; the results indicated that the improved scheme simplified the moulding process and the riser elimination process, enabled effective feeding in the casting, and greatly reduced shrinkage defects. Sun et al. [16] developed a die casting process for automotive aluminium alloys, and the results demonstrated that optimizing the process parameters significantly reduced the shrinkage cavities and defects in the castings. Li et al. [17] analyzed the influence of riser designs on the feeding effects of gray cast iron pieces. The simulation results indicated that changing the height and diameter of riser necks can greatly improve riser feeding effects and reduce shrinkage defects in the casting products. This study used mold flow analysis to design a casting system for an exhaust manifold and adjust the process parameters so as to effectively enhance the quality and yield of exhaust manifold castings.

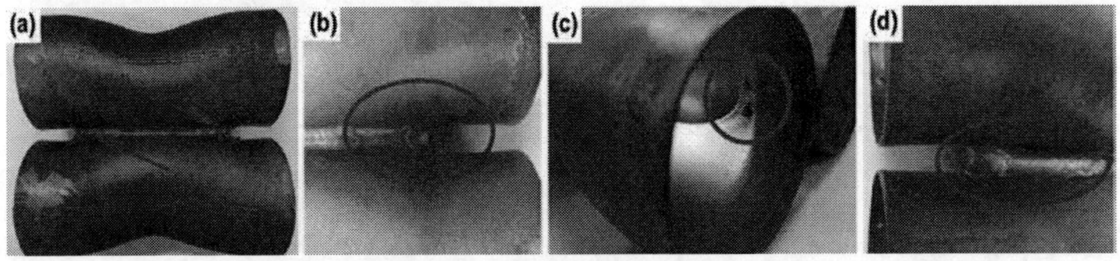

Fig. 1. Welded repairs on exhaust manifold; (a) locations of repairs on exhaust manifold; (b)-(d) welding and penetration marks of repaired defects circled in red.

2 Experimental and numerical methods

2.1 Process parameter settings of investment casting

We chose SUS304 stainless steel for this study, which is resistant to corrosion and high temperatures and has good mechanical properties. As a result, it is a suitable material for exhaust manifolds. We imported the physical properties of the SUS304 stainless steel into the metal mold flow analysis, which include density (ρ): 7,930 kg/m3, thermal conductivity (K): 14.39 W/m·K, specific heat (S): 494 J/kg·K, latent heat (L): 290 kJ/kg, liquidus temperature (T_l): 1,454 °C, and solidus temperature (T_s): 1,399 °C, as shown in Table 1. We designed the process parameters of the initial scheme and the improved scheme based on the material characteristics. These properties and parameters include the liquidus and solidus temperatures, casting material ingredients, ceramic shell temperature $T_{ceramic}$ (°C), casting temperature $T_{casting}$ (°C), pouring speed $V_{pouring}$ (cm/s), and ceramic shell thickness δ (mm), as shown in Table 2.

Table 1. Thermophysical properties of SUS304 stainless steel [9].

ρ (kg/m3)	K(W/m·K)	S(J/kg·K)	L(kJ/kg)	T_l (°C)	T_s (°C)
7930	14.39	494	290	1454	1399

Table 2. Process parameters of casting schemes.

	Initial scheme	Improved scheme
Casting material	SUS 304	
Shell mold material	Zircon sand	
Mesh number	3,000,000	
$T_{ceramic}$ (°C)	1050	1050
$T_{casting}$ (°C)	1600	1600
$\delta(mm)$	6.0	
$V_{pouring}$ (cm/s)	7	9

2.2 Molten metal flow analysis and method

We performed numerical simulation and analysis for exhaust manifolds using the AnyCasting software package. The simulation process was divided into three major parts: pre-processing (physical model selection, meshing, and parameter setting), computation processing (flow field calculation, temperature field calculation, stress field calculation, and defect prediction analysis), and post-processing (result analysis). Furthermore, the continuity equation, the Navier-Stokes equations, energy equation, and volume function equation can be used to calculate the filling and solidification processes of the liquid metal and the changes in temperature, speed, and pressure, as shown in Eqs. (1)-(6) [1,8-10].

(1) Continuity equation:

$$\frac{\partial\rho}{\partial t}+\frac{\partial(\rho u)}{\partial x}+\frac{\partial(\rho v)}{\partial y}+\frac{\partial(\rho w)}{\partial z}=0 \tag{1}$$

(2) Navier-Stokes equations:

$$\rho\left(\frac{\partial u}{\partial t}+u\frac{\partial u}{\partial x}+v\frac{\partial u}{\partial y}+w\frac{\partial u}{\partial z}\right)=-\frac{\partial p}{\partial x}+\rho g_x+\mu\left(\frac{\partial^2 u}{\partial x^2}+\frac{\partial^2 u}{\partial y^2}+\frac{\partial^2 u}{\partial z^2}\right) \tag{2}$$

$$\rho\left(\frac{\partial v}{\partial t}+u\frac{\partial v}{\partial x}+v\frac{\partial v}{\partial y}+w\frac{\partial v}{\partial z}\right)=-\frac{\partial p}{\partial y}+\rho g_y+\mu\left(\frac{\partial^2 v}{\partial x^2}+\frac{\partial^2 v}{\partial y^2}+\frac{\partial^2 v}{\partial z^2}\right) \tag{3}$$

$$\rho\left(\frac{\partial w}{\partial t}+u\frac{\partial w}{\partial x}+v\frac{\partial w}{\partial y}+w\frac{\partial w}{\partial z}\right)=-\frac{\partial p}{\partial z}+\rho g_z+\mu\left(\frac{\partial^2 w}{\partial x^2}+\frac{\partial^2 w}{\partial y^2}+\frac{\partial^2 w}{\partial z^2}\right) \tag{4}$$

(3) Energy conservation equation:

$$\rho\frac{\partial H}{\partial t}+\rho H\nabla\cdot\vec{u}-\nabla\cdot(k\nabla T)-q=0 \tag{5}$$

(4) Volume function equation:

$$\frac{\partial F_V}{\partial t}+u_1\frac{\partial F_V}{\partial x}+u_2\frac{\partial F_V}{\partial y}+u_3\frac{\partial F_V}{\partial z}=0 \tag{6}$$

In this study, defect prediction was performed using the Niyama criterion [9,18], which predicts microscopic defects based on crystal formation within the castings.

$$G/\sqrt{R}<C_{Niyama} \tag{7}$$

In the Niyama criterion, a $G/\sqrt{R}$ value less than the threshold value means a greater probability of shrinkage porosity, whereas a $G/\sqrt{R}$ greater than the threshold value means a smaller probability of shrinkage porosity [18].

In Eqs. (1)-(7), u, v, and w denote the speed vectors along the X, Y, and Z axes; ρ represents density; μ is the dynamic viscosity of the liquid metal; g_x, g_y, and g_z are the gravity acceleration vectors; t denotes time; q represents volume heat flux; H is enthalpy; G denotes the temperature gradient; R represents to cooling rate, and C_{Niyama} is the cooling rate.

3 Results and discussions

3.1 Mold flow analysis of initial scheme for exhaust manifold

The filling and solidification processes of the liquid metal are crucial to casting quality. We therefore incorporated AnyCasting numerical simulation and analysis to examine the cavity filling process, the directions of solidification and cooling, turbulent flows, and poured short, as shown in Fig. 2. As shown in Fig. 2 (a), the liquid metal flows vertically through the gate and fills the horizontal runner. It then enters the exhaust manifold via the ingate and encounters a surface, which creates impact and causes splashes and turbulent flows. Thus, the liquid metal cannot flow steadily into the exhaust manifold. The portions circles in red in Fig. 2(b) show that the surfaces of the exhaust manifold have already solidified, while the pipe junctions and runners have not. This means that the walls of the pipe junctions are thicker and have greater geometric changes. Furthermore, the solidification process of the pipe junction walls creates isolated residual melts on both the left and right sides. Figure 3 displays the probabilities and distributions of defects in the initial scheme. A darker red indicates a higher probability of shrinkage porosity defects, whereas a darker blue means a lower probability of shrinkage porosity defects. Figures 3(a) and 3(c) show larger areas with high probability of shrinkage porosity defects on top of the exhaust manifold near the ingates and at the bottom. Furthermore, the parts circled in red in Fig. 3(b) show a lot of gray areas at the pipe junctions, which mean poured short defects.

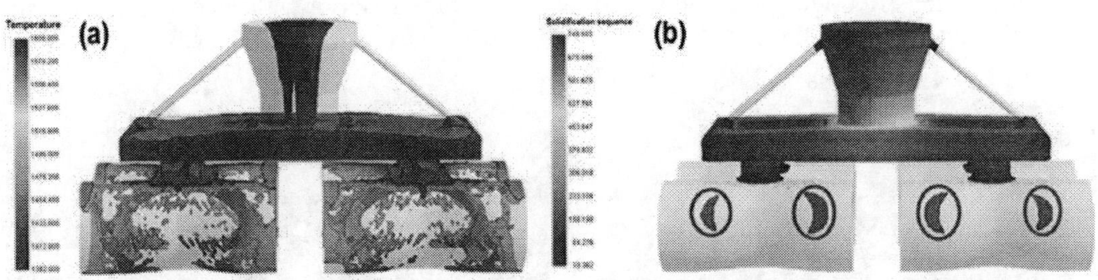

Fig. 2. Analysis of filling and solidification processes of liquid metal in initial scheme: (a) filling sequence (t=2.38 s); (b) solidification sequence (t=69.4 s).

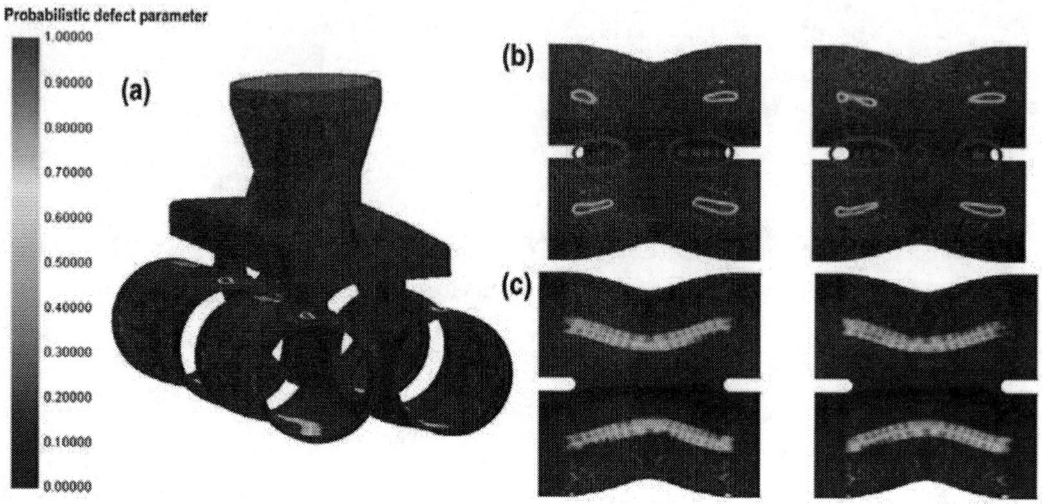

Fig. 3. Probabilistic defect distribution of initial scheme: (a) pattern tree; (b) top view of exhaust manifold; (c) bottom view of exhaust manifold.

3.2 Design and mold flow analysis of improved scheme for exhaust manifold

Based on the simulation results of the initial scheme, we had the four ingates filled at the same time in the improved scheme and installed two thin boards (L52×W50×H4.5) on both sides of the pipe junctions for solidification feeding. This solved the problems of shrinkage porosity defects, poured short, and turbulent flows. As shown in Fig. 4(a), the liquid metal flowed swiftly toward the thin boards and the bottoms of the exhaust manifold. The splash issue was much improved in the improved scheme, and the flow field of the filling process was much steadier. This prevented air from being captured within cavities and the liquid metal, thereby preventing casting defects caused by the captured air. As shown in Fig. 4(b), the thin boards prompted the feeding of the portions to the left and right of the pipe junctions, and the results show that the isolated residual melt can be effectively concentrated on the board to feed the portions to both sides of the pipe junctions. Figures 5(a) and 5(c) present significantly reduced probabilities of defects at the top and bottom of the exhaust manifold. As shown with the parts circled in red in Fig. 5(b), the installation of the thin boards effectively enabled the filling of the pipe junctions, thereby eliminating the poured short defects in the initial scheme and improving the quality of the casting.

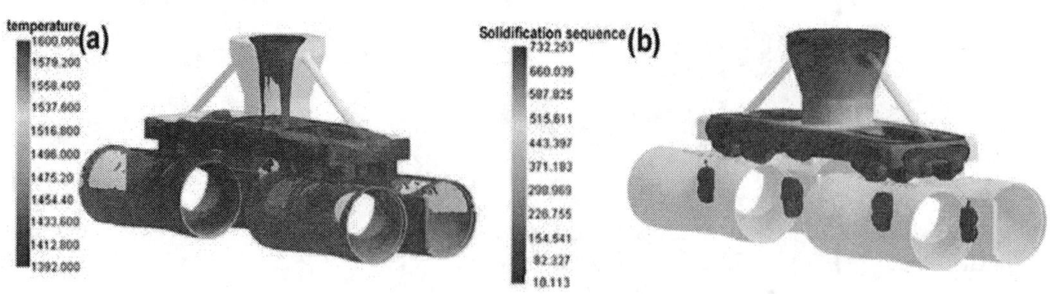

Fig. 4. Analysis of filling and solidification processes of liquid metal in improved scheme: (a) filling sequence (t=2.45 s); (b) solidification sequence (t=147.3 s).

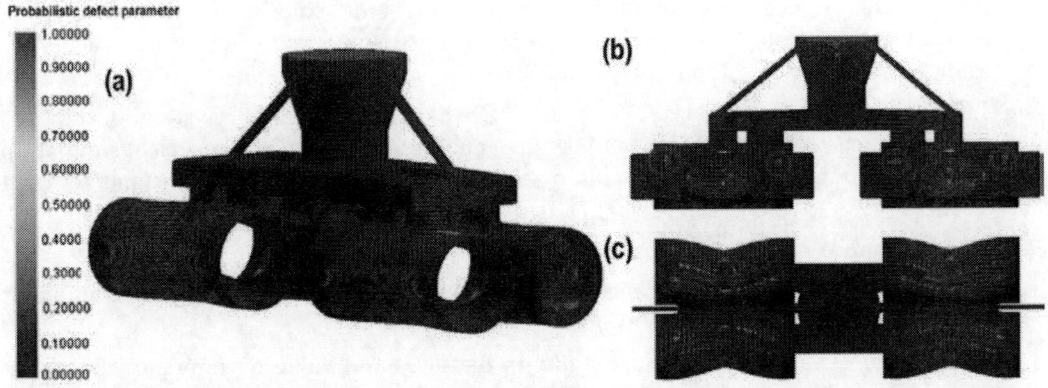

Fig. 5. Probabilistic defect distribution of improved scheme: (a) pattern tree; (b) sectional view of exhaust manifold; (c) bottom view of exhaust manifold.

4 Conclusions

Increasing the number of ingates and altering their dimensions can effectively improve the stability of the early filling process and reduce turbulent flows. The addition of thin boards at the pipe junctions of the exhaust manifold achieved feeding effects during the solidification process, thereby reducing shrinkage and poured short defects. The simulation results revealed significantly lower probabilities of defects in the improved scheme than in the initial scheme, particularly around the ingates, at the bottom of the exhaust manifold, and at the pipe junctions, which means greater exhaust manifold quality.

Acknowledgment

The authors gratefully acknowledge the support provided for this research by the Ministry of Science and Technology, R.O.C. under grants MOST 106-2221-E-020 -014 and MOST 106-2622-E-020 -004 -CC3.

References

1. G.F. Mi, X.Y. Liu, K.F. Wang, J.T. Niu, Numerical simulation of low pressure die-casting aluminum wheel, China Foundry, Vol. 48, No. 1(2009)
2. P.H. Huang, W.J. Wu, C.H. Shieh, Compute-aided design of low die-casting process of A356 aluminum wheels, Applied Mechanics and Materials, Vol. 864, pp. 173-178(2017).
3. P.H. Huang, W.J. Wu, C.H. Shieh, Numerical simulations of low pressure die-casting for A356 aluminum rims, Materials Science Forum, Vol. 893, pp. 276-280(2017).
4. P.H. Huang, Y.T. Chen, B.T. Wang, An effective method for separating casting components from the runner system using vibration-induced fatigue damage, Int J Adv Manuf Technol, Vol. 74, pp. 1275-1282 (2014).
5. P.H. Huang, M.J. Guo, A study on the investment casting of 17-4PH stainless steel helical impeller of centrifugal pump, Materials Research Innovations, 19, S9, 77-81 (2015) DOI 10.1179/1432891715Z.0000000001924
6. P.H. Huang, J.Y. Luo, S.C. Hung, C.J. Lin, H.H. Cheng, Optimal pouring system design for investment casting of cladding thin-plate heater using metallic mold flow Analyses, Applied Mechanics and Materials, Vol. 627, pp. 46-49(2014)

7. P.H. Haung, C.J. Lin, Computer-aided modeling and experimental verification of optimal gating system design for investment casting of precision rotor, Int J Adv Manuf Technol, Vol. 79, pp. 997-1006(2015)
8. J.K. Kuo, P.H. Huang, H.Y. Lai, J.R. Chen, Optimal gating system design for investment casting of 17-4PH stainless steel enclosed impeller by numerical simulation and experimental verification, International Journal of Advanced Manufacturing Technology, 92, 1093-1103 (2017) DOI: 10.1007/s00170-017-0198-0
9. User manual of AnyCasting, version 6.0
10. Y.Y. Li, Prediction of shrinkage defect in steel casting for marine engine cylinder cover by numerical analysis (2007)
11. X.G. Zhao, Z.X. Xiao, J.Z. Guo, Study on the threshold value of simulated shrinkage porosity for a large steel casting, Foundry, Vol. 64, No. 5, pp. 425-429(2015)
12. J.H. Huang, Y. Cao, H.Q. Wang, W.T. Zheng, X.G. Yuan, Casting process design and optimization of martensite stainless steel pumping impeller, Foundry, Vol. 64, No. 4, pp. 317-320(2015).
13. H.B. Ma, H.J. Chuan, Effect of structure of back cover on casting defect and performance, Foundry, Vol. 64, No. 10, pp. 1042-1045(2015)
14. M.Y. Yang, Y.L. Chi, FEM Numerical simulation of temperature and defects prediction algorithm of casting solidification, Foundry, Vol. 66, No. 1, pp. 44-49(2017)
15. S. Li, H.Y. Peng, Solidification process simulation and process optimization for the EPC process of the tail roller, Foundry, Vol. 64, No. 9, pp. 861-863(2015).
16. J. Sun, S.X. Xu, J. Tang, F.F. Wang, Z.Q. Han, Development of squeeze casting technology for manufacturing aluminum alloy subframe for automobile, Foundry, Vol. 64, No. 1, pp. 17-21(2015)
17. S. Li, R.H. Shen, X.G. Yuan, Effect of riser parameters on gray iron feeding, Foundry, Vol. 64, No. 11, pp. 1116-1119(2015)
18. L.W. Pan, L.J. Zheng, H. Zhang, W.L. Gao, Applicability of shrinkage porosity prediction for casting with Niyama criterion, Journal of Beijing University of Aeronautics and Astronautics, Vol. 37, No. 12, pp. 1534-1540(2011)

MATEC Web of Conferences **185**, 00009 (2018) https://doi.org/10.1051/matecconf/201818500009
ICPMMT 2018

Artificial intelligence in safety system

Li-Shan Chen[1], *Yen-Ming* Tseng[2,*], and *Xiao-Na* Lin[1]
[1]School of Management, Fujian University of Technology, No.3 Xueyuan Road, Fuzhou University Town, 350108, Fuzhou City, Fujian Province, China

[2]College of Intelligence Robot, Fuzhou Polytechnic, No.8 Lianrong Road, Fuzhou University Town, 350108, Fuzhou City, Fujian Province, China

Abstract. This research aims to study learning environment, and let the learning environment become smart. Swarm intelligence, cloud computing, and active Ultra-High Frequency RFID were used on it. We built friendly human-computer-interface software for users to use as pad phone. The Extensible Markup Language (XML) and C sharp language were used in this research. If the users begin to search, the kernel safety learning system automatically communicates with other RFID readers by agents, and the agents can search the closer camera for users. This study's result has successfully implemented to Chin-Huo educational organization, and it would be helpful for the paterfamilias to hold all situations about their children at Chin-Huo educational organization. Paterfamilias can understand their children's learning, going to Chin-Huo and leaving Chin-Huo through personal computers, or notebooks simultaneously or asynchronously by the computer-mediated communication. That will be great help in the grip of whole after-school remedial education, teaching and learning situation.

1 Introduction

In many applications, the environmental context for and drivers of movement patterns are just as important as the patterns themselves [1]. Kopf et al. [2] use a standard technique to estimate scene depth, but assign depths to image gradients rather than pixels. Digital artifacts created via transformational technologies often embody implicit knowledge that must be correctly interpreted to successfully act upon the artifacts [3]. With continued advances in communication network technology and sensing technology, there is astounding growth in the amount of data produced and made available through cyberspace.

Sun and Chen [4] [5] apply adopts the Windows Media Player along the RTP/RTSP protocol in order to embed the mobile information system into the users' pad phone, and provides a solution (including hardware solutions) to promote campus safety environment. Chen combines the swarm intelligence and Web Services to transform a conventional library system into an intelligent library system having high integrity, usability, correctness, and reliability software for readers [6]. L. S. Chen, and S. L. Chen [7] built the intelligent system and developed a knowledge base of the computer-parts.

*Corresponding author: swk1200@qq.com

This research used the swarm intelligence, cloud computing, and active Ultra-High Frequency RFID for learning environment, and built friendly human computer interface software for users to use as pad phone. We program the system and software with Extensible Markup Language (XML) and C sharp language.

2 Literature review

2.1 For cloud computing

Hayes et al. [8] show how leveraging an underlying platform-as-a-service (PaaS) cloud model can provide integration with web service patient monitoring systems while providing high availability, scalability, and security. The emergence of cloud computing has led enterprises to rethink fundamentally how they organize their content assets. In particular, IT managers are challenged to decide what content should be managed in the cloud and what should not [9]. Lian, Yen and Wang [10] investigate the critical factors that will affect the decision to adopt cloud computing technology in developing countries, specifically in Taiwan's hospital industry. Ranjan and Zhao [11] advance the management and delivery of services in large, heterogeneous, uncertain, and evolving cloud computing environments.

2.2 For swarm intelligence

The explosive growth of mobile apps has given rise to the challenge of finding out interesting apps for users. Recommender systems are employed to meet this challenge. However, as the lack of user and app data, the development of recommender systems for mobile apps is still at a slow pace [12]. The multi-level thresholding is one of the most important issues in image segmentation. It is a time consuming problem, i.e. finding appropriate threshold values could take exceptionally long computational time [13]. Swarm robotics is an approach to collective robotics that takes inspiration from the self-organized behaviors of social animals. Through simple rules and local interactions, swarm robotics aims at designing robust, scalable, and flexible collective behaviors for the coordination of large numbers of robots [14].

2.3 For RFID

Broekmeulen, and Donselaar [15] suggest a replenishment policy for perishable products which takes into account the age of inventories and which requires only very simple calculations. Zhou [16] takes a different perspective by modeling item-level information visibility in general. Delgado, Ros, and Vila [17] present a system that is able to process the information provided by a Tagged World to identify user's behavior and to produce alarms in dangerous situations. Abad, Palacio, Nuin, Zárate, Juarros, Gómez, and Marco [18] present important advantages regarding conventional traceability tools and currently used temperature data loggers such as more memory, reusability, no human participation, no tag visibility needed for reading, possibility of reading many tags at the same time and more resistance to humidity and environmental conditions. Lee, and Chan [19] propose a genetic algorithm to determine such locations in order to maximize the coverage of customers. Also, the use of RFID is suggested to count the quantities of collected items in collection points and send the signal to the central return center.

3 Research methods

This research used the swarm intelligence, cloud computing, and active Ultra-High Frequency RFID for learning environment, and built friendly human computer interface software for users to use as pad phone. The Extensible Markup Language (XML) and C sharp language were used in this research. The framework of RFID system was shown as Figure 1, and its searching process was shown as Figure 2. In Figure 1, there were eight RFID Readers and eight IP-Cameras for the safety learning system. In Figure 2, the agents generator would make agents which could search the optimum path of RFID Reader.

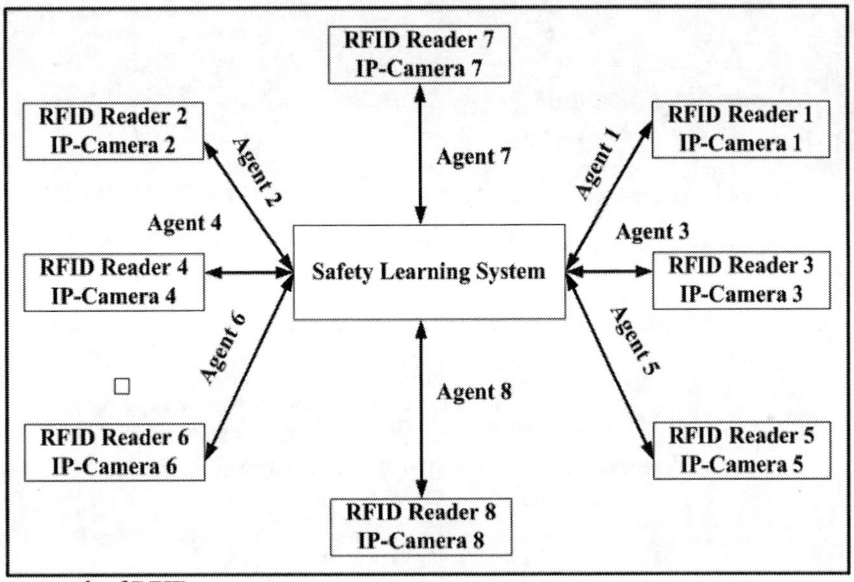

Fig. 1. Framework of RFID system.

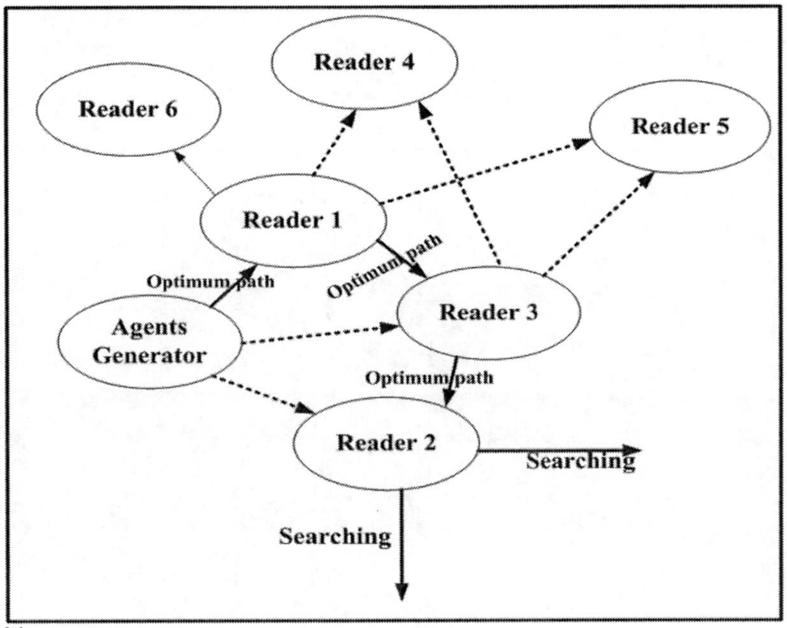

Fig. 2. Searching process.

4 Research results

The safety system has been successfully developed, as shown in Figures 3-6. Figure 3 is the login frame. Figure 4 is welcome frame. There are eight areas in this safety system; (1) The left side of the gate, (2) The right side of the gate, (3) The eastern side of the house, (4) House Back, (5) Classroom 1, (6) Classroom 2, (7) Off ice, (8) Leisure area, as shown in Figure 5. Figure 6 is the searching results of classroom 1.

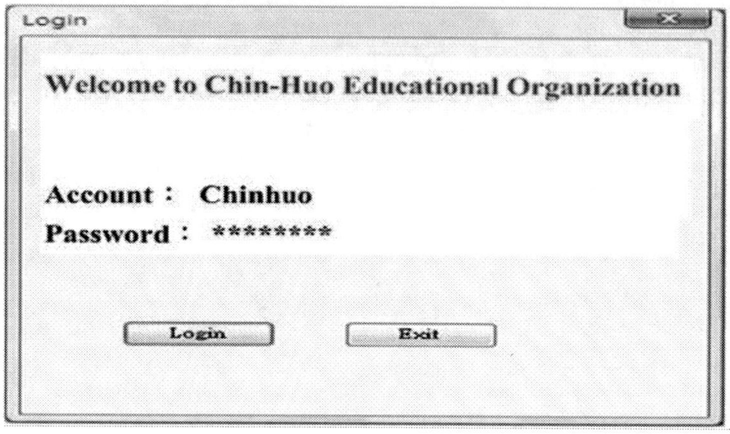

Fig. 3. Login frame.

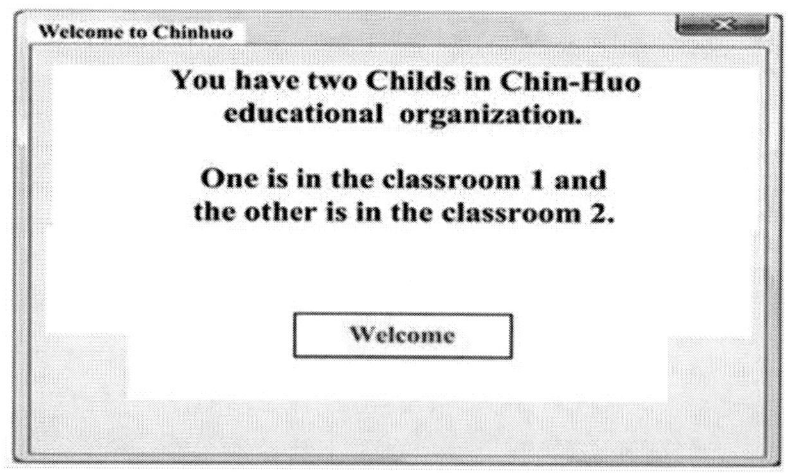

Fig. 4. Welcome frame.

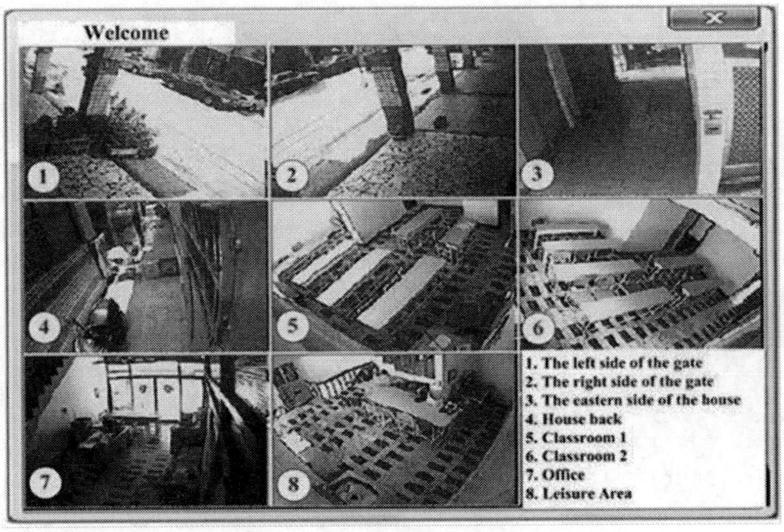

Fig. 5. Eight areas.

Fig. 6. Searching results (Classroom 1).

5 Conclusions

This research aims to study learning environment, and let the learning environment become smart. Swarm intelligence, cloud computing, and active Ultra-High Frequency RFID were used on it. We built friendly human-computer-interface software for users to use as pad phone. The Extensible Markup Language (XML) and C sharp language were used in this research. The paterfamilias' time in operating the instrument was saved. Even someone who does not have the professional knowledge. This research also develops friendly human computer interface software for users to use as personal computers or notebooks. The size of the software is 30 kilobits; therefore, the software is not a liability for the users' tools. The result has been successfully implemented on Chin-Huo educational organization and it would be helpful for the paterfamilias to hold all situations about their children at one educational organization. It will be of great help in the grip of whole after-school remedial education, teaching and learning.

References

1. S. Bleischa, M. Duckhama, A. Galtonb, P. Laubec, J. Lyond. IJGIS, Vol. 28, 363-382 (2014)
2. J. kopf, F. Langguth, D. Scharstein, R. Szeliski, M. Goesele. ACM, TOG Homepage archive, Vol. 32, No. 199(2013)
3. P.M. Leonardi, D. E. Bailey, MIS Quarterly, 32 (2), 411-436 (2011)
4. S.Y. Sun, L.S. Chen, IJOI, Vol. 6, 63-67 (2014)
5. S.Y. Sun, S.Y.,L. S. Chen, IJOI, Vol. 6, 167-178 (2013)
6. L.S. Chen. LCATS , Vol.34, 1-10 (2010)
7. L.S. Chen. LCATS, Vol.32, 127-141 (2009)
8. G. Hayes, K. El-Khatib, C. McGrego, CLIEE, Vol. 260,1149-1158 (2014)
9. I. Gonzenbach, C. Russ, J.V. Brocke, ECMISRP,145-161 (2014)
10. J.W. Lian, D.C. Yen, Y.T. Wang, IJIM, Vol. 34, 28-36 (2014)
11. R. Ranjan, L. Zhao, JS, Vol. 65, 154-184 (2013)
12. X. Xia, X. Wang, X. Zhou, T. Zhu, LNIEE, Vol. 274, 405-412 (2014)
13. A. Marciniak, M. Kowal, P. Filipczuk, J. Korbicz, ISTMDAISC, Vol. 230, 301-311 (2014)
14. M. Brambilla, E. Ferrante, M. Birattari, M. Dorigo, SI , Vol. 7, 1-41(2013)
15. R. A.C.M. Broekmeulen, K. H. V. Donselaar, COR, 36 (11), 3013-3018(2015).
16. W. Zhou, EJOR, 198 (1), 252-258 (2009)
17. M. Delgado, M. Ros, M. A. Vila, ESWA, 36 (6), 9899-9906 (2015)
18. E. Abad, F. Palacio, M. Nuin, A. G. Zárate, A. Juarros, J. M. Gómez, S. Marco, JFE, 93 (4), 394-399 (2014)
19. C.K. M, T. M. Lee, T.M. Chan, ESWA, 36 (5), 9299-9307 (2015)

Design and development of an automatic gas poisoning prevention and ventilation system

Ho-Chih Cheng[1], *Min-Chie* Chiu[1,*], *Che-Min* Chiu[2] , and *Cheng-Ying* Yang[1]

[1]Department of Mechanical and Automation Engineering, Chung Chou University of Science and Technology, No. 6, Lane 2, Sec. 3, Shanjiao Rd., Yuanlin City, Changhua County 510, Taiwan, R.O.C.
[2]Institute of Biomedical Engineering, National Tsing Hua University, No. 101, Section 2, Kuang-Fu Rd., Hsinchu City 300, Taiwan, R.O.C.

Abstract. As both carbon monoxide (CO) and carbon dioxide (CO_2) exit during a fire, they are obvious indicators for the need to be alert to the possibility of fire. Another problem concerning poisoning by carbon monoxide (CO) is asphyxiation that occurs in houses during winter. The development of an automatic gas poisoning prevention system in conjunction with a ventilation function using carbon monoxide/carbon dioxide sensors would prove beneficial, necessary. As presented here, this system includes two gas sensors, an alarm, a ventilation device, a motor, and a rain-protection louver. The louver is manipulated by a motor. Two thresholds of gas concentration are preset inside the microcontroller via a PC. The louver is opened by the motor as the first threshold of gas is reached. Additionally, an alarm system is triggered and the ventilation fan starts up if the second threshold of gas concentration is reached. Consequently, image-monitoring via the PC is established using an IPCAM.

1 Introduction

Fires are disasters that quickly result in tremendous loss of human life. In order to decrease the danger of fire, a prevention system is essential. The first automatic smoke detector based on an ionization current was invented in 1940 [1]. Gupta [2], in 1985, developed an automatic fire detection system. He adopted an automatic fire detection (AFD) system to differentiate the reliability between the presence and absence of fire. He found that not all fires emit smoke. However, combustion gases emitted from fires still existed. Therefore, the combustion gases were regarded as an early warning indicator for fires [3]. The gases closely related to the combustion process include carbon monoxide (CO), carbon dioxide (CO_2) and nitrogen oxides (NOx). In 2000, Hagen and Milke [4] indicated that a system of gas sensors was capable of detecting fires faster than a system of two ordinary smoke detectors. Based on an infrared photoacoustic technique, Nebiker and Pleisch [5, 6], in 2001 and 2001, detected the gas concentration of carbon dioxide using a detector. As disturbances like steam, fog, dust pollution, and condensed water may cause false alarms

*Corresponding author: minchie.chiu@msa.hinet.net

[7], the sensor system used to identify an extinguished fire using a video-based system as well as vision observing needed to be immune to ambiguous signals. According to an investigation by Krüll et al. [8] and Tobera et al. [9], because of the high occurrence of hydrogen during an open fire, a H2-Sensor [0–10ppm] was implemented as a fire indicator. The main features of this semiconductor gas sensor (GTE GSME) were observed to have a very fast response time and to be highly sensitive [10]. Derbel [11], in 2004, improved the fire detectors by using gas sensors and neural networks. Cestari et al. [12], in 2005, analyzed 32 fire tests and 11 nuisance tests using the output from ionization, photoelectric and carbon monoxide (CO) detectors, and a thermocouple. Their results revealed that fire sensitivity and nuisance immunity were directly related to three sets of parameters including (a) a increase in temperature rise and production of CO, (b) CO and ionization detection, and (C) a temperature increase and CO and ionization detection.

In order to increase the reliability of aircraft smoke detectors, and to reduce the response time, Chen et al., in 2007, proposed a fire detection system based on the simultaneous measurements of carbon monoxide, carbon dioxide, and smoke [13]. Mulrooney et al. [14], in 2007, developed an optical fiber based system suitable for monitoring the presence of carbon dioxide (greenhouse gas) emitted from the exhaust system of a motor vehicle. Later, they designed a mid-infrared optical fiber sensor used to detect carbon monoxide exhaust emissions [15]. Their analysis revealed that the optical fibre was capable of monitoring the presence of carbon dioxide and carbon monoxide in the exhaust system of gas engines. According to an investigation from Gutmacher et al. [16] in 2012, CO reached the gas sensors over 4 min before any smoke could be detected by the optical smoke detectors. Based on IEEE 802.15.4 and/or Ethernet, Becher et al. [17], in 2010, detected evaporating hazardous material released from moving sources using a gas sensor network which is a gas-sensing platform communicating via a TCP/IP protocol and a wireless sensor network.

As mentioned above, carbon monoxide (CO) and carbon dioxide (CO_2) which are closely related to combustion processes are essential indicators in fire prevention and need to be monitored online. Moreover, from the prospective of health, carbon monoxide (CO) has also been recognized as a risk factor for adverse cardiovascular outcomes. Min et al. [18], in 2009, proved that the effect of two specific MetS components (blood glucose and triglycerides) was related to CO exposure. Moreover, because gas (carbon monoxide (CO) and carbon dioxide (CO_2)) poisoning often occurs in houses in Taiwan during winter, the development of a gas poisoning prevention system is necessary. In order to avoid fires and gas poisonings, an automatic gas poisoning prevention and ventilation system using carbon monoxide/carbon dioxide sensors is proposed. Here, two gas sensors, an alarm, a ventilation device, a motor, and a rain-protection louver are adopted. The louver is opened by the motor when the first threshold of the gas is reached. Moreover, a remotely monitor alarm system is triggered if the second threshold of gas concentration is reached. Simultaneously, forcing ventilation is started.

2 Automatic gas poisoning prevention and ventilation system

The automatic gas poisoning prevention and ventilation system is depicted in Fig. 1. As indicated in Fig. 1, the hardware of the gas detection and forced ventilation system is composed of two gas sensors (carbon monoxide (CO) and carbon dioxide (CO_2)), a microcontroller, an alarm, a ventilation device, a motor, and a rain-protection louver. Each part is controlled by a microcontroller – PIC18F4520 shown in Fig. 2. A motor linked to the microcontroller can be rotated clockwise/ counter-clockwise to open/close the louver. Two gas sensors (carbon monoxide (CO) and carbon dioxide (CO_2) linked to the microcontroller are adopted to detect the gases' concentration. The concentration values of carbon monoxide (CO) and carbon dioxide (CO_2) will be wirelessly sent from the microcontroller to the server PC. The algorithm of gas detection and actuation is illustrated in Fig. 3. As indicated in Fig. 3, the gas ppm is fed back online to the server PC when the server PC sends an inquiry of gas concentrations (carbon monoxide (CO) and carbon dioxide (CO_2)) to the microcontroller. The louver is opened via the motor when either carbon monoxide's ppm or carbon dioxide's ppm reaches the first warning threshold; if not, the louver is closed and the inquiry of gas concentrations is continued. Thereafter, a second check of the gas concentrations of the second threshold is performed. The on-site alarm is triggered and the ventilation fan is started when one of the gas ppms exceeds the value of the second threshold. Also, an alarm notification at the interface of the server PC is initiated; if not, both the on-site alarm and ventilation fan are turned off. The alarm notification of the server PC vanishes.

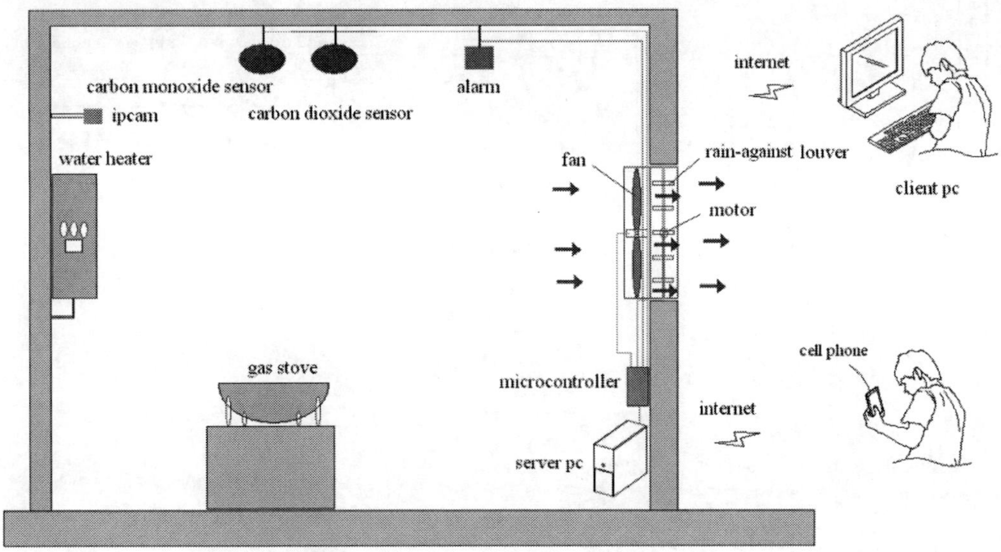

Fig. 1. A remote gas detection/force ventilation/monitoring system.

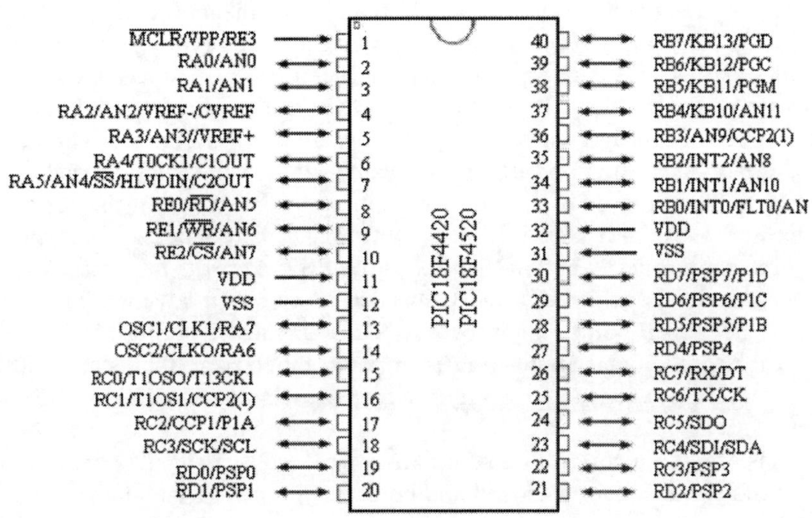

Fig. 2. The diagram of the microcontroller – PIC18F4520.

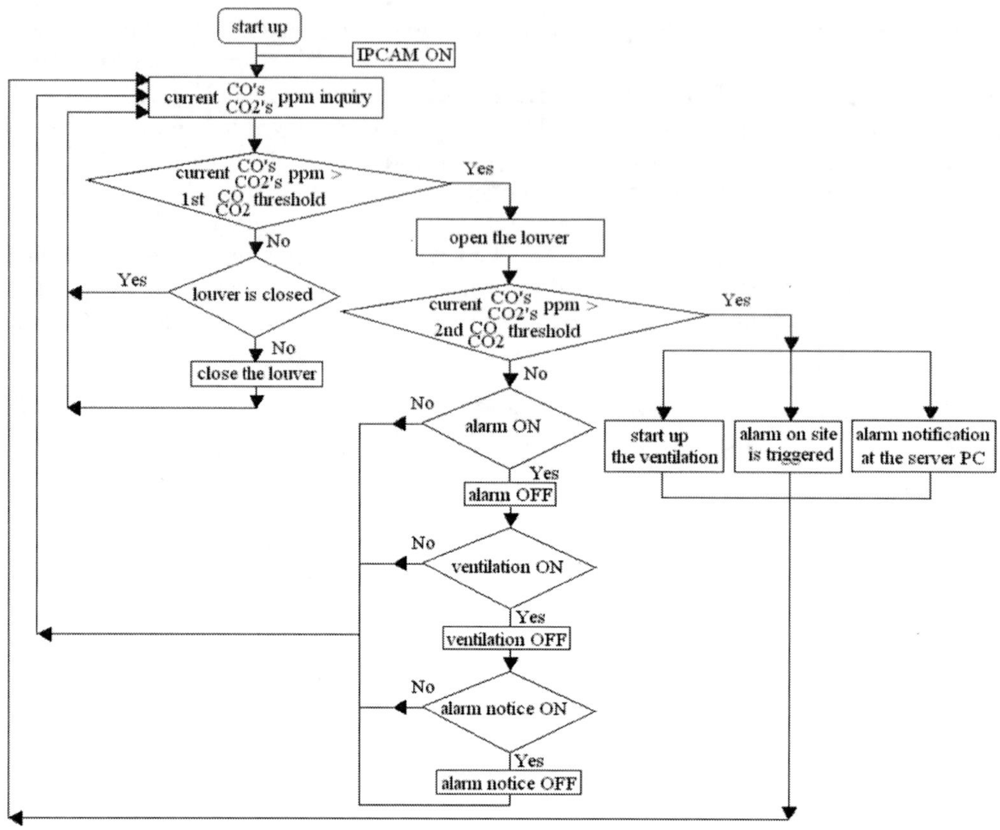

Fig. 3. The algorithm for gas detection and actuation.

3 Human/machine interface

In order to monitor/control the poisoning environment online, a server PC is wirelessly connected to the microcontroller (single chip) using both an RS-232 and a blue tooth protocol. Two gas sensors used to detect gas concentrations (carbon monoxide (CO) and carbon dioxide (CO_2)) are connected to the microcontroller. The detected values are fed back to the server PC via the blue tooth and shown on the interface dialogue. The server PC's interface which is programmed by a VB6.0 is depicted in Fig. 4. In addition, in order to visualize the on-site situation, an IPCAM connected to the wireless AP router is also installed on site. The process of setting and running an interface on the server PC is also illustrated in Fig. 5. As indicated in Fig. 5(a), to make the connection between the server PC and the microcontroller, a button that opens the RS232 port is initiated. As illustrated in Fig. 5(b), the thresholds of the CO and CO_2 need to be set in advance before starting the automatic gas detecting/monitoring process. Also, as shown in Fig. 5(c), the automatic function of gas detection and monitoring is initiated by clicking the "start" button. As can be seen in the figure, the alarm's green light which signals a safe condition for the detecting zone is preset.

Three kinds of system controlling situations (@1S, @2S, and @3S) are exhibited online. For the @1S status, the louver is closed and both the ventilation fan and the alarm are OFF. This indicates that the environment is safe. Second, for the @2S status, the louver is opened and both the ventilation fan and the alarm are OFF. This indicates that the gas concentration has reached the first threshold (warning) and the natural convection via the louver opening is required. Consequently, the @3S status indicates that the louver is opened and both the ventilation fan and the alarm are ON. This situation appears when the gas concentration reaches the second threshold (danger) and forced convection is initiated using the ventilation fan. The above three conditions (@1S, @2S, and @3S) can also be individually manipulated by clicking the buttons of @1S, @2S, and @3S on the server PC/ client PC interface.

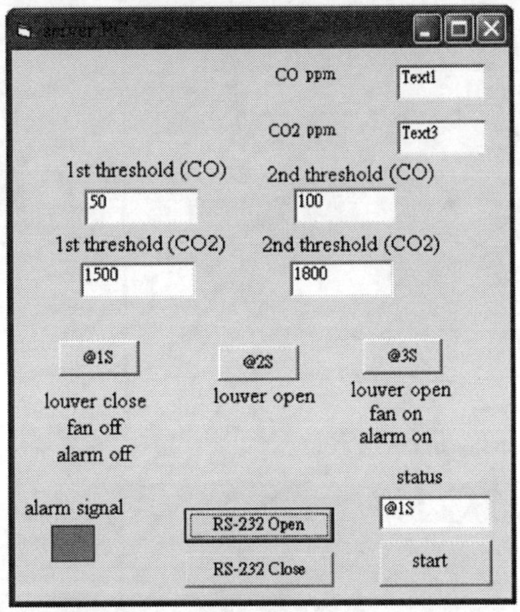

Fig. 4. Interface of the server PC programmed by VB.

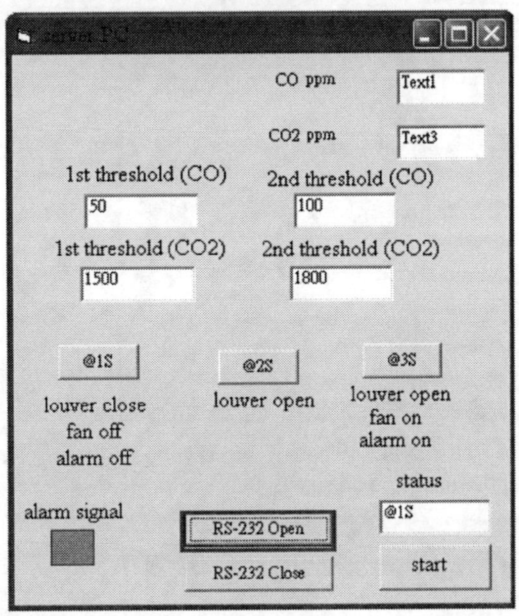

Fig. 5(a). Starting the gas monitoring/controling system (server PC) [connected to the microcontroller].

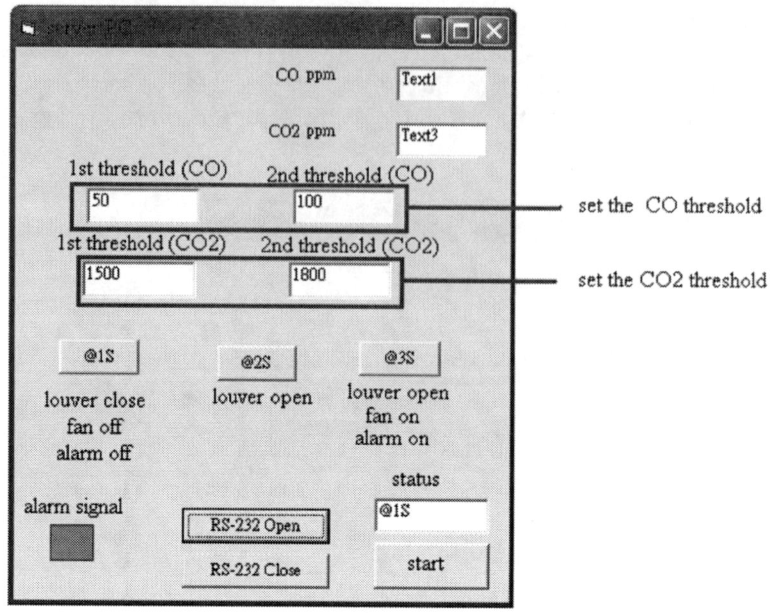

Fig. 5(b). Setting the thresholds of the CO and CO_2.

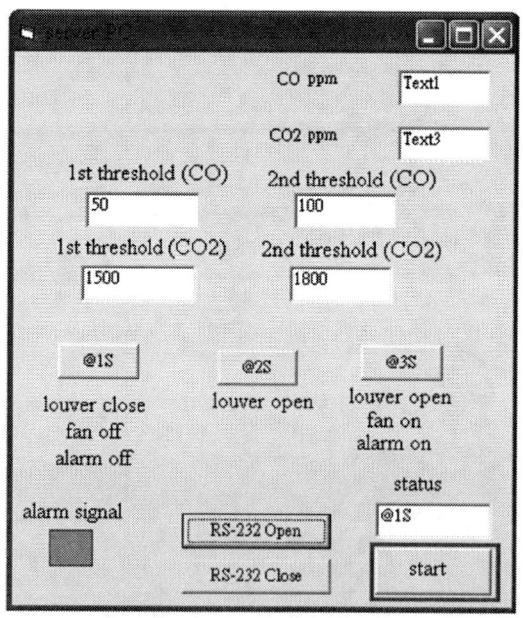

Fig. 5(c). Initiating the data transmission between the server pc and the microcontroller.
Fig. 5. Setting and running the interface on server PC.

4 System communication design [19, 20]

For a remote gas monitoring/controlling system, the communication between the server port and the client port is necessary. In this paper, the TCP/IP communication between the server PC and the client PC is adopted using Winsock with the VB program. The corresponding communication diagram interface between the server PC and the client PC is illustrated in Fig. 6.

(1) Listening at the server port: The condition "Listen" for the server port (server PC) is

performed while the server PC is waiting for the connection from the client port (client PC) at the ready condition of "Listen." In addition, one WINSOCK object in the server port will respond to one client only. One WINSOCK is responsible for the "Listen" function when the "Local Port" is preset as the port of "Listen."

(2) Connection at the client port (client PC): A connection request is sent via the WINSOCK object. The WINSOCK will assign a "Port Number" (at "Listen" condition) for the server port (server PC) at the WINSOCK 's "Remote Port."

(3) Reception at the server port (server PC): The server port (server PC) receives a ConnectionRequest command from the client port (client PC) when the client port submits a request of "Connect." The client port's connection request will be accepted by the server port by using the "Accept" via a new WINSOCK.

(4) Sending data from the client port (client PC): The data sent from the client port to the server port will be initiated using the "SendData" command when the internet connection between the server port and the client port is ready.

(5) Receiving data at the server port (server PC): DataArrival from the client port (client PC) will be received on the server port when the client port delivers data to the server port. The server port can read the data from the client port by using "GetData."

(6) Receiving data at the client port (client PC): The client port will acknowledge DataArrival when the server port submits data to the client port. The client port will then receive the data via "GetData."

(7) Close: The connection will be ended by clicking "Close" when the communication work is completed.

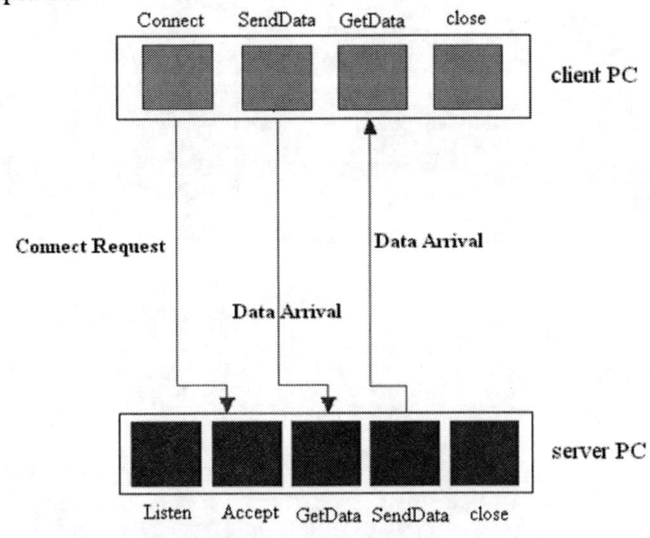

Fig. 6. Communication diagram between the server PC and the client PC.

5 Results and discussions

The automatic gas poisoning prevention and ventilation system using carbon monoxide/carbon dioxide sensors has been introduced. The gas concentration can be automatically detected via server PC/client PC control ports. Two thresholds (1st threshold in the warning situation and the 2nd threshold in the danger situation) are specified before automatic gas detection and ventilation controlling is initiated. Immediate gas concentrations are detected and shown on the interface of the server PC and the client PC. A demonstration of the automatic gas poisoning prevention and ventilation system is depicted in Fig. 7(a)-(n). As indicated in Fig. 7(a), the CO and CO_2 are generated in the kitchen. In Fig. 7(b), the gas concentration increases gradually. As shown in Fig. 7(c), the gas concentration reaches the first threshold (warning level). As illustrated in Fig. 7(d), a command from the server PC/client PC is sent to the microcontroller to actuate the motor to open the louver. As can be seen in Fig. 7(e), gas is diluted by natural convection via the louver opening. However, as shown in Fig. 7 (f), because the increase in gas is larger than gas dilution rate, the gases accumulate. As indicated in Fig. 7(g), the gas concentrations reach the second threshold (danger status). The on-site alarm is then triggered. Concurrently, the alarm flashing red is reflected onto the interfaces of the server PC and the client PC. Subsequently, as illustrated in Fig. 7(h), the forced ventilation is started. The gas concentration shown in Fig. 7(i) decreases when the forced convection is ongoing. As indicated in Figs. 7(j)-(l), the alarm is terminated and the ventilation fan turned off when the gas concentration dilutes to the 1st threshold. The warning signal on the interface turns from red to green. Furthermore, the gas concentration continues to dilute to a level below the 1st threshold. Therefore, the louver is closed. The prototype automatic gas poisoning prevention and ventilation system has been introduced and shown in Fig. 8.

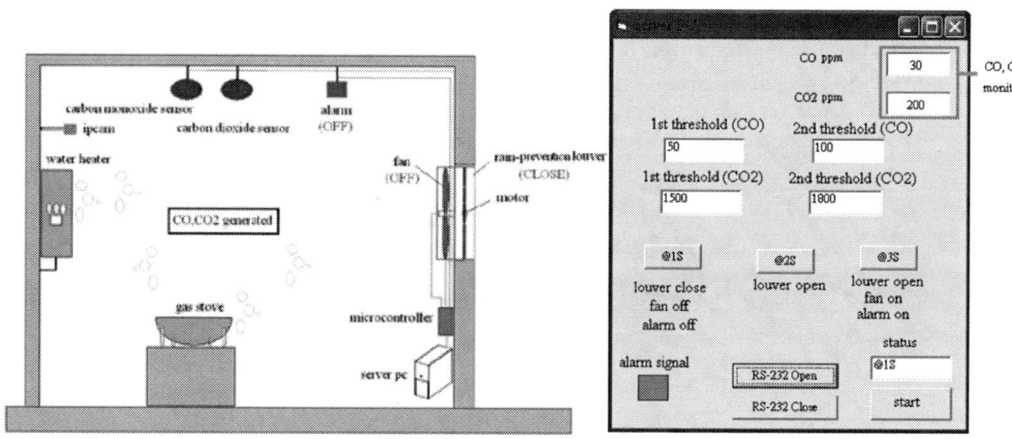

Fig. 7(a). Gas generated.

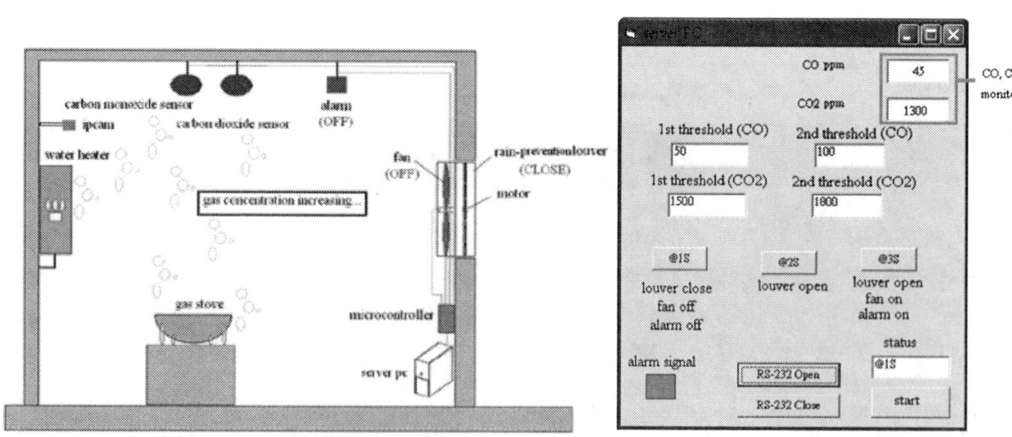

Fig. 7(b). Gas concentration increases.

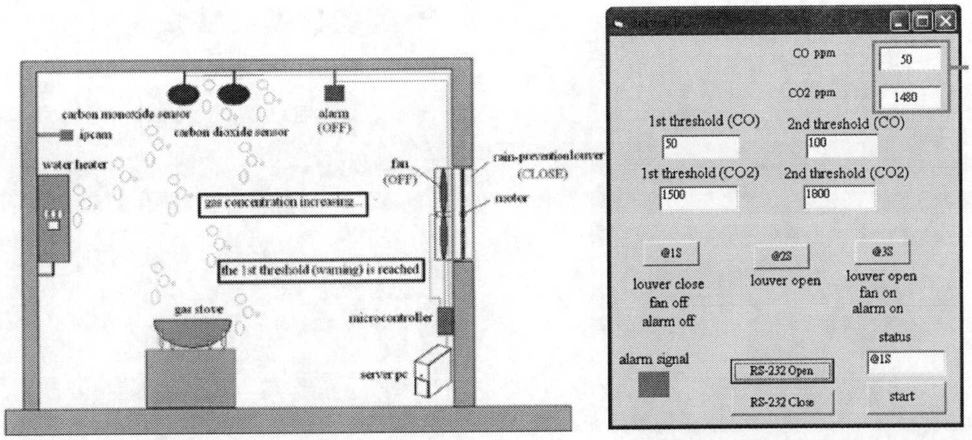

Fig. 7(c). First gas threshold is reached.

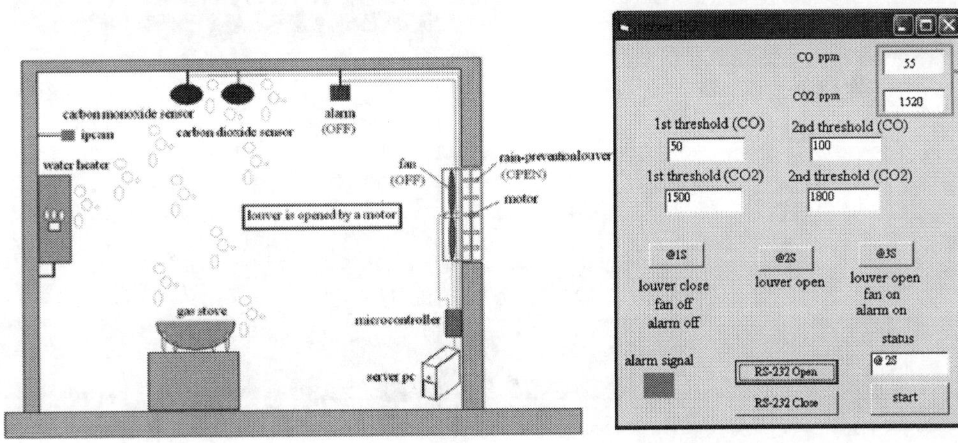

Fig. 7(d). Louver is opened.

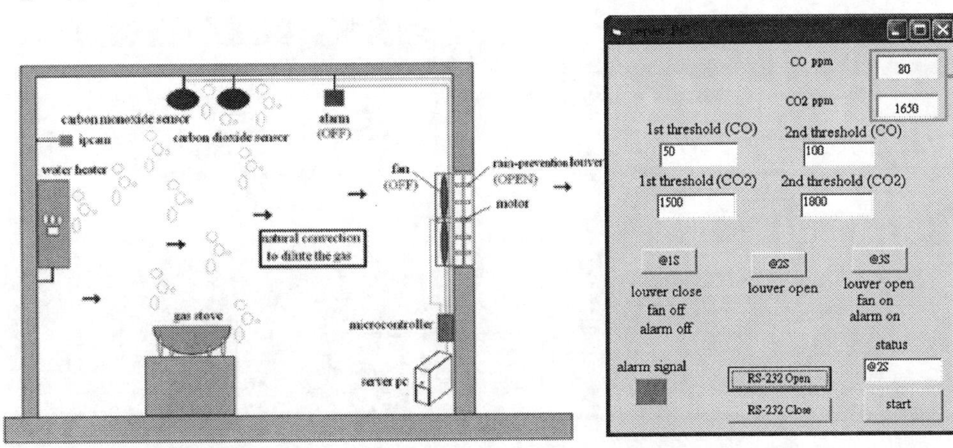

Fig. 7(e). Natural convection to dilute the gas.

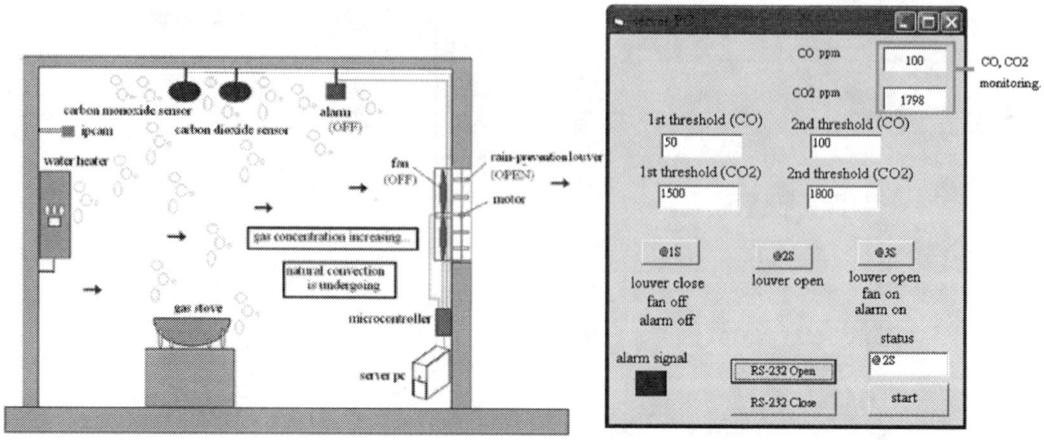

Fig. 7(f). Gas concentration increases (ongoing natural convection).

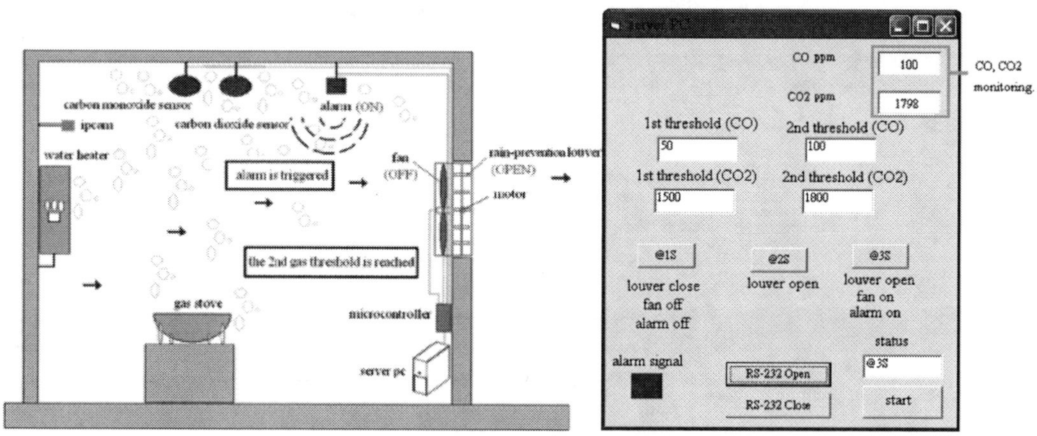

Fig. 7(g). Second gas threshold is reached (alarm is triggered).

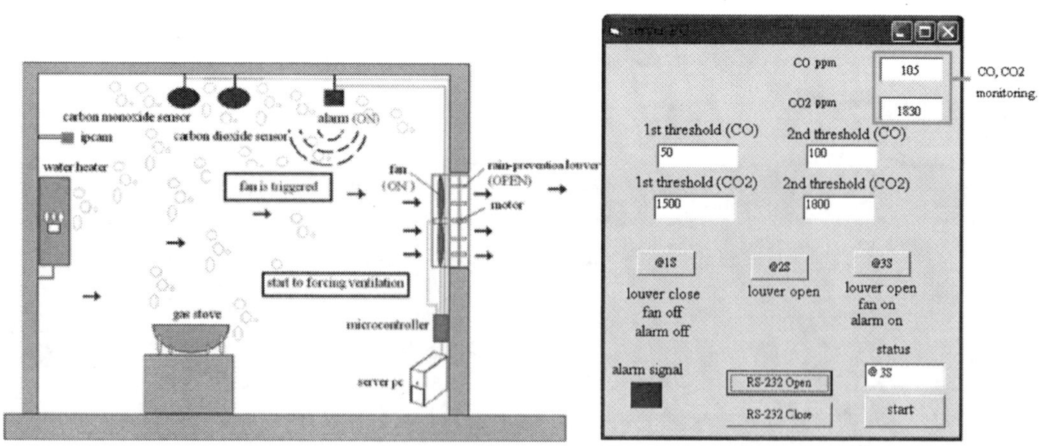

Fig. 7(h). Fan is triggered (starts forced ventilation).

 https://doi.org/10.1051/matecconf/201818500010

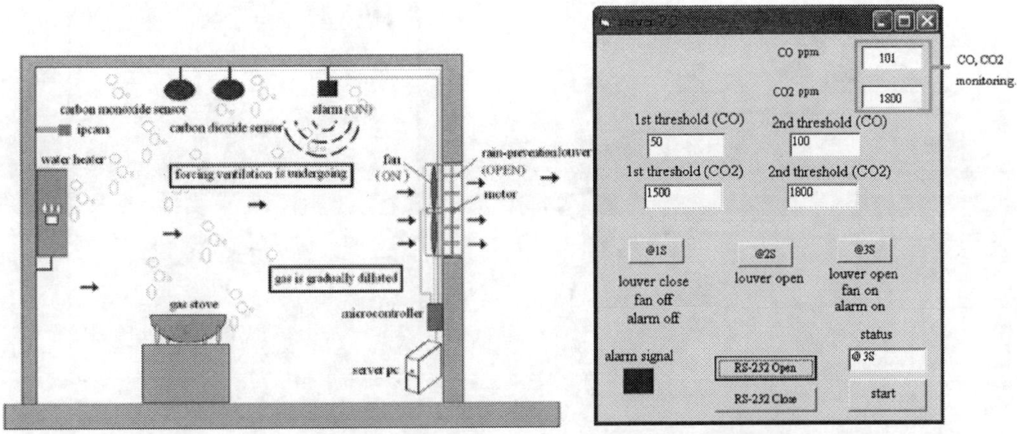

Fig. 7(i). Gas concentration is gradually dilluted (ongoing forced ventilation).

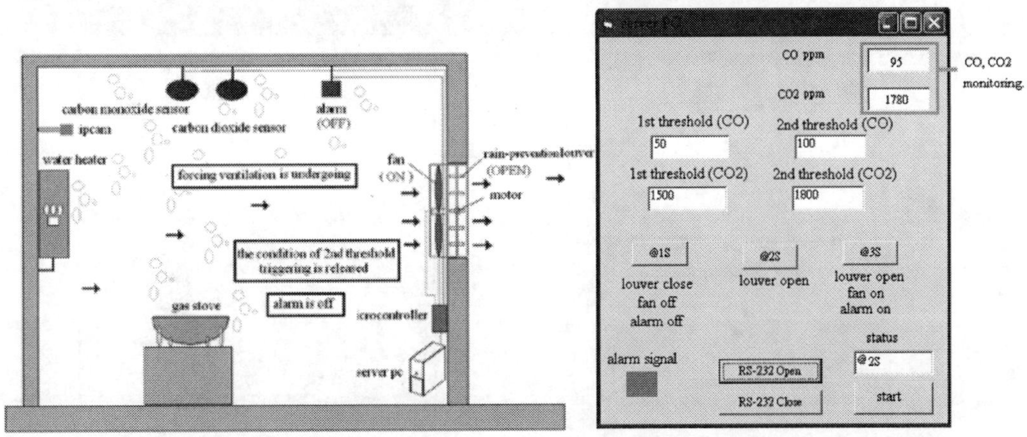

Fig. 7(j). Second threshold's warning condition is set off (alarm goes off).

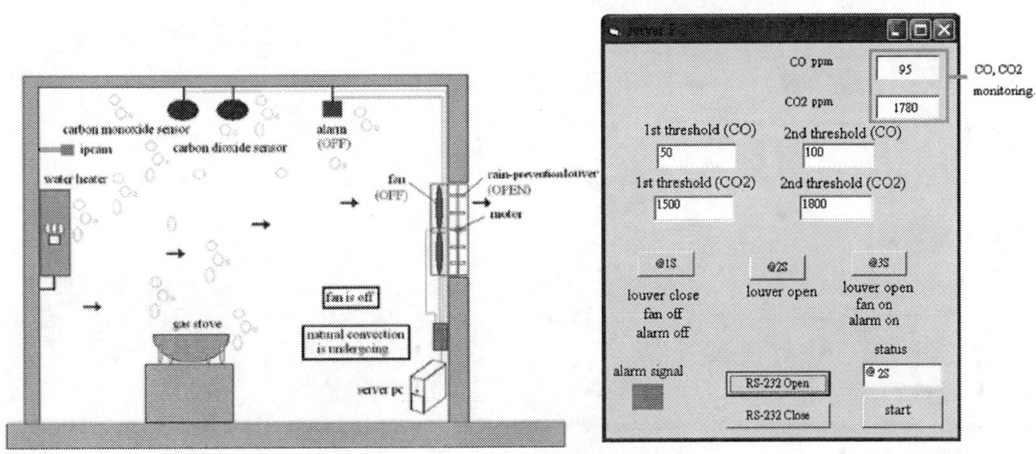

Fig. 7(k). Gas is continuously diluted (ongoing natural convection).

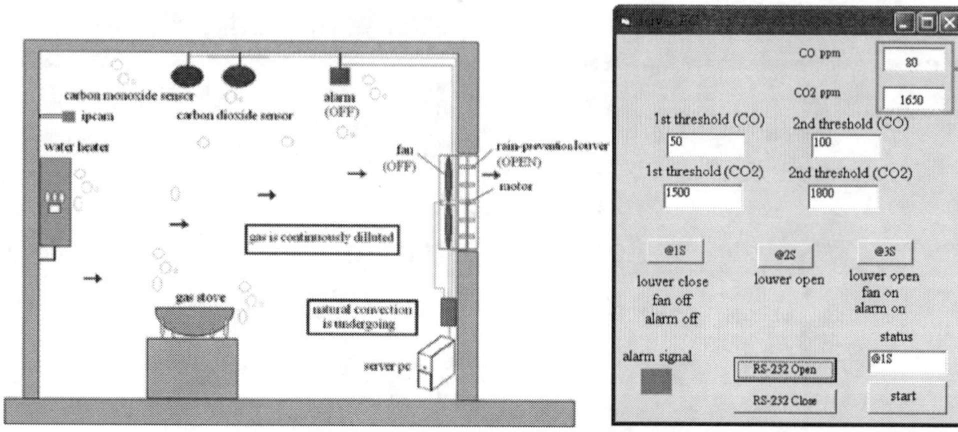

Fig. 7(l). Second threshold's warning condition is set off (fan is turned off).

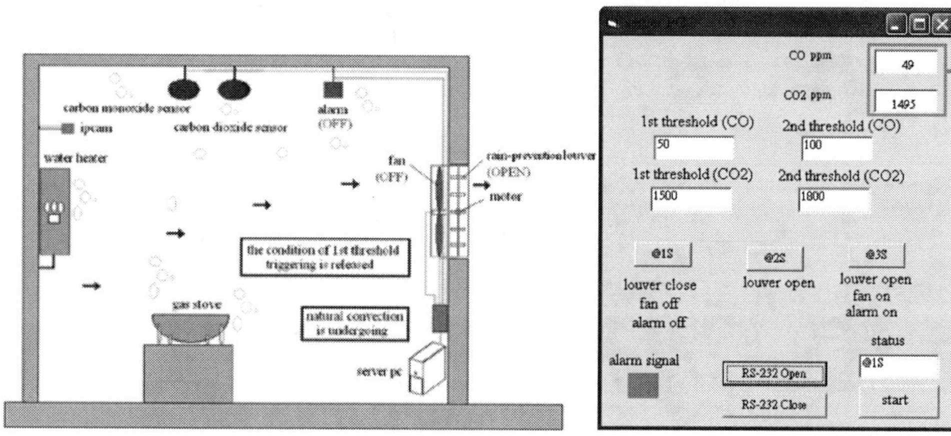

Fig. 7(m). First threshold's warning condition is set off.

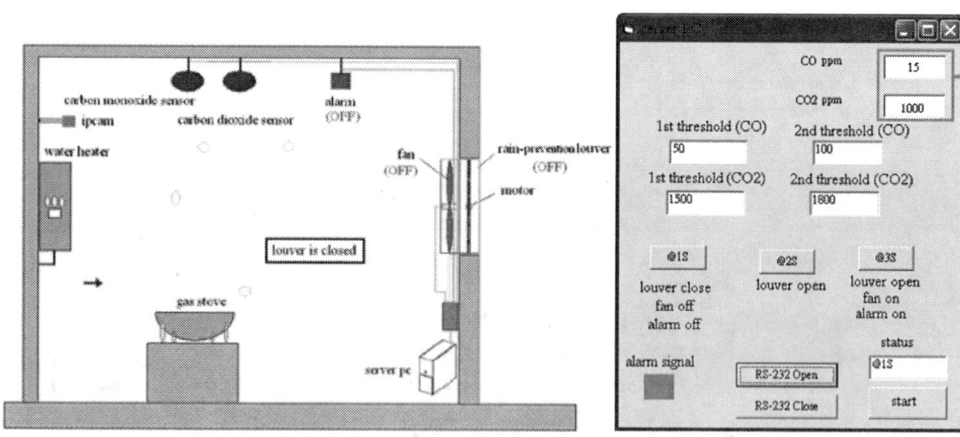

Fig. 7(n). Louver is closed.
Fig. 7. Process of automatic remote gas detection/forced ventilation/monitoring system.

Fig. 8(a). Gas sensors, microcontroller, motor, and ventilation fan.

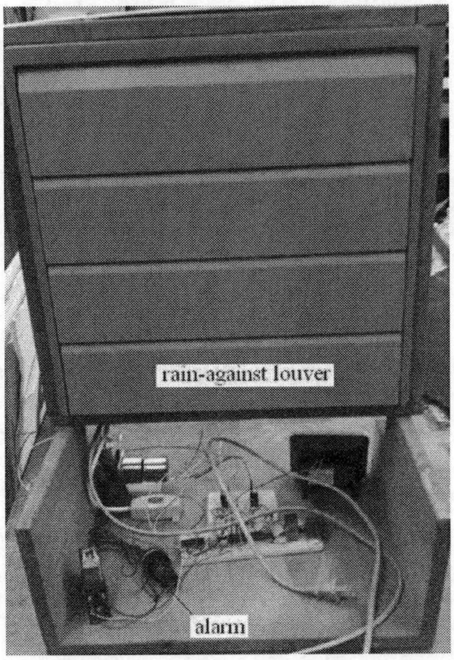

Fig. 8(b). louver

Fig. 8. Prototype of an automatic gas poisoning prevention and ventilation system.

6 Conclusions

It has been shown that a wireless automatic gas poisoning prevention and ventilation system can be remotely monitored and manipulated during a gas detecting process. Using the RS-232 communication standard and blue tooth protocol, the server PC is wirelessly connected to the microcontroller (single chip). The microcontroller connects to two gas sensors, an alarm device, a motor (used to open/close the louver), and a ventilation fan. To remotely implement the gas poisoning prevention and ventilation system, human/machine interfaces in both the server PC port and the client PC port programmed by VB6.0 has been established. Three system operating modes (@1S: louver CLOSE and ventilation fan OFF and alarm OFF; @2S: louver OPEN and ventilation fan OFF and alarm OFF; @3S: louver OPEN and ventilation fan ON and site alarm ON and interface alarm ON) are designed into the interface. Current gas concentrations can be automatically monitored and controlled online by clicking "start" on the interface. Two thresholds of gas concentrations marked in both the warning level and danger level need to be specified before the system is initiated. In order to remotely monitor the on-site situation, a wireless USB AP router wirelessly connected to the IPCAM is adopted. The image can be sent back to the server PC via the Wi-Fi protocol.

Consequently, a prototype of the automatic gas poisoning prevention and ventilation system has been demonstrated.

References

1. H.J. Tobler, Ionisationsfeuermelder: Technische Und Wirtschaftliche Bedeutung, Bulletin SEV/VSE, **23(6)**, 13-15 (2006)
2. Y.P. Gupta, Automatic Fire Detection Systems: Aspects of Reliability, Capability and Selection Criteria, Fire Safety J., **8(2)**, 105-117 (1985)
3. U. Hoefer, D. Gutmacher, Fire Gas Detection, a Procedia Engineering, **47**, 1446-1459 (2012)
4. B.C. Hagen, J.A. Milke, The Use of Gaseous Fire Signatures as a Mean to Detect Fires, Fire Safety J., **34(1)**, 55-67 (2000)
5. P.W. Nebiker, R.E. Pleisch, Photoacoustic Gas Detection for Fire Warning, Fire Safety J., **36(2)**, 173-180 (2001)
6. P.W. Nebiker, R.E. Pleisch, Photoacoustic Gas Detection for Fire Warning, Fire Safety J., **37**, 429-436 (2002)
7. W. Krüll, R. Tobera, I. Willms, H. Essen, N. V. Wahl, Early Forest Fire Detection and Verification Using Optical Smoke, Gas and Microwave Sensors, Procedia Engineering, **45**, 584 -594 (2012)
8. W. Krüll, I. Willms, R. Tobera, B. Wiggerich, Early Forest Fire Detection and Suppression – An Integrated Approach, *14th International Conference on Automatic Fire Detection*, AUBE '09, Duisburg, Germany, (2009)
9. R. Tobera, W. Krüll, I. Willms, Optical Smoke and Gas Sensors as An Additional Method for Early Wildfire Verification, *14th International Conference on Automatic Fire Detection*, AUBE '09, Duisburg, Germany (2009)
10. GTE Industrieelektronik GmbH, http://www.adicos.de
11. F. Derbel, Performance Improvement of Fire Detectors by Means of Gas Sensors and Neural Networks, Fire Safety J., **39(5)**, 383-398 (2004)
12. L. A. Cestari, C. Worrell, J.A. Milke, Advanced Fire Detection Algorithms Using Data from the Home Smoke Detector Project, Fire Safety J., **40(1)**, 1-28 (2005)
13. S.J. Chen, D.C. Hovde, K.A. Peterson, A.W. Marshall, Fire Detection Using Smoke and Gas Sensors, Fire Safety J., **42(8)**, 507-515 (2007)
14. J. Mulrooney, J. Clifford, C. Fitzpatrick, E. Lewis, Detection of Carbon Dioxide Emissions from a Diesel Engine Using a Mid-infrared Optical Fibre Based Sensor, Sensors and Actuators A: Physical, **136(1)**, 104-110 (2007)

15. J. Mulrooney, J. Clifford, C. Fitzpatrick, P. Chambers, E. Lewis, A Mid-infrared Optical Fibre Sensor for the Detection of Carbon Monoxide Exhaust Emissions, Sensors and Actuators A: Physical, **144(1)**, 13-1 (2008)
16. D. Gutmacher, U. Hoefer, J. Wöllenstein, Gas Sensor Technologies for Fire Detection, Sensors and Actuators B: Chemical, **175**, 40-45 (2012)
17. C. Becher, P. Kaul, J. Mitrovics, J. Warmer, The Detection of Evaporating Hazardous Material Released from Moving Sources Using a Gas Sensor Network, Sensors and Actuators B: Chemical, **146(2)**, 513-520 (2010)
18. J. Y. Min, D. Y. Paek, S. I. Cho, K. B. Min, Exposure to Environmental Carbon Monoxide May Have a Greater Negative Effect on Cardiac Autonomic Function in People with Metabolic Syndrome, Science of the Total Environment, **407(17)**, 4807-4811 (2009)
19. H. C. Cheng, M. C. Chiu, The Design of An Automatic Wind Electricity Monitoring System for a Battery Charge Process, J. of Information & Optimization Sciences, **34(6)**, 373-388 (2013)
20. H. C. Cheng, M. C. Chiu, Automatic Solar Electrical Monitor for a Battery Charge Process Using a Network Remote Control System, Applied Mechanics and Materials, **336-338**, 1211-1216 (2013).

An assessment of transmission efficiency on a hard disk's vibration

Ying-Chun Chang [1], *Ho-Chih* Cheng [2], *Min-Chie* Chiu [2,*], and *Min-Shiang* Chang [1]

[1]Department of Mechanical Engineering, Tatung University, No.40, Sec. 3, Zhonshan N. Rd., Taipei City 104, Taiwan, R.O.C.

[2]Department of Mechanical and Automation Engineering, Chung Chou University of Science and Technology, No. 6, Lane 2, Sec. 3, Shanjiao Rd., Yuanlin City, Changhua County 510, Taiwan, R.O.C.

Abstract. At present, there is a trend to design a hard disk with a larger capacity, a faster revolution speed, and a precise mechanism. Therefore, the influence of data transmission with respect to vibration is obvious. To reduce interference of the vibrational impact on the data transmission efficiency, vibrational abatement on the hard disk using the damping material becomes crucial. Because of the complicated relationship for the damping coefficient and spring constant to the hardness (D1, D2, and D3) of the damping material installed under the hard disk, it is difficult to theoretically assess an optimal hardness combination of the damping material. Therefore, an alternative way by using an experimental study in conjunction with Taguchi method, an Artificial Neural Network (ANN), and a GA Method is proposed. In this paper, a hard disk is placed on a vibration tester that is an analogue to a dynamic vibrational circumstance induced by a vibrational base. The data transfer rate of the hard disk will be detected by using IOmeter software under various base-excitation accelerations, tilted angles of the hard disk, and targeted frequencies. To reduce the vibrational impact on the data transmission efficiency, an assessment of an optimal three-layer damping material installed under the hard disk using the Taguchi method, the Artificial Neural Network (ANN), and the GA Method is proposed. Before the optimization of the damping material is performed, the required experimental sets (the hardness for three layers of damping material) of the data transmission testing with respect to various design parameters will be determined by using the Taguchi method. The ANN, a simplified objective function (OBJ), will be established by inputting the hardness of three layers of damping material and their related data transmission efficiency at three targeted frequencies. Thereafter, the optimal hardness for three layers of the damping material will be obtained using a genetic algorithm (GA). Consequently, the optimal hardness of the three-layer damping material with respect to various tilted angles and target frequencies will be assessed.

*Corresponding author: minchie.chiu@msa.hinet.net

1 Nomenclature

This paper is constructed on the basis of the following notations:
bit: bit length of chromosome
B_0, B_i, B_{ij}, B_{ijk}: the coefficient of the node function in the ANN
chrm: the length of the chromosome
c_i: the damping coefficient of the i-th layer damping material (N-m s^{-1})
c_{sys} : damping coefficient of vibration system (N-m s^{-1})
CPM: the product of the penalty function
D_1, D_2, D_3: design parameters of hardness for damping material (Shore, HS)
$f_{i1}(D_i)$: the implicit function of the damping coefficient for the i-th layer damping material with hardness of D_1
$f_{i2}(D_i)$: the implicit function of the spring constant for the i-th layer damping material with hardness of D_1
FSE: the deviation of mean square
h: the unit's number in a hidden layer
IOPS: transmission rate (Input/Output per second)
$iter_{max}$: maximum iteration during GA optimization
k_i: the spring constant of the i-th layer damping material
k_{sys}: the spring constant of vibration system
m: mass of the hard disk (kg)
M: the number of the design parameters
N: the number of training data
N_p: the total possible searching number ($=2^M$)
pc: crossover ratio
pm: mutation ratio
pop: number of population
Q: the number of the network's coefficients
r: the ration of ω/ω_n
x_i, x_j, x_k: the input data in the ANN
y_k: the output value in the ANN
k_p: the penalty function in the ANN
Y : input amplitude for the exciting base (m)
$\hat{y}_i$: the required data in the ANN
y_i: the predicted data for ANNM
σp^2 : the error variation in the ANN
ς : damping ratio of a vibration system
ω : angular velocity of the vibrational base
ω_n : natural angular velocity of the vibrational system

2 Introduction

Because of a larger capacity, a faster revolution speed, and a precise mechanism designed for a hard disk, the vibrational impact on data transmission becomes obvious. In order to reduce the influence of the data transmission on a hard disk, a vibrational control for the hard disk is necessary. Chiou and Hung (1993) simulated a floppy disk using a fixed–free circular plate rotating with a constant speed. Tsai *el al.* (2007) developed an active vibration-isolated system for a notebook's hard disk by modifying the connection between the hard disk and the primary board using both the Piezoelectric actuator and sensor within a space-constrained situation. They also verified the above vibration abatement effect by analyzing the natural frequency and the mode shape via ANASYS and an experimental test. Shiu (2008) investigated the influence of the disk's data transmission with respect to various vibrational frequencies, vibrational amplitudes and tilted angles of the disk. Because of the wide application of neural network techniques used in the parameter design during the manufacturing process (Chen, 2006; Tai, Chi, 2008), Chang (2011) advanced Shiu's study (2008) and assessed the optimization of the hardness on the damping material using the neural network in conjunction with the GA optimizer. However, only one-layer of the damping material was applied in vibration reduction. No effective experimental plan used in reducing the number of experimental tests is proposed. Therefore, in order to enhance vibration control and reduce the number of experimental tests, an assessment of the vibrational impact on the data transmission efficiency, an assessment of an optimal three-layer damping material (shown in Fig. 1) installed under the hard disk using a Taguchi method, an Artificial Neural Network (ANN), and a GA Method is proposed.

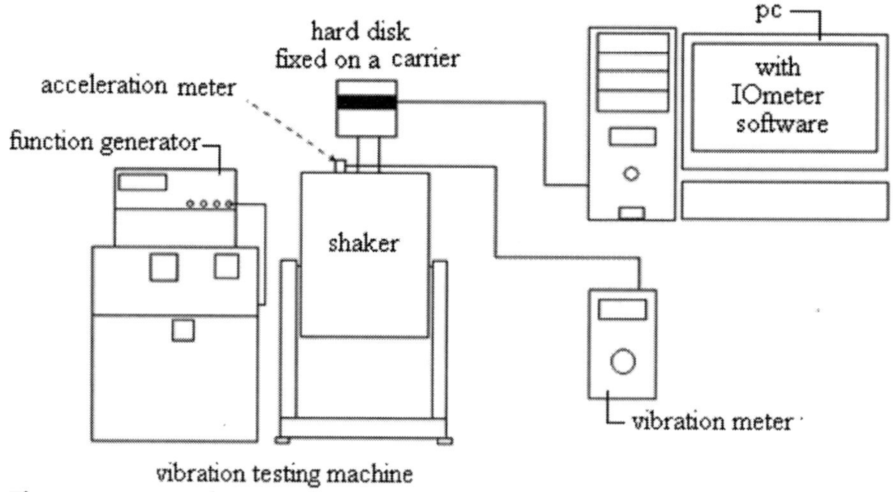

Fig. 1. The measurement of IOPS Base-excitation vibrating system for a hard disk.

3 Mathematical model in vibration

For a three-layer damping material installed under the hard disk with tilted angle of 0 °, the equivalent spring/damping system is depicted in Fig.2. The related free-body diagram for a one-mass vibrational system is also shown in Fig. 3. Based on Newton's second law, the equation of motion is

$$m\ddot{z}_1 + c_{sys}(\dot{z}_1 - \dot{z}_o) + k_{sys}(z_1 - z_o) = 0 \tag{1a}$$

$$\frac{1}{c_{sys}} = \frac{1}{c_1} + \frac{1}{c_2} + \frac{1}{c_3} \;;\; \frac{1}{k_{sys}} = \frac{1}{k_1} + \frac{1}{k_2} + \frac{1}{k_3} \;;\; c_1 = f_{11}(D_1) \;;\; k_1 = f_{12}(D_1) \;;$$

$$c_2 = f_{21}(D_2); k_2 = f_{22}(D_2); c_3 = f_{31}(D_3); k_3 = f_{32}(D_3); \tag{1b}$$

where D_1, D_2 , and D_3 are the hardness with respect to three kinds of damping material. Assuming that the base is harmonically excited yields

$$z_o = Y\sin(\omega \cdot t) \tag{2}$$

The relative displacement z is set at

$$z = z_1 - z_o \tag{3}$$

Plugging Eqs.(2) and (3) into Eq. (1) yields

$$m\ddot{z} + c_{sys}\dot{z} + k_{sys}z = m\omega^2 Y \sin(\omega \cdot t) \tag{4}$$

The solution to Eq. (4) is

$$z(t) = Z\sin(\omega t - \phi) \tag{5a}$$

$$Z(\omega) = \frac{mY\omega^2}{\sqrt{\left(k - m\omega^2\right)^2 + c_{sys}^2\omega^2}};$$

$$= Y\frac{r^2}{\sqrt{\left(1-r^2\right)^2 + \left(2r\zeta\right)^2}} \tag{5b}$$

$$r = \frac{\omega}{\omega_n};\ \phi = \tan^{-1}\left(\frac{c_{sys}\omega}{k - m\omega^2}\right); \tag{5c}$$

$$\varsigma = \frac{c_{sys}}{2\sqrt{k_{sys}m}} = \frac{\dfrac{c_1\cdot c_2\cdot c_3}{c_1\cdot c_2 + c_1\cdot c_3 + c_2\cdot c_3}}{2\sqrt{\left(\dfrac{k_1\cdot k_2\cdot k_3}{k_1\cdot k_2 + k_1\cdot k_3 + k_2\cdot k_3}\right)m}} \tag{5d}$$

$$= \frac{\dfrac{f_{11}(D_1)\cdot f_{21}(D_2)\cdot f_{31}(D_3)}{f_{11}(D_1)\cdot f_{21}(D_2) + f_{11}(D_1)\cdot f_{31}(D_3) + f_{21}(D_2)\cdot f_{31}(D_3)}}{2\sqrt{\dfrac{f_{12}(D_1)\cdot f_{22}(D_2)\cdot f_{32}(D_3)}{f_{12}(D_1)\cdot f_{22}(D_2) + f_{12}(D_1)\cdot f_{32}(D_3) + f_{22}(D_2)\cdot f_{32}(D_3)}\cdot m}}$$

Because the damping coefficients and spring constants with respect to three kinds of damping material are related to the implicit functions ($f_{11}, f_{12}, f_{21}, f_{22}, f_{31}, f_{32}$) of the hardness ($D_1$, D_2, and D_3), it is difficult to exactly find a best hardness combination of the three damping material. Therefore, an alternative way by using a Taguchi method, an Artificial Neural Network (ANN), and a GA Method is proposed.

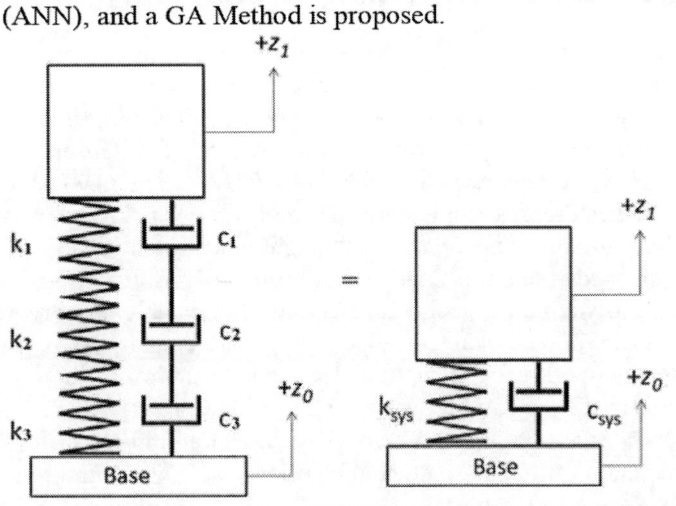

Fig. 2. The equivalent spring/damping system of IOPS base-excitation vibrating system for a hard disk (tilted angle: 0°).

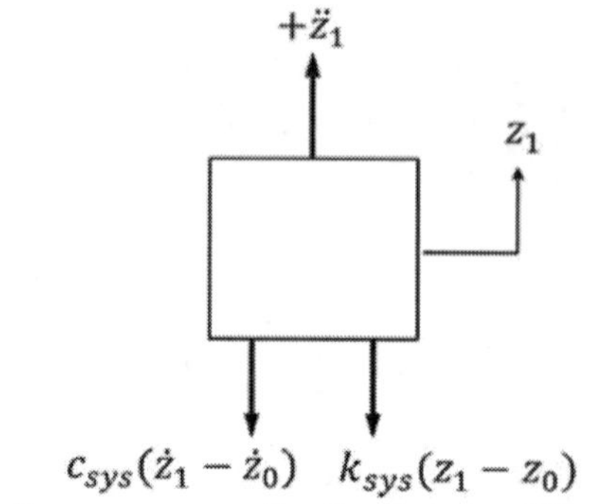

Fig. 3. The free-body diagram for a one-mass vibrational system(tilted angle: 0 °).

4 Taguchi method

The Taguchi design is an experimental method used to improve quality by using appropriate design parameters determined by practical operations on a site or computer simulation (2000). To efficiently reduce the number of conventional experiments, the orthogonal array (Chang, 2000; Wei, 2002) using design parameters (control factors) in columns and standard (level) quantities in rows is presented. By analyzing the parameters' sensitivity to quality, the optimal parameters can be approached by the Taguchi Method (Lan, Chiu, Yeh, 2008). The steps in establishing the analysis of engineering quality using Taguchi Method include the following:
A. Selection of quality.
B. Judging the ideal functions of quality.
C. Listing all the factors that will influence quality.
D. Deciding the control factors and the related levels.
E. Selecting an appropriate orthogonal table and arranging a complete experimental plan for the control factors and the related levels.
F. Performing experimental work and recording experimental results.
G. Data analysis and comparison.

5 Artificial neural network model (ANNM)

Artificial Neural Networks (ANNs) have an advantage in establishing complex non–linear relationships for complicated problems. They may be viewed as universal approximators; however the main disadvantage of this approach is that detected dependencies are hidden. To overcome this drawback, the Group Method of n Data Handling (GMDH) was developed by Ivakhnenko (1971). The GMDH is a self-organized adaptive model. The interconnections between the layers of neurons are simplified, and an automatic algorithm for structure design and weight adjustment is established. Based on the GMDH's feed-forward networks and short-term polynomial transfer functions, the coefficients of the polynomial transfer functions are obtained using a regression technique. The regression technique is then combined with the emulation of the self-organizing activity for neural network (NN) structural learning. In addition, the input variable set in each layer is created. The assembled number is $p!/[(p-\square r)!r!]$ where p is the number of input variables and r is normally set to be two. As indicated in Fig. 4, the polynomial neural network is composed of an input layer, a hidden layer — Σ（Summation）, and an output layer（product）where the hidden layer is the weight summation and the output layer is the product of the input and weighted value (Patrikar, Provence, 1996). The jth output —z_{jk} is

$$z_{jk} = \sum_{i=0}^{n} W_{ij} X_{ij} \tag{6}$$

The total output of the neural network is expressed as

$$y_k = \prod_{j=1}^{h} z_{jk} \tag{7}$$

where h is the unit's number in a hidden layer.
Combining Eqs.(4)(5) yields

$$y_k = B_0 + \sum_{i=1}^{n} B_i x_i + \sum_{i=1}^{n}\sum_{j=1}^{n} B_{ij} x_i x_j + \sum_{i=1}^{n}\sum_{j=1}^{n}\sum_{k=1}^{n} B_{ijk} x_i x_j x_k + \ldots\ldots\ldots\ldots \tag{8}$$

where y_k is the output value, x_i, x_j, x_k is the input data, and B_0, B_i, B_{ij}, and B_{ijk} are the coefficient of the node function.

To obtain the ANNM, the experimental data of the hardness (D1, D2, and D3) for three layers of damping material installed under the hard disk and the output data (IOPS) is used. The trained ANNM can be achieved using both the training data bank and the polynomial calculation in conjunction with the PSE standard (deviation of mean square).
PSE has the form

$$\text{PSE} = \text{FSE} + k_p \tag{9}$$

$$\text{FSE} = \frac{1}{N}\sum_{i=1}^{N} (\hat{y}_i - y_i)^2 \tag{10}$$

where FSE is the deviation of the mean square, k_p is the penalty function, N is the number of training data, $\hat{y}_i$ is the required data, and y_i is the predicted data for the ANNM.
The penalty function k_p is

$$k_p = \text{CPM}\frac{2\sigma p^2 Q}{N} \tag{11}$$

where CPM is the product of the penalty function, σp^2 is the error variation, and Q is the number of the network's coefficients. The related steps of the ANNM construction are shown in Fig. 5. The predicted IOPS can be obtained by inputting arbitrary design data. The ANNM, an OBJ function, works in conjunction with the GA optimizer during the optimization process.

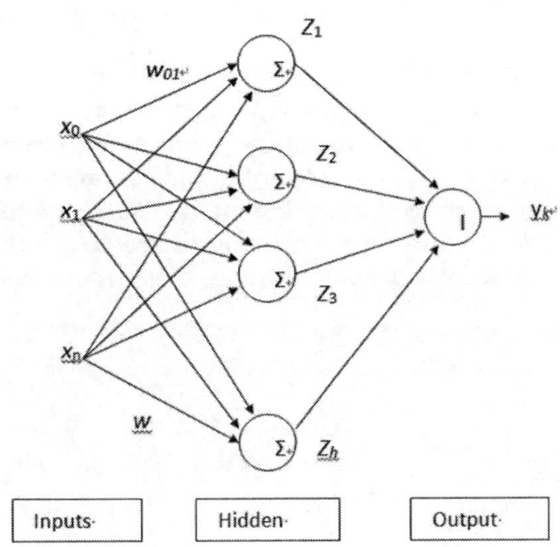

Fig. 4. Structure of the artificial neural network.

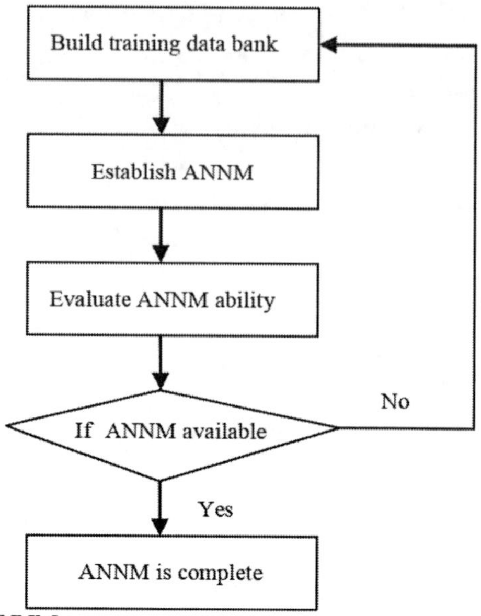

Fig. 5. The steps in the ANNM.

6 Genetic algorithm

The concept of Genetic Algorithms, was first formalized by Holland (1975) and later extended to functional optimization by Jong (1975). It involves the use of optimization search strategies patterned after the Darwinian notion of natural selection. For the optimization of the objective function (*OBJ*), the design parameters of $X1$, $X2$, …, X_m were determined. Initialization is carried out by setting the population size (pop), the length of the chromosome (chrm), the crossover ratio (pm), the mutation ratio (pc), the maximum iteration ($iter_{max}$), the selection method(elitism), the parameter numbers, and the searching ranges of the parameters. Each candidate parent will be selected by the coding/decoding transformation and the fitness (*OBJ*) calculation. The precision (M) of the parameter search is

$$M = \frac{P_{max} - P_{min}}{N_p - 1} \tag{12}$$

where N_p ($=2^m$) is the total possible searching number, m is the number of the design parameters, P_{max} is the maximum range of the parameter, and P_{min} is the minimum range of the parameter. The tournament selection will be adopted as the elitism mechanism in the GA optimization. In addition, the uniform crossover shown in Fig. 6 is applied in the optimization process. Moreover, as indicated in Fig. 7, the mutation scheme is also used to widen the range of the chromosome. The operations in the GA method are pictured in Fig. 8. The process was terminated when a number of generations exceeded a pre-selected value of $iter_{max}$.

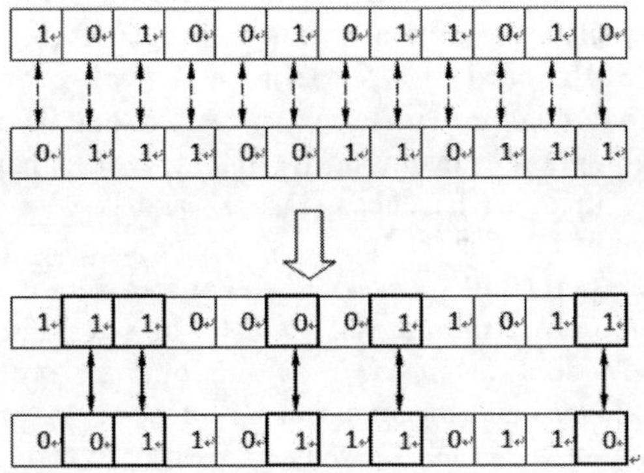

Fig. 6. Mechanism of uniform crossover.

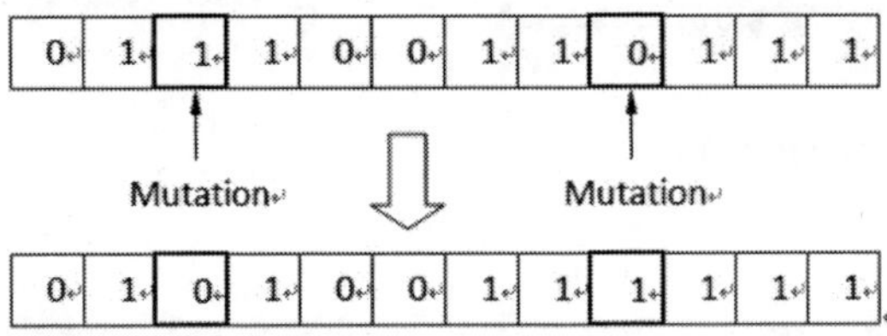

Fig. 7. Mechanism of mutation.

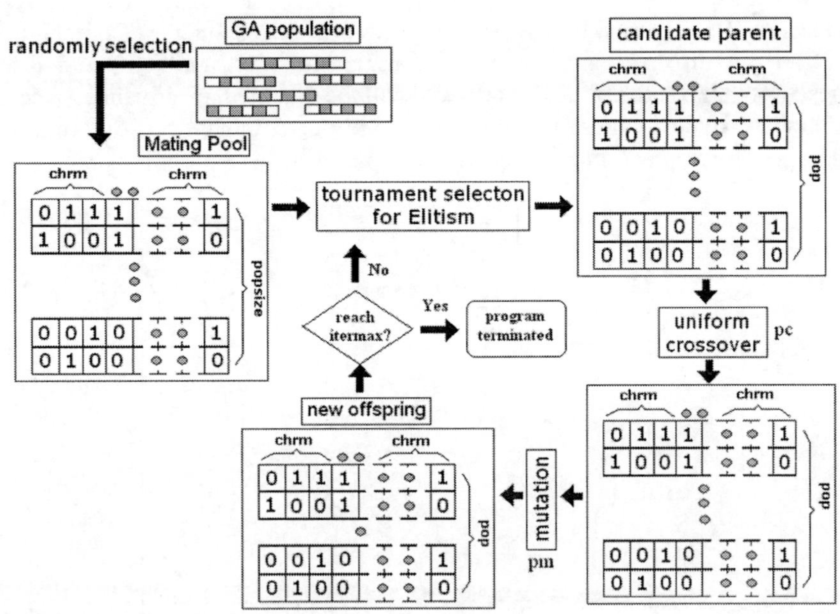

Fig. 8. The block diagram of the GA optimization on plenums.

7 Case study

In this paper, a three-layer damping material installed under the hard disk shown in Fig. 1 is established. An acceleration of 0.2 g is added in the base-excitation vibration. There are three kinds of hard disk with tilted angles (0°, 5°, and 10°) adjusted during the experimental test. To improve the transmission efficiency of a hard disk within a base-excitation environment, a three-layer damping material with various combination of hardness is adopted. In addition, to realize the critical frequencies (target frequencies) that reduce the transmission efficiency of the hard disk, experimental work of an IOPS measurement at three tilted angles without damping material is carried out in advance. Thereafter, for optimal design with the highest transmission efficiency on a hard disk during the base excitation situation, three kinds of hardness (D_1, D_2, and D_3) for three layers of damping material bundled and installed under the hard disk are chosen as the input data. The related experimental transmission rate (IOPS) serves as the output data in the ANNM. In addition, to verify the ANNM correction, training data will be inputted into the ANNM. Using the ANNM as an OBJ and adopting the GA in the optimal process, the best hardness combination of the three-layer damping material with respect to three kinds of tilted angles at the target frequency will be assessed. Moreover, in order to efficiently reduce the number of conventional experiments, the Taguchi design is introduced. The orthogonal array with respect to three tilted angles (0 °, 5 °, and 10°) and three critical frequencies (140 Hz, 150 Hz, and 160 Hz) will be established by using design parameters (control factors) in columns and standard rows (level) of quantities.

8 Results and discussion

8.1 Results

In order to realize the critical frequency (target frequency) that reduces the transmission efficiency of the hard disk, a hard disk without added damping material is tested using three kinds of tilted angles. The results shown in Fig. 9 indicate that the IOPS at the ranges of 120Hz~180Hz is fairly low. Therefore, three target frequencies of 140 Hz, 150 Hz, and 160 Hz are selected. Before establishing the ANNM and based on the Taguchi method, an orthogonal array with respect to three tilted angles (0 °, 5 °, and 10°) and three critical frequencies (140 Hz, 150 Hz, and 160 Hz) will be planned in advance. According the orthogonal tables, related experimental work for various combinations of damping material and tilted angles at various critical frequencies is planned.

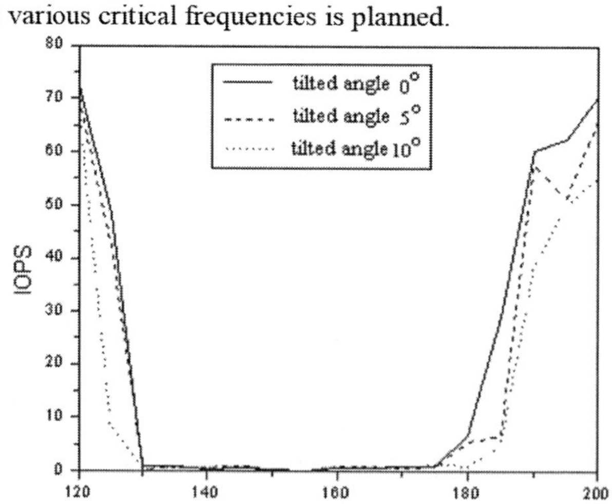

Fig. 9. The hard disk's IOPS with respect to three kinds of tilted angles without adding damping material.

8.1.1 IOPS at a Hard Disk's Tilted Angle of 0°

Using damping material of $D_1=D_2=D_3=50$, the IOPS with respect to acceleration at the target frequency of 140 Hz is shown in Fig. 10. Fig. 10 indicates that the IOPS has the maximum value of acceleration at 0.08 (g). Similarly, adopting damping material of $D_1=D_2=D_3=50$, the IOPS with respect to acceleration at the target frequency of 150 Hz is shown in Fig. 11. Fig. 11 indicates that the IOPS has the maximum value of acceleration at 0.03 (g). Also, using damping material of $D_1=D_2=D_3=50$, the IOPS with respect to acceleration at the target frequency of 160 Hz shown in Fig. 12 indicates that the IOPS has the maximum value of acceleration at 0.06 (g). Therefore, the experimental work on the transmission measurement with respect to three target frequencies 140 Hz, 150 Hz, and 160 Hz under a tilted angle of 0° and accelerations of 0.08, 0.03, and 0.06 (G) is performed and shown in Tables 1~3.

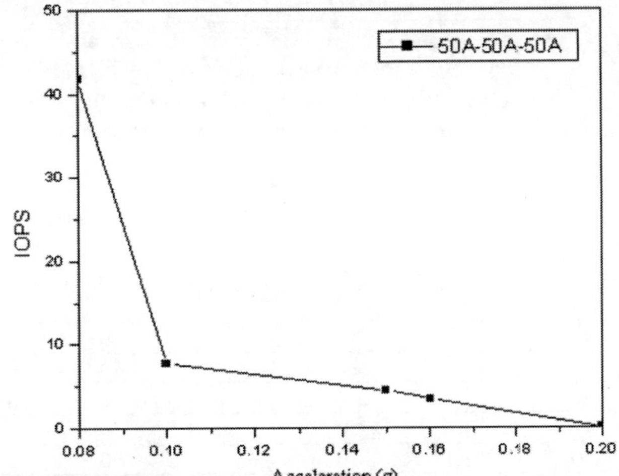

Fig. 10. The hard disk's IOPS with respect to acceleration at the tilted angle (0°) and the target frequency (140 Hz) using damping material (D1=D2=D3=50).

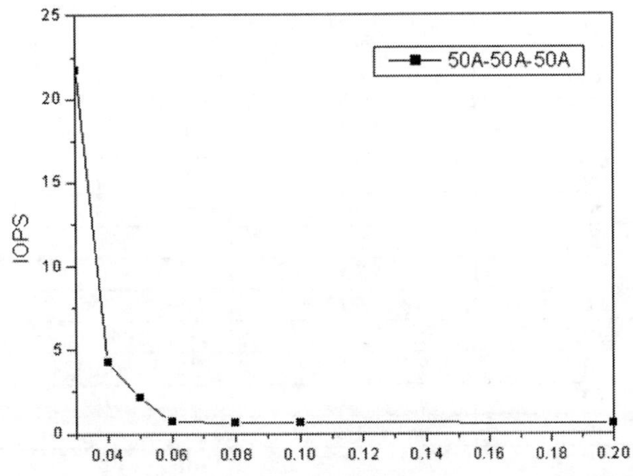

Fig. 11. The hard disk's IOPS with respect to acceleration at the tilted angle (0°) and the target frequency (150 Hz) using damping material (D1=D2=D3=50).

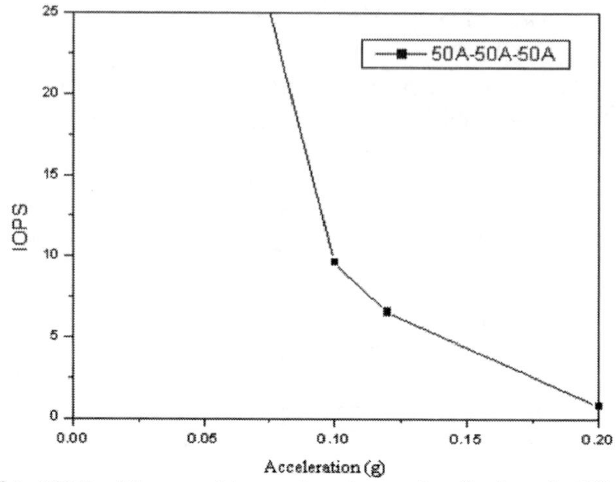

Fig. 12. The hard disk's IOPS with respect to acceleration at the tilted angle (0°) and the target frequency (160 Hz) using damping material (D1=D2=D3=50).

Table 1. Experimental results of a hard disk with a three-layer damping material (hardness of D_1, D_2, and D_3) at a tilted angle of 0° and 140 Hz.

Item	Parameter 1 (D_1)	Parameter 2 (D_2)	Parameter 3 (D_3)	IOPS
1	30	40	50	21.97232
2	30	60	70	55.29873
3	30	70	40	9.57526
4	30	40	60	7.86317
5	30	50	70	3.30748
6	40	50	60	60.14307
7	40	70	30	57.00203
8	40	40	40	58.85171
9	40	30	50	59.83321
10	40	70	60	14.17967
11	50	60	70	56.15763
12	50	30	60	57.31137
13	50	70	50	54.46577
14	50	40	60	56.48192
15	50	30	70	60.81673
16	60	70	50	57.26186
17	60	40	50	3.23626
18	60	60	60	58.67868
19	60	30	50	34.40444
20	60	40	70	8.46796
21	70	30	40	1.70073
22	70	50	60	60.57304
23	70	70	70	54.00194
24	70	40	60	58.30641
25	70	50	30	61.19392

Table 2. Experimental results of a hard disk with a three-layer damping material (hardness of D_1, D_2, and D_3) at a tilted angle of 0° and 150 Hz.

Item	Parameter 1 (D_1)	Parameter 2 (D_2)	Parameter 3 (D_3)	IOPS
1	30	40	50	4.55979
2	30	60	70	18.63423
3	30	70	40	2.17793

4	30	40	60	6.44938
5	30	50	70	18.16737
6	40	50	60	27.91441
7	40	70	30	1.60061
8	40	40	40	9.94398
9	40	30	50	55.45931
10	40	70	60	53.85391
11	50	60	70	39.70867
12	50	30	60	35.34186
13	50	70	50	23.01421
14	50	40	60	21.13465
15	50	30	70	49.11832
16	60	70	50	40.18326
17	60	40	50	53.69491
18	60	60	60	20.15838
19	60	30	50	52.57962
20	60	40	70	21.34739
21	70	30	40	21.94138
22	70	50	60	35.19878
23	70	70	70	43.44021
24	70	40	60	54.94346
25	70	50	30	26.09167

Table 3. Experimental results of a hard disk with a three-layer damping material (hardness of D_1, D_2, and D_3) at a tilted angle of 0° and 160 Hz.

Item	Parameter 1 (D_1)	Parameter 2 (D_2)	Parameter 3 (D_3)	IOPS
1	30	40	50	32.46401
2	30	60	70	4.35431
3	30	70	40	3.74602
4	30	40	60	47.84439
5	30	50	70	20.38102
6	40	50	60	4.89399
7	40	70	30	5.44983
8	40	40	40	16.99811
9	40	30	50	1.85630
10	40	70	60	20.50971
11	50	60	70	3.38123
12	50	30	60	8.46935
13	50	70	50	1.01219
14	50	40	60	1.56729
15	50	30	70	1.42908
16	60	70	50	45.36873
17	60	40	50	10.61518
18	60	60	60	42.03457
19	60	30	50	47.61785
20	60	40	70	13.42458
21	70	30	40	1.83411
22	70	50	60	2.34113
23	70	70	70	15.12661
24	70	40	60	15.65359
25	70	50	30	44.52777

Using the experimental data from the ANNM and applying the GA optimizer, the optimization of the combined hardness (D_1, D_2, and D_3) is performed. The related GA control parameters are shown in Table 4. The optimal results of the design parameters (D_1=D_2=D_3=50) at the target frequencies is shown in Table 5. To meet the real hardness specifications of the damping material, real design data that is close to the optimal data are adopted. The real design data and related IOPS for three target frequencies at a tilted angle of 0° are depicted in Table 6. Moreover, the original design data (without optimization) and related IOPS are illustrated in Table 7. Comparing the transmission efficiency before and after optimization is performed, results in targeted frequencies of 140 Hz, 150 Hz, and 160 Hz are plotted in Figs. 13~15.

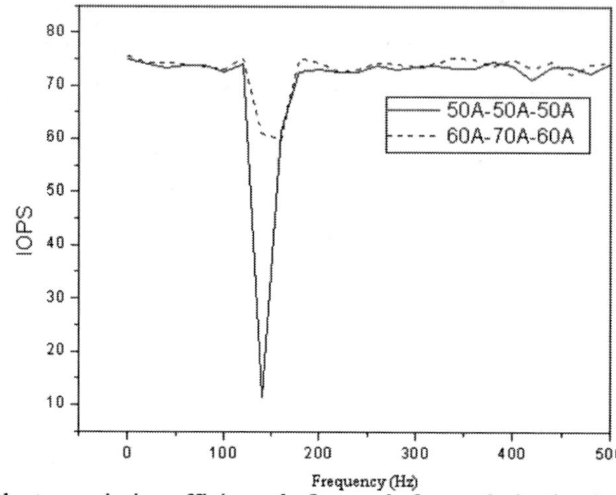

Fig. 13. Comparing the transmission efficiency before and after optimization is performed at the target frequency of 140 Hz (tilted angle of 0°).

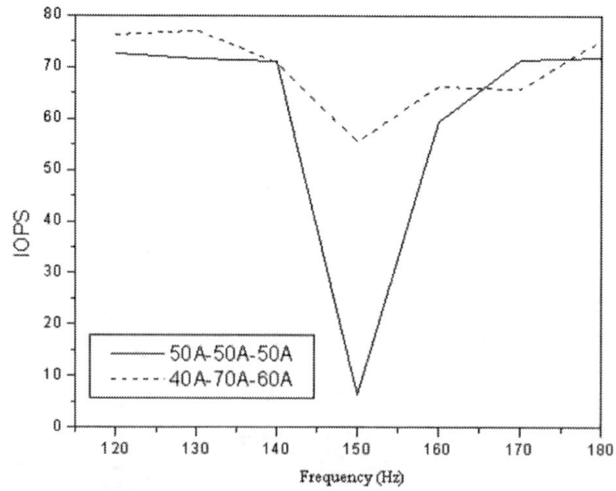

Fig. 14. Comparing the transmission efficiency before and after optimization is performed at the target frequency of 150 Hz (tilted angle of 0°).

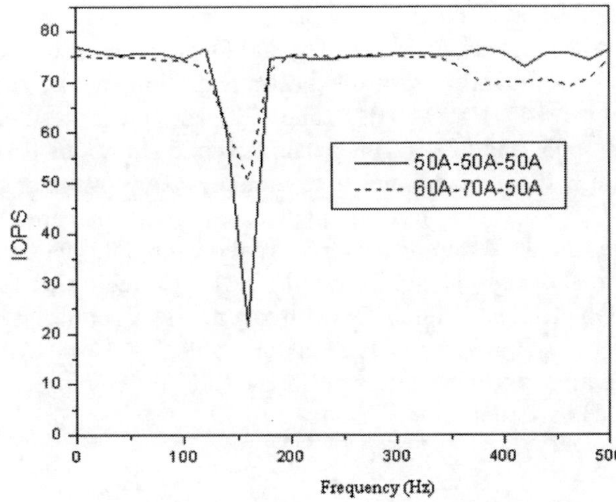

Fig. 15. Comparing the transmission efficiency before and after optimization is performed at the target frequency of 160 Hz (tilted angle of 0°).

Table 4. Selected GA parameters during shape optimization.

GA parameters	Value (or condition)
design variables	3
chrm	40
pop	100
elitism	(tournament)
crossover	(uniform crossover)
pc	0.8
pm	0.05
$iter_{max}$	1000

Table 5. Optimal results of a hard disk with a three-layer damping material (hardness of D_1, D_2, and D_3) at three target frequencies (tilted angle = 0°).

Target frequency (Hz)	Optimal design parameter			OBJ
	Parameter 1 (D_1)	Parameter 2 (D_2)	Parameter 3 (D_3)	IOPS
140	57.5589	73.5525	62.3321	74.48192
150	38.6027	67.3665	59.2364	64.3681
160	57.3395	68.1867	52.0627	56.3769

Table 6. Practical results of a hard disk with a three-layer damping material (hardness of D_1, D_2, and D_3) at three target frequencies (tilted angle = 0°).

Target frequency (Hz)	Practical design parameter			IOPS
	Parameter 1 (D_1)	Parameter 2 (D_2)	Parameter 3 (D_3)	
140	60	70	60	61.06
150	40	70	60	55.45
160	60	70	50	50.57

Table 7. Original IOPS (D_1=D_2=D_3=50) at three target frequencies without optimization (tilted angle = 0°).

Target frequency (Hz)	Practical design parameter			IOPS
	Parameter 1(D_1)	Parameter 2 (D_2)	Parameter 3 (D_3)	
140	50	50	50	11.36
150	50	50	50	6.36
160	50	50	50	21.43

8.1.2 IOPS at a Hard Disk's Tilted Angle of 5°

Similarly, the optimization of hardness (D_1, D_2, and D_3) for a hard disk at a tilted angle of 5 ° is performed. The experimental work on the transmission measurement with respect to three target frequencies 140 Hz, 150 Hz, and 160 Hz under a tilted angle of 5 ° and accelerations of 0.08, 0.03, and 0.06 (G) is performed and shown in Tables 8~10. Using the experimental data from the ANNM and applying the GA optimizer, the optimal design parameters (D_1=D_2=D_3=50) at the target frequencies is shown in Table 11. To meet the real hardness specification for the damping material, real design data that is close to the optimal data is adopted. The real design data and related IOPS for three target frequencies at a tilted angle of 5 ° are depicted in Table 12. Moreover, the original design data (without optimization) and related IOPS are illustrated in Table 13. Comparing the transmission efficiency before and after optimization is performed, results in targeted frequencies of 140 Hz, 150 Hz, and 160 Hz are plotted in Figs. 16~18.

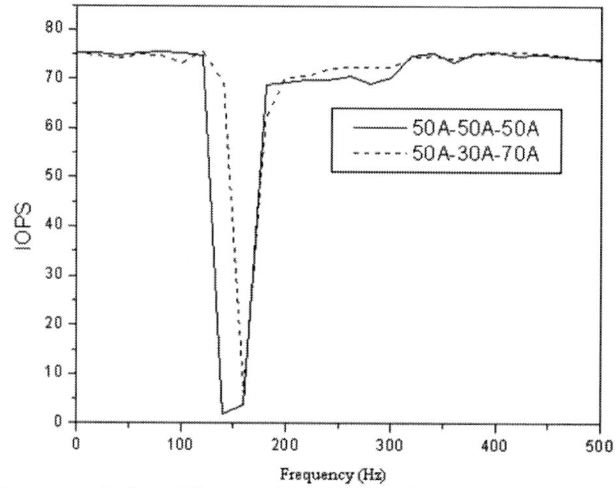

Fig. 16. Comparing the transmission efficiency before and after optimization is performed at the target frequency of 140 Hz (tilted angle of 5°).

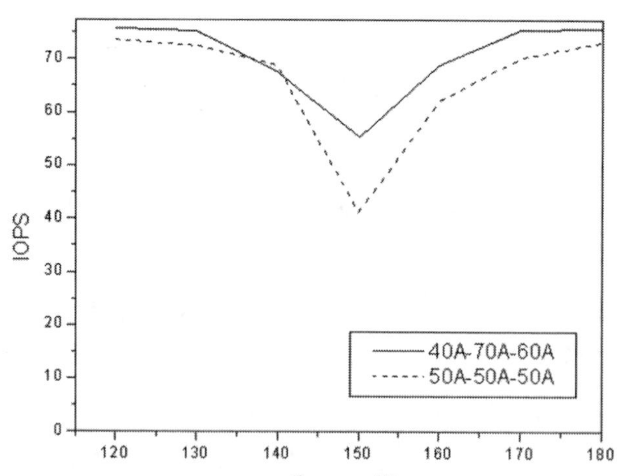

Fig. 17. Comparing the transmission efficiency before and after optimization is performed at the target frequency of 150 Hz (tilted angle of 5°).

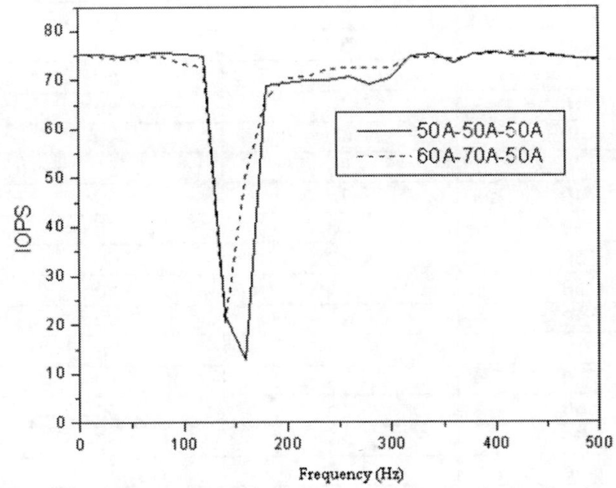

Fig. 18. Comparing the transmission efficiency before and after optimization is performed at the target frequency of 160 Hz (tilted angle of 5°).

Table 8. Experimental results of a hard disk with a three-layer damping material (hardness of D_1, D_2, and D_3) at a tilted angle of 5° and 140 Hz.

Item	Parameter 1 (D_1)	Parameter 2 (D_2)	Parameter 3 (D_3)	IOPS
1	30	40	50	3.09138
2	30	60	70	46.66624
3	30	70	40	8.46956
4	30	40	60	17.57293
5	30	50	70	8.92974
6	40	50	60	15.87552
7	40	70	30	58.84662
8	40	40	40	53.74312
9	40	30	50	27.87799
10	40	70	60	11.23359
11	50	60	70	65.64529
12	50	30	60	33.51206
13	50	70	50	59.55521
14	50	40	60	64.06865
15	50	30	70	65.92546
16	60	70	50	63.33052
17	60	40	50	3.85896
18	60	60	60	56.78851
19	60	30	50	33.13341
20	60	40	70	1.08436
21	70	30	40	1.08953
22	70	50	60	11.36135
23	70	70	70	59.00901
24	70	40	60	60.67589
25	70	50	30	52.10223

Table 9. Experimental results of a hard disk with a three-layer damping material (hardness of D_1, D_2, and D_3) at a tilted angle of 5° and 150 Hz.

Item	Parameter 1 (D_1)	Parameter 2 (D_2)	Parameter 3 (D_3)	IOPS
1	30	40	50	47.52115
2	30	60	70	32.82668
3	30	70	40	6.44015

4	30	40	60	4.52681
5	30	50	70	26.76372
6	40	50	60	5.16036
7	40	70	30	8.50283
8	40	40	40	22.62648
9	40	30	50	33.25664
10	40	70	60	66.28423
11	50	60	70	20.91662
12	50	30	60	28.80622
13	50	70	50	4.21008
14	50	40	60	22.10794
15	50	30	70	5.63342
16	60	70	50	13.67966
17	60	40	50	51.90243
18	60	60	60	44.63756
19	60	30	50	55.86545
20	60	40	70	48.08131
21	70	30	40	24.77616
22	70	50	60	8.86914
23	70	70	70	13.55824
24	70	40	60	24.37131
25	70	50	30	38.42977

Table 10. Experimental results of a hard disk with a three-layer damping material (hardness of D_1, D_2, and D_3) at a tilted angle of 5° and 160 Hz.

Item	Parameter 1 (D_1)	Parameter 2 (D_2)	Parameter 3 (D_3)	IOPS
1	30	40	50	27.22337
2	30	60	70	5.98962
3	30	70	40	2.66482
4	30	40	60	19.90615
5	30	50	70	9.59281
6	40	50	60	6.80697
7	40	70	30	1.56727
8	40	40	40	3.11403
9	40	30	50	2.17442
10	40	70	60	23.17627
11	50	60	70	17.51487
12	50	30	60	7.00062
13	50	70	50	1.10011
14	50	40	60	1.81187
15	50	30	70	2.81924
16	60	70	50	50.95341
17	60	40	50	17.82112
18	60	60	60	40.61928
19	60	30	50	44.27424
20	60	40	70	10.56677
21	70	30	40	1.82861
22	70	50	60	1.70727
23	70	70	70	47.45103
24	70	40	60	23.71135
25	70	50	30	21.09823

Table 11. Optimal result of a hard disk with a three-layer damping material (hardness of D_1, D_2, and D_3) at three target frequencies (tilted angle = 5°).

Target frequency (Hz)	Optimal design parameter			OBJ
	Parameter 1 (D_1)	Parameter 2 (D_2)	Parameter 3 (D_3)	IOPS
140	47.2256	33.4489	66.3825	72.8632
150	42.8749	70.3235	61.8612	57.3342
160	62.0962	70.9005	45.3971	64.8916

Table 12. Practical results of a hard disk with a three-layer damping material (hardness of D_1, D_2, and D_3) at three target frequencies (tilted angle = 5°).

Target frequency (Hz)	Practical design parameter			IOPS
	Parameter 1 (D_1)	Parameter 2 (D_2)	Parameter 3 (D_3)	
140	50	30	70	66.22
150	40	70	60	55.36
160	60	70	50	50.02

Table 13. Original IOPS (D_1=D_2=D_3=50) at three target frequencies without optimization (tilted angle = 5°).

Target frequency (Hz)	Practical design parameter			IOPS
	Parameter 1 (D_1)	Parameter 2 (D_2)	Parameter 3 (D_3)	
140	50	50	50	6.43
150	50	50	50	41.24
160	50	50	50	12.12

8.1.3 IOPS at a Hard Disk's Tilted Angle of 10°

Equally, the optimization of hardness (D_1, D_2, and D_3) for a hard disk at a tilted angle of 10 ° is performed. The experimental work for the transmission measurement with respect to three target frequencies 140 Hz, 150 Hz, and 160 Hz under a tilted angle of 10 ° and accelerations of 0.08, 0.03, and 0.06 (G) is performed and shown in Tables 14~16. Using the experimental data from the ANNM and applying the GA optimizer, the optimal design parameters (D_1=D_2=D_3=50) at the target frequencies are shown in Table 17. To meet real hardness specifications for the damping material, real design data that is close to the optimal data is adopted. The real design data and related IOPS for the three target frequencies at a tilted angle of 10 ° are depicted in Table 18. Moreover, the original design data (without optimization) and related IOPS are illustrated in Table 19. Comparing the transmission efficiency before and after optimization is performed, results in frequencies of 140 Hz, 150 Hz, and 160 Hz are plotted in Figs. 19~21.

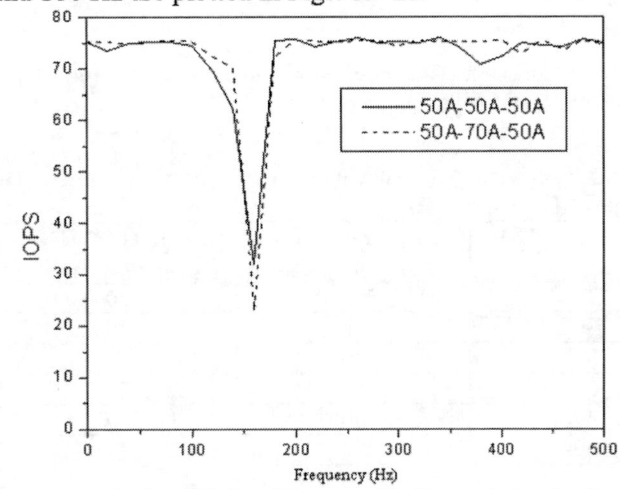

Fig. 19. Comparing the transmission efficiency before and after optimization is performed at the target frequency of 140 Hz (tilted angle of 10°).

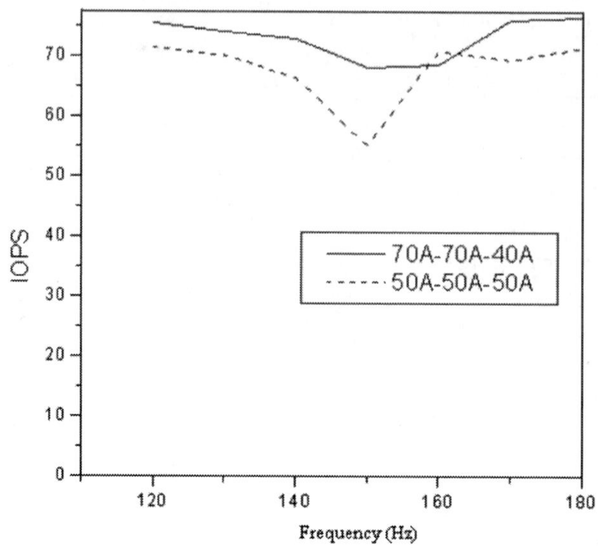

Fig. 20. Comparing the transmission efficiency before and after optimization is performed at the target frequency of 150 Hz (tilted angle of 10°).

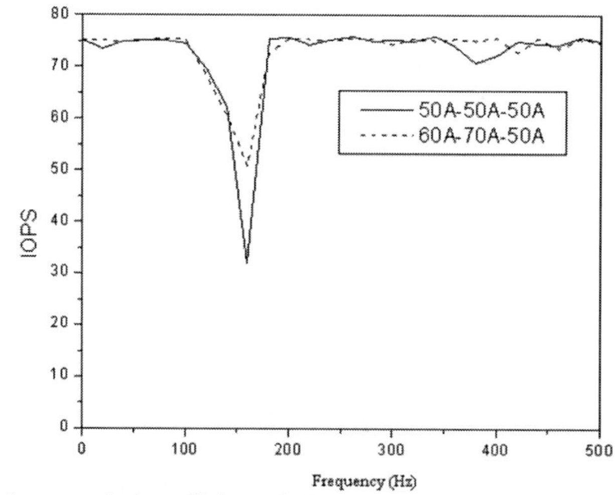

Fig. 21. Comparing the transmission efficiency before and after optimization is performed at the target frequency of 160 Hz (tilted angle of 10°).

Table 14. Experimental results of a hard disk with a three-layer damping material (hardness of D_1, D_2, and D_3) at a tilted angle of 10° and 140 Hz.

Item	Parameter 1 (D_1)	Parameter 2 (D_2)	Parameter 3 (D_3)	IOPS
1	30	40	50	1.80081
2	30	60	70	48.68469
3	30	70	40	13.81347
4	30	40	60	3.19666
5	30	50	70	3.00747
6	40	50	60	3.30326
7	40	70	30	65.93559
8	40	40	40	29.20518
9	40	30	50	53.84428
10	40	70	60	10.00949

11	50	60	70	65.52351
12	50	30	60	10.81129
13	50	70	50	66.19763
14	50	40	60	39.31431
15	50	30	70	40.71456
16	60	70	50	55.72275
17	60	40	50	57.18641
18	60	60	60	54.34626
19	60	30	50	44.93873
20	60	40	70	3.17489
21	70	30	40	1.68952
22	70	50	60	59.49621
23	70	70	70	47.39454
24	70	40	60	53.81705
25	70	50	30	46.89436

Table 15. Experimental results of a hard disk with a three-layer damping material (hardness of D_1, D_2, and D_3) at a tilted angle of 10° and 150 Hz.

Item	Parameter 1 (D_1)	Parameter 2 (D_2)	Parameter 3 (D_3)	IOPS
1	30	40	50	45.32489
2	30	60	70	40.98301
3	30	70	40	2.81179
4	30	40	60	8.98616
5	30	50	70	17.39083
6	40	50	60	43.29303
7	40	70	30	8.76413
8	40	40	40	12.17231
9	40	30	50	44.13115
10	40	70	60	56.33476
11	50	60	70	28.36053
12	50	30	60	41.23089
13	50	70	50	4.53785
14	50	40	60	7.09557
15	50	30	70	21.23833
16	60	70	50	56.01829
17	60	40	50	50.85596
18	60	60	60	34.41528
19	60	30	50	52.28129
20	60	40	70	15.39809
21	70	30	40	60.10214
22	70	50	60	26.79049
23	70	70	70	21.32682
24	70	40	60	48.12752
25	70	50	30	29.67769

Table 16. Experimental results of a hard disk with a three-layer damping material (hardness of D_1, D_2, and D_3) at a tilted angle of 10° and 160 Hz.

Item	Parameter 1 (D_1)	Parameter 2 (D_2)	Parameter 3 (D_3)	IOPS
1	30	40	50	25.32489
2	30	60	70	2.57491
3	30	70	40	3.11967
4	30	40	60	27.58612
5	30	50	70	13.67468
6	40	50	60	11.70491
7	40	70	30	1.55053

8	40	40	40	3.84822
9	40	30	50	1.73401
10	40	70	60	3.99291
11	50	60	70	5.18783
12	50	30	60	7.23593
13	50	70	50	1.86749
14	50	40	60	1.71731
15	50	30	70	1.83482
16	60	70	50	47.65246
17	60	40	50	22.72884
18	60	60	60	49.75337
19	60	30	50	40.21922
20	60	40	70	12.46766
21	70	30	40	1.41149
22	70	50	60	1.38483
23	70	70	70	49.78942
24	70	40	60	4.23209
25	70	50	30	37.41532

Table 17. Optimal result of a hard disk with a three-layer damping material (hardness of D_1, D_2, and D_3) at three target frequencies (tilted angle = 10°).

Target frequency (Hz)	Optimal design parameter			OBJ
	Parameter 1 (D_1)	Parameter 2 (D_2)	Parameter 3 (D_3)	IOPS
140	46.3258	71.0052	54.6587	70.3894
150	72.2311	67.5274	41.5288	67.2318
160	56.1108	73.5341	54.2323	59.5564

Table 18. Practical result of a hard disk with a three-layer damping material (hardness of D_1, D_2, and D_3) at three target frequencies (tilted angle = 10°).

Target frequency (Hz)	Practical design parameter			IOPS
	Parameter 1 (D_1)	Parameter 2 (D_2)	Parameter 3 (D_3)	
140	50	70	50	70.14
150	70	70	40	68.11
160	60	70	50	50.85

Table 19. Original IOPS (D_1=D_2=D_3=50) at three target frequencies without optimization (tilted angle = 10°).

Target frequency (Hz)	Practical design parameter			IOPS
	Parameter 1 (D_1)	Parameter 2 (D_2)	Parameter 3 (D_3)	
140	50	50	50	62.14
150	50	50	50	55.16
160	50	50	50	32.03

8.2 Discussion

As deduced in Eq.(5), in case of $\omega^2 << \omega_n^2$, $Z(\omega)$, the displacement of the disk, is simplified as

$$Z(\omega) = Yr^2 \tag{13}$$

If the frequency is $\omega^2 >> \omega_n^2$, $Z(\omega)$ will become

$$Z(\omega) = Y \tag{14}$$

Whereas, in case of $\omega^2 = \omega_n^2$ (resonance), $Z(\omega)$ yields

$$Z(\omega)=\frac{Y}{2\varsigma_{sys}}=\frac{\sqrt{\left(\frac{k_1\cdot k_2\cdot k_3}{k_1\cdot k_2+k_1\cdot k_3+k_2\cdot k_3}\right)m\cdot Y}}{\frac{c_1\cdot c_2\cdot c_3}{c_1\cdot c_2+c_1\cdot c_3+c_2\cdot c_3}}=\frac{\sqrt{\left(\frac{f_{12}(D_1)\cdot f_{22}(D_2)\cdot f_{32}(D_3)}{f_{11}(D_1)\cdot f_{21}(D_2)+f_{11}(D_1)\cdot f_{31}(D_3)+f_{21}(D_2)\cdot f_{31}(D_3)}\right)m\cdot Y}}{\frac{f_{11}(D_1)\cdot f_{21}(D_2)\cdot f_{31}(D_3)}{f_{11}(D_1)\cdot f_{21}(D_2)+f_{11}(D_1)\cdot f_{31}(D_3)+f_{21}(D_2)\cdot f_{31}(D_3)}} \quad (15)$$

As mentioned in Eqs.(13)~(15), for a tilted angle of 0 °, the displacement of the hard disk induced by the external based-excitation vibration is tightly related to the forcing based-excitation frequency, the system's natural frequency, the amplitude of the base-excitation (Y), and the hardness combination of three damping material. Here, the amplitude of the base-excitation (Y) is related to the base acceleration. If it is in the case of the resonant condition (for a specified bas-excitation motion), the displacement of the hard disk will decrease when the system's damping ratio increases. Also, the transmission efficiency will be improved if the induced displacement of the hard disk is small. Therefore, the finding of a best combination for three damping material is essential. Moreover, for a tilted angle with non-zero degree, because of the unbalance force installed on the spring/damping system, the induced displacement of the hard disk will be expected to increase.

As indicated in Fig. 4, the transmission efficiency at the ranges of 120Hz~180Hz is very low. Therefore, the three target frequencies of 140 Hz, 150 Hz, and 160 Hz are selected and used in the hardness optimization for the three-layer damping material. As indicated in Tables 6~7, for a hard disk vibrating at a tilted angle of 0 ° and accelerating at 0.08(G), the transmission efficiency (IOPS) will be improved from 11.36 to 61.06 at 140 Hz using the hardness combination of (60, 70, 60). In addition, the transmission efficiency (IOPS) will be enhanced from 6.36 to 55.45 at 150 Hz using the hardness combination of 40, 70, and 60. Also, the transmission efficiency (IOPS) will increase from 21.43 to 50.57 at 160 Hz using the hardness combination of 60, 70, and 50. Additionally, as indicated in Tables 12~13, for a hard disk vibrating at a tilted angle of 5 ° and accelerating at 0.03(G), the transmission efficiency (IOPS) will be improved from 6.43 to 66.22 at 140 Hz when using the hardness combination of 50, 30, and 70. Also, the transmission efficiency (IOPS) will be enhanced from 41.24 to 55.36 at 150 Hz when using the hardness combination of 40, 70, and 60. In addition, the transmission efficiency (IOPS) will increase from 12.12 to 50.02 at 160 Hz using the hardness combination of 60, 70, and 50. Moreover, as indicated in Tables 18~19, for a hard disk vibrating at a tilted angle of 10 ° and accelerating at 0.06(G), the transmission efficiency (IOPS) will be improved from 62.14 to 70.14 at 140 Hz when the hardness combination is 50, 70, and 50. Also, the transmission efficiency (IOPS) will be enhanced from 55.16 to 68.11 at 150 Hz when the hardness combination is 70, 70, and 40. Consequently, the transmission efficiency (IOPS) will increase from 32.03 to 50.85 at 160 Hz when the hardness combination is 60, 70, and 50.

9 Conclusions

As can be seen in Eqs.(13)~(15), the induced displacement of the hard disk is closely related to the forcing based-excitation frequency, the system's natural frequency, the acceleration (Y), and the hardness combination of three damping material. Also, the tilted angle with unbalanced force will largely influence the hard disk's motion. In order to depress the vibration effect to the hard disk's data transmission, a best combination for three damping material, an appropriate base acceleration, and a fitful tilted angle are necessary.

Experimental work shown in Fig. 4 indicates that the transmission efficiency at the ranges of 120Hz~180Hz is very low when using the acceleration of base-excitation at 0.2 (G). To reduce the vibrational impact on data transmission efficiency, a three-layer damping material installed under the hard disk is adopted. An assessment of the damping material's hardness (D_1, D_2, and D_3) is performed by using the Artificial Neural Network Model (ANNM) and the GA Method. Before the optimization of the damping material is performed, the required experimental sets for data transmission testing will be determined by using the Taguchi method. Here, the ANNM, a simplified objective function (OBJ), is established by inputting the hardness (D_1, D_2, and D_3) of three layers of damping material and their related data transmission efficiency (IOPS) at a targeted transmission frequency. Results reveal that the transmission efficiency of a hard disk at three tilted angles can be efficiently improved at targeted frequencies within a base-excitation vibrating environment using the GA optimization. For a hard disk with a tilted angle of 0 °, the increments of IOPS at the target frequencies (140Hz, 150 Hz, 160 Hz) are 49.7, 49.09, and 29.14. In addition, for a hard disk tilted at 5 °, the increments of IOPS with respect to the target frequencies (140Hz, 150 Hz, 160 Hz) are 50.79, 14.12, and 37.9. Also, for a hard disk tilted at 10 °, the increments of IOPS with respect to the target frequencies (140Hz, 150 Hz, 160 Hz) are 8.0, 12.95, and 18.82.

Consequently, the assessment of the three-layer damping material's hardness using the Taguchi method, the GA optimization, as well as the ANN's model is quite efficient.

References

1. J. Van der Geer, J.A. J. Hanraads, R.A. Lupton, The art of writing a scientific article. Journal of Scientific Communications, **163**, 51-59 (2000)
2. C.C. Chang, Optimum isolation design for enhancing transmission efficiency of a hard disk drive, Master Thesis, Tatung University, Taiwan (2011)
3. J.C. Chang, The application of Taguchi method in the finite element analysis for the optimal structural design of compressed sheet in huge extended plate, Master Thesis, National Chung Hsing University, Taiwan (2000)
4. H.M. Chen, Shape optimization of rubber dome of key shell fragment using polynomial network and genetic algorithm, the 23th Chinese Society of Mechanical Engineer, 129-135 (2006)
5. W.J. Chiou, S.C. Huang, Modeling and vibration simulation of a spinning PC hard disk, Proceedings of the 10th National Conference of the CSME, 667-677 (1993)
6. J.H. Holland, Adaptation in natural and artificial system, Ann Arbor, MI, The University of Michigan Press (1975)
7. A.G. Ivakhnenko,. Polynomial theory of complex system, IEEE Trans. Syst. Man. Cyber., **1(4)**, 364-368 (1971)
8. D. Jong, Analysis of the behavior of a class of genetic adaptive systems, PhD Dissertation, The University of Michigan Press (1975)

9. T.S. Lan, M.C. Chiu, L.J. Yeh, An approach to rib design of injection molded project using finite element and Taguchi method, Information Technology Journal, **7(2)**, 299-305, (2008)
10. H.H. Li, Taguchi method-theory and experiment on quality design, Gau-Li, Taipei, (2000)
11. A. Patrikar, J. Provence, Nonlinear system identification and adaptive control using polynomial networks, Math. Comput. Modeling, **l23(1/2)**, 159-173 (1996)
12. S.J. Shiu, Transmission rate enhancement of disk drive system subjected to base excitation at different inclined angles, Master Thesis, Tatung University, Taiwan (2008)
13. C.C. Tai, Y.L. Chi, The influence of processing parameters on the flatness for a contact surface of plastic connectors, Journal of Advanced Engineering, **3(2)** (2008)
14. M.S. Tsai, W.H. Yuan, J.M. Chang, Active vibration control of hard-disk drives using PZT actuated suspension systems, The 13th International Congress on Sound and Vibration, Cairns, Australia (2007)
15. H.S. Wei, Analysis for the optimum conditions in derrick hook by using Taguchi method, Master Thesis, National Ping Tung University of Science and Technology, Taiwan (2002)

Numerical investigation on the forming behaviour of stainless/carbon steel bimetal composite

Zhou Li[1], *Jingwei* Zhao[1], *Qingfeng* Zhang[2], *Sihai* Jiao[2], and *Zhengyi* Jiang[1,*]

[1]School of Mechanical, Materials, Mechatronic and Biomedical Engineering, University of Wollongong, Wollongong NSW 2522, Australia
[2]Baosteel Research Institute (R&D Centre), Baoshan Iron & Steel Co., Ltd., Shanghai 200431, China

Abstract. Bimetal composites have wide applications due to their excellent overall performance and relatively low comprehensive cost. The aim of this study is to investigate the forming behaviour of stainless/carbon steel bimetal composite during stamping by finite element method (FEM). In this work, the bonding interface of bimetal composite sheet was assumed to be perfect without delamination during the plastic forming process for simplicity. Uniaxial tensile tests on base metal (carbon steel) and compositing metal (stainless steel) were first carried out, respectively, in order to obtain the tensile properties of each of the component materials required in the forming simulation. Processing variables, including the layer stacking sequence, relative thickness ratios of two layers and friction were considered, and their effects on the distributions of circumferential stress and thickness strain were analysed. The bimetal composite sheet was set as the eight-node solid elements in the developed FEM model, which is effective for evaluating the distributions of circumferential stress and thickness strain, and predicting the high-risk region of necking during the stamping of bimetal composites. The simulation results can be used as an evaluation indicator of the capability of forming machine to ensure the bimetal composite can be safely formed.

1 Introduction

With the increasing demands for the multi-functionality of industrial products, many research efforts in materials science and engineering have focused on developing engineering materials of better mechanical capabilities and low comprehensive cost, such as bimetal composites [1]. Some common fabrication processes, such as hemming, bending, and large plastic forming, are adopted to further manufacture those bimetal and laminated metal composites [2]. In this case, many research works have been studied on the behaviours of bimetal composite sheets during the forming process.

Kapinski [3] presented the initial concepts of deep drawing process for bimetal elements of sheet materials. The inertia forces, plastic hardening behaviour of deformed material and dynamic stress were utilised in the problems of plasticity, and the experimental results were

[*]Corresponding author: jiang@uow.edu.au

provided as well. Parsa et al. [4] studied the behaviour of two-layer aluminium-stainless steel laminated sheets during the direct drawing and reverse redrawing processes, i.e. the first and second drawing stages, by the finite element analysis and the experimental approach. Their studies indicated that the contact of stainless steel with the punch leads to the maximum drawing ratio in direct redrawing, but the aluminium should contact the punch in order to obtain the highest drawing ratio in reverse redrawing. Morovvati et al. [5] used the experimental and finite element investigation on wrinkling of circular two-layer metal sheets in the deep drawing process, and the influences of different blank holder forces on winkling were analysed. They also studied the relationship between the required forming force and the drawing radio. Likewise, Atrian et al. [6] studied the effects of different parameters, such as lubrication, blank-holder force and the diameter of the composite blank, on the deep drawing of steel/brass laminated sheets. However, in the previous works, those studied bimetal or laminated metal composites were thin sheets, but less people paid attention to the formability of the medium and thick bimetal composite.

In this study, the FEM investigation was conducted for the forming behaviour of the medium and thick bimetal sheets without the blank holder. The effects of some processing variables, including the layer stacking sequence, relative thickness ratios of two layers and friction on the forming force and the material flow were investigated and discussed. In order to predict the high-risk region of necking during the stamping process, the distributions of circumferential stress and thickness strain in the bimetal sheets were determined using the FE simulation.

2 Tensile properties of bimetal composite

A bimetal composite sheet, consisting of the duplex ferritic-austenitic stainless steel (2205) layer with the thickness of 3 mm and the carbon steel (AH36) layer with the thickness of 8 mm, was bonded by the hot rolling as schematically shown in Fig. 1a. Chemical compositions of two essential components are shown in Table1. After roll bonding, the bonding interface was assumed to be thin and perfect without the delamination during the stamping process for simplicity in this work, and the morphology of bonding zone is shown in Fig. 1b.

Table 1. Chemical compositions of parent and clad metal plates.

Steel	Element (wt %)							
	C	Si	Mn	P	S	Cr	Mo	Ni
2205	0.03	1.00	2.00	0.03	0.02	22-23	3.0-3.5	4.5-6.5
AH36	0.18	0.50	0.90-1.60	0.035	0.035	0.20	0.08	0.40

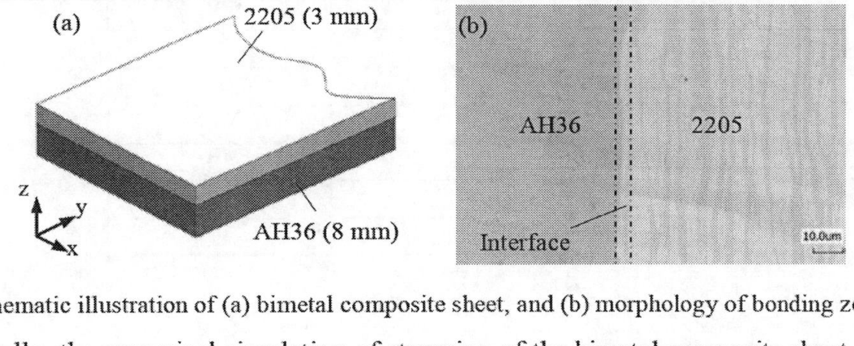

Fig. 1. Schematic illustration of (a) bimetal composite sheet, and (b) morphology of bonding zone.

Generally, the numerical simulation of stamping of the bimetal composite sheet can be conducted if the mechanical properties of its individual component layers and their thickness values were obtained. Therefore, the tensile tests on individual metal layer cut from the bimetal composite sheet by Spark Erosion Cutting (SEC) were carried out first, in order to obtain their individual mechanical properties. All uniaxial tensile tests were conducted on a universal testing machine Instron-8801 with crosshead speed of 0.5 mm/min, and the measurement of longitudinal deformation was carried out using an extensometer of gauge length 50 mm. The results of tensile tests were shown in Fig. 2, which indicated experimental stress-strain curves for the bimetal plates with the relative thickness ratio 3:8, the 2205 stainless steel layer and the AH36 carbon steel layer. It is to note that the strength of these layers in the bimetal sheet was much different.

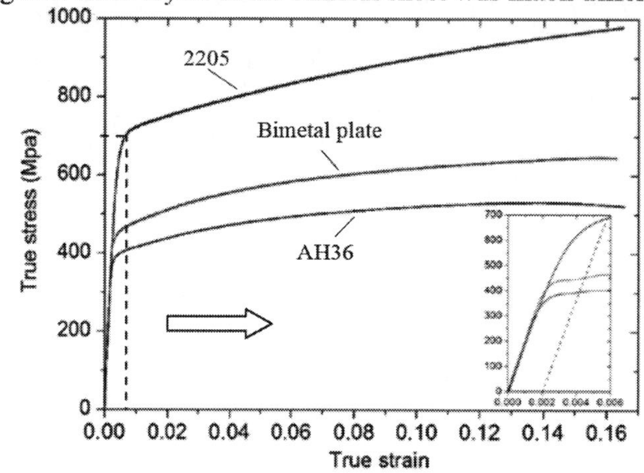

Fig. 2. Tensile properties of bimetal composite and its component layers.

3 FE simulation

The stamping was performed under two types of the sheet stacking sequence, i.e. the relative position of two layers, namely SC and CS. In the SC situation, the 2205 stainless steel layer was in contact with the punch and the AH36 carbon steel layer was positioned underneath. In the CS case, it was a reverse situation, i.e. the carbon steel of the bimetal sheet was in contact with the punch. The stamping die and the experimental procedures were illustrated in Fig. 3, and the dimensions of the major parts of the die set were shown in Table 2 as well.

Table 2. Dimensions of parts in the stamping die.

Parameters	Dimensions (mm)
Punch diameter	138
Punch profile radius [1]	14.5
Punch profile radius [2]	24.5
Die diameter	160
Die profile radius[1]	14
Die profile radius[2]	25
Clearance between punch and die	11

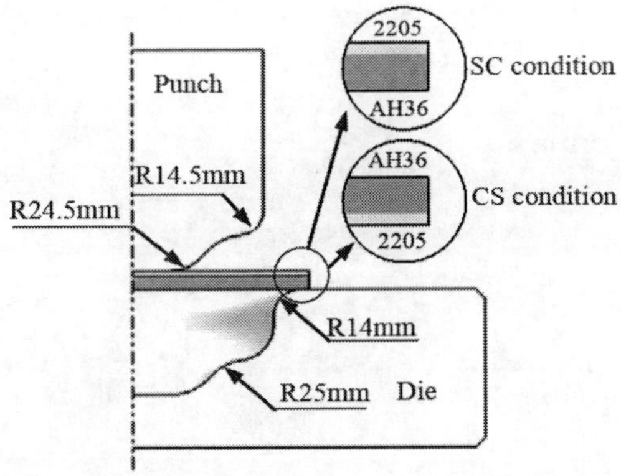

Fig. 3. Forming die set and experimental procedure.

The behaviour of the stamping of bimetal composite plate was more complicated than that of monolithic metal sheets, due to the mechanical properties of its component layers about the plastic yield stress and Young's moduli, relative thickness ratios of each layer and sheet stacking situations. Therefore, the FE simulation can be introduced to determine the situation of bimetal sheets stamping using the commercial software ABAQUS. The initial diameter of sheet model is 177 mm calculated by an inverse technique in the forming simulation software AUTOFORM, and the press speed of punch is 3.25 mm/s assuming the rate independent behaviour without considerable discrepancies. In addition, the punch stroke is 65 mm according to the initial distance between the punch and die. As the bonding interface was assumed perfect, the bonding interface was set by a coherent node connection, and the model of composite plate was built together with different material properties in two layers obtained from previous tensile tests. The material was set as homogeneous and rate-independent following the von Mises yield criterion. An eight-node linear solid element C3D8R was used to mesh the plates, and there are 13824 elements in the FE model of bimetal plate, as shown in Fig. 4. In addition, the punch and die were set as a rigid body to reduce the computational time, and simulation conditions are shown in Table 3.

Table 3. FE simulation conditions.

Items	**Parameters**
Initial sheet diameter	177 mm
Punch speed	3.25 mm/s
Punch stroke	65 mm
Element type	C3D8R
Element quantity	13824

https://doi.org/10.1051/matecconf/201818500012

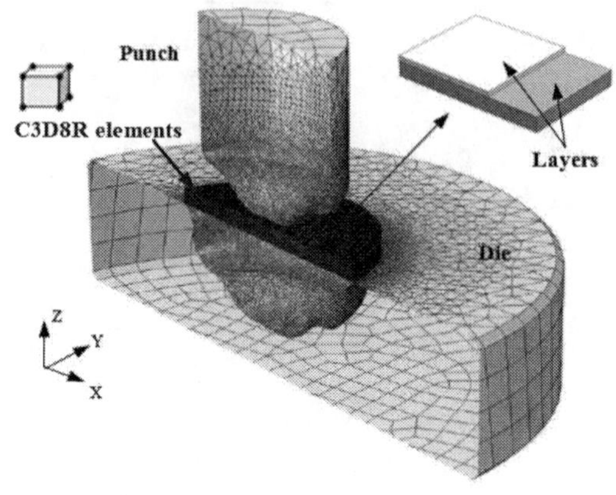

Fig. 4. FE model of bimetal plates stamping process.

4 Results and discussions

The effects of several conditions, such as the stacking sequence, relative thickness ratios and friction conditions, in the stamping process were presented and discussed. First, the layer stacking sequence and relative thickness ratios of two layers are of particular interest for the stamping behaviour of bimetal plates. To study the different influences of the plate stacking sequence and relative thickness ratios, four different simulations were preformed, i.e. the SC condition with the thickness ratios 3:8 and 1:1(total 11 mm thick), and CS condition with the thickness ratios 8:3 and 1:1. The load-stamping stroke curves were obtained from four different conditions under the dry condition with the frictional coefficient of 0.29 between tools and stamping plates, as shown in Fig. 5. There are three stages in the stamping process, i.e. the range of 0-40 mm (I stage), 40-60 mm (II stage) and >60 mm (III stage). It is noted that the load values of four different conditions were almost the same in the initial I stage, but the difference began to appear and increased in the II stage. The first change position of load for all conditions appeared when the small profile corner of the punch (point A shown in Fig. 5) contacted the deformed plates, and the second change point of force appeared when the bigger profile corner of the die (point C shown in Fig. 5) contacted the deformed plates.

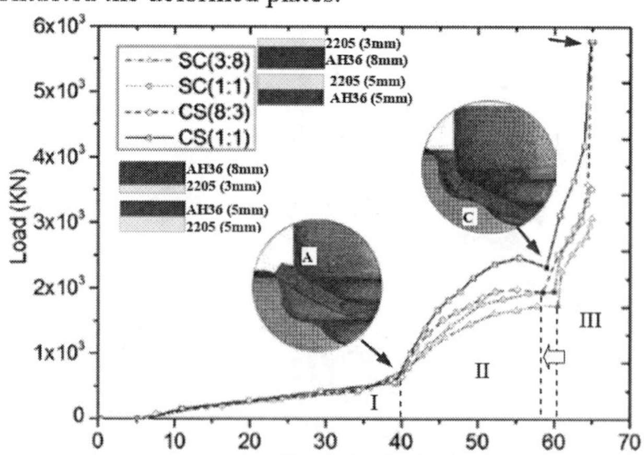

Fig. 5. Load-punch stroke curves in four different conditions.

It can be seen from the stage II of Fig. 5 that the forces of CS conditions all were larger than those of SC conditions with the same ratios of thickness and the former maximum forces were almost twice larger than that of the latter. Under the same stacking condition, the load increased with increasing the proportion of stainless steel thickness, i.e. the force of CS (1:1) condition was larger than that of CS (8:3) condition and this trend was the same for two SC conditions. In addition, it is worth noting that the beginning time of the stage III of stamping process in SC conditions was behand the time of CS conditions. The reasons were that the low carbon steel (AH36) with better plasticity was positioned in the underneath layer at large deformations resulting in the larger thickness reduction of the plate.

The friction condition between tools and stamping plates can affect the values of force and the formability of plates. Therefore, in order to observe the influence of friction, the load and stamping displacement curves in the SC condition with two previous thickness ratios were shown in Fig. 6, when the frictional coefficients were assumed to be μ=0.1 and 0.29 under lubricant and dry conditions, respectively [6]. Fig. 6 indicates that the friction coefficient did not affect the load in the initial stage I of stamping process, because the contact area was small and the contact pressure was not sufficient at this stage. However, the load decreased in the stage II for both two different ratios with a smaller friction coefficient. The effect of the friction generally can be observed apparently at large deformations in the stamping process [7].

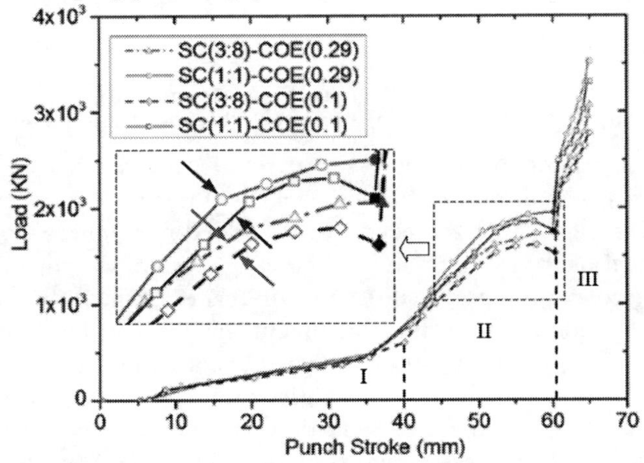

Fig. 6. Load-punch stroke curves with different friction coefficients

In addition to the load, the evaluation of material flow for bimetal plates stamping is important to reduce the defect or failure of forming products. The study of circumferential stress and thickness strain of the deformed component can determine the material flow situation and predict the occurrence of possible defects or even fracture [8, 9]. There are two areas of profile radius in the punch (A and B corners shown in Fig. 7a) and one area of profile radius (C corner shown in Fig. 7a), and the deformed plate was subjected to higher circumferential stress in some transition regions between those corners, which may cause the fracture in adjacent region. It is also shown in Fig. 7b that the circumferential stress distribution of SC condition (3:8) and μ=0.1 was zero for inside and outside layers near these areas. This trend was similar in other conditions.

 https://doi.org/10.1051/matecconf/201818500012

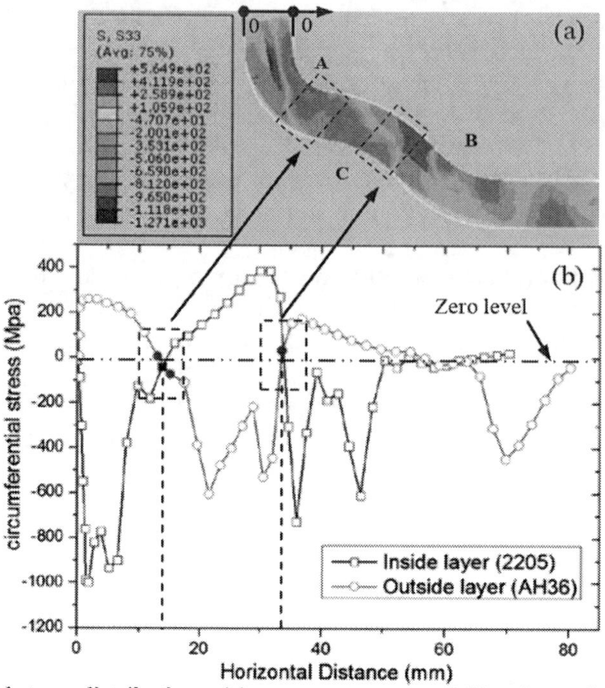

Fig. 7. Circumferential stress distributions: (a) stress contours, and (b) values of two layers.

Thickness strain values of both layers and total plates were also obtained from FE results. These values were determined in the thickness direction by measuring the thickness change of various regions from the centre of deformed component. Total and individual thickness strain distributions of SC and two friction conditions with the same thickness ratio (3:8) were shown in Fig. 8. It is shown that the small friction coefficients in comparison to dry condition made the component thickness to be uniform, i.e. the small total thickness reduction occurred, but the effect was not sufficient. The thickness of inner layer (2205) in both conditions fluctuated slightly from the centre to the edge of A corner of the punch, but increased by almost 15% in the A corner region due to the compression situation. In contrast, the thickness strain of outer layer (AH36) reached its low values in the A and B corner regions, and thus the same situation was for the total thickness strain. It was found that a lower thickness strain occurred in the underneath layer at a higher risk to fail. In addition, the negative thickness strain of total and individual layers at the centre of component was caused by the biaxial tensile stress state at this region [6].

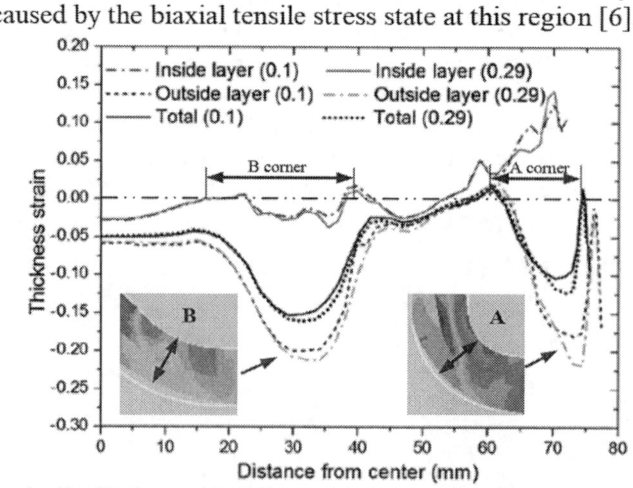

Fig. 8. Thickness strain distributions with different friction conditions.

5 Conclusions

Forming characteristics of 2205 stainless/AH36 carbon steel bimetal composite during stamping were investigated based on the FE numerical simulation. The following conclusions can be obtained:

(1) The stacking sequence of layers has an influence on the load of punch, which means that the load is higher when the strong material (2205) is positioned in the underneath layer, but its effect is rather small in the initial stamping process.

(2) It has been found that, in all conditions, the punch load increases with the increase of the proportion of 2205 stainless steel thickness. Similarly, the influence is not sufficient at the beginning of the plastic deformed stage of bimetal plates.

(3) Friction coefficients affect the load in the second stage of stamping process in SC conditions, and the load decreases for different thickness ratios with a smaller friction coefficient.

(4) Distributions of the circumferential stress and thickness strain near the punch arears of profile radius indicate that the high risk of fracture appeared in these areas.

The authors wish to acknowledge the financial support from Baosteel-Australia Joint Research and Development Centre (BAJC) under project of BA-16009. The first author is greatly thankful for the scholarship support (IPTA) from University of Wollongong and scholarship support from China Scholarship Council (CSC).

References

1. K. Yilamu, et al., J Mater Process Tech, **210**, 272-278 (2010)
2. B.X. Liu, et al., Mater Sci Eng: A, **679**, 172-182 (2017)
3. S. Kapinski, J Mater Process Tech, **60**, 197-200 (1996)
4. K.Y. Mohammad Habibi Parsaa, Norio Takakura, Int J Mech Sci, **43**, 2331-2347 (2001)
5. M.R. Morovvati, Fatemi, A., and Sadighi, M., Int J Adva Manu Tech, **54**, 113-121 (2011)
6. A. Atrian and Fereshteh-Saniee, F., Compos B: Eng, **47**, 75-81 (2013)
7. F. Fereshteh-Saniee, Pillinger, I., and Hartley, P., J Mater Process Tech, **153**, 151-156 (2004)
8. F. Dehghani and Salimi, M., In J Adva Manu Tech, **82**, 163-177 (2016)
9. J.G. Liu and Xue, W., T Nonferr Metal Soc, **23**, 964-969 (2013)

A tuned vibration absorber constituted of shape memory alloy wires for vibration reduction of platform structures: design and implementation

Yun-Ting Liao[1], *Jia-Hong* Lin[2], and *Chun-Ying* Lee[2,*]

[1]Graduate School of Information, Production and Systems, Waseda University, Kitakyushu-shi, Fukuoka 808-0135, Japan
[2]Graduate Institute of Manufacturing Technology, National Taipei University of Technology, Taipei 10608, Taiwan

Abstract. Machinery can suffer from mechanical vibrations since resonance may be generated from time-varying external excitations under different operation conditions. These detrimental vibrations may significantly influence the device's performance, effectiveness and reliability in operation. In this paper, an innovative, simple and high-efficiency tuned vibration absorber (TVA) consisting of shape memory alloy (SMA) wires, which is referred to a wire-type tuned vibration absorber (WTVA), is proposed to reduce the induced vibration. Experiments are carried out using a six-degree-of-freedom platform which is designed to simulate the frame of precision machinery in practical applications. With the equivalent stiffness of SMA wires adjusted by the controlled electric current, the frequency tunability of WTVA can be achieved. When the natural frequency of WTVA tuned in with the disturbance frequency, the experimental results demonstrate that the efficiency in vibration reduction of the platform is drastically increased even with considerable weight difference between WTVA and the platform. Moreover, the tunable frequency span also increases greatly due to the new design of WTVA and the material characteristics of SMA wires.

1 Introduction

Due to advances in industry, the demands for product precision and reliability have become more stringent. Most of the machinery is operated under the influence of rotating drivers, such as motors and fans, which will incite the oscillations from their rotational imbalances. In some cases, the excitation frequency from devices during operation matches its natural frequency and induces the so-called resonant vibration. The resonance will usually cause excess noise, and influence not only the accuracy but also the service life of machinery. In more severe cases, it may lead to rapid catastrophic failure. Hence, many experts and scholars focus much of their efforts in how to effectively avoid and reduce vibration, especially under resonance. Basically, every physical system has more than one

[*]Corresponding author:leech@ntut.edu.tw

resonant frequency. While the machinery is operated under varied rotating speeds, resonance at the different frequencies may be generated due to varied excitation frequencies. Therefore, the requirements of robust vibration reduction for machinery must be adaptive and tunable.

Traditional solutions to vibration reduction that involve structural modifications are often time consuming and expensive. Blocking the problematic frequencies in variable nature limits the operation of the system by the user. One possible solution is an installation of a dynamic vibration absorber (DVA). The DVA was first invented by Frahm [1] in 1909. Basic operation principle of DVA is to vibrate out of phase with target structure, thereby applying a counteracting force. Furthermore, there are numerous ways of reducing and preventing vibration, for example, by changing the stiffness of a structure, by increasing damping using viscoelastic materials, or by using control technology. Three types of control methods can be categorized: passive [2], active [3] [4] and semi-active. The passive vibration absorber is designed beforehand by choosing adequate stiffness and/or damping, and the vibration caused by specific natural frequency of primary structure can be reduced by attaching the DVA. Usually, the passive DVA which aims at a specific and single resonant frequency is used on the well-defined system. For instance, Heo et al. [2] developed a passive DVA with fixed parameter to reduce the vibration and noise of disk drives at a certain rotating speed. However, vibration was only suppressed at a specific disk speed. Therefore, the passive DVA cannot cope with variable systems. In order to overcome this difficulty, Chang et al. [3] proposed the active DVA to suppress and reduce the vibration by voice coil motor as the actuator. To achieve active control, the actuator plays the key role in the active DVA. The active DVA always uses an actuator (usually pneumatic or hydraulic) to generate the actually equal and opposite external force. Chang et al. [4] further designed a controller to achieve a better performance of the active DVA with tunable frequency. Although the active DVA can effectively reduce the vibration with multiple speeds, the components and actuator of the active DVA are complex and expensive.

In this paper, a semi-active DVA is considered and designed. Since the parameter of semi-active DVA can be adjusted, it is the so-called adaptively tuned vibration absorber (TVA). TVA is more efficient than passive DVA and almost as effective as active DVA [5]. Most of all, it is more stable and energy conserving. The concept of semi-active configuration is to replace the active force generators with continually adjustable elements which can vary and/or shift the rate of energy dissipation in response to instantaneous condition of motion. The most important requirement of TVA is its adjustable parameter can directly correspond to the extent of required frequency. The adjustable characteristic can be realized by altering the equivalent stiffness which directly influences the natural frequency. Since conventional materials such as steel, copper and aluminium with fixed physical properties cannot meet the requirement of adjustable nature, intelligent materials with different kinds of adjustable material properties were introduced to design or applied to the TVA in previous literatures, such as viscoelastic solids [6, 7], shape memory alloys (SMAs) [8-11], ferromagnetic shape memory materials [12], piezoelectric materials [13, 14], electro-rheological fluid[15, 16], magnetorheological fluid [17-19] and shape memory polymer [20] etc. However, most designs only employ single tuned frequency in vibration reduction. The tunable frequency span is still quite limited.

The major purpose of this paper is to design an innovative, simple and high-efficiency TVA. A TVA constituted of the SMA wires, which is referred to a wire-type TVA (WTVA), is designed to reduce the vibration of the primary structure. The extremely large and recoverable strain characteristics of the SMA imply that it can satisfy the requirement of tunable frequency via the external straining of the WTVA. The primary structure is presented as a platform with six degrees of freedom in loading deformation and employed to mimic the practical frame of machinery. Since the practical devices are mostly operated with multiple and variable speeds, more than one resonant frequency therefore is liable to be excited and further generate considerable vibration. To naturalize the vibration energy of the platform under resonance, the natural frequency of the WTVA must be tuned, by adjusting its axial force, as close as possible to the resonant frequency of platform. Moreover, the design of WTVA has to provide multiple tunable natural frequencies in order to extend its operational frequency span.

2 Tuned vibration absorber & shape memory alloy

2.1 Tuned vibration absorber

The simplest form of a TVA is an auxiliary mass that is attached to the vibrating system by a spring only. TVA is very robust in reducing vibration if a system is excited by a periodically varying force of constant frequency, especially when this frequency is very close to the resonant frequency. Theoretical formulation on the TVA can be found in many vibration textbooks [21]. Thus, derivation of its dynamic equations was not presented herein for brevity.

In this paper, the purpose is to design a WTVA with high efficiency for vibration reduction, extensively tunable frequency span with simple and compact configuration. Figure 1 illustrates the schematic diagram of the proposed WTVA. This WTVA consists of a spring element made of two Ni-Ti SMA wires and an auxiliary mass placed at the mid-span. When the auxiliary mass moves in lateral direction, the equivalent stiffness can be varied by adjusting the axial force of the wires. The adjustment screw, which moves the slider platform and the movable end of the SMA wire, provides the first static pretension setting. Moreover, the power supply provides the control electric current to the SMA wires and generates an additional tension by heating. Heating SMA wires controls the dynamically tunable tension and subsequently the stiffness of TVA.

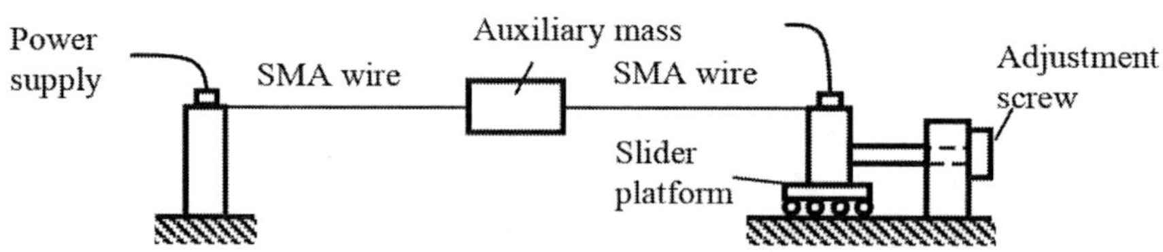

Fig. 1. Schematic diagram of the proposed WTVA.

2.2 Mathematical modelling of the WTVA

To derive the natural frequency which can be tuned by controlling axial force, a mathematical formulation is built to facilitate design procedure. Since their weights are considerably negligible comparing with auxiliary mass, the SMA wires are assumed to be weightless. Moreover, deformation of the auxiliary mass is much smaller than that of wires under loading. Only the wire's stiffness is considered in the overall stiffness of WTVA. WTVA can therefore be assumed to be a single degree of freedom system on a plane. Figure 2 shows the free-body diagram (FBD) of wire under oscillation. The SMA wires are assumed to be uniform and the tension P in the wires is constant. $w(x, t)$ represents the transverse displacement of the wire. ρ and θ, respectively, represent the mass per unit length and the slope of the deflected wire. Taking the equation of motion in lateral axis, the dynamic differential equation is derived as

$$\frac{\partial}{\partial x}(P\frac{\partial w(x,t)}{\partial x})+f(x,t)=\rho\frac{\partial^2 w(x,t)}{\partial t^2} \tag{1}$$

If the external force $f(x, t) = 0$, then the free vibration equation could be obtained as

$$c^2\frac{\partial^2 w}{\partial x^2}-\frac{\partial^2 w}{\partial t^2}=0\ \text{, where } c=\sqrt{\frac{P}{\rho}} \tag{2}$$

The free vibration equation, Eq. (2), can be solved by method of separation of variable. The solution is written as the product of functions $W(x)$ and $T(t)$, which depend on coordinate x and time t, respectively:

$$w(x,t)=W(x)T(t) \tag{3}$$

By solving the separated second-order ordinary differential equations, these functions are given by

$$W(x)=A\cos\left(\frac{\omega x}{c}\right)+B\sin\left(\frac{\omega x}{c}\right) \tag{4}$$

$$T(t)=C\cos(\omega t)+D\sin(\omega t) \tag{5}$$

For the proposed WTVA, the wires are fixed at both ends, so one of the boundary conditions is $W(0) = 0$. The general solution of Eq. (3) can be written as

$$w(x,t)=B\sin\frac{\omega x}{c}(C\cos\omega t+D\sin\omega t) \tag{6}$$

Corresponding to the position of auxiliary mass, which is at $x = L/2$ (L represents the total length of wires) with the mass m, the boundary condition should be satisfied as

$$m\left.\frac{\partial^2 w}{\partial t^2}\right|_{x=\frac{L}{2}}=2P\left.\frac{\partial w}{\partial x}\right|_{x=\frac{L}{2}} \tag{7}$$

Substituting Eq. (6) into Eq. (7), we can find

$$\omega\tan\frac{\omega L}{2c}=-\frac{2P}{mc} \tag{8}$$

Thus, the relationship between axial force P and natural frequency ω of WTVA is obtained. The related parameters in Eq. (8) are the mass of auxiliary mass m, the total length L and mass per unit length ρ of the SMA wires, which will be first defined when designing the WTVA. Hence, Eq. (8) provides the design formulation that the tunable frequency can be realized by adjusting the axial force P.

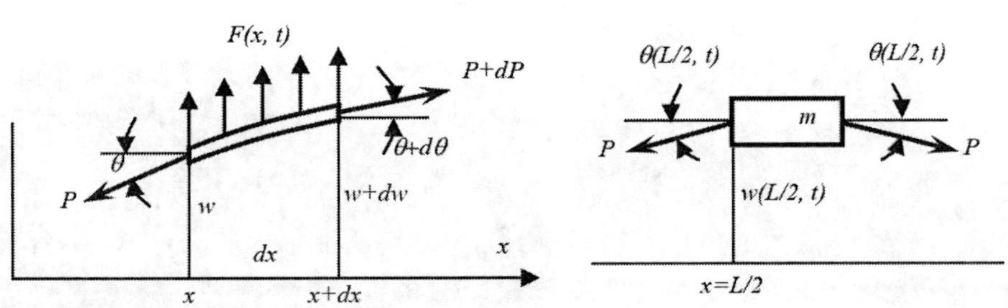

Fig. 2. Free body diagrams of the SMA wire and the auxillary mass in the WTVA under external loading.

2.2 Shape memory alloy

SMA is a type of intelligent material which has been shown to exhibit extremely large, recoverable strains (around 10%) and it is these temperature- and/or stress-controlled properties which allow SMA to be utilized in many exciting and innovative engineering applications. In addition, SMA actuators also have attractive features such as high power to weight ratio, silence operation, no electromagnetic interference, and compact in size [22]. Two main types of SMAs are copper-aluminium-nickel and nickel-titanium (NiTi) alloys. NiTi-based SMAs are much more preferable for most applications because of their better mechanical properties compared with other existing SMAs [23]. Rustighi et al. [8] tuned the natural frequency by employing the tunable elastic modulus (stiffness) of Ni-Ti SMA with temperature. A change in the tunable frequency of 21.4% with 120 seconds of response time was obtained. However, 120 seconds of response time is only suitable for applications where the forcing frequency changes slowly with time, i.e. its practical application is limited. In spite of the attractive material property of SMA, several types of TVA consisting of other shape memory materials are proposed. For example, Lee et al. [20] studied the TVA which employs the hybrid shape memory materials in cantilevered configuration. By tuning the stiffness of the proposed TVA, more than 45% of tunable frequency span is obtained, and approximately 50% of vibration reduction is achieved with 70 seconds of response time. As shown in previous researches [8, 9, 20], the characteristics such as efficiency of vibration reduction, tunable frequency span and response time are still inferior for certain applications. Even in the active DVA which is proposed by Chang et al. [4], the vibration reduction is only improved by approximately 50% with tunable frequency. Thus, using different configuration design to improve the TVA performance is attempted in this study.

3 System descriptions

3.1 The dynamic characteristics of platform

The platform which is designed to simulate the practical frame for machinery is shown in Fig. 3. Under operation of the machinery, vibration will be generated due to slight imbalance in the rotor of the motor. An AC servo motor is used on the platform to excite the platform by adjusting the speed of motor. For example, the motor can drive a spindle in various speeds to satisfy the different machining needs. A pedestal is designed to place the WTVA. One side of the WTVA is fixed on a movable holder in order to adjust the initial pretension of the WTVA by screw. The other side of the WTVA is fixed on the fixed holder.

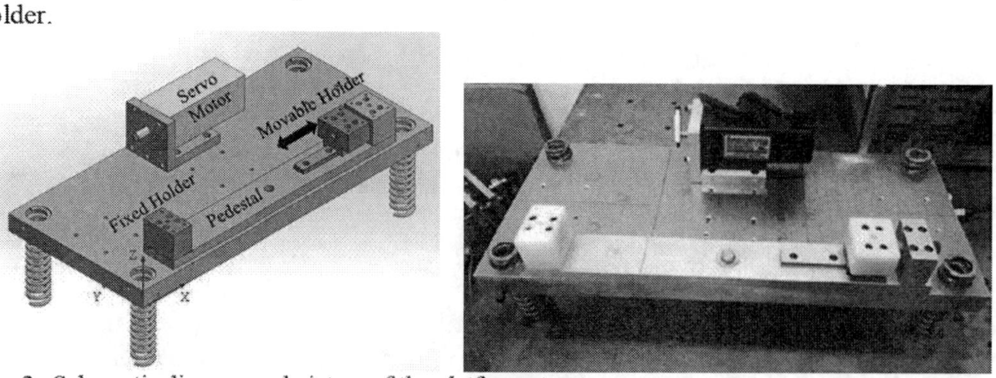

Fig. 3. Schematic diagram and picture of the platform structure.

Before designing the parameters of the WTVA, the dynamic properties of the target platform must be confirmed by simulation and experimental verification. The simulation model is established and simulated by SolidWorks Simulation® in advanced. Moreover, the simulation results are verified by experiment. The experimental confirmation utilizes an impact hammer to perform the impulse modal testing. The comparison between simulation and experimental results is shown in Table 1. The modal shapes and corresponding natural frequencies of the first six modes of the platform are presented. Considering there are still a lot of details simplified in simulation model, the simulation error is reasonable and acceptable. In this paper, the speed of motor is deliberately regulated to stimulate the resonant frequency and imitate the practical device which suffers the resonance.

Table 1. Natural frequencies and corresponding modal shapes of the platform.

Mode	Natural Frequency (Hz) SolidWorks Simulation®	Natural Frequency (Hz) Experimental Result	Error(%)	Modal Shape
1st	18.16	17.25	+5.28	*y*-axis translation
2nd	18.44	18.25	+1.04	*x*-axis translation
3rd	29.76	30.25	-1.62	*z*-axis rotation
4th	29.91	31.50	-5.05	*z*-axis translation
5th	36.71	42.25	-13.1	*x*-axis rotation
6th	50.52	51.50	-1.90	*y*-axis rotation

According to the configuration of the pedestal chosen for placing the WTVA, the controllable modes of the platform are 1st, 4th and 5th modes. The other modes are not included for vibration reduction demonstration in this paper. However, they are easy to implement. For instance, the WTVA just needs to be replaced along the Y-axis for controlling other modes.

3.2 Design of the WTVA

For the design of the proposed WTVA, its specific parameters, described in Eq. (8), should be determined according to the natural frequencies of the target platform. The diameter (d), total length (L) and density per unit length (ρ) of Ni-Ti SMA wire are 0.6 mm, 90 mm and 1.84E-6 kg/mm, respectively. The mass of rectangular-shaped auxiliary mass (m) is 0.054 kg, which is in the order of hundredth of the platform. In order to measure the magnitude of axial force which is generated from SMA wires, two strain gages are attached on opposite sides of the rectangular auxiliary mass block. By measuring the strain of the mass block under loading, the corresponding axial force generated from the SMA wires can be directly obtained.

To control the tunable frequency of WTVA, the operating procedures are explained in the following. At the first step, a pretension is exerted on the movable holder to deform the SMA wires by tightening the adjustment screw, as shown in Fig. 1. After exerting the initial pretension, the movable holder is fixed. The deformed WTVA is then heated by applying electric current, so that the SMA wires are prone to return to their original length because of shape memory effect. Since both ends are fixed, the SMA wires are restrained to return to their original lengths. Therefore, a further internal axial force is generated, attaining a large controllable and tunable frequency capability.

Before being installed on the platform, the WTVA must be first characterized to determine the relationship among pretension, control electric current and corresponding natural frequency. Figure 4 presents the natural frequency of the WTVA under applied electric current and various pretensions. By adjusting the pretension, the working frequency span of WTVA can be roughly adjusted to cover the target frequency. It is seen in Fig. 4(a) that the formulation predicts the natural frequency of the WTVA with good accuracy. With the control of electric current, the tunable frequency span of WTVA is found to be approximately from 15 Hz to 45 Hz, as presented in Fig. 4(b), which represents a change of about 200% from the lowest frequency. Moreover, the little difference between the heating and cooling control demonstrates insignificant nonlinear effect and makes the control simpler. Compared with previous researches [8, 19, 20], the tunable frequency span is drastically improved due to the design of this WTVA.

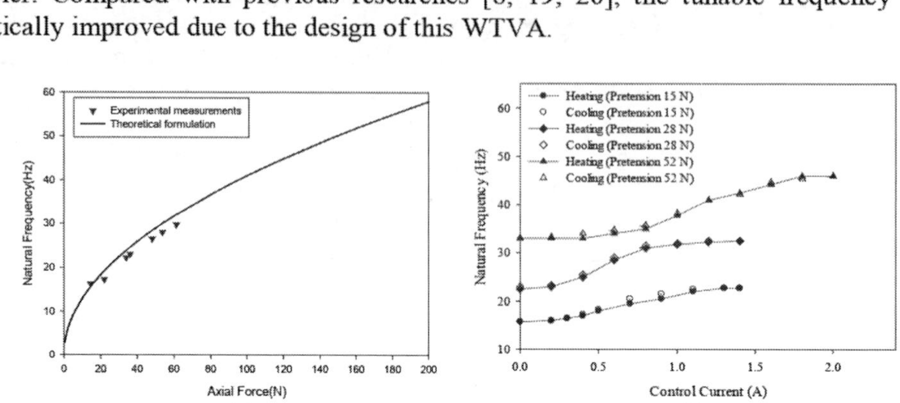

(a) (b)

Fig. 4. The natural frequency of the WTVA under different applied pretensions: (a) natural frequency controlled by applied axial force, (b) natural frequency further controlled by applied electric current.

4 Experimental results and discussion

4.1 Experimental setup

WTVA is installed onto the platform as shown in Fig. 5. The DC power supply provides the control electric current to SMA wires in order to precisely adjust the natural frequency of the WTVA. The AC servo motor can excite the platform in different vibration frequencies by controlling its speed. A tri-axial accelerometer is utilized to measure the vibration and the effectiveness of vibration reduction can be evaluated. Both signals from tri-axial accelerometer and DC power supply are recorded and further processed by the dynamic signal analyser. According to its installed location, the WTVA can be employed to control the 1st mode (17.25 Hz), the 4th mode (31.5 Hz) and 5th mode (42.25 Hz) of the platform.

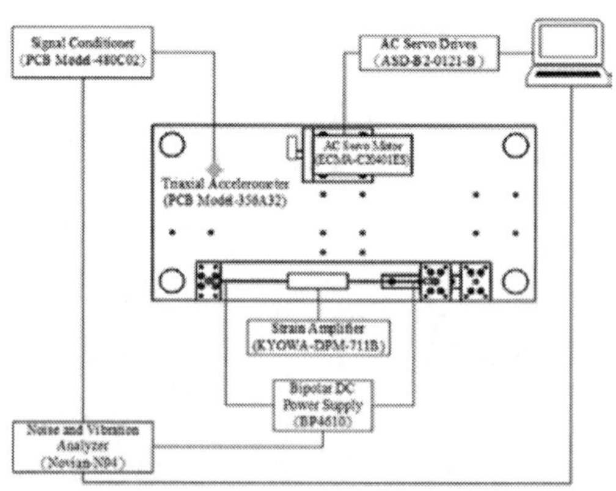

Fig. 5. Experimental setup for using WTVA in vibration control of a platform structure.

4.2 Performance of vibration reduction

For vibration reduction of the platform's 1st mode, the speed of the motor is first adjusted to reach the resonant frequency of 17.25 Hz (1035 rpm). A 0.4-A electric current with 15 N of pretension is the closest to the driving frequency of the platform, as shown in Fig. 4. The measured vibration result is shown in Fig. 6. The power supply is adjusted from 0 A to 0.4 A at 3 s after the beginning of data recording for the heating situation shown in Fig. 6(a). It is seen that the amplitude of the platform reaches the steady state after 30 s. As the results before and after control show, the vibration reduction rate reaches approximately 84%. Apart from the heating procedure, the response time of cooling procedure is also considered since the rotating speed of the devices could alter in the practical application. To demonstrate the control effectiveness during cooling process, the current of the WTVA is initially adjusted to 1.3 A to offset from its resonant frequency. Afterwards, the control current on the WTVA is adjusted from 1.3 A to 0.4 A at same instant as heating example. The result presented in Fig. 6(b) shows the effect of vibration reduction can achieve around 77.6%.

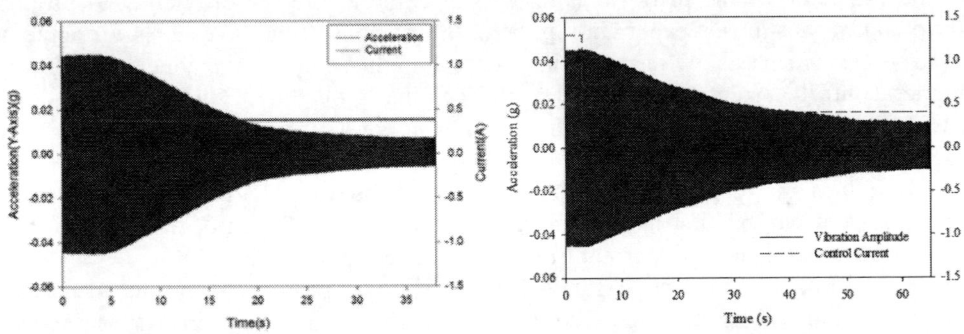

(a) (b)

Fig. 6. Experimental results of 1st mode vibration reduction at 1035-rpm speed: (a) heating, (b) cooling.

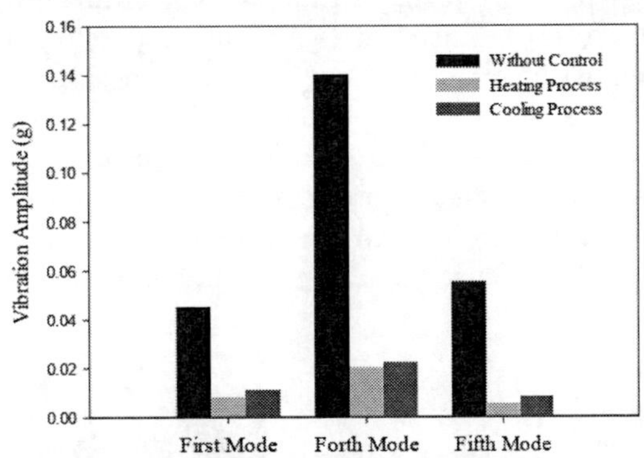

Fig. 7. Experimental results for vibration reduction of the platform.

The vibration control for the 4th and 5th modes of the platform with WTVA is also investigated. It should be noted that different modes have different natural frequencies and modal shapes. Nevertheless, the steps are same as those in the experiment for the first mode. Only the motor speed, pretension and electric current of the WTVA are different. The pretensions for controlling the 4th and 5th modes are respectively adjusted to 28N and 52N at the beginning. The electric currents for precisely tuning the natural frequency of WTVA are 0.8 A and 1.3 A, respectively. The speeds of AC servo motor are separately tuned to1890 rpm (31.5 Hz) and 2535 rpm (42.25 Hz) to excite the resonance of platform. Fig. 7 summarizes the measured results to compare the performance of vibration reduction. The vibration amplitude of the platform can greatly reduce up to 90% even with this WTVA having considerably small weight fraction comparing with platform. The results demonstrate that the proposed WTVA not only can considerably increase the capability of vibration reduction compared with previous research, but also extend the controllable frequency span for practical application.

4.3 Improvement for steady-state time response

A long response time of the WTVA can influence its practical applications. Generally, vibration control requires timely response. It is evident that the response time of steady-state during cooling process is usually longer than heating, since the cooling speed of the SMA wires depends on the relative temperature between SMA wires and surrounding under natural convection. For time-varying systems, the control robustness is improved by decreasing the response time of the systems.

The 1st mode of the platform is chosen to demonstrate the efficiency of proposed speed-up approaches in this paper. In Fig. 8(a), the response time takes 27 s for achieving the steady-state which is slow response in practical application. As the concept for heating, the higher control current can accelerate the heating procedure. Therefore, the control current for heating process is doubled from 0.4 Ampere to 0.8A, and kept constant for 3 seconds. The result is showed in Fig. 10. The response time is reduced to 5 seconds, which is around 1/6th of the response time compared with Fig. 8(a). The result also reveals that the larger current can be first used to accelerate the heating response, then adjust to and keep the corresponding control current of natural frequency of the WTVA.

The approach which is employed to accelerate response time in the cooling procedure is to apply a fan to forcibly cool down the SMA wires. In a similar fashion to the experimental processes, the result obviously shows that the response time can be greatly reduced by force-cooling device, as shown in Fig. 8 Compared with the response time without forced cooling, and the response time is shortened by 50%.

In this paper, the natural frequency of the WTVA is tuned by manually controlling the electrical current. In order to increase the efficiency of the vibration reduction, an adaptive control system must be employed since the natural frequency of the WTVA must be as close as possible to the target's. Adaptive control system could not only increase the efficiency, but also appropriately adjust the heating current to improve the time response. Moreover, the vibration energy of the platform is absorbed to the WTVA. Meanwhile, the vibration amplitude of WTVA becomes extremely huge. Therefore, the vibration energy of the WTVA can be harvested and reused by design of an energy harvesting device.

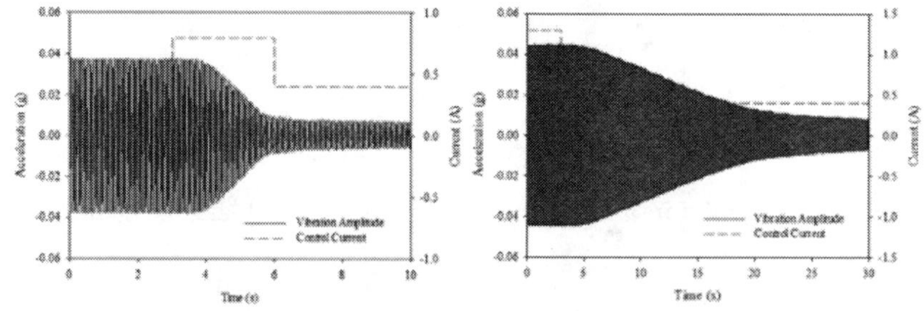

(a) (b)

Fig. 8. The results of improved steady-state time response for 1st mode of platform: (a) with the regulation of electric current in heating procedure, (b) with forced cooling device in cooling procedure.

5 Conclusions

This paper proposes the feasibility of the wire-type vibration absorber (WTVA) with simple and tiny architecture, which consists of the shape memory alloy (SMA) wires and the auxiliary mass with strain gages. The tunable span of the WTVA is found to be approximately from 15 to 45 Hz, which represents a change from the lowest frequency of about 200%. The frequency variation is greatly increased with considerable weight difference compared with previous research. In addition, the experimental results confirm that the vibration amplitude of platform is effectively reduced, while natural frequency of the WTVA is adjusted to track the specific frequency of the platform. The maximum efficiency of vibration reduction achieves around 90%.

In this paper, the approaches for shortening response time are also proposed for practical application. In the heating procedure, the response time is greatly reduced by increasing the heating control current at beginning, since more power input can accelerate the thermal heating to achieve quickly the target temperature of the WTVA. As the result, the response time in heating procedure is improved. In the cooling procedure, the force-cooling device is utilized to accelerate the cooling speed of the WTVA. The results prove that the proposed approaches can reduce response time of the WTVA effectively.

References

1. H. Frahm, *Device for damping vibrations of bodies*, US Patent No.989, 958 (1909).
2. J.W. Heo, J. Chung, J.M. Park, IEEE T Consum Electr, **48**, 874-878 (2002)
3. C.S. Chang, T.S. Liu, Jpn J Appl Phys, **45**, 1120-1123 (2006)
4. C.S. Chang, T.S. Liu, IEEE T Magn, **43**, 799-801 (2007)
5. A.E. Mansour, R.C. Ertekin, ISSC 2003-15th International Ship and Offshore Structures Congress, **1**, 242-246 (2003)
6. R. Fosdick, Y. Ketema, J Appl Mech, **65**, 17-24 (1998)
7. Y. Ketema, J Sound Vib, **216**, 133-145 (1998)
8. E. Rustighi, M.J. Brennan, B.R. Mace, Smart Mater Struct, **14**, 19-28 (2004)
9. K. Williams, G. Chiu, R. Bernhard, J Sound Vib, **249**, 835-848 (2002)
10. B. Tiseo, A. Concilio, S. Ameduri, A. Gianvito, J Theor Appl Mech, **48**, 135 (2010)
11. M.A. Savi, A.S. dePaula, D.C. Lagoudas, J Intel Mat Syst Str, **22**, 67-80 (2011)
12. N.N. Sarawate, M.J. Dapino, J Intel Mat Syst Str, **20**, 1625-1634 (2009)
13. N.W. Hagood, A. vonFlotow, J Sound Vib, **146**, 243-268 (1991)
14. P. Bonello, M. J. Brennan, S.J. Elliott, Smart Mater Struct, **14**, 1055-1065 (2005)
15. S.B. Choi, Y.-K. Park, J Sound Vib, **172**, 428-432 (1994)
16. C. Hirunyapruk, M.J. Brennan, B.R. Mace, W. Li, Smart Mater Struct, **19**, 1-10 (2010)
17. Y. Sun, M. Thomas, J Vib Control, **17**, 1253-1264 (2011)
18. H.X. Deng, X.L. Gong, J Intel Mat Syst Str, **18**, 1205-1210 (2007)
19. J.-H. Koo, M. Ahmadian, J Intel Mat Syst Str, **18**, 1137-1142 (2007)
20. C.Y. Lee, C.C. Chen, T.H. Yang, C.J. Lin, J Intel Mat Syst Str, **23**, 1725-1734 (2015)
21. J.P. DenHartog, *Mechanical Vibration*, Dover Pub., 1985
22. M. Follador, M. Cianchettim, A. Arienti, C. Laschi, Smart Mater Struct, **21**, 1-10 (2012)
23. J.M. Jani, M. Leary, A. Subic, M.A. Gibson, Mater Design, **56**, 1078-1113 (2014)

Analyses of contact forces and kinetic motion on heavy load ball-screw

Chin-Chung Wei[1,*] and *Wen-Hsien* Kao[2]

[1]Department of Power Mechanical Engineering, National Formosa University, No.64 Wenhui Rd. Huiwei Township, Yunlin County, 63205, Taiwan R.O.C.
[2]Institute of Mechatronoptic Systems, Chienkuo Technology University, Changhua, No. 1, Chieh Shou N Rd., Chang Hua 500, Taiwan R.O.C.

Abstract. Effects of contact angle and groove factor of a heavy load ball-screw are discussed with the variation of contact forces at eight ball circulations. Contact forces are varying as a sinusoidal function of each circulation owing to the variation of phase angle. With the increase of contact angles, contact forces at each ball circulation are decreased and variation in each ball circulation. The decrease of the contact forces means that the contact stresses of contact areas are reduced. Fatigue life of raceways can thus be extended. Low groove factor can reduce skidding speed and friction coefficient. By the analyzing results, optimal transmission performance can be achieved in a heavy-loaded ball-screw.

1 Introduction

Heavy load ball-screw is wieldy used in plastic injecting mold machine, applying in injecting and load-lock axes. Non-preload is applied in a heavy load ball-screw. Operating conditions are low rotational speed and extremely high axial load, contact pressure on a contact area of ball-raceway can reach 1.8 GPa. Therefore, contact forces on each circulation of balls will not be equal owing to the contact mechanism. This will lead to serious wear occurred at the ball circulation which is nearby the flange of nut. The aim of the study is established a kinematic simulation model of multi-cycles ball-screw without preload in order to estimate contact forces on each circulation. This can help in the heavy loaded ball-screw design on uniform contact forces. Finite element model is made by a commercial software, Mash-Part, as shown in the Fig.1. Effective balls are 128 and set in the raceways, contact behavior is set as Hertz contact situation. An axial load is applied to a side of the flange of the nut.

The kinetic study of a ball-screw is started by Lin et al.[1]. They established fundamental coordinates to describe the motion of a ball's center, which is revoluted around a screw. Contact and kinetic behaviors among ball, and raceways of nut and screw are thus be obtained for discussion of geometrical parameters. Wei and Lin[2] considered effects of the

*Corresponding author: ccwei@nfu.edu.tw

centrifugal force of the ball and lubrication at contact areas. Kinetic motion and contact behavior were introduced by a preloaded two-circulation ball-screw numerical model[3]. Sliding velocity and contact force will bring the wear occurred at the contact area between ball and raceway. A two body wear model considering the roughness effect was introduced in the ball-screw contact mechanism [4], the density of roughness peak and sliding speed are major factors to the wear of a preloaded ball-screw using in the normal operating conditions. Sliding roll ratio and position of pure rolling point are major factors to the kinetic motion and wear [5,6]. Contact mechanism and kinetic motion of a heavy load ball-screw are different to a normal situation. Non-preload is applied, and extreme high contact force has existed on the contact area. The study is present a theoretical model on the analyses of contact forces and kinetic motion of a heavy load ball-screw. A commercial software Mesh-Part is also used in the calculating of contact forces and kinetic behavior related to the contact angle and groove factor are discussed in the presented study.

2 Analysing methodology

The center position of any ball o′ in a ball-screw can be determined by an absolute coordinate (x,y,z), rotational coordinate (x′,y′,z′), and moving coordinate (t,n,b) as shown in the Fig.1. Contact coordinates (Xo, Yo, Zo) and (Xi, Yi, Zi) located on the contact ellipses at ball-nut and ball-screw contact areas, respectively. α_o and α_i are contact angles of ball-nut and ball-screw, respectively.

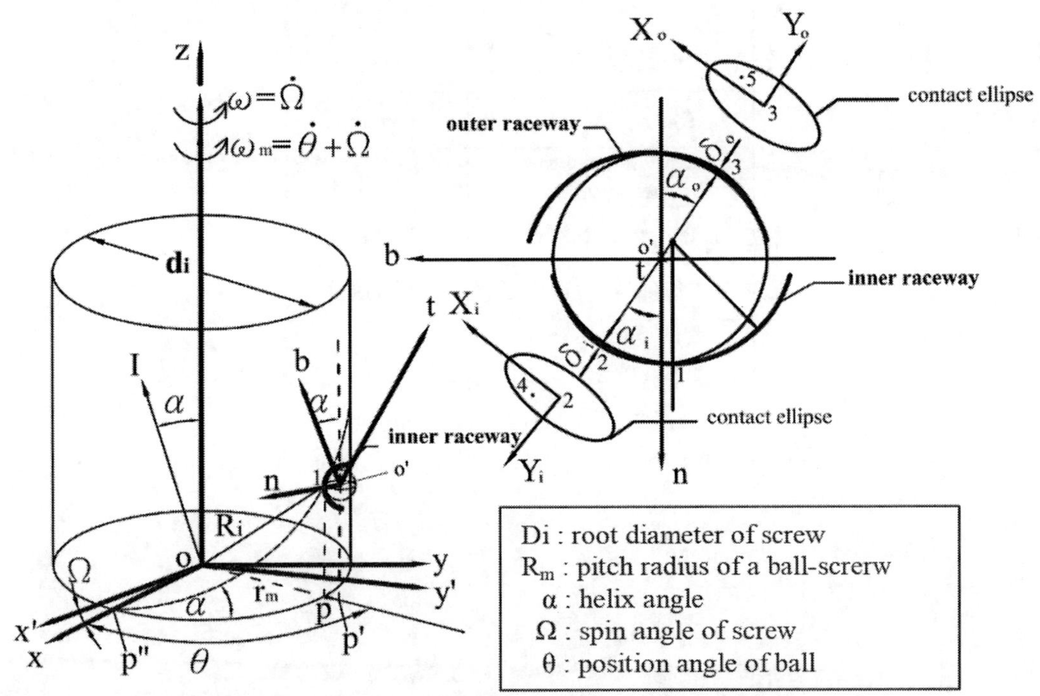

Fig. 1. Absolute, rotational and moving coordinates for a screw and any ball.

A finite element model has been established by a commercial software Mash Part, as shown in the Fig.2. Eight circulations are shown in the Fig. as L1, L2, L3, L4, R1, R2, R3, and R4. The axial load is applied to the top side of the nut, so the force is translated from the up nut to the down nut. Numerical parameters of a heavy load ball-screw are listed in the Tab.1.

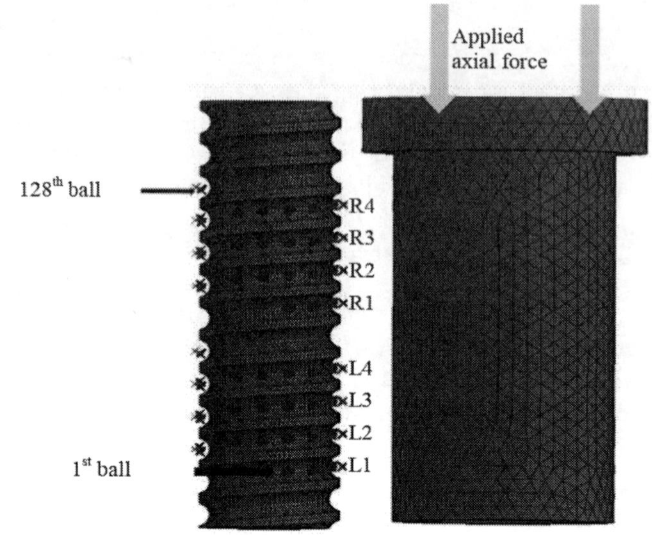

Fig. 2. FEM model for a double nut ball-screw.

Table 1. Parameters of a double-nut 8 circulations heavy load ball-screw.

Parameters	Value
Lead, L	25 mm
Dia. of a ball, D_b	19.05 mm
Dia. of pitch, D_m	106 mm
Curvature of the raceway, $r_{o,i}$	10.3 mm
Effective ball number	128
Axial load	236 kN
Rotational speed of screw	554 rpm

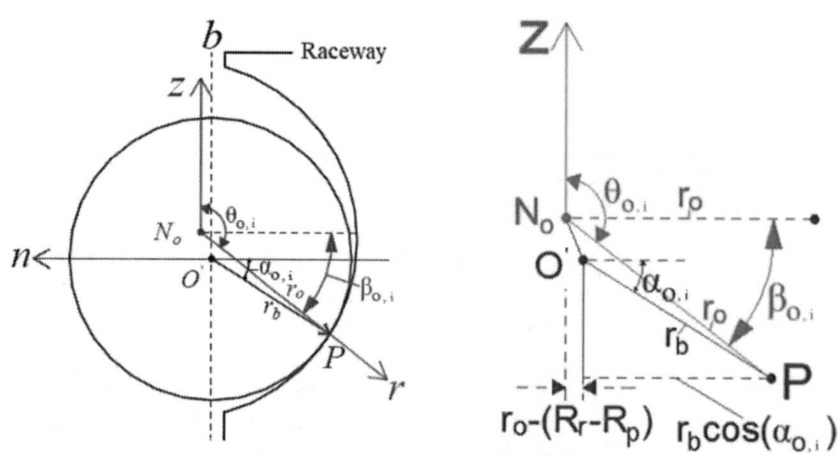

Fig. 3. Geometry of contact area.

Contact geometry is shown in the Fig.3. The contact area is varying with the variation of operating conditions. While setting the boundary on the FEM model, the contact area is assumed as a contact stripe [4]. In the Fig.3, the center of the raceway is No, and the contact distribution of angle is defined as $\theta_{o,i}$, which is written as

$$\theta_{o,i} = cos^{-1}\left\{\frac{(r_b \times cos\alpha_{o,i}) + [r_{o,i} - (R_r - R_p)]}{r_{o,r}}\right\} \pm \frac{\pi}{2} \tag{1}$$

where, the subscript o and i denote the ball-nut and ball-screw contact sides. r_b is the radius of a ball. $r_{o,i}$ is the radius of the raceway of the nut and screw, respectively. R_r and R_p are radii of screw's root and pitch, respectively. $\alpha_{o,i}$ is the contact angles of the raceway of the nut and screw, respectively.

By the past study of Wei[2], the orbital and spinning speeds of a ball along the screw, ω_m and ω_R are written as:

$$\omega_m = \frac{\omega}{1 - \frac{(1+\gamma' C_{\alpha_o})(C_{\alpha_i} + S_{\alpha_i})}{(1-\gamma' C_{\alpha_i})(C_{\alpha_o} + S_{\alpha_o})}} \tag{2}$$

$$\omega_R = \frac{-\omega_m(1+\gamma' C_{\alpha_o})\sqrt{2}C_\alpha}{\gamma'(C_{\alpha_o} + S_{\alpha_o})} \tag{3}$$

where, $\gamma' = \frac{r_b}{r_m}$, ω is the rotational speed of the screw.

The value of the sliding speed between ball and raceway can be separated as tangential direction (t-axis direction) and vertical direction (X-direction), respectively. The t-direction is parallel to the moving direction of ball center. The X-direction is orthogonal to the t-direction. These two sliding speeds can be written as :

$$V_{St} = r_b(\omega_t - \omega S_\alpha) \tag{4}$$

$$V_{SX} = d(\omega_m - \omega) - r_b\left[(\omega_b - \omega C_\alpha)C_{\alpha_i} - \omega_n S_{\alpha_i}\right] \tag{5}$$

The partial speed of ω_R at t, n and b directions are written as

$\omega_t = \omega_R C_\beta S_{\beta'}$; $\omega_n = -\omega_R S_\beta$ and $\omega_b = -\omega_R C_\beta C_{\beta'}$, the angle β and β^' are spinning angles at orthogonal directions, respectively. The sliding speed at the X-direction is due to the spinning and yawing motion of a ball. It means the skidding motion between ball and raceway.

3 Results and discussions

A heavy load ball-screw is applied an extremely high axial load on the nut. Each circulation balls are carried high contact force, and contact forces are greatly increased with the increase of ball numbers, as shown in Fig.4. The black line separates right and left nuts, contact forces at the left nut are much lower than the right nut. Contact forces are varying as a sinusoidal curve due to the effect of the phase angle of each ball is different. The increase of contact angles can decrease the contact forces.

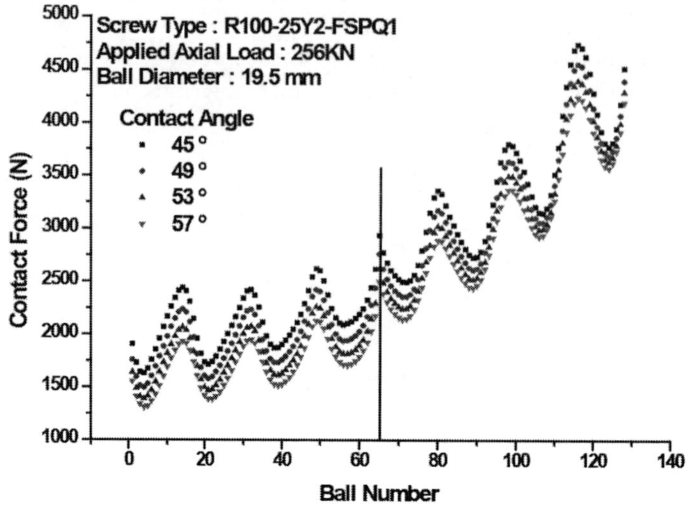

Fig. 4. Contact forces distribution V.S. contact angles.

Contact forces are averaged at each circulation and shown in the Fig. 5. The difference of contact forces at each contact angle condition is similar, but the latest circulation is not the same. This is due to the applied axial load is acted on the right end of the nut, the latest circulation is the first part to bear the force and then decay with the circulations. Therefore, the right nut bears much higher contact forces than the left nut. The 7th and 8th circulations, R3 and R4, thus bear similar contact forces. In the figure, the increase of contact angle can effectively reduce the contact force at each circulation. Because higher contact angle can let the radial part contact force become smaller. But too large contact angle will greatly increase the contact stress of top raceway, it will lead to fatigue failure happened on it.

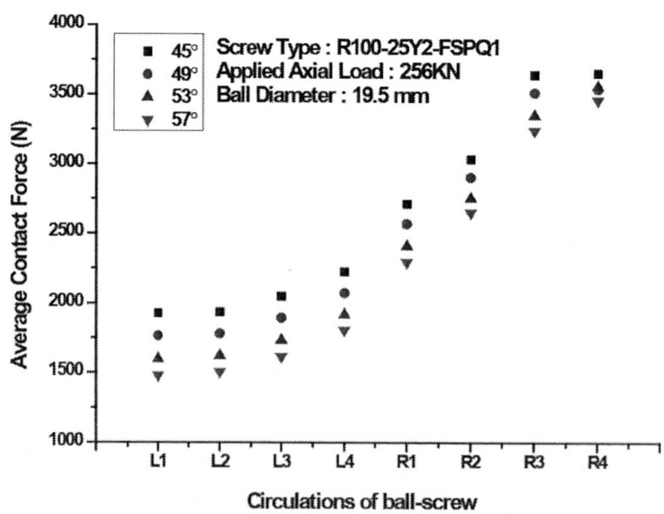

Fig. 5. Average contact forces vary with contact angles and circulations.

The design of groove factor is very important to the contact mechanism of a heavy load ball-screw. The groove factor is well known as the ratio of the radius of the raceway to the diameter of a ball. When the groove factor is near to 0.5, the contact area between ball and raceway has the largest value. The contact angle is also varied with the variation of the groove factor, as shown in the Fig.6. The decrease of the groove factor increases the contact angle and reduces the contact forces, especially at contact areas of the left nut.

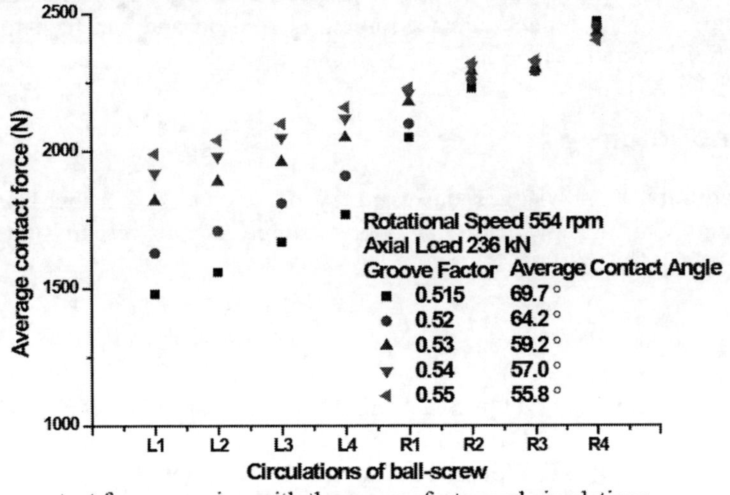

Fig.6. Average contact forces varying with the groove factor and circulations.

Friction coefficient increases with the increase of the groove factors due to the sliding speed are also increased. It's interesting in the increases of sliding speeds along t- and X-direction is different. The value of sliding speed at t-direction is greater than X-direction, but the rising slope is not so large. The rising of sliding speed at X-direction with the increase of the groove factor means the skidding motion on contact area becomes serious. So the friction coefficient is also rising with the increase of the groove factor.

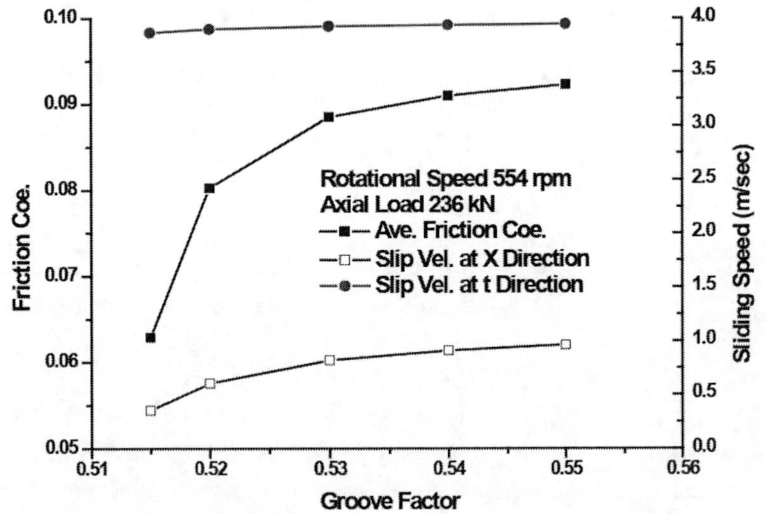

Fig.7. Frition coefficient varies with the groove factor and sliding speeds.

4 Conclusions

The study uses a commercial software Mash-Part and creates a theoretical method for analyzing contact and kinetic behavior to a heavy load ball-screw under different raceway geometry conditions. The contact angle and groove factor were discussed for the lowest contact forces to multi-circulations. Increasing the contact angle or reducing the groove factor can effectively decrease contact forces of each circulation. The friction coefficient is dominated by the sliding speed at X-direction, which is along the raceway and vertical to the moving direction of ball centre. The speed means the skidding motion between ball and raceway. Lower groove factor has lower sliding speed and friction coefficient. The aim of the study indicates how to reduce contact force and friction and can help to design high transmission efficiency and fatigue life heavy-load ball-screw.

5 Acknowledgment

The author gratefully acknowledges the Ministry of science and technology (MOST) and Hiwin Technology Co. in Taiwan, R.O.C., whose supported this research under grant Most-106-2262-E-150-001-CC2.

References

1. M.C. Lin, B. Ravani, S. A. Velinsky, ASME J. Mech. Des, 116, 849 (1994)
2. C.C. Wei, J.F. Lin, J. Mech. Des,125(4), 717 (2004)
3. C.C. Wei, J.F. Lin, and J. H. Horng, Tri. Int., 42, 1816 (2009)
4. C.C. Wei, R.S. Lai, Mech. and Machine Theory, 46, 880 (2011)
5. C.C. Wei, W.L. Liou, and R. S. Lai, Wear, 292-293, 111 (2012)
6. J.Z. Hu, M. Wang, T. Zan, Mech. and Machine Theory, 79, 158(2014)

Based on FP LDs with split band technique in two-channel fiber optical CATV transport system

Wen-Shing Tsai[1,*], *Hong-Yu* Dai[1], and *Chung-Yi* Li[2]

[1]Department of Electrical Engineering, Ming Chi University of Technology, Taipei, 24301 Taiwan
[2]Institute of Electro-Optical Engineering, National Taipei University of Technology, New Taipei City, 10608 Taiwan

Abstract. This study proposes a two-channel fiber optic cable television (CATV) transport system based on Fabry–Perot laser diodes (FP-LDs) with a split band technique. To reduce the interference between channels, we apply the split band technique to two channels with different frequencies by using two converters at each moment. In this two-channel transmission scheme, composite second-order (CSO) and composite triple-beat (CTB) distortions induced by the systems are possibly confined in unused channels and result in good transmission performances. Through a 40 km standard single-mode fiber transmission, excellent performances of carrier-to-noise ratio (≥50 dB), CSO (≥70 dB), and CTB (≥72 dB) are obtained using the proposed fiber optic CATV transport systems. The proposed systems employing FP-LDs with a split band technique are also simpler and more cost effective than conventional externally modulated systems.

1 Introduction

For fiber optical CATV transport systems, the maximum fiber link using cascaded erbium-doped fiber amplifiers (EDFAs) is limited by RF parameters such as carrier to noise ratio (CNR), composite second-order (CSO), and composite triple beat (CTB) [1, 2]. These parameters can be seriously degraded by stimulated Brillouin scattering (SBS), stimulated Raman scattering (SRS), and fiber dispersion effects in 1550-nm fiber optical CATV transport systems. Several ways have been proposed to solve these problems such as employing an externally modulated transmitter and differential detection through two fiber links [3], using external light injection locking or sideband filtering techniques [4,5], the CSO and CTB performances can be improved [6-8]. Nevertheless, an expensive externally modulated transmitter as well as sophisticated differential detection, external light injection locking, and sideband filtering techniques are required. And further, it is difficult to obtain good CNR, CSO, and CTB transmission performances owing to full channel loading. In this letter, a structure of directly modulated fiber optical CATV transport system based on Fabry-Perot laser diodes (FP LDs) with split band technique is proposed and experimentally demonstrated. Under such two-channel transmission scheme, CSO and CTB

*Corresponding author: wst@mail.mcut.edu.tw

distortions induced by systems are dramatically reduced. Over a 40-km single-mode fiber (SMF) transmission, excellent performances of CNR (≥50 dB), CSO (≥70 dB), and CTB (≥72 dB) are obtained in the proposed directly modulated fiber optical CATV transport systems. To compare with a conventional externally modulated fiber optical CATV system, the proposed system reveals a prominent merit with simpler and more economic advantages.

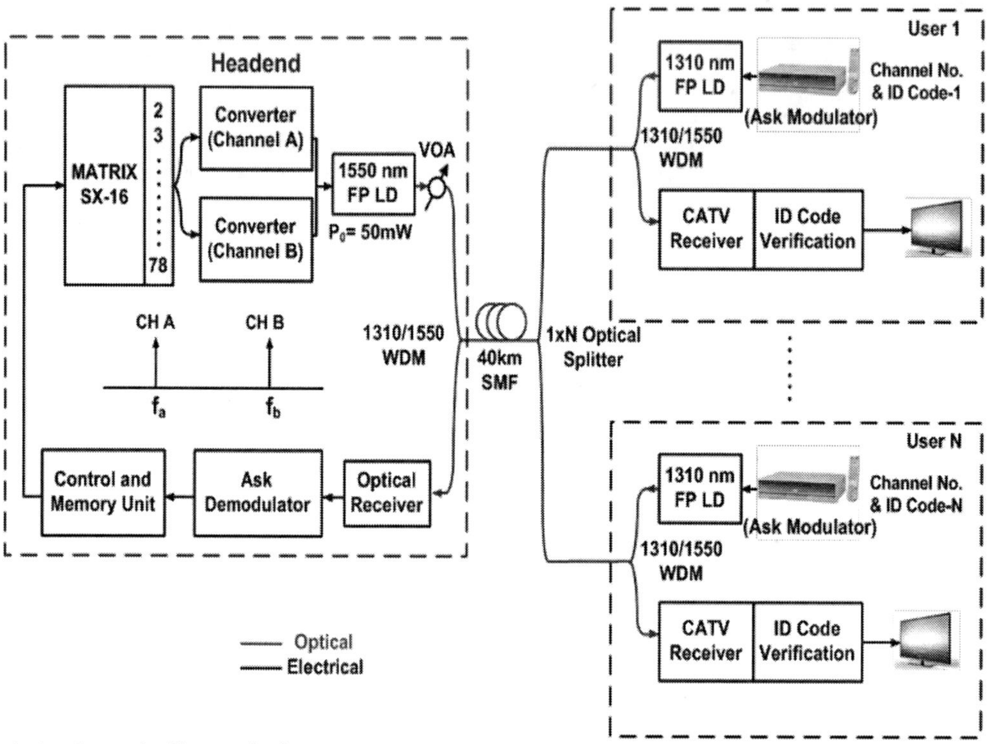

Fig. 1. A schematic fiber optical CATV transport system based on FP LDs with two-channel transmission.

The schematic system of fiber optical CATV transport system based on FP LDs with split band technique is shown in Fig. 1. This system can be used by N users simultaneously. Each user can uplink the identification (ID) code of desired channel to the headend through direct modulating the 1310 nm FP LD via a channel selector such as set-top box, station selector etc. The headend picks the selected channel through the ID code uploaded by user and downloads to user via direct modulating the 1550 nm FP LD. As the user receives the channels transmitted from the headend, an ID code verification will be excuted to verify whether the channel is the user desired. Meanwhile, to avoid the interference of other channels, we choose two channels with different frequencies by using two converters at each moment which avoid the expensive equipment and complex systems that are required by the system of full channel loading [3-9].

2 Experimental setup

Fig. 2 presents two different fiber optical CATV transport systems. Fig. 2(a), referred to as system I, shows the conventional 77-channel fully loading externally modulated fiber optical CATV transport system. Fig. 2(b), referred to as system II, shows our proposed directly modulated fiber optical CATV transport system with two-channel transmission. System I employs an EDFA to deliver optical signal over a 40-km SMF transmission. System II employs a high power FP LD to transmit optical signal over a 40-km SMF transmission. In system I, the EDFA has an output power of 16 dBm and a noise figure of 4.5 dB at an input power of 0 dBm. For the externally modulated transmitter with an optical modulation index (OMI) of 3.5%, its SBS-suppression ability is about +16 dBm. Thereby, the optical power launched into the fiber should be kept at ≤+16 dBm to avoid the degradation induced by the SBS effect. The FP LD used in system II has an operating wavelength of 1550 nm, an operation current of 500 mA, and a relative intensity noise (RIN) of about -130 dB/Hz. The OMI per channel of the FP LD transmitter is 4.8%, and the optical output power of the FP LD transmitter is 50 mW.

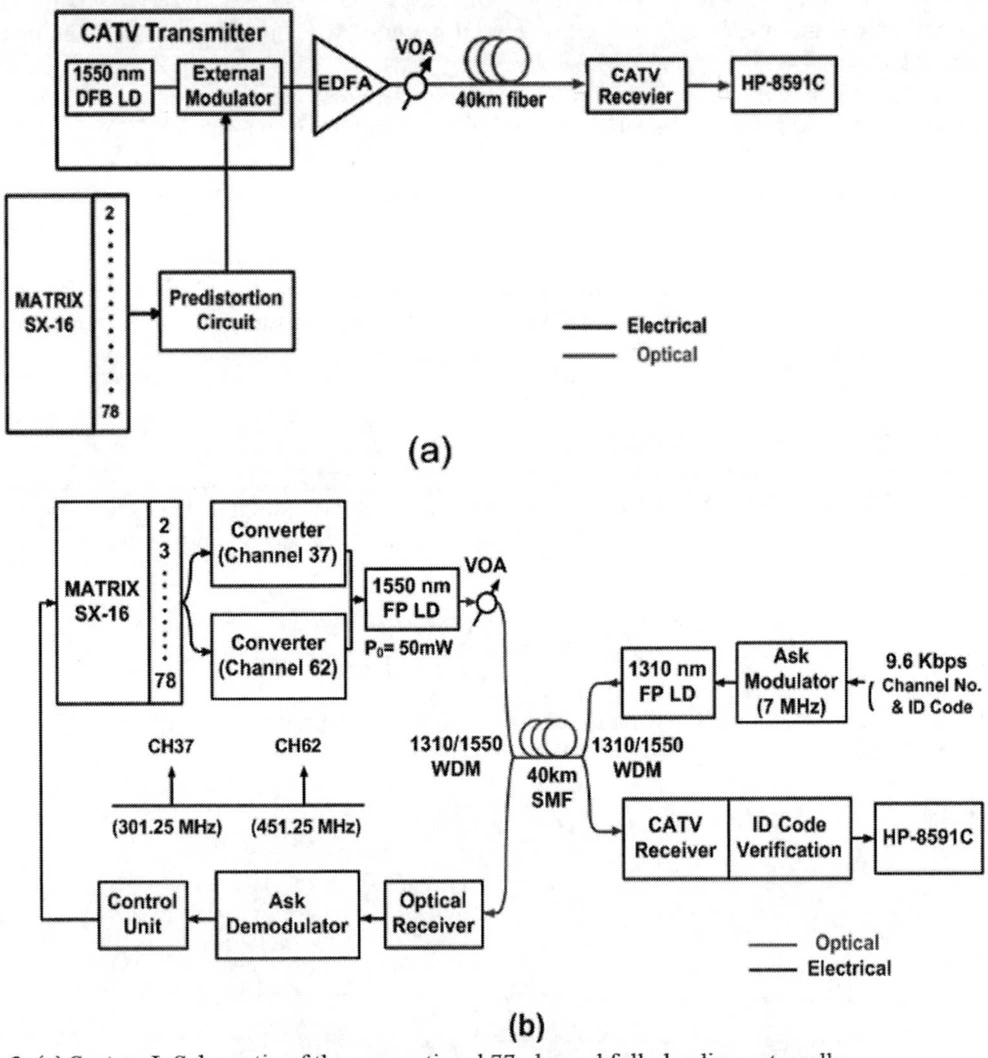

Fig. 2. (a) System I: Schematic of the conventional 77-channel fully loading externally modulated systems. (b) System II (new proposed): Schematic of the NTSC 77-channel systems based on FP LDs with two-channel transmission.

In system I, as shown in Fig. 2(a), channels 2-78 (55.25-547.25 MHz) generated from a multi-carrier generator (Matrix SX-16) are fed into an external modulator. The input power of the EDFA is kept at 0 dBm, the fiber transmission loss is about 0.4 dB/km, and each EDFA has a 16-dB link budget. Therefore, the optical power amplified by an EDFA has a sufficient power budget for transmission over a 40-km SMF link. The variable optical attenuator (VOA) is set after the EDFA, which results in less distortion as the optical power launched into the fiber. Over a 40-km SMF transmission, the optical signal is fed into an optical receiver. The optical input power level of the optical receiver is adjusted as 0 dBm, and the RF output level of the optical receiver is kept ab over 30 dBmV per channel. All CATV parameters (CNR, CSO, and CTB) are measured and analyzed using an HP-8591C CATV analyzer after a 40-km SMF transmission.

In system II, as shown in Fig. 2(b), channels 2-78 are split by a 1×2 RF splitter, went through two separate converters, recombined the CATV signals using a 2×1 RF combiner, and directly fed into a 1550-nm FP LD transmitter. It should be noted that system I is a uni-directional transmission system, whereas system II is a bi-directional transmission system. The control code (9.6 Kbps) for channel number selection and ID code at the optical node is modulated by an amplitude shift keying (ASK) modulator, and the upstream carrier frequency is located at 7 MHz. By using two-way transmission, system II just only transmit two channels to the optical node with two separate converters at the headend. The converters will convert the selected channels into channel 37 and channel 62 with ID code by a control unit. The function of the control unit is to verify the ID code of user and pick out the channel selected by the user. Over a 40-km SMF transport, the optical signal is launched into a CATV receiver with ID code verification for down-link transmission. For optimum performance of the CATV receiver, the optical input power level of the optical receiver is kept at 0 dBm, and the RF output level of the optical receiver is kept above 30 dBmV per channel. CNR, CSO, and CTB parameters are measured and evaluated using an HP-8591C CATV analyzer over a 40-km SMF link.

3 Results and discussion

Figs. 3-5 show the measured CNR, CSO and CTB values under NTSC channel number of systems I and II, as well as the schematic CNR, CSO and CTB definitions. CNR can be expressed as:

$$CNR = 10\log\left[\frac{\text{Carrier level}}{\text{Noise level}}\right] \tag{1}$$

It can be seen that the CNR value of system II is lower than that of system I about 3 dB. The CNR performance is given by [10-11]:

$$\mathrm{CNR} = \left(\mathrm{CNR}_{\mathrm{RIN}}^{-1} + \left(\mathrm{CNR}_{\mathrm{th}}^{-1} + \mathrm{CNR}_{\mathrm{shot}}^{-1}\right)\right)^{-1} \tag{2}$$

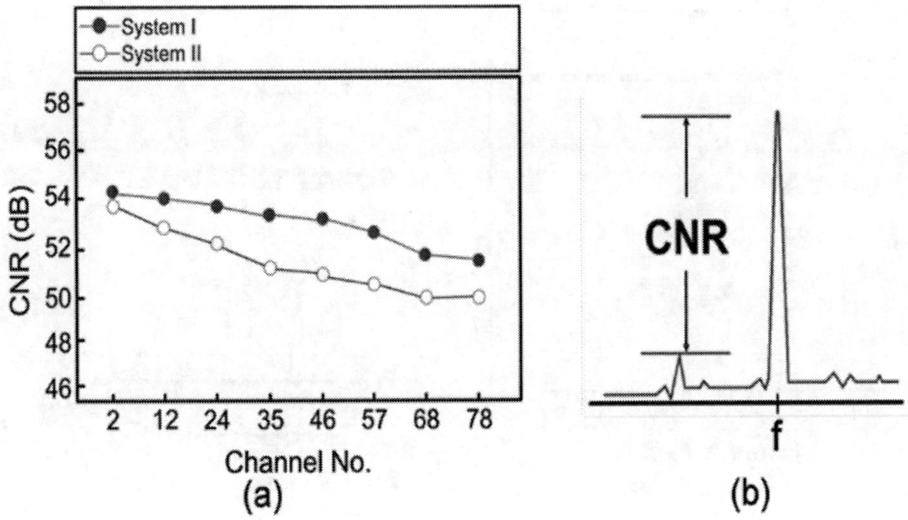

Fig. 3. (a)The measured CNR value under NTSC channel number of systems I and II, and (b) the schematic CNR definition.

where CNR_{RIN} results from the RIN of LD; CNR_{th} (resulting from thermal noise) and CNR_{shot} (resulting from shot noise) are associated with the optical receiver. The CNR value of system II can be improved by using a FP LD with lower RIN value (≤-135 dB/Hz), because higher CNR value can be obtained from a LD with lower RIN value. However, the CNR value of system II is equal to or higher than the threshold value (≥50 dB), indicating the least quality of demand.

By contrast, the CSO and CTB performances of the system II can be significantly improved. CSO and CTB are given by:

$$CSO = 10\log\left[\frac{\text{(Carrier level)}}{\text{(Second order beat distortion level)}}\right] \tag{3}$$

$$CTB = 10\log\left[\frac{\text{Carrier level}}{\text{Triple order beat distortion level}}\right] \tag{4}$$

The CSO and CTB values of the system II are higher than 70 and 72 dB. These improvements can be attributed to the use of two-channel transmission scheme to reduce the CSO and CTB distortions induced by systems. The CSO and CTB distortions can be stated by [12]:

$$\mathrm{CSO} = 10\log\left[\frac{\mathrm{mD}\lambda_c^2\mathrm{Lf}}{4\mathrm{c}}\sqrt{16(\Delta\tau)^2 + \frac{4\lambda_c^4\mathrm{L}^2\pi^2\mathrm{f}^6}{\mathrm{c}^2}}\right] + 10\log \mathrm{N}_{\mathrm{CSO}} + 6 \tag{5}$$

$$CTB = 10\log\left[\frac{9m^2D^2\lambda_c^4L^2f^2}{4c}\left(4(\Delta\tau)^2 + 4\pi^2 f\right)\right] + 10\log N_{CTB} + 6 \tag{6}$$

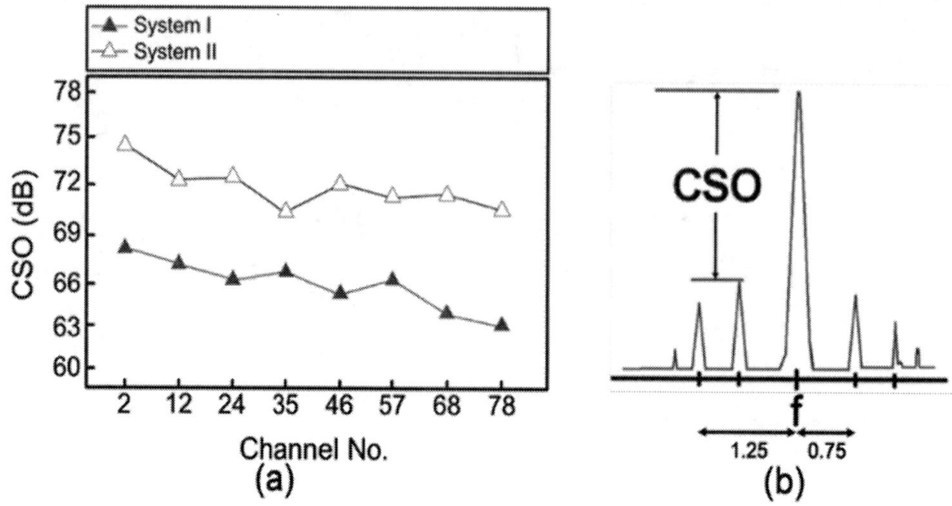

Fig. 4. (a)The measured CSO value under NTSC channel number of systems I and II, and (b) the schematic CSO definition.

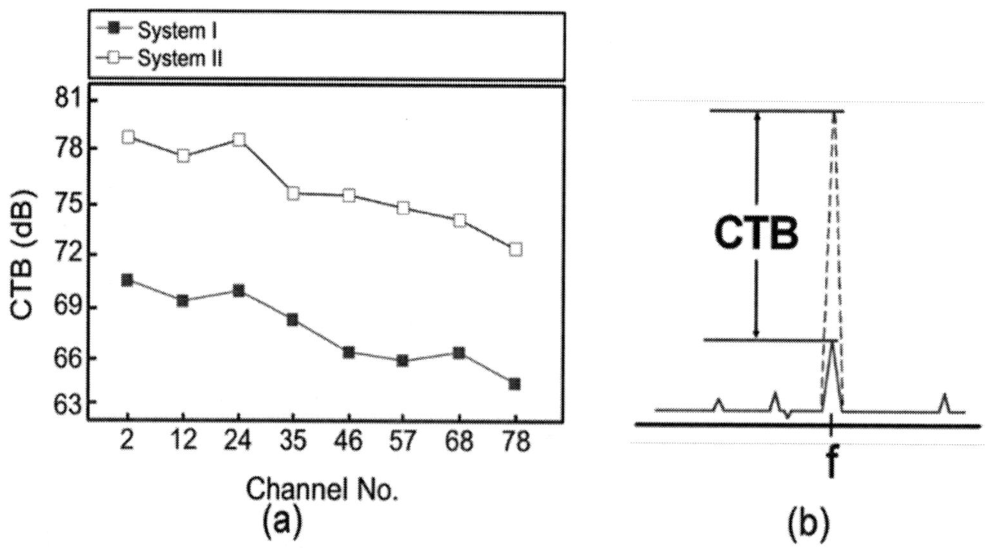

Fig. 5. (a)The measured CTB value under NTSC channel number of systems I and II, and (b) the schematic CTB definition.

where m is the OMI, D is the dispersion coefficient, λ_c is the optical carrier wavelength, L is the fiber length, f is the RF frequency, $\Delta\tau$ is the fiber dispersion, NCSO and NCTB are the product counts of CSO and CTB. By using two-channel transmission scheme, a very smaller N_{CSO} and N_{CTB} can be obtained from just only two channels. Thereby, a major part of CSO and CTB distortions will be automatically deleted in such two-channel transmission scheme. For instance, the carrier frequencies of CH37 and CH62 are 301.25 and 451.25 MHz, we can derive the second-order intermodulation distortions (IMD2):

$$451.25\text{ MHz (CH62)} - 301.25\text{ MHz (CH37)} = 150\text{ MHz} \quad (7)$$

$$301.25\text{ MHz (CH37)} + 451.25\text{ MHz (CH62)} = 752.5\text{ MHz} \quad (8)$$

The carrier frequencies of CH19 and CH117 are 151.25 and 751.25 MHz, it can be seen that the CSO distortions induced by CH37 and CH62 are located at CH19 (151.25 MHz – 150 MHz = 1.25 MHz) and CH117 (752.5 MHz – 751.25 MHz = 1.25 MHz). The above results tell us that CSO distortions induced by CH37 and CH62 are located at the unused channels of CH19 and CH117, by which far from the carrier frequencies of CH37 and CH62. By using two-channel transmission scheme, system II can dramatically remove a major part of CSO distortions. Similarly, we can also derive the third-order intermodulation distortions (IMD3):

$$301.25\,\text{MHz}\,(\text{CH37}) * 2 - 451.25\,\text{MHz}\,(\text{CH62}) = 151.25\,\text{MHz} \tag{9}$$

$$451.25\,\text{MHz}\,(\text{CH62}) * 2 - 301.25\,\text{MHz}\,(\text{CH37}) = 601.25\,\text{MHz} \tag{10}$$

The carrier frequencies of CH19 and CH87 are 151.25 and 601.25 MHz, it also can be seen that the CTB distortions induced by CH37 and CH62 are located at CH19 and CH87. The above results tell us that CTB distortions induced by CH37 and CH62 are located at the unused channels of CH19 and CH87, by which far from the carrier frequencies of CH37 and CH62. By using two-channel transmission scheme, system II can dramatically remove a major part of CTB distortions.

4 Conclusions

In this paper, we presented a fiber optical cable television (CATV) transport system using Fabry-Perot laser diodes (FP LDs) with split band technique. Unlike conventional full channel transmission, the 77-channel NTSC system selects only two channels using two converters at each moment, indicating a two-channel form transmission. For the two-channel form with split band technique transmission system, both IMD2 and IMD3 distortions are located in the unused frequency band. Distortion-induced interferences can be obviously suppressed. The values of both CSO and CTB for two-channel transmission are much higher than those of conventional full channel transmission. The value of CNR for the proposed system is less by about 2 dB than that of the conventional system because the direct modulation scheme has insufficient amplifier output power and less power in receiver. In addition, the performance of this system can be improved further using a push-pull preamplifier circuit that eliminates other high-order harmonic distortions to obtain excellent CSO and CTB values. The proposed system can instead of requiring the use of externally modulated 1550-nm transmitter, sophisticated SBS suppression, and differential detection techniques, only uses a high-power FP laser diode to transmit signals over 40 km with good performance in terms of CNR (≥50 dB), CSO (≥70 dB) and CTB (≥72 dB). Compared to externally modulated 1550-nm transmission systems, our proposed system has the distinctive advantages of simplicity and cost effectiveness, and is very suitable to local village and town communications.

References

1. W.S. Tsai, H.L. Ma, H.H. Lu, Y.P. Lin, H.Y. Chen, S.C. Yan, Opt. Exp. **18,** 26077-26083 (2010)
2. W.Y. Lin, C.H. Chang, P.C. Peng, H.H. Lu, and C.H. Huang, Opt. Exp. **10,** 10301-10307 (2010)
3. H.S. Su, C.Y. Li, W.Y. Lin, H.H. Lu, C.H. Chang, P.Y. Wu, Y.P. Lin, C.Y. Chen, Opt. Exp. **19,** 26928-26935 (2011)

4. H.H. Lu, A.S. Patra, S.J. Tzeng, H.C. Peng, W.I. Lin, IEEE Photon. Technol. Lett. **20,** 351-353 (2008)
5. C.Y. Li, H.S. Su, C.Y. Chen, H.H. Lu, H.W. Chen, C.H. Chang, C.H. Jiang, Opt. Exp. **19,** 14000-14007 (2011)
6. W.H. Chen, and W.I. Way, IEEE/OSA J Lightw. Technol. **22,** 1679-1693 (2004)
7. H.H. Lu, IEEE Photon. Technol. Lett. **14,** 1478-1480 (2002)
8. M.R. Phillips, K.Y. Wu, and F.X. Villarruel, IEEE/OSA J Lightw. Technol. **26,** 2647-2652 (2008)
9. C.H. Yeh, C.W. Chow, Y.F. Wu, F.Y. Shih, J.H. Chen, and C.L. Pan, IEEE Photon. Technol. Lett. **23,** 1627-1629 (2011)
10. P.Y. Wu, H.H. Lu, C.L. Ying, C.Y. Li, and H.S. Su, IEEE/OSA J Lightw. Technol. **29,** 2422-2427 (2011)
11. W.I. Way, *Broadband Hybrid Fiber/Coax Access System Technologies* (New York: Academic, 1998)
12. S. Ovadia, *Advances in red VCSEL technology, Broadband Cable TV Access Networks: Form Technologies to Application s* (Prentice-Hall, 2001)

The effect of thrust force in drilling composite materials using step core-ball drill

Hsing-Ming Teng and *Chung-Chen* Tsao*

Tahua University of Science and Technology, Department of Mechatronic Enginering, 30743 Chunglin Hsinchu, Taiwan, ROC

Abstract. Composite materials have gained increasing popularity over the past few decades due to their superior mechanical properties, such as high strength-to-weight ratio, fighting against high temperature and corrosion resistance. The assembly of enormous aeronautical components and structures require the machining of composite materials. Drilling is the most important hole- making process in the final assembly. When drilling composite materials, a number of defects are generating. Delamination caused by drilling thrust has been showed as one of the most problematic defects after drilling composite laminates. With a pressing need for decreased delamination, many studies are turning more and more toward tool geometry and machining parameters. Drilling of composite plates using a step core-ball drill, which is a special drill to improve the chip flow and reduces the thrust force at the exit of hole, is investigated in this study. The experimental results found that the step core-ball drill was efficient in drilling of carbon fiber reinforced plastic (CFRP) and did not produce loading on the drill exit at the proper drilling conditions. The results obtained from this study feeding back for fundamental research efforts could steer future studies on the drilling composite materials in the most promising direction.

1 Introduction

Composite materials have rapidly increasing popularity over the past few decades because their superior mechanical properties such as high strength-to-weight ratio, fighting against high temperature and corrosion resistance, especially the percentage of carbon fiber reinforced plastics (CFRPs) has reached up to 40% in the new generation of aircrafts manufacturing [1]. The Dreamliner 787 of Boeing's commercial aircraft, which has 50% composite (including plastics and carbon fiber) of its primary aircraft structure, is constructed in tubular segments using a molding process that can eliminate 1,500 aluminum sheets and 50,000 fasteners during final assembly [2-3]. Drilling is mostly used for hole-making process in composite parts for final assembly operations. Drilling-induced defects, such microcracking, fiber pullout, delamination, splintering and burr, reduces the hole quality of the machined parts. However, delamination is highly concern during drilling composites due to its inhomogeneity and anisotropic nature. Miner and Mackey point out that not only for new concepts of tooling but also for different realms of cutting conditions to solve the drilling-induced damages in drilling composite materials [4-5]. In general,

*Corresponding author: aetcc@tust.edu.tw

drilling-induced delamination easily occurs both at the entrance (peel-up) and the exit (push-out) sides of the workpiece. Carbon fiber reinforced plastics (CFRPs), which are one of the interest subjects in machining composite materials, are known to be difficult to machine due to the abrasive nature and low thermal conductivity. Many studies have been shown that different techniques can increase drilled quality of composites by changing the configuration of the process such as feed rate, spindle speed, tool geometry and material, coolant type, coated tool, back-up plate, etc. Feed rate and tool geometry, however, are the two most important cutting parameters affecting the quality of drilled holes, and feed rate and drilling-induced thrust force are closely related [6-29]. It has been recommended to set feed high to avoid the peel-up defect while the drill just entrances the workpiece. That action made the high risk on delamination at the exit side of the hole. Additionally, the influence of drilling speed on the hole quality is shown to be minimal than feed rate; therefore, high drilling speed is adopted for better productivity.

The twist drill, which has quite a complicated tool geometry, is a traditionally use tool for hole-making in industrial practice. The chisel edge of the twist drill, which has a null cutting speed at the center point, pushes the uncut thickness (last lamina) as it approach the exit side in drilling. The action of the chisel edge on the thrust force increases with the feed rate and may account for over 80% of the total force needed to drill a hole. The exit side delamination occurs when thrust force exceeds the inter-laminar bond strength of the composite material. Step core-ball drill, which has a half ball geometry on the end of outer cutting tool, is a special drill to improve the chip flow and reduces the thrust force at the exit of hole in drilling of composite plates, as shown in Fig. 1. Additionally, plated diamond bit plays an increasingly important role in civil, geology, petroleum, construction and other industrial applications. Some researches and enterprises now pay close attention by advantages of plated diamond bit such as high bonding strength, high grit protrusion and large space for chip [30-33]. An experimental analysis for thrust forces associated with the various diamond grits and drilling parameters using step core-ball drill in drilling of composite materials was investigated in the present study. An innovative tool, however, can be consulted in efficiency and quality of drilling composite materials in domestic industry in the future.

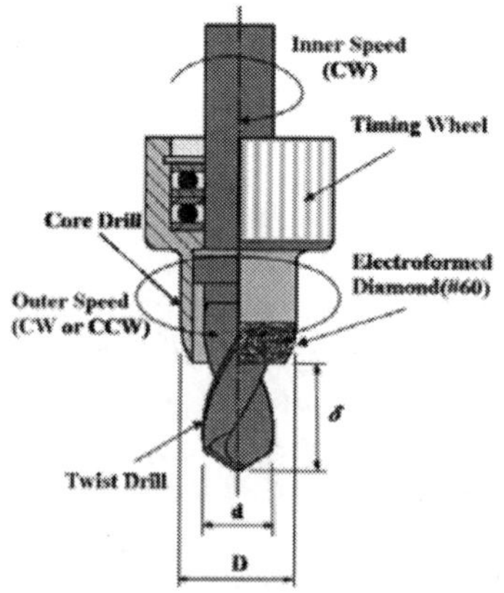

Fig. 1. Schematic of step core-ball drill.

2 Experimental setup and method

The composite laminates were fabricated using the prepreg 41C30126 fabric carbon fiber/epoxy matrix (Formosa Plastics Corporation) by autoclave molding. The stacking sequence of the laminates was $[0/90]_{8S}$. The carbon fiber reinforced plastic (CFRP) laminates were approximately 4.0 mm thick and 16 plies were laid up. Specimens of 60×60 mm^2 were cut, using a water-cooled diamond table saw. Drilling tests were conducted using a 3-axis LEADWELL V30 vertical machining center with a 5.5-kW spindle drive, for which the thrust force was measured using a Kistler 9273-type piezoelectric dynamometer. The force signals were transmitted to a Kistler 5011 multichannel charge amplifier and stored on a P4 personal computer. The experimental setup to measure thrust forces for step core-ball drilling is shown in Fig. 2. The proper fixture was held rigidly on top of the dynamometer. The outer drill diameter, plated length and core-ball tool thickness of the step core-ball drills are 10, 12 and 1 mm, respectively. The end of the core-ball tool plated with #40 (0.38 mm), #50 (0.27 mm) and #60 (0.25mm) diamond grits on the cutting surface, which are different in grain size, were used in all tests. The internal part of the step core-ball drills is the twist drill. Tungsten carbide twist drills of 5.6 and 6.8 mm diameter were used to obtain the entire drilling-thrust. All experiments were performed twice, under dry drilling conditions at spindle speeds of 800, 1,000 and 1,200 rpm for outer drills, respectively, spindle speed of 1,214 rpm for inner drill, and feed rates of 15, 18 and 20 mm/min. In this study, the rotational direction of the inner drill is always clockwise (CW), and the rotational direction of outer drill is always counter clockwise (CCW). The different rotational direction between inner drill and outer drill is shown in Fig. 3.

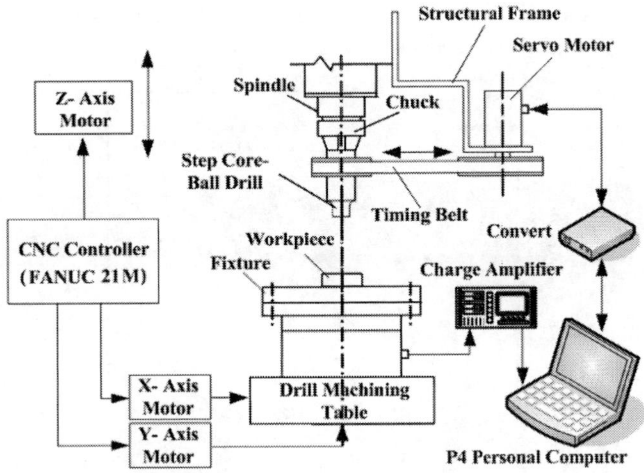

Fig. 2. Schematic diagram of experimental set-up.

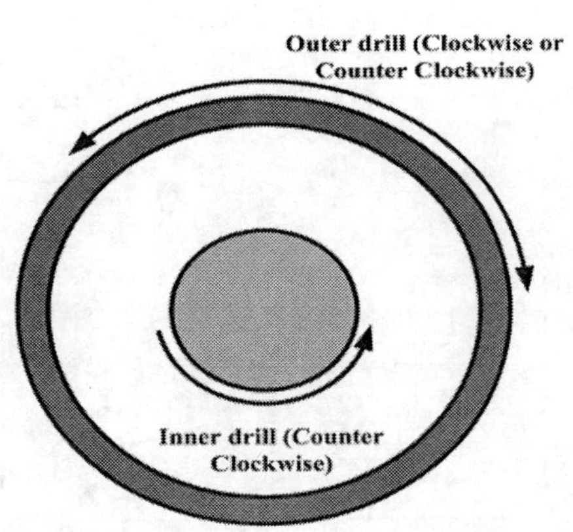

Fig. 3. The different rotational direction between inner drill and outer drill for step core-ball drill (Bottom View).

3 Results and discussions

Fig. 4 depicts the thrust force of step core-ball drill with a 1000 rpm spindle speed at various plated diamond grits and feed rates in drilling CFRP. From Fig. 4, the thrust forces of step core-ball drill with various plated diamond grits increases as the feed rate increases. Numerous papers have recognized that thrust force (or delamination) increases with feed rate [5, 8-10, 28]. For minimizing drilling-induced damage, low feed rate is preferred in drilling of composite laminates. The thrust force with #60 plated diamond grit is larger than thrust force with #40 plated diamond grit as shown in Fig. 4. This result may be explained by the fact that the thrust force at the exit section when the #60 plated diamond grit core drill was used became large due to the variation of the total contact area between plated diamond grits and machined surface. From these experiments, it may be concluded that the larger plated diamond grit of core drill is appropriate for the CFRP drilling without fuzzing and delamination. The study experimentally explains how the plated diamond grit significantly affects the thrust force in drilling composite materials. Fig. 5 depicts the thrust force of step core-ball drill with an 18 mm/min feed rate at various plated diamond grits and rpm spindle speeds in drilling CFRP. From Fig. 5, the thrust forces of step core-ball drill with various plated diamond grits decreases as the spindle speed increases, which might be explained by the fact that the epoxy resin of the CFRP was damaged by the heat generated due to the increase of cutting speed. Also, numerous papers have proved that thrust force (or delamination) decreases with spindle speed.

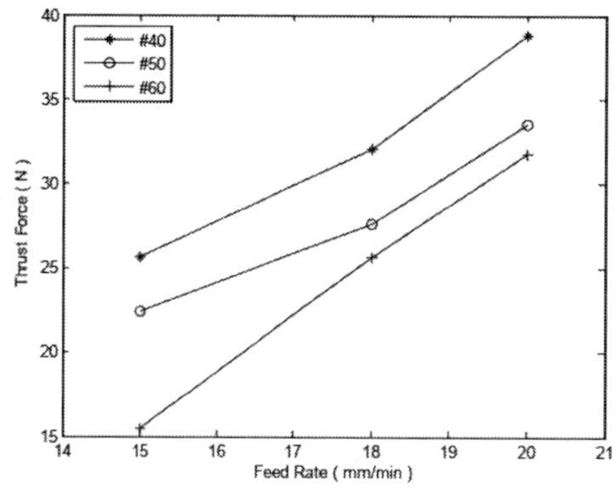

Fig. 4. Thrust force of step core-ball drill with a 1000 rpm spindle speed at various plated diamond grits and feed rates in drilling CFRP.

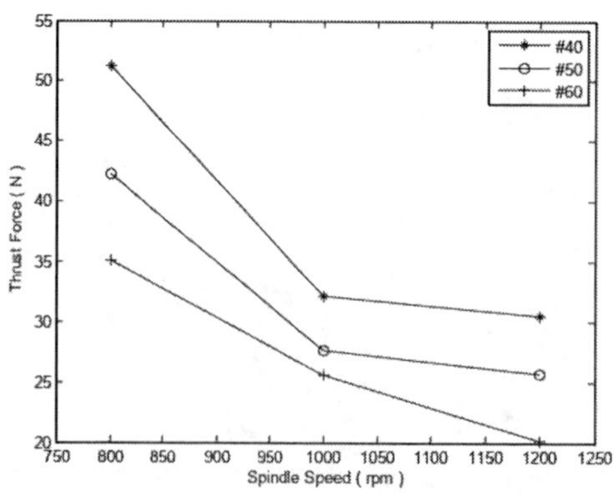

Fig. 5. Thrust force of step core-ball drill with an 18 mm/min feed rate at various plated diamond grits and spindle speeds in drilling CFRP.

4 Conclusions

An experimental analysis for thrust forces associated with the various plated diamond grits and drilling parameters using step core-ball drill in drilling of composite materials is investigated in the present study. The results reveal that, the thrust force of the step core-ball drill with #40 plated diamond grit is lower than those with #60 plated diamond grit. In other words, the larger plated diamond grit can cause larger thrust force at the step core-ball drill exit in the workpiece. However, a increase in feed rate at the end of drilling using step core-ball drill in drilling of composite materials will cause the cracking around the exit edge of the hole. It was also stated that the larger the feeding load is, the larger the thrust force is.

Acknowledgment

The research is partially supported by Ministry of Science and Technology, Taiwan, ROC, under contract MOST-105-2221-E-233 -001.

References

1. Y. Karpat, O. Bahtiyar, B. Değer, B. Kaftanoğlu, CIRP Ann. **63**(1), 81 (2014)
2. http://www.dailymail.co.uk/sciencetech/article-2041863/Boeing-787-Dreamliner-reality-carbon fibre-plane-delivered-Japan.html
3. https://www.thoughtco.com/boeings-787-dreamliner-820385
4. L.H. Miner, EI Segundo California, Soc. Plast. Eng. 85 (1980)
5. B.A. Mackey, *Proc. 37th Conf. Soc. Plas. Ind.*, Reinf. Plast./Compos. Institute, 1 (24D), (1982)
6. W. König, C. Wul, P. Grass, H. Willerscheid, CIRP Ann. **34**(2), 537 (1985)
7. I. El-Sonbaty, U.A. Khashaba, T. Machaly, Compos. Struct. **63**, 329 (2004)
8. H. Hocheng, C.K.H. Dharan, ASME J. Eng. Ind. **112**, 236 (1990)
9. H. Hocheng, C.C. Tsao, J. Mater. Process. Technol. **140**, 335 (2003)
10. H. Hocheng, C.C. Tsao, J. Mater. Process. Technol. **167**, 251 (2005)
11. H. Hocheng, C.C. Tsao, Int. J. Mach. Tools Manuf. **46,** 1403 (2006)
12. C.C. Tsao, Int. J. Mach. Tools Manuf. **46**, 1653 (2006)
13. M. Fernandes, C.Cook, Int. J. Mach. Tools Manuf. **46,** 70 (2006)
14. M. Fernandes, C. Cook, Int. J. Mach. Tools Manuf. **46**, 76 (2006)
15. S. Jain, D.C.H. Yang, ASME J. Eng. Ind.. **116,** 475 (1994)
16. U.A. Khashaba, M.A. Serif, M.A. Elhamid, Compos. Part A Appl. Sci. Manuf. **38**, 61 (2007)
17. W. König, P. Grab, CIRP Ann. **38**(1), 119 (1989)
18. A. Koplev, A. Lystrup, T. Vorm, Compos. **14**, 371 (1983)
19. R. Piquet, B. Ferret, F. Lachaud, P. Swider, Compos. Part A Appl. Sci. Manuf. **31**, 1107 (2000)
20. C.C. Tsao, H. Hocheng, Int. J. Mach. Tools Manuf. **44**, 1085 (2004)
21. C.C. Tsao, H. Hocheng, Int. J. Mach. Tools Manuf. **45**, 1261 (2005)
22. C.C. Tsao, Int. J. Mach. Tools Manuf. **46**, 1653 (2006)
23. C.C. Tsao, H. Hocheng, Int. J. Mech. Sci. **49**, 983 (2007)
24. C.C. Tsao, Mater. Des. **29**, 1740 (2008)
25. C.C. Tsao, Y.C. Chiu, Int. J. Mach. Tools Manuf. **51**, 740 (2011)
26. C.C. Tsao, H. Hocheng, Y.C. Chen, CIRP Ann. **61(1)**, 91 (2012)
27. M.S. Won, C.K.H. Dharan, ASME J. Manuf. Sci. Eng.**124**, 242 (2002)

28. C.C. Tsao, W.W. Wu, Int. J. Computer. Integr. Manuf. **27**, 979 (2014)
29. U.A. Khashaba, A.A.El-Keran, J. Mater. Process. Technol. **249**, 415 (2017)
30. Y.Q. Xing, J.X. Deng, G D. Zhang, Z. Wu, F.F. Wu, J. Manuf. Process. **26**, 31 (2017)
31. W.Y. Song, F.B. Liu, C.G. Bu, G.D. Zhou, Procedia Eng. **73**, 48 (2014)
32. I. Sultana, Z. Shi, M.H. Attia, V. Thomson, Procedia CIRP **45**, 23 (2016)
33. P.C. Priarone, M. Robiglio, R. Melentiev, L. Settineri, Procedia CIRP **66**, 181 (2017)

Vibration modes and sound characteristic analysis for different sizes of singing bowls

Bor-Tsuen Wang [1,*], *Chun-Lang* Tsai [1], and *Ying-Hui* Wu [2]

[1]Department of Mechanical Engineering, National Pingtung University of Science and Technology, Pingtung, 912, Taiwan
[2]Machinery Division, National Nei-Pu Senior Agricultural Industrial Vocational School, Pingtung, 912, Taiwan

Abstract. The singing bowl is used not only for the instrument of Buddhism but also for musical therapy. This work aims to investigate the correlation of vibration modes and percussion sound for singing bowls. A typical singing bowl is first selected to perform finite element analysis (FEA) for theoretical modal analysis (TMA) as well as experimental modal analysis (EMA). Modal parameters of singing bowl, including natural frequencies and mode shapes, can be obtained from analysis and experiment, respectively. Singing bowl FE model can then be updated and verified by adjusting material properties and used to predict structural vibration modes. The percussion sound of singing bowl is also measured to obtain its sound spectrum. The peak frequency response of singing bowl sound can be interpreted and contributed from circular vibration modes of the bowl. With the knowledge of sound generation mechanism for the singing bowl, this work also studies the percussion sound characteristics of seven different sizes of singing bowls. Results show the fundamental frequency and overtone frequencies of singing bowl percussion sound are higher for the smaller size. Interestingly, that the peak resonant frequencies have near the integer ratio relationship makes the singing bowl revealing harmony sound effects. The radiated sound spectrum can be well calibrated and predicted for different sizes of singing bowls. This work shows the analytical and experimental approaches in studying the singing bowl percussion sound that strongly correlated to structural vibration modes and can be adopted for future development of singing bowls.

1 Introduction

Singing bowls also known as Tibetan singing bowls from Himalayan can produce harmony sound. Singing bowls are not only used for the instrument of Buddhism but also for musical therapy. This work aims to investigate the vibration characteristics of a singing bowl by finite element analysis (FEA) and experimental modal analysis (EMA).

Wang *et al.* [1] presented structural and vibro-acoustic analysis on the harmonic sound plate made by steel which vibration modes and acoustic sound pressure modes can be

[*]Corresponding author:wangbt@mail.npust.edu.tw

identified and related to the percussion sound. Wang *et al.* [2] further explored the possibility of using glass material to manufacture the harmonic glass plate (HGP). With the use of FEA for design aid and EMA for the verification, the design of HGP can be achieved from the steel to the glass. The numerical analysis and experimental techniques are crucial for obtaining structural vibration modes as well as sound characteristics.

Bells are similar to singing bowls and have drawn much attention in scientific study. Pan [3] examined the ancient Chinese musical bell which is the almond-shape cross section rather than circular bell. The almond-shape bell can produce two-tone effect for structural vibration mode properties. Lee *et al.* [4] investigated the correlation between percussion sound and vibration modes of Korean bell. With the aid of FEA, they analysed bell's vibration mode shapes and predicted the percussion sound. The optimum percussion location can be designed to obtain the lasting longer sound. McLachlan *et al.* [5] adopted FEA to design the bell that can generate harmonic sound. The geometry of bell was optimized to obtain the special sound quality of bell. McLachlan [6] indicated the major parameters affecting the bell's percussion sound are the thickness, height, radius, and curvature shape and cone angle of bell. Fletcher *et al.* [7] measured the sound spectrum for different bells applied with different levels of impact forces. The percussion sound responses are shown with different decay effects.

Different musical instruments with different sound generation mechanisms are of interest. Myers *et al.* [8] showed the sound generation mechanism for brass instrument. Different geometry variation and effects on radiated sound frequencies were investigated. For guitars, other than the string vibration the radiated sound is amplified through the resonance body with sound hole. Ezcurra [9] applied FEA to calibrate the material properties of sound board and examined vibration mode shapes of sound board to observe the generated sound spectrum. Boullosa [10] conducted sound and vibration measurement on several guitars to compare the radiated sound characteristics.

Wang *et al.* [11] presented correlation study on a singing bowl between vibration modes and percussion sound spectrum. The ring vibration modes can be known dominating the percussion sound response, and different striking methods may result in a slight different sound effect. This work aims to study a series of seven singing bowls from the large size to small ones. Theoretical modal analysis (TMA) on a typical singing bowl is first performed by FEA. EMA on the singing bowl is also carried out to verify vibration modes, including natural frequencies and mode shapes. The percussion sound spectrum can then be measured and interpreted by vibration modes of the singing bowl. Finally, the sound spectrums for the seven different sizes of singing bowls are studied to calibrate the differences, such as the fundamental frequency and overtone frequencies.

2 Analyses and verification of singing bowl vibration modes

Fig. 1 shows the typical singing bowl A with the height of 128mm and the radius of 260mm. Both FEA and EMA will be performed, respectively, on the singing bowl A to obtain its natural frequencies and corresponding mode shapes. Fig. 2 shows the procedure for model verification (MV). The purpose of MV is to calibrate the finite element (FE) model of singing bowl that can be equivalent to the real structure.

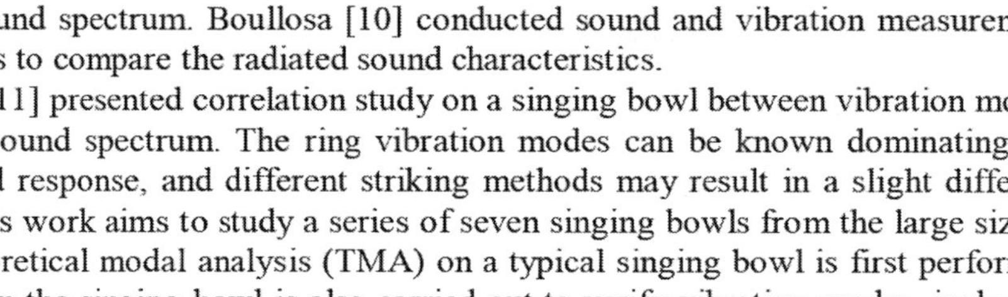

(a) picture (b) FE model

Fig. 1. Typical singing bowl A.

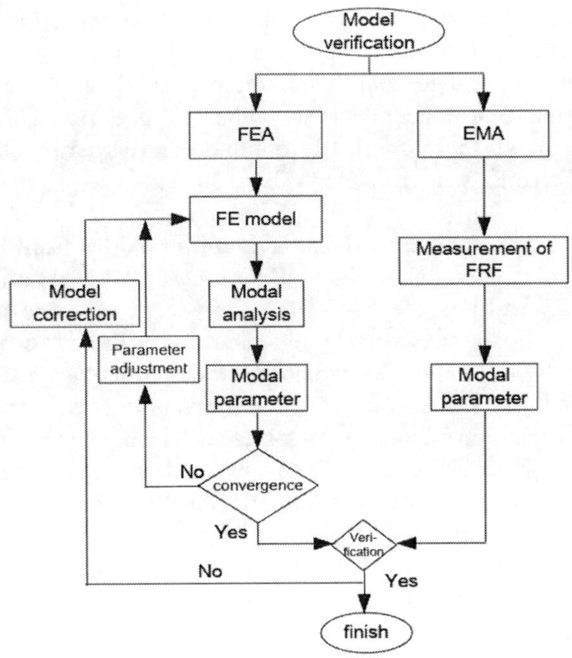

Fig. 2. Procedures for model verification (MV).

The first step in MV is to perform FEA on the singing bowl. The FE model for the singing bowl as shown in Fig. 1(b) is built and performed TMA to numerically obtain structural modal parameters, including natural frequencies and mode shapes. The singing bowl FE model is constructed by 8-node brick elements, SOLID187 in ANSYS. The singing bowl is made of copper alloy, which material properties are density 4802.6 kg/m^3, Young's modulus 39.37 GPa and Poisson ratio 0.3. The material properties are calibrated after MV. Table 1 shows the convergence analysis of singing bowl FE models. Model (D) with 26,348 elements and 53,226 nodes is adequate for reasonable and accurate analysis in comparison of errors of natural frequencies between Models (D) and (C), within 0.2% up to 2000 Hz.

Table 1. Convergence analysis of singing bowl FE models.

	Model (A)	Model (B)	Model (C)	Model (D)	% B vs. A	% C vs. B	% D vs. C
elements	6,689	10,131	18,399	26,348	-	-	-
nodes	13,637	20,595	37,217	53,226	-	-	-
F-07 (Hz)	132.0	130.9	130.2	130.1	0.84	0.54	0.08
F-09 (Hz)	369.9	366.4	364.2	363.8	0.96	0.60	0.11
F-11 (Hz)	699.2	692.3	687.2	686.3	1.00	0.74	0.13
F-13 (Hz)	1107.4	1095.7	1086.5	1084.8	1.07	0.85	0.16
F-15 (Hz)	1583.7	1565.2	1550.5	1547.9	1.18	0.95	0.17
F-17 (Hz)	2117.1	2090.2	2069.3	2065.1	1.29	1.01	0.20

The second step is to perform EMA on the real singing bowl. In order to obtain real structure's modal parameter, we use the impact hammer to excite the singing bowl in free boundary condition and the accelerometer to pick up the vibration response. Fig. 3(a) shows the experimental setup for performing EMA on the singing bowl, while Fig. 3(b) reveals the 24 measurement points on the ring surface of singing bowl. The accelerometer is fixed at Point 1 to avoid possible nodal points of vibration modes, and the impact hammer is applied to excite the singing bowl at the 24 grid points. The structural frequency response functions (FRFs), $H_{ai,fj}(f)$ or simply $H_{ij}(f)$, between the acceleration at the i-th location and the impact force at the j-th location can be measured. The 24 sets of FRFs can be imported to the curve-fitting software ME'scopeVES to determine experimental modal parameters, including natural frequencies, mode shapes and modal damping ratios.

The third step of MV is to compare the obtained modal parameters from FEA and EMA, respectively. The important concept is that if the FE model can be equivalent to the real structure of singing bowl, the numerically obtained and experimentally determined modal parameters should be consistent with each other.

Table 2 summarizes the comparison of natural frequencies obtained from FEA and EMA, respectively. One can observe the axisymmetric modes from FEA, such as modes F-07 and F-08, because of the circular shape of singing bowl. The natural frequencies of two axisymmetric modes are the same from FEA; however, only one mode is obtained from EMA for mode O-01. For modes such as F-11 and F-12, the corresponding experimentally extracted modes are O-03 and O-04, which natural frequencies are not exactly the same. This can be explained that the singing bowl is not in a perfect axisymmetric shape.

For those modes that can be put aside for comparison, the physical meanings of mode shapes should be the same. Table 2 also shows the error percentage of natural frequencies between FEA and EMA. The root mean square (RMS) error for all comparable modes is 4.27%. Only the natural frequency error of the first mode is 7.43%, and other modes are within 5%. This indicates the equivalency of FE model to the real structure of singing bowl.

As one can see in Table 2, the modes can be categorized into two types of modes, including the (z,θ) modes and (r,θ) modes. Those modes are in a reasonable sequence from lower order of modes to higher order, such as (z,θ)=(1,2), (z,θ)=(1,3), and etc.

Table 3 shows the theoretical and experimental mode shapes for (z,θ) modes. One can observe both the (z,θ) mode shapes from FEA and EMA agree reasonably to each other. The (z,θ) modes of singing bowl are the ring modes that only the upper ring surfaces will vibrate. For the (z,θ) mode, z is for the vertical direction always with one partition, i.e. z=1, and θ is for the partition number in the circumferential direction.

Table 4 shows theoretical mode shapes for the (r,θ) modes that are the local mode effects due to the bottom surfaces of singing bowl. That the (r,θ) modes are not experimentally measured is the cause of measuring points only on the ring surface as shown in Fig. 3(b). The physical meaning of (r,θ) modes can be characterized for the bottom surface of singing bowl where r is for the radial direction, and θ is for the circumferential direction. It is noted that there is no axisymmetric mode for (r,θ)=(r,0) as shown in Table 2.

Next section will demonstrate that the percussion sound of singing bowl is dominated by the ring modes, i.e. the (z,θ) modes, and the (r,θ) modes related to the bottom surface vibration of singing bowl do not contribute to the sound radiation, because the singing bowl is generally struck at the ring surface.

(a) Instrumentation setup

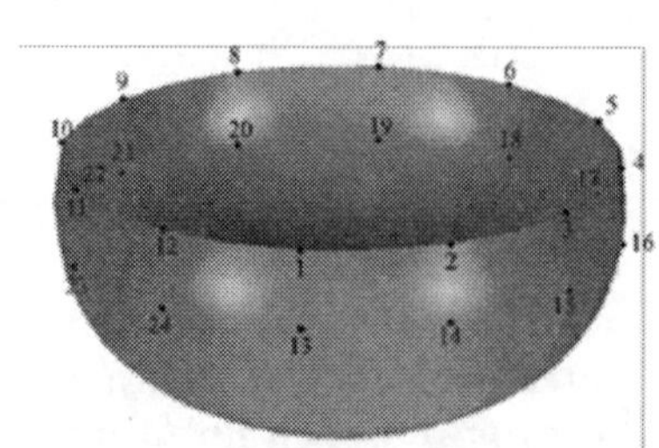

(b) Grip points for EMA

Fig. 3. Experimental setup for EMA on the singing bowl.

Table 2. Comparison of natural frequencies between FEA and EMA.

FEA			EMA			Natural frequency error (%)
Mode No.	Natural frequency (Hz)	Physical meaning of mode shapes	Mode No.	Natural frequency (Hz)	Physical meaning of mode shapes	
F-07	130.1	$(z, \theta)=(1,2)$	O-01	121.1	$(z, \theta)=(1,2)$	7.43
F-08	130.1					
F-09	363.8	$(z, \theta)=(1,3)$	O-02	361.7	$(z, \theta)=(1,3)$	0.58
F-10	363.8					
F-11	686.3	$(z, \theta)=(1,4)$	O-03	700.0	$(z, \theta)=(1,4)$	-1.96
F-12	686.4		O-04	712.5		-3.66
F-13	1084.8	$(z, \theta)=(1,5)$	O-05	1123.4	$(z, \theta)=(1,5)$	-3.44
F-14	1085.0		O-06	1135.2		-4.42
F-15	1547.9	$(z, \theta)=(1,6)$	O-07	1614.1	$(z, \theta)=(1,6)$	-4.10
F-16	1548.1		O-08	1621.9		-4.55
F-17	2065.1	$(z, \theta)=(1,7)$	O-09	2152.3	$(z, \theta)=(1,7)$	-4.05
F-18	2065.3		O-10	2169.5		-4.80
F-19	2142.8	$(r, \theta)=(1,0)$	-	-	-	-
F-20	2579.3	$(r, \theta)=(1,1)$	-	-	-	-
F-21	2579.3		-	-	-	-
F-22	2627.9	$(z, \theta)=(1,8)$	O-11	2752.3	$(z, \theta)=(1,8)$	-4.52
F-23	2628.0					
F-24	2737.4	$(r, \theta)=(2,0)$	-	-	-	-
F-25	2908.7	$(r, \theta)=(2,1)$	-	-	-	-
F-26	2908.7		-	-	-	-
F-27	3111.7	$(r, \theta)=(1,2)$	-	-	-	-
F-28	3111.7		-	-	-	-
F-29	3205.7	$(r, \theta)=(3,1)$	-	-	-	-
F-30	3205.8		-	-	-	-
F-31	3228.5	$(z, \theta)=(1,9)$	O-12	3357.8	$(z, \theta)=(1,9)$	-3.85
F-32	3228.8		O-13	3382.0		-4.53
F-33	3245.8	$(r, \theta)=(3,0)$	-	-	-	-
F-34	3319.9	$(r, \theta)=(2,2)$	-	-	-	-
F-35	3319.9		-	-	-	-
F-36	3414.9	$(r, \theta)=(1,3)$	-	-	-	-
F-37	3414.9		-	-	-	-
F-38	3486.4	$(r, \theta)=(3,2)$	-	-	-	-
F-39	3486.4		-	-	-	-
F-40	3511.0	$(r, \theta)=(4,0)$	-	-	-	-
RMS for Natural frequency error (%)						4.27

Table 3. Comparison of mode shapes between FEA and EMA for modes (z, θ).

(z, θ)	(1,2)	(1,3)	(1,4)	(1,5)
FEA	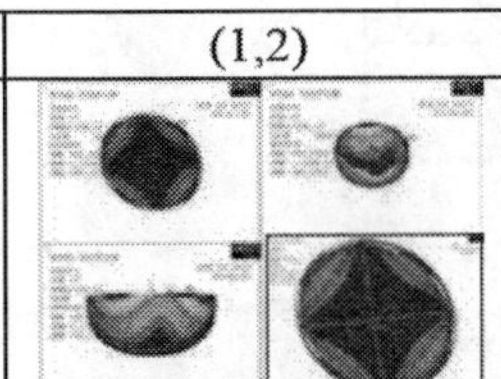 F-07_130.1 Hz	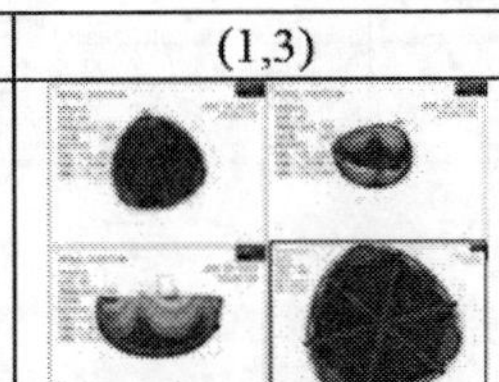 F-09_363.8 Hz	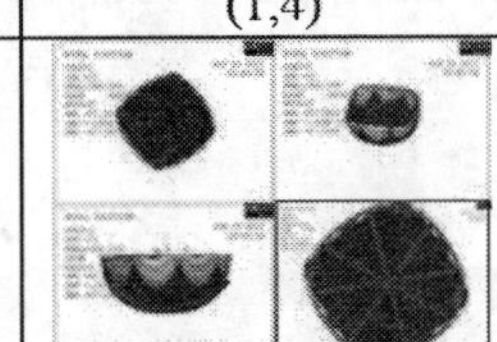 F-11_686.3 Hz	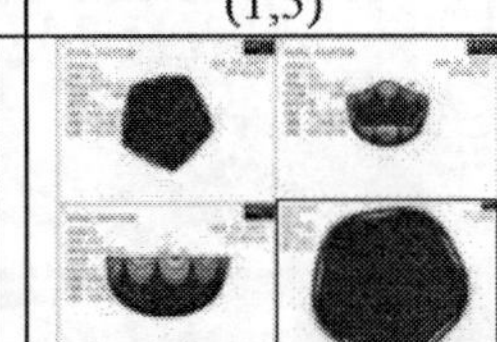 F-13_1084.8 Hz
EMA	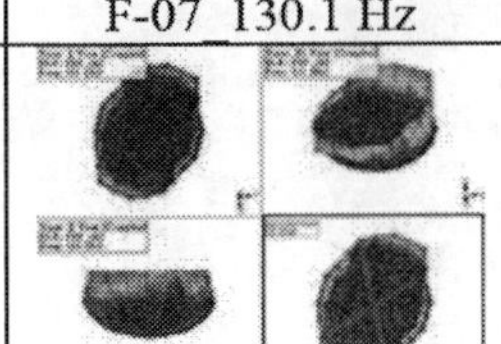 O-01_121.1 Hz	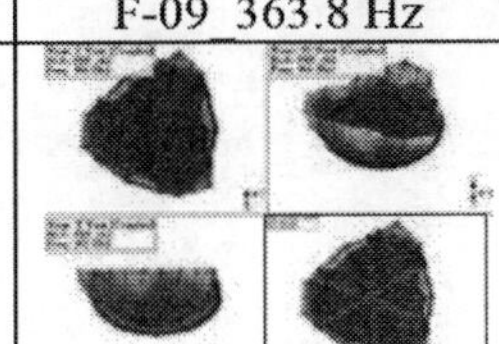 O-02_361.7 Hz	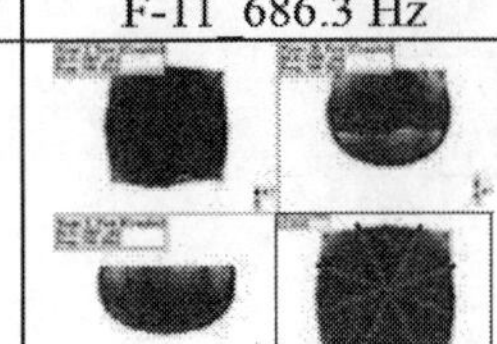 O-03_700.0 Hz	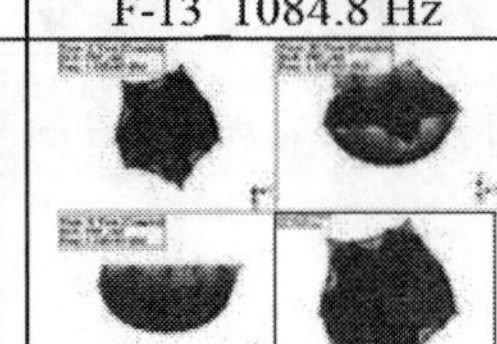 O-05_1123.4 Hz

Table 3. Comparison of mode shapes between FEA and EMA for modes (z, θ) (Continued).

(z, θ)	(1,6)	(1,7)	(1,8)	(1,9)
θ FEA	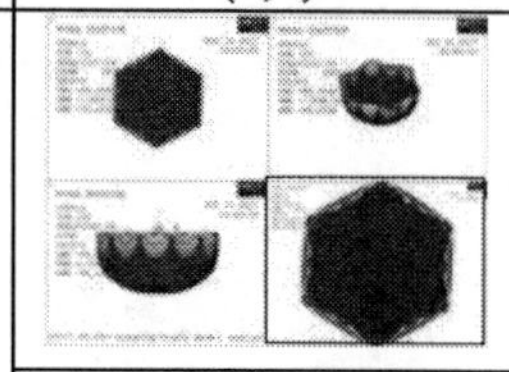	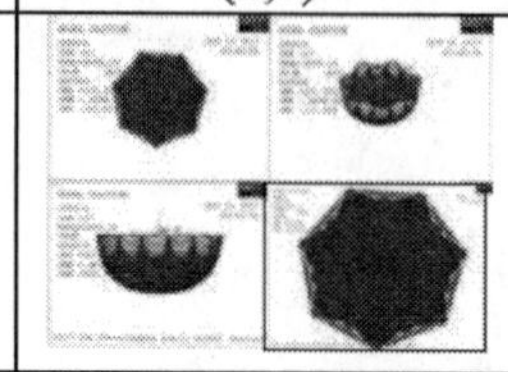	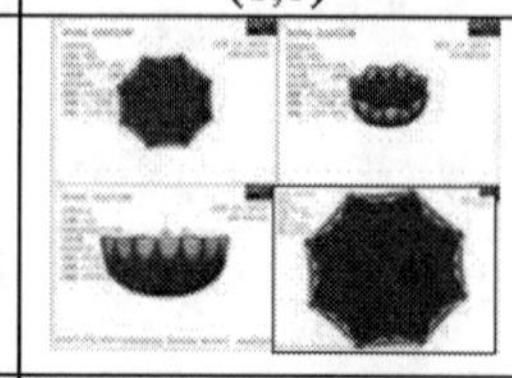	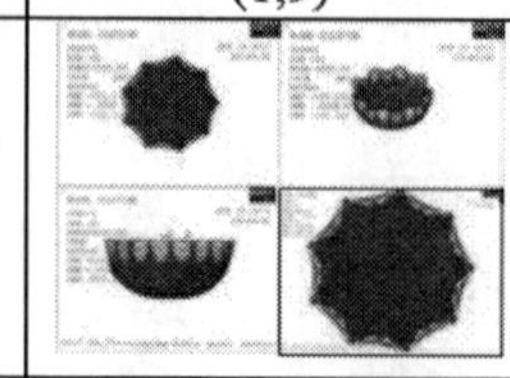
	F-15_1547.9 Hz	F-17_2065.1 Hz	F-22_2627.9 Hz	F-31_3228.5 Hz
EMA	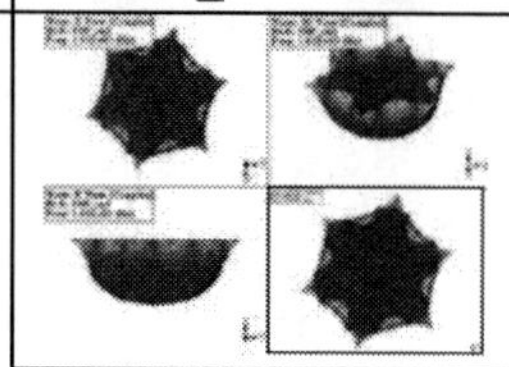	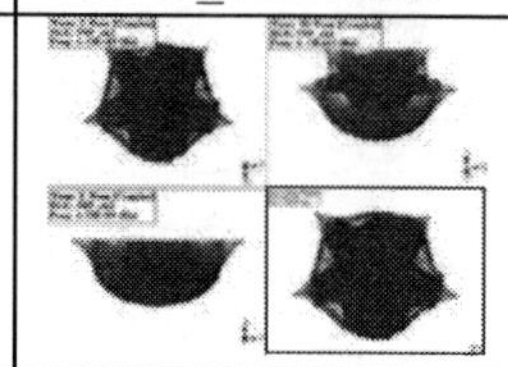	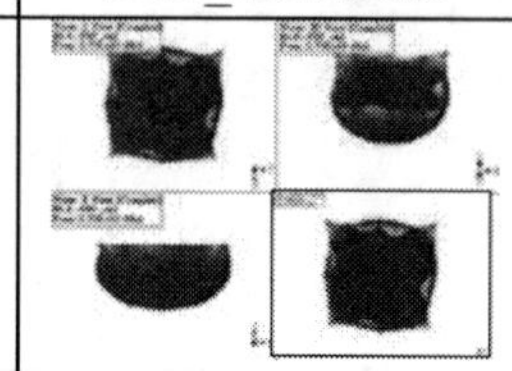	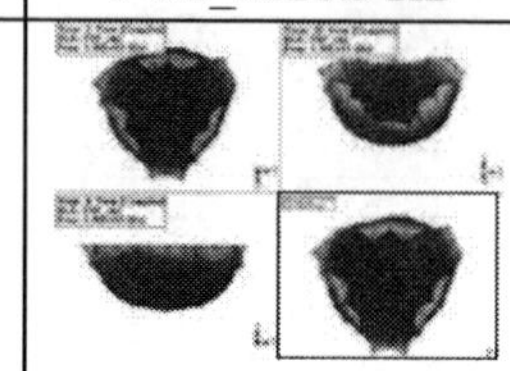
	O-07_1614.1 Hz	O-09_2152.3 Hz	O-11_2752.3 Hz	O-12_3357.8 Hz

Table 4. Theoretical mode shapes for modes (r, θ) from FEA.

r \ θ	0	1	2	3
1	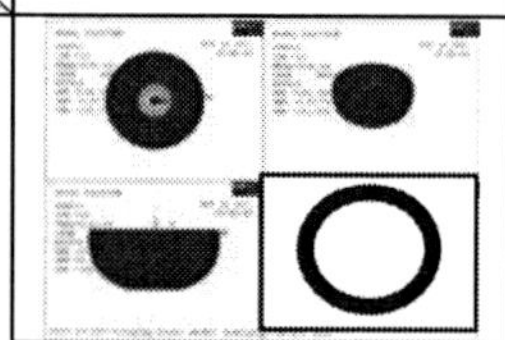 F-19_2142.8 Hz (r,θ)=(1,0)	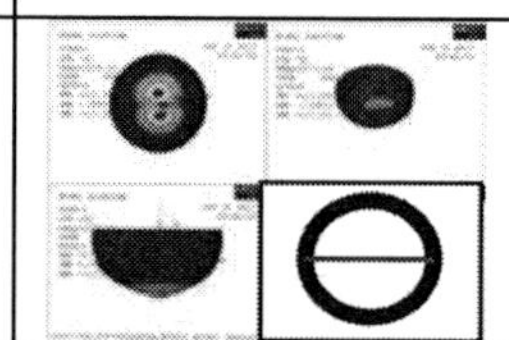 F-20_2579.3 Hz (r,θ)=(1,1)	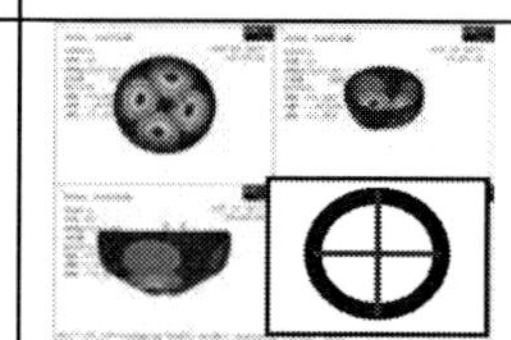 F-27_3111.7 Hz (r,θ)=(1,2)	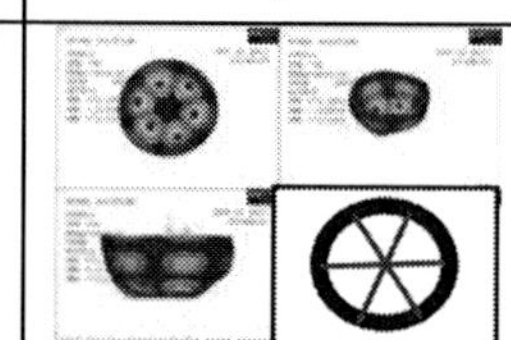 F-36_3414.9 Hz (r,θ)=(1,3)
2	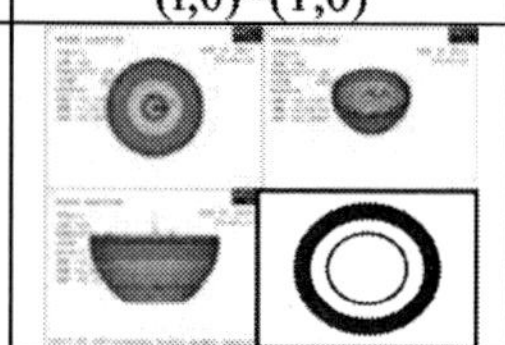 F-24_2737.4 Hz (r,θ)=(2,0)	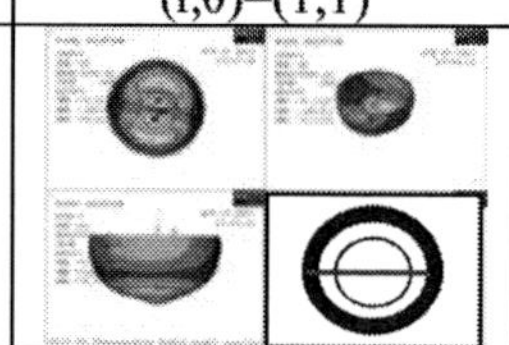 F-25_2908.7Hz (r,θ)=(2,1)	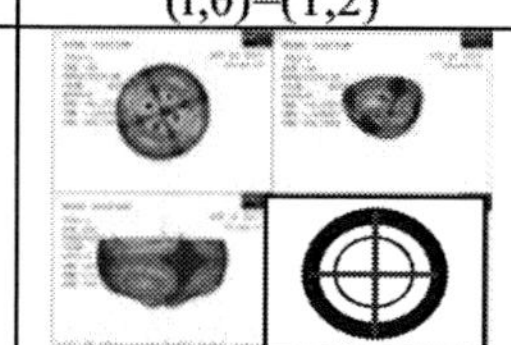 F-34_3319.9 Hz (r,θ)=(2,2)	
3	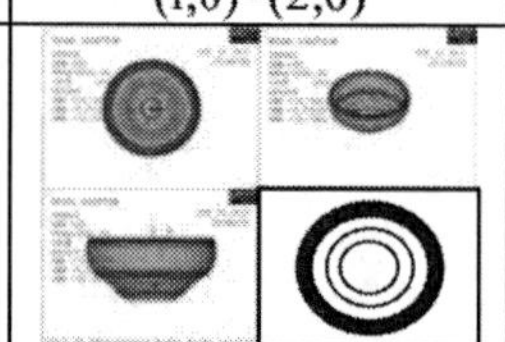 F-33_3245.8 Hz (r,θ)=(3,0)	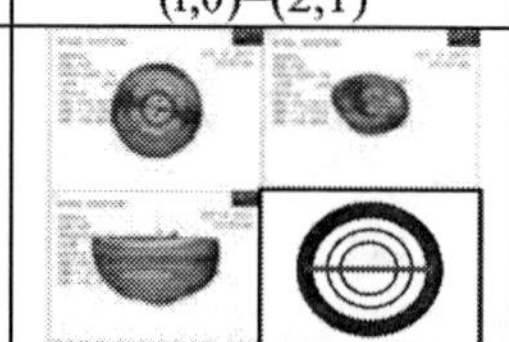 F-29_3205.7 Hz (r,θ)=(3,1)	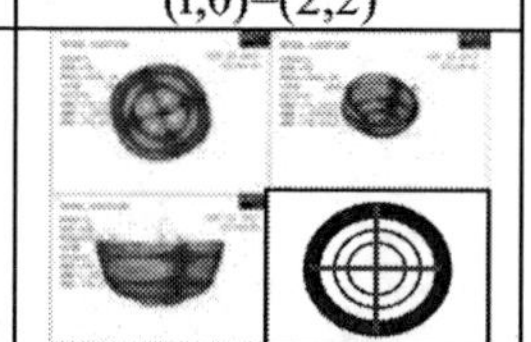 F-38_3486.4Hz (r,θ)=(3,2)	
4	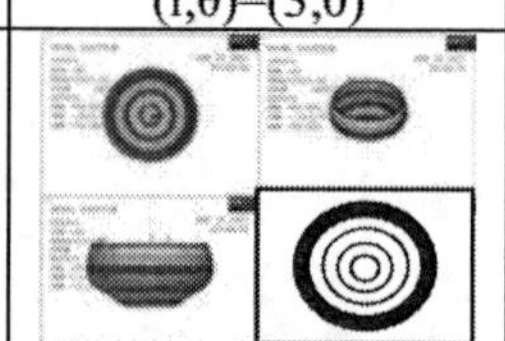 F-40_3511.0Hz (r,θ)=(4,0)			

3 Correlation of percussion sound to vibration modes

Section 2 shows the theoretical analysis of vibration modes and experimental verification of FE model for the singing bowl. The structural mode shapes have been physically interpreted for all modes that include the (z, θ) modes and (r, θ) modes. The interest is to know how the vibration modes affect the percussion sound of singing bowl.

The microphone is positioned near the singing bowl away from the edge about 20cm as shown in Fig. 3(a). The singing bowl is stuck by the soft tip stick at the top of ring surface. The time response of sound pressure is measured and performed spectral analysis to obtain the percussion sound spectrum.

Fig. 4 shows the sound spectrum of singing bowl A. On the top of sound spectrum for each peak resonance, the corresponding mode shapes from FEA and EMA are depicted as shown. One can observe that those peak resonances are related to the ring modes, i.e. the (z, θ) modes, and there is no (r, θ) modes in sound spectrum. This is the cause that the (r, θ) modes are not excited for striking on the ring surface.

Since the percussion stick is with a soft tip, the lower modes of vibration can be induced mostly. The sound responses at lower modes are higher, and the sound pressure level become descending at higher modes. The first peak resonance frequency about 118.4Hz can be identified as the fundamental frequency and the rest peak resonance frequencies are the overtones that affect the perceived sound quality.

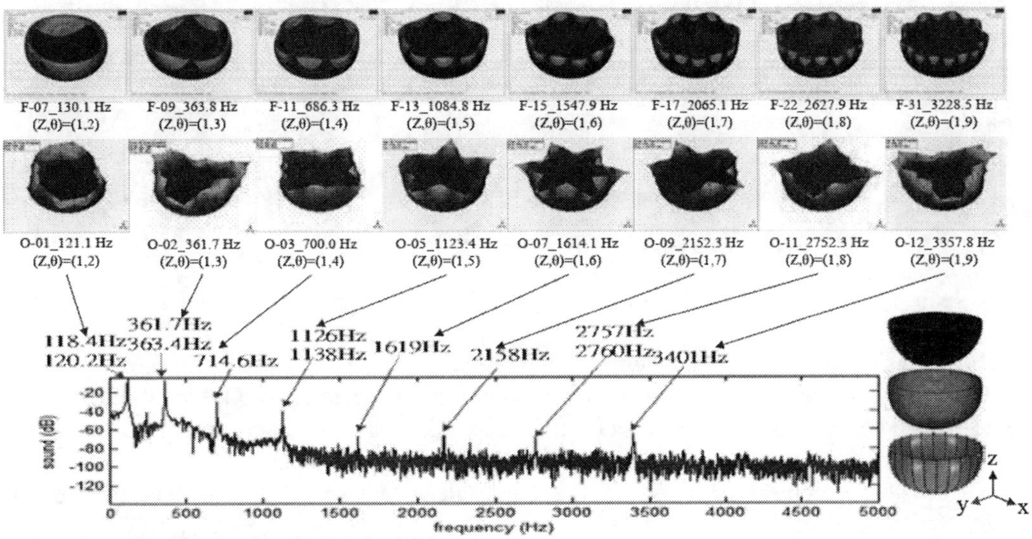

Fig. 4. Sound spectrum for singing bowl A.

4 Percussion sound characteristics for different singing bowls

Fig. 5 shows the picture of seven different sizes of singing bowls and number from 1 to 7. These singing bowls are designed as a set of bowls. The largest one is Number 1 with the radius of 358mm and the height of 148mm. The interest is how these different sizes of singing bowls sound with the same strike excitation. As we realize the sound generation mechanism for the singing bowl A that is similar to those singing bowls in Fig. 5 in term of geometry characteristics. Therefore, we can examine the sound spectrum and evaluate the sound quality of these singing bowls.

Fig. 5. Different sizes of singing bowls number 1 to 7.

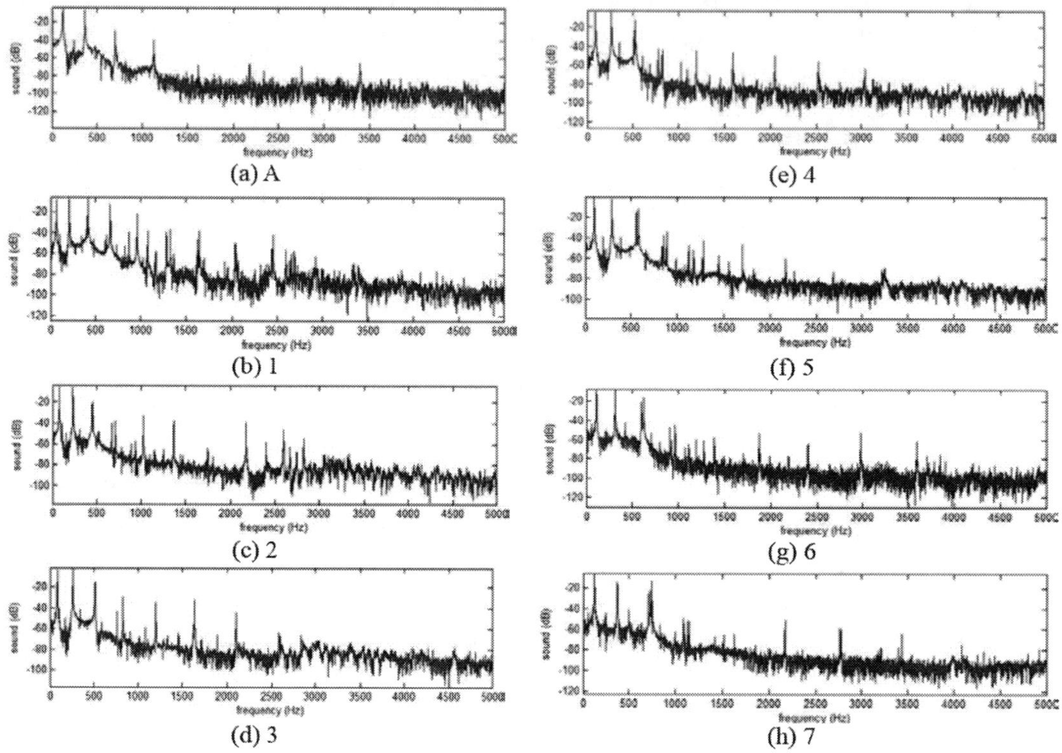

Fig. 6. Sound spectrum of singing bowl A and different sizes of singing bowls number 1 to 7.

The percussion sound measurement is preceded as discussed in Section 3. Sound spectrum for all singing bowls can be obtained and shown in Fig. 6. One can observe that the sound spectrums among these singing bowls are similar to each other; however, the peak response frequencies are different. As discussed, those peak resonance responses come from the structural vibration modes and will reveal the different perceived sound quality.

Table 5 summarizes the peak resonance frequencies from Fig. 6 for singing bowl A and Number 1 to 7. The first peak resonance frequency can be the (z, θ)=(1,2) mode and known as the fundamental frequency. The higher modes may also contribute to the radiated sound and known as the overtone frequencies. The frequency ratios in Table 4 are obtained by taking the overtone frequencies divided by the fundamental frequency. It is interesting to note that the frequency ratios for all singing bowls appear the values near integer ratios, especially 3 and 6 mostly. The harmonic sound effect is evident, and this is why the perceived sound quality for these singing bowls is felt harmony.

Table 5. Peak resonance frequencies for different singing bowls.

Bowl NO.	**A**		**1**		**2**		**3**		**4**		**5**		**6**		**7**	
Mode NO.	**Peak Frequency (Hz)**	**Frequency ratio**	**Peak Frequency (Hz)**	**Frequency ratio**	**Peak Frequency (Hz)**	**Frequency ratio**	**Peak Frequency (Hz)**	**Frequency ratio**	**Peak Frequency (Hz)**	**Frequency ratio**	**Peak Frequency (Hz)**	**Frequency ratio**	**Peak Frequency (Hz)**	**Frequency ratio**	**Peak Frequency (Hz)**	**Frequency ratio**
S-01	118.4	**1.00**	66.76	**1.00**	75.76	**1.00**	82.14	**1.00**	89.4	**1.00**	99.56	**1.00**	107.4	**1.00**	126	**1.00**
S-02	120.2	1.02	67.63	1.01	225.8	2.98	83.59	1.02	263.3	**2.95**	289.1	2.90	109.1	1.02	128.9	1.02
S-03	361.7	**3.05**	206.1	**3.09**	227.6	**3.00**	256.9	**3.13**	509.1	5.69	290.2	**2.91**	313.2	**3.15**	370.6	2.94
S-04	363.4	3.07	207.8	3.11	441.5	5.83	504.7	6.14	513.5	5.74	555	5.57	596.5	**5.99**	375.3	**2.98**
S-05	714.6	**6.04**	403.4	**6.04**	444.4	5.87	507.9	6.18	526.5	5.89	561.3	5.64	609.8	6.12	710	5.63
S-06	1126	9.51	414.8	6.21	451.6	6.76	513.7	6.25			577.9	5.80	623.7	6.26	722.7	5.74
S-07	1138	9.61	665	**9.96**	454.8	6.81	761.6	9.27			580.5	5.83	626.4	6.29	741.3	5.88
S-08	1619	13.67	960.1	14.38	667	**9.99**	827.5	**10.07**			851.6	8.55			750.3	**5.95**
S-09	2158	18.23	1079	16.16	671.6	10.06	1203	14.65			886.1	8.90				
S-10	2757	23.29	1330	**19.92**	708.2	10.61	1637	19.93								
S-11	2760	23.31			1026	15.37										
S-12	3401	28.73			1369	20.51										
S-13					2169	32.49										

Table 6. Fundamental frequencies for different singing bowls.

Bowl NO.	**1**	**2**	**3**	**4**	**5**	**6**	**7**
Fundamental Frequency(Hz)	66.76	75.76	82.14	89.4	99.56	107.4	126
Scale	C2	D2	E2	F2	G2	A2	B2
Scale Frequency(Hz)	65.41	73.42	82.41	87.31	98.00	110.00	123.47
Frequency Difference (Hz)	1.35	2.34	-0.27	2.09	1.56	-2.60	2.53
Frequency Error (%)	2.06	3.19	-0.33	2.39	1.59	-2.36	2.05

By examining the fundamental frequencies of the seven singing bowls from 1 to 7, Table 6 summarizes the fundamental frequencies of these singing bowls and compares the fundamental frequencies with musical note's standard frequencies. It is noted that C2-B2 are followed the scientific notation system for musical notes Do-Si. Note that the frequency differences are within 2 to 3 Hz, and the frequency error is about 2-3%. The set of seven singing bowls is intended for the design to produce musical notes from C, D up to B that is an octave of musical notes, i.e. Do, Re up to Si.

5 Conclusions

This work conducts theoretical analysis and experimental approach in examining the percussion sound characteristics of a set of singing bowls. A typical singing bowl is first studied to numerically and experimentally determine structural modal parameters. Base on the comparison of natural frequencies and corresponding mode shapes, the FE model of singing bowl can be validated and used to assist on characterize the vibration mode characteristics of singing bowls. The ring modes of singing bowl are the most dominated modes for percussion sound radiation. The local modes at the bottom side of singing bowl can also be identified and do not attribute to generate the percussion sound. The set of seven singing bowls is calibrated to have the fundamental frequencies from C to B of an octave of musical notes. In particular, the harmonic sound effect with the integer frequency ratios of near 3 and 6 is found and revealed a good perceived sound quality.

The authors thanks The Sanxi Collection Company for providing the singing bowls that were studied in this work. This work is also partially supported by Ministry of Science and Technology, Taiwan, under the contract number: MOST 105-2622-E-020-001 -CC3.

References

1. B.T. Wang, C.H. Huang, Y.H. Wu, *The 23nd International Congress on Sound and Vibration (ICSV23)*, Paper No.:440 (2016)
2. B.T. Wang, Y.L. Tsai, P.H. Chen, H.Y. Chang, K.T. Yu, *20th International Congress on Sound & Vibration,* Paper No.: 431 (2013)
3. J. Pan, *The Journal of the Acoustical Society of America*, **84**, S1, 135 (2009)
4. J.M. Lee, S.H. Kim, S.J. Lee, J.D. Jeong, H.G. Choi, *Journal of Sound and Vibration*, **257**, 779-790 (2002)
5. N. McLachlan, B.K. Nigjeh, A. Hasell, *The Journal of the Acoustical Society of America*, **114**, 1, 505-511 (2003)
6. N. McLachlan, *75th Acoustical Society of America Conference,* 1-8 (2004)
7. N.H. Fletcher, W.T. McGee, A.Z. Tarnopolsky, *The Journal of the Acoustical Society of America*, **131**, 3, 1437-1444 (2002)
8. A. Myers, R.W. Pyle, J. Gilbert, D.M. Campbell, J. P. Chick, S. Logie, *The Journal of the Acoustical Society of America*, **131**, 678-688 (2012)
9. A. Ezcurra, *Journal of Sound and Vibration*, **194**, 640-644 (1996)
10. R.R. Boullosa, *Applied Acoustics*, **63**, 311-322 (2002)
11. B.T. Wang, Y.C. Lan, C.H. Huang, Y.S. Wu, C.H. Chang, *2013 Annual Meeting and 26th Symposium of Taiwan Acoustical Association*, Paper No.: C5 (2013). (in Chinese)

A novel flexible tool with rotary ultrasonic spindle for fine grinding of tungsten carbide

Albert Wen-Jeng Hsue* and *Yi-Zhong* Zheng

Department of Mold and Die Engineering, National Kaohsiung University of Science and Technology, Taiwan

Abstract. Tungsten carbide is a typical difficult-to-cut material by conventional machining processes. In this paper, a novel design of flexible abrasives tool combined with a rotary ultrasonic machining (RUM) spindle is conducted to reduce the labor force significantly. The newly designed flexibility of tool-tip is aimed at preventing overcutting from the CNC grinding. The grinding conditions with resulted surface morphology of the tungsten steel were investigated through Taguchi design of experiment and ANOVA analysis. The machining capability of the novel flexible tool is compared with conventional tools through specific grinding paths under proper operational conditions.

1 Introduction

As a very strong mould material, tungsten carbide is usually difficult to cut by traditional machining processes. Its hardness was usually risen up to 85 to 91 HRA. At the other hand, the machined surface roughness by most of the CNC machining cannot reach mirror-like quality; and they must rely on secondary polishing for very period in the normal processing. In this paper, a novel flexible abrasives tool combined with a rotary ultrasonic machining (RUM) spindle is conducted to reduce the labour force for polishing significantly. The newly designed flexibility of tools is aimed at preventing overcutting from the general CNC machine grinding. The grinding conditions with resulted surface morphology of the tungsten steel were investigated and analyzed. The suitable operating conditions are studied through Taguchi design of experiments and ANOVA analysis.

The ultrasonic assisted machining (UAM) was consider as a suitable option for the processing of difficult-to-cut materials in the past decades [1]. In the recent years, the rotary ultrasonic machining (RUM) spindle was further proposed to successfully assist the machining processes of hard and brittle materials [2-3]. It provides much better properties than UAM in both increasing tool life and reducing the cutting force. Therefore, it is emerging in the production fields up to date. For example, Shen et al. [4] revealed that RUM can increase the material removal rate of ceramics, although there were micro cracks remained at the cutting edge. But the self-dressing of tool abrasives by using the RUM spindle seems trivial since the impact effect of rotary ultrasonic causing the small breakage

*Corresponding author:albert.hsue@kuas.edu.tw

of abrasives. Guo et al. [5] verified that the surface integrity of cutting edge of brittle materials conducted by RUM is better than that by UAM. They also confirmed the superior quality of RUM over conventionally done by the RAM ultrasonic machining.

For fundamental insights of grinding, Ismail et al. [6] proposed a concept of effective radius to describe the region of proper geometrical transcription and better surface roughness conducted with RUM assisted grinding. The quantity of remaining abrasives resident within the circle of that effective radius is an essential index to evaluate the grinding effect of RUM assistance. It is also a good index for evaluation of worn-out or tool life.

Wang et. al. [7] proposed a concept of system matching for RUM type of grinding of titanium in 2014. They reveal the interaction between main factors, and the analysis technology for improving surface integrity under the system matching conditions. Wang et. al. [8] also investigated the grinding of brittle materials further with this system concept in the same year. They further establish a predictive model for the grinding force and the surface integrity with RUM effect. Before that, Hu et. Al. [9] have proposed a report on modeling of material removal rate in rotary ultrasonic machining. They simulate the effects of main factors on the material removal rate by establishing a removal model through design of experiments (DOE) and ANOVA analysis.

In the lab, Hsue and Wang also proposed a series of RUM drilling and grinding experiments on brittle materials, such as ZrO2, Sapphire and Na-Ca glass through Taguchi-ANOVA analysis in the 2015 [10]. They also investigated the machining resistance response, tool life and the effects of the paths (cycloid VS. linear) on the removal efficiency and the surface integrity. It confirmed that RUM provided much longer tool life, smaller machining force and lower roughness compared to the conventional grinding processes. However, the material removal rate (MRR) depends on not only the ultrasonic assistance, but also the operational conditions, and even the machining paths. In which, cycloid paths provided not only higher MRR but also better roughness under the assistance of RUM.

This research is aimed at experiments with a newly designed flexible tool and contribution to the polishing processes of a mould made of tungsten carbide. The maximum MRR and the minimum three-dimensional surface roughness with reasonable power consumption were set as the targets. Also, the polishing resistance reflected by the RUM machining processes are investigated.

First of all, the Taguchi's L16 orthogonal experiments are conducted using the rough abrasives of # 400 meshes tool. The major contribution factors for enhancing material removal rate are investigated. The optimized combination of operational conditions is also analysed by ANOVA and compared with two kinds of tool holders. Secondary, the fundamental polishing investigation by using #1000 abrasives in grain size of 13 μm and proper cutting depth combined with the RUM spindle are conducted. Their polishing forces vs. feed rate, and the spindle speed are also investigated and analysed. For further improvement of finishing processes, effects of cycloid paths with and without flexible tool under RUM are compared to conventional linear paths.

2. Experiment setup and design of the flexible tool

2.1 Experimental setup

To investigate the effect of the novel tool for RUM grinding, we adopted a RUM spindle built on a machnie tool, VU-5, by a local machine builder, Tongtai machine & tool Co. The ultrasonic power consumption of its rotary spindle can be adjusted from 0% to 100%, therefroe, with certain power and without ultrasnoic vibrataion can be compared in the experiments. As shown in Figure 1, the main component of the experimental setup is described as followed. In which, (1) indicates the RUM spindle, (2) : flexible tool holder and the diamond abrasive tool, (3) : workpiece, (4) : fixture jig for workpiece, (5) : six axes force sensor, (6) : main clamper vise mounted on machine table, (7) : Reinyshaw non-contact tool calibrator. In which, the fixture platelet (4) milled to a square cavity in advanced i saimed at holding workpiece and connected to the force sensor. The 6 axial force sensor (5) with a protection acrylic box was then mounted on the main vise (6). The cutting force exerted on workpiece was measured and amplified through a signal conditioner and then acquired by a notebook PC through DAQ with USB interface. To protect the entire instrument setup from machining fluid flushing, they are covered with a sheet of plastic cloth as shown in Figure 1(a).

As shown in Figure 1(b), this research adopts tungsten carbide in type C01 as workpiece sample, which was fabricated by Cheng-Chang carbide Co., Ltd. Its grain size distributes from 1.0 to 2.0 μm, with the density of 15.0 g/cm^3, hardness of 90-90.5 HRA. It is typically a sort of strong mold material for fastening components such as drawing and forging of the bolt heads. Typically, tungsten-carbide is a chemical compound containing equal parts of tungsten and carbon atoms. In its most basic form, tungsten carbide is a fine gray powder, but it can be pressed and formed into shapes for use in industrial machinery, cutting tools, moulds, other tools and instruments and etc. Tungsten carbide is approximately twice as strong as steel, with a strong Young's modulus. It is so tough with comparable hardness with corundum (α-Al_2O_3) and can only be polished and finished with abrasives of superior hardness such as cubic boron nitride (CBN) or diamond powder, and wheels.

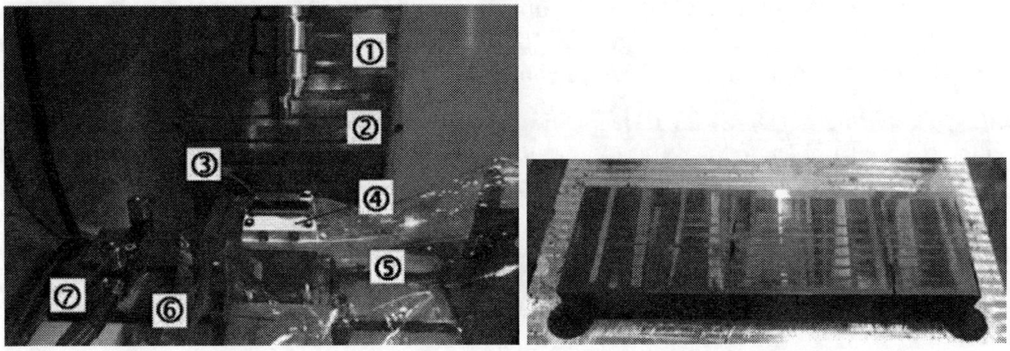

(a) (b)

Fig. 1. (a)Setup of the RUM spindle with flexible abrasive tool, (b) Tungsten carbide after grinding testing.

2.2 Design of experiment - preparing and planning

This research investigates the MRR and surface integrity of RUM grinding with the newly developed flexible RUM tool. In the rough machining stage, main factors include feedrate, RUM consumption power, spindle speed, and grinding depth. The surface terrain of work after tests is also investigated. Preparing processes adopted a CBN ceramic milling tool with 16 mm diameter, succeeded with a 6mm diameter of #230 tool and another #400 abrasive tools for rough grinding.

The assembly of the proposed flexible tool holder is depicted in a conceptual plot as shown in Figure 2, and the customized abrasive tool with diameters 12mm and 6mm are shown in Figure 3. The flexible tool holder involvoes a screw nut, two sheets of spring ring, an innner shaft for holding tool with inner threads, linear sleeve bearing for holding the innner shaft, fix-pin for fixing the innner shaf to the tool holder's shell. Then, the tool holder combined with customized abrasive tool was adopted to the spindle of the machine tools.

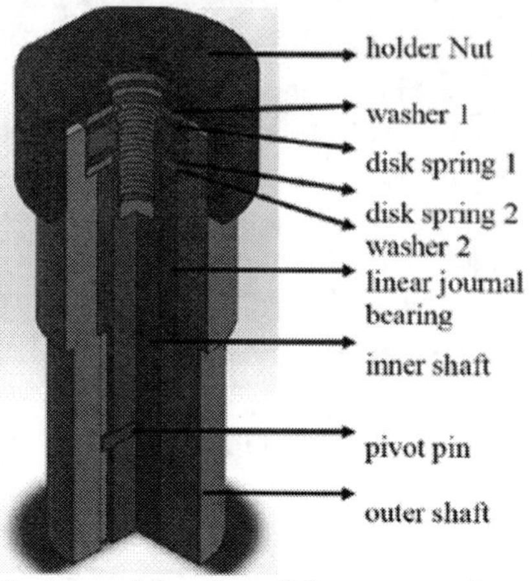

Fig. 2. Assembly plot of the proposed flexible tool holder.

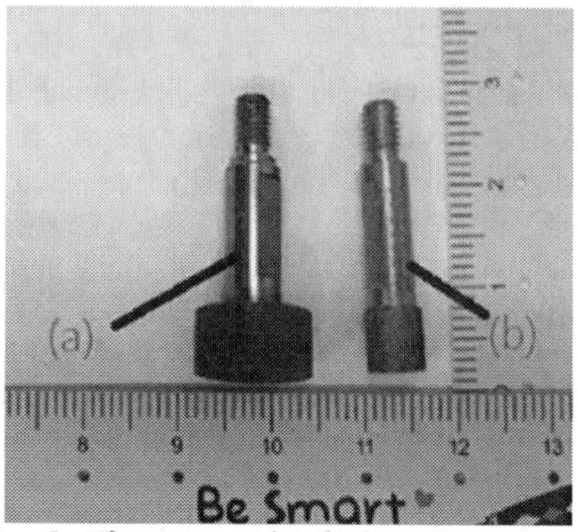

Fig. 3. The customized abrasive tool with diameters of (a) 12 mm and (b) 6 mm.

Preliminary plans from the roughing to finish grinding were first designed. Rough grinding was first conducted with abrasive tools of mesh #230 (62μm) and #400 (38μm grain), combined with surveying factors include tool types and grinding depth. The spindle speed is fixed at 5 krpm, feed rate at 10 mm/min, and with RUM in 100% power or without RUM (0%), and grinding depths from 2, 4, 6, 8 to 10μm. For finishing stage, spindle speeds was conducted from 1k to 8k rpm with mesh #1000 abrasive tool (13μm grain) with fixed depth 3μm and fixed feedrate of 10 mm/min. And then, machining feedrate was conducted from 5, 10, 20, 40 to 80 mm/min with fixed 5k rpm and the fixed depth 3μm.

Design of experiment (DOE) for RUM grinding with the proposed flexible tool holder was conducted based on Taguchi-ANOVA method. The middle-finishing experiment through L16 orthogonal array adopted a 6mm diameter tool with mesh #400 abrasives. The surface integrity was measured by a KEYENCE VK-X200 laser microscope, and the removal volume was estimated from the machined slot based upon the measured profiles.

Table 1. L16 orthogonal array for midterm-finishing of Tungsten carbide by DOE.

Test No	*Spindle (rpm)*	*Feedrate (mm/min)*	*depth (μm)*	*Tool-hoolder*	*RUM*
1	4000	10	3	Fixed	on
2	4000	20	6	Fixed	on
3	4000	30	9	Flexible	off
4	4000	40	12	Flexible	off
5	5000	10	6	Flexible	off
6	5000	20	3	Flexible	off
7	5000	30	12	Fixed	on
8	5000	40	9	Fixed	on
9	6000	10	9	Fixed	off
10	6000	20	12	Fixed	off
11	6000	30	3	Flexible	on
12	6000	40	6	Flexible	on
13	7000	10	12	Flexible	on
14	7000	20	9	Flexible	on
15	7000	30	6	Fixed	off
16	7000	40	3	Fixed	off

Design of experiment for final finishing through L9 orthogonal array adopted an abrasive tool with #1000 mesh grits. The contribution of dominant factors and the optimum combination of the operation conditions are derived from Taguchi-ANOVA analysis. The optimization combination is then carefully verified by a set of confirmed experiments.

Table 2. L9 DOE orthogonal array for final finishing by flexible tool.

Test No	*Spindle (rpm)*	*Feedrate (mm/min)*	*Power of Ultrasonic generator (%)*
1	4000	9	0
2	4000	12	30
3	4000	15	100
4	5000	9	30
5	5000	12	100
6	5000	15	0
7	6000	9	100
8	6000	12	0

3. Experiment results and discussions

3.1 Response of surface integrity to grinding depth

The first set of rough experiments adopted a 6 mm abrasive tool with #230 grits (about 62μm). The fixed spindle speed at 5k rpm results in a tangential velocity of 1.57 m/s, and feedrate is fixed at 10 mm/min. Effect of grinding depth with RUM and without RUM, associated with fixed and flexible abrasive tool is compared and discussed as follows.

Tree dimensional surface roughness (Sa) with the fixed and the novel flexible tools are conducted and compared, respectively. With the fixed tools, the surface roughness decades with the grinding depth, no matter if the RUM is applied. But the minimum seems appeared at the 6μm depth. The fixed tool with RUM produced results no better than that without RUM. Although the roughness trend looks similar for those with flexible tools, however, the flexible RUM seems much more sensitive than fixed tools and produced better roughness than that without RUM, but the deviations at shallow depth (which is less than or equal to 4μm depth) are relatively large. The minimum roughness achieved Sa 0.48μm at 8μm depth is concluded.

Effect of grinding depth on the profiles is conducted from profile measurement through Mitutoyo SJ-400 instrument. As shown in Figure 4, the material removal rate can be estimated from the cross-section area of approximated trapezoid in machined slot. The fixed tool with RUM, as shown in Figure 4(a), produced deeper and more integrity profile than that with flexible tool in Figure 4(b).

The MRR wth RUM and No-RUM are calculated from the product of grinding length and this cross-section area as shown in Figure 4. The removal ability of both fixed tool and the flexible rool looks integrity. The MRR is in a positive proportional relationship to grinding depth for both cases. But with the fixed tools, as shown in Figure 5(a), the MRR of both wth RUM or No-RUM produced no obvious difference except lower depth under 4 μm. However, from the Figure 5(b), the MRR without RUM is almost 3 times of that with RUM, revealing the flxiblity of novel tool combined with RUM producees less removal volume rate.

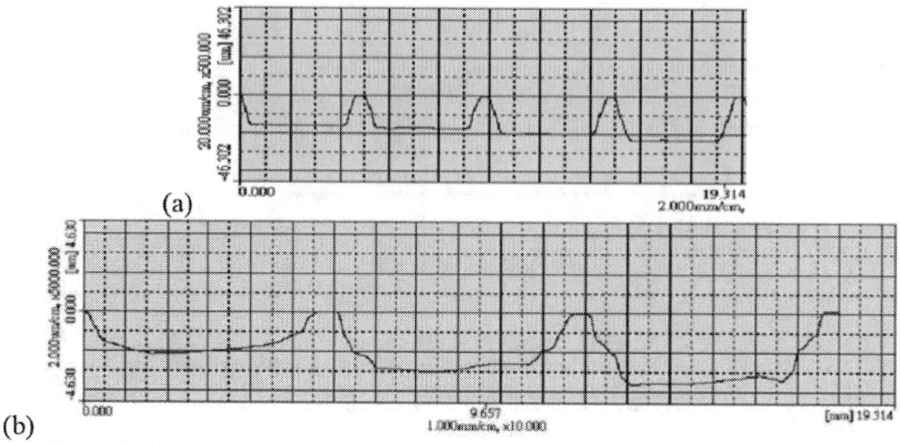

Fig. 4. Effect of grinding depth on machined profiles without RUM, but with (a) the fixed tools 4, 6, 8, 10μm, (b) the flexible tools 4, 6, 8μm.

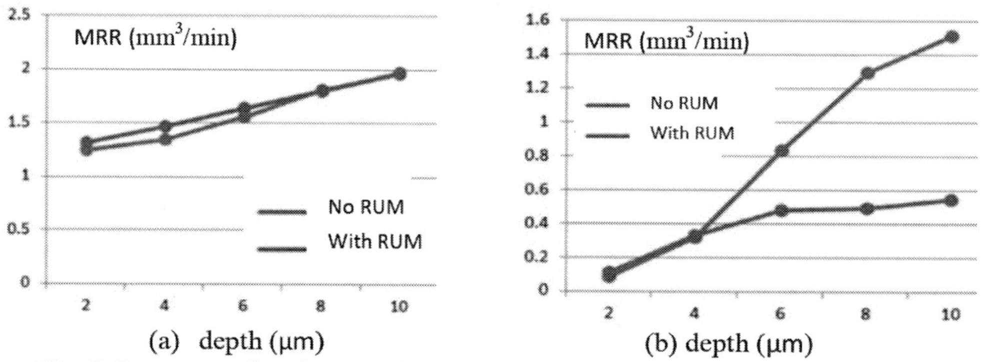

Fig. 5. Response of surface roughness to grinding depth with (a) the conventional fixed tools, and (b) the flexible tools.

As shown in Figure 6, the grinding force is in a positive proportional relationship to grinding depth for both cases. No matter flexible tool or fixed tool types, as shown in Figures 6 (a), the grinding force of with RUM can reduce the grinding force significantly compared to that without RUM. From the Figure 6(a), the resistance force without RUM is almost 2.3 times of that with RUM conducted in depth of 10um for conventional. At the other hand as shown in Figure 6(b), it is almost 3 times difference for flexible tool, revealing the effect of RUM on saving tool [illegible]ng.

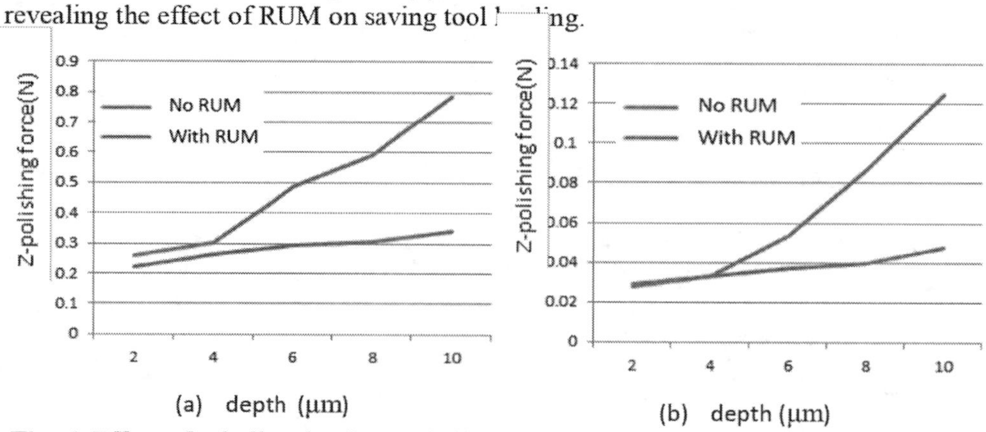

Fig. 6. Effect of grinding depth on grinding force with (a) the conventional fixed tools, and (b) the flexible tools.

Therefore, the Z-axis polishing force increases with the increasing grind-depth, and decreases with the spindle speed. Generally, the polishing force by using RUM is smaller than that by no-RUM and the difference gets larger as the grinding depth gets deeper.

3.2 Response of surface roughness to the spindle speed and the feed rate

The experiments since from this section are all conducted with the novel flexible tools with 6 mm diameter and the abrasive diamond grains of #1000. The experimental aims at the response of surface roughness and profile to the spindle speed, through 1krpm to 7k rpm, or tangential velocity of 19 m/s to 151 m/s. Its grinding depth is fixed at 3μm associated with 10mm/min of feedrate, and the machining path is a single round trips in a linear way (go forth and come back to its origin). As shown in Figure7, the 3D surface roughness reveals decreasing trends as the spindle speed increases, no matter if rotary ultrasonic wave is applied or not. However, with ultrasonic assistance its Sa is apparently depressed both at the lowest speed and at higher speed above 132 m/s. The best Sa 0.25μm appears at 132 m/s, which is about 24% improved from the case without ultrasonic assistance.

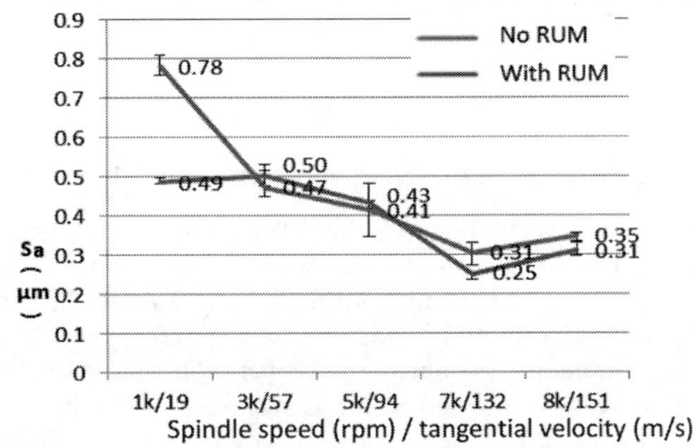

Fig. 7. Response of surface roughness to the spindle speed.

On the other hand, the response of surface roughness and profile to the feedrate is conducted with fixed spindle speed at 5k rpm (94 m/s). As shown in Figure 8, As the feedrate increases from 5 mm/min to even 40mm/min, there is a decreasing trends for Sa, no matter with rotary ultrasonic assistance or not. In general, the ground surface roughness with ultrasonic is better than that without ultrasonic. Therefore, the best 3D surface roughness appears at 20 mm/min for Sa 0.56μm with ultrasonic assistance.

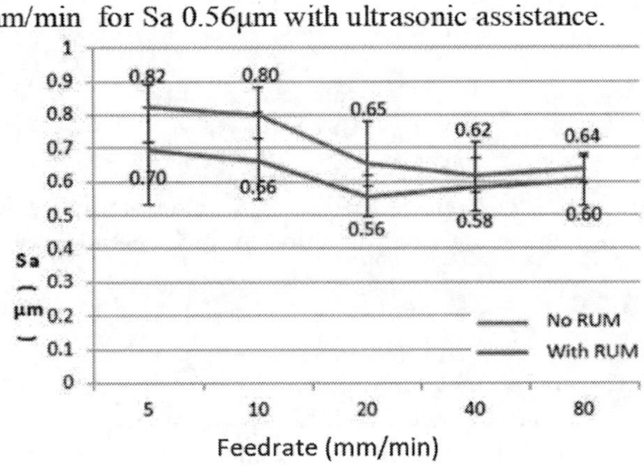

Fig. 8. Response of surface roughness to the machining feed rate.

3.3 Finishing experiments and comparison on polishing paths

As shown in Figure 9, by using #400 grit abrasives tools, the S/N ratio of MRR for a mid-term finishing experiments is obtained according to the L16 array in Table 1. From the chart of S/N ratio, the dominant contribution is evaluated and ranked as Grinding depth, Spindle speed, Feed rate, and level of RUM. The optimum combination of operation conditions for the highest MRR is 5k rpm, feed rate at 5 mm/min, 12μm depth, with RUM and the flexible tool. From the first two columns, the higher feedrate associated with the speed above 5k rpm causes poor MRR. The reason is due to the collapse of the diamond abrasive from the tool-tip in those conditions. Since removal volume is estimated from the product of the actual cross-section as shown in Figure 4. And saving the grinding efficiency, the worn out of tool tip reduces the removal rate significantly.

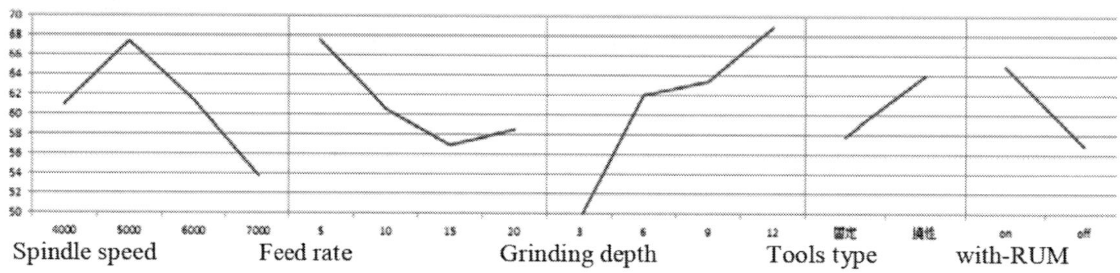

Fig. 9. S/N ratio of MRR for a midterm finishing (#400 grit) from L16 array.

By using 10 mm diameter tools with abrasives of #1000 grits, as shown in Figure 10, the S/N ratio of surface roughness for final finishing experiments in a L9 array (Table 2) is depicted. The grinding depth is fixed as 3μm, with flexible tool. According to ANOV analysis, the rank of dominant contribution to the roughness is Spindle speed, Ultrasonic power level, and Feedrate. The optimum combination of operation conditions for the lowest roughness is 5k rpm, 9 mm/min, and 100% full RUM power, and it contributed to Sa 0.454μm. The confirm experiment revealed a Sa 0.418μm, deviates 7.6% from the prediction result. Observation from Figure 10 reveals above 5~6 k rpm, the tool worn out cancels the advantage of high spindle speed on surface roughness. Moreover, higher ultrasonic power level seems improve the surface roughness a little bit.

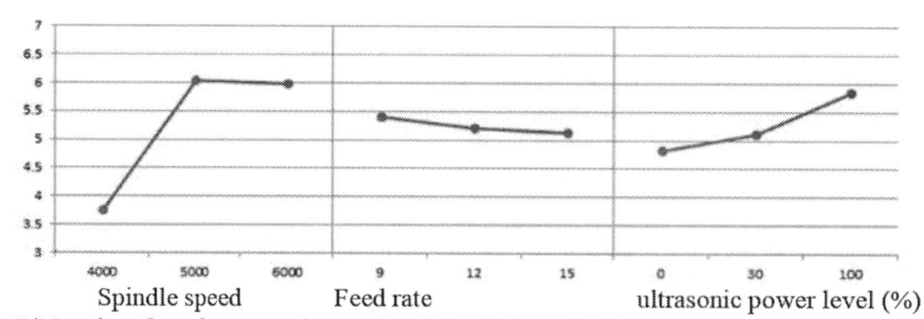

Fig. 10. S/N ratio of surface roughness for final finishing in linear path (flexible tool)

Another strategy to adopt cycloid path as shown in Figure 11, the Taguchi Finishing L9 experiments reveal similar trends as that in Figure 11, except the effect of ultrasonic power. The optimum combination of operation conditions for the lowest roughness is 5k rpm, feedrate 9 mm/min, without RUM power, and it contributed to Sa 0.172μm according to the ANOVA analysis. However, confirm experiment revealed a Sa 0.185μm, deviates 7% from the prediction result.0.172μm, which is much smaller than that roughness by linear path as shown in Figure 10.

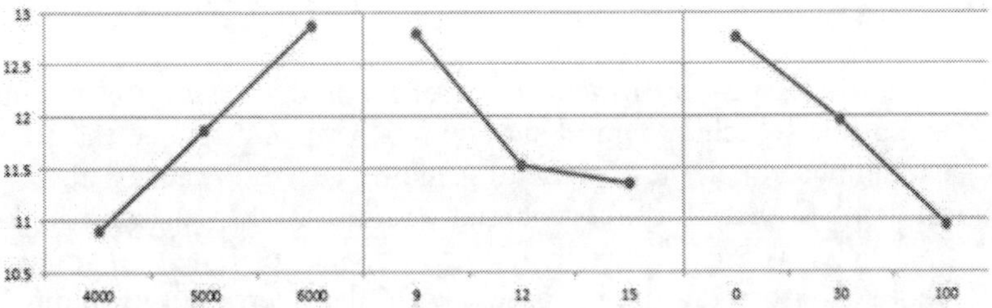

(a)Spindle speed (b)Feed rate (c) ultrasonic power level (%)

Fig. 11. S/N ratio of surface roughness for final finishing in cycloid paths.

Therefore, the L16 Taguchi experiments are conducted by using the tool with rough abrasives of # 400 meshes. The major contribution factors for enhancing material removal rate (MRR) are grinding depth > spindle speed > feed rate. The optimized combination by ANOVA is the set of spindle speed 5000 rpm, feed rate 5mm/min, grinding depth of 12 μm, with flexible tool holder, and with ultrasonic assistance. Its highest MRR can reach up to 0.72mm3/min. For the quality of surface roughness (SR), the main contribution factors are kinds of tool holder> grinding depth> spindle speed. The optimized parameters are under spindle speed 5k rpm, 5 mm/min, depth 3μm, with a flexible tool holder, and without ultrasonic assistance. Its minimum roughness reaches up to Sa 0.44μm in a good machinability.

The better performance of the carbide conducted through the cycloidal flexible grinding with rotary ultrasonic spindle was shown in Figure 13. The predicted surface roughness Sa is computed as 0.172μm. While the confirmation experimental obtaining from the DOE by Taguchi method are Sa roughness of 0.185μm, with residual error 7.6% comparing with that predicted surface roughness. And the flatness is 0.29μm, respectively. As shown in Figure 13 (b), along the measuring of 15.44mm length on the surface, there remained trivial revolution marks of tool path. The terrain caused by tool marks was risen from -0.41-0.20μm. It was caused by the tool inclination and the vibration of the abrasives on the tool tip.

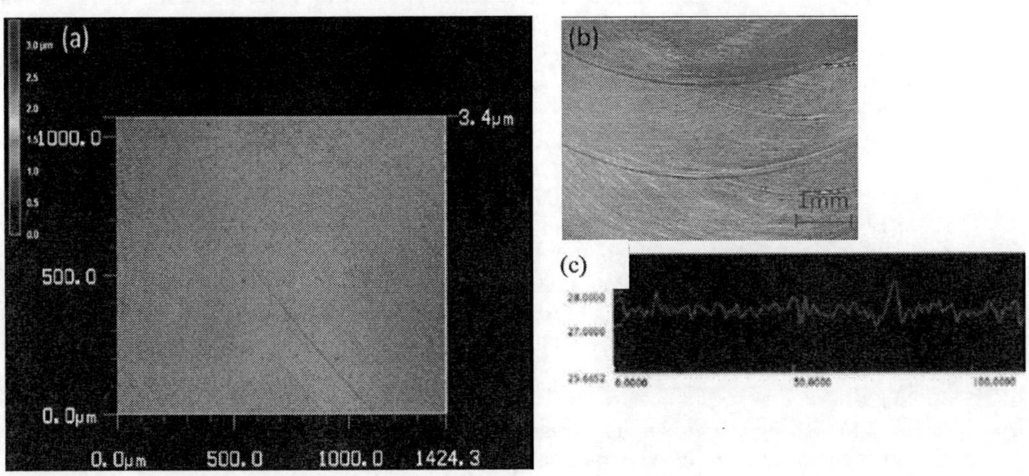

Fig. 12. (a) Surface of tungsten carbide after cycloidal path polishing with (b) macro view, and (c) its profile remained.

4. Conclusions

Form above discussion, it is confirmed the novel flexible abrasives tool combined with a RUM spindle is feasible to achieve good surface integrity, and reduce the labor force with regular CNC machining. For the fundamental grinding survey by using abrasives of 13μm grits and 3μm of depth at that 7k rpm combined with the RUM, surface roughness reaches about Sa 0.25μm, which is 24% better than conventional no-RUM polishing. It is worthy noticed that Z-axis resistant force increases with the increasing grinding depth, and decreases with the spindle speed. Generally, the polish force by using RUM is smaller than that by no-RUM, and their difference gets larger as the polish depth gets deeper. The surface roughness also decreases with the increasing spindle speed and the deceleration of the feed rate, respectively.

For further finish polishing, Taguchi design and ANOVA analysis reveal that for cycloid path, with feed rate of 9 mm/min, 6000 rpm and without RUM can reach the best flatness of 0.29μm, and the lowest surface roughness of Sa 0.185μm with reasonable tool life. Based upon the above results, therefore, this paper confirms the feasibility of the newly designed flexible tool and contributes this hybrid polishing strategy of the tungsten carbide.

Acknowledgement

Financial support from both Tongtai machine & tool Co., Ltd and National Kaohsiung University of Science and Technology, and the technology support from Dr. H.P. Chen and Mr. M.J. Chang are gratefully acknowledged.

References

1. P. Legge, Ultrasonic drilling of ceramics, Ind. Diam. Rev. **24**, 20-24 (1964)
2. Mohsen Ghahramani Nik, Mohammad R. Movahhedy, Javad Akbari, Ultrasonic-assisted grinding of Ti6Al4V alloy, Procedia CIRP, 353-358 (2012)
3. Hu Gong, F.Z. Fang, X.T. Hu, 2010, Kinematic view of tool life in rotary ultrasonic side milling of hard and brittle materials, Int. J. of Mach. Tools and Manuf. **50(3)**, 303-307 (2010)
4. J.Y. Shen, J.Q. Wang, B. Jiang, X.P. Xu, Study on wear of diamond wheel in ultrasonic vibration-assisted grinding ceramic, Wear **332–333**, 788-793 (2015)
5. B. Guo, Q. Zhao, Ultrasonic vibration assisted grinding of hard and brittle linear micro-structured surfaces, Prec. Eng. **48**, 98-106 (2016)
6. M. Fauzi Ismail, K. Yanagi, H. Isobe, Geometrical transcription of diamond electroplated tool in ultrasonic vibration assisted grinding of steel, Int. J. Mach. Tools & Manuf. **62**, 24-31 (2014)
7. Taghi Tawakoli, Bahman Azarhoushang, Influence of ultrasonic vibrations on dry grinding of soft steel, Int. J. Mach. Tools & Manuf. **48**, 1585-1591 (2008)
8. Y. Wang, B. Lin, X. Cao, S. Wang, An experimental investigation of system matching in ultrasonic vibration assisted grinding for titanium, J. Mater. Proc. Tech. **214**, 1871-1878 (2014)
9. Y. Wang, B. Lin, S. Wang, X. Cao, Study on the system matching of ultrasonic vibration assisted grinding for hard and brittle materials processing, Int. J. Mach. Tools & Manuf. **77**, 66-73 (2014)
10. P. Hu, J.M. Zhang , Z.J. Pei, C. Treadwell, Modeling of material removal rate in rotary ultrasonic machining: designed experiments, J. Mater. Proc. Tech. **129**, 339-244. (2002)
11. C.J. Wang, Albert W.J Hsue, The optimal combination of operational conditions and the tool wear characteristics of direct coupled rotary ultrasonic machininng, Proceedings of the Conference ICPMMT 2015, Kenting, Taiwan, May 21-24 (2015)

Light and eco-friendly structure design of an instant noodle bowl

Jiunn Fang[1] and *Po-Chuan* Chen[2,*]

[1]Department of Aerospace and Systems Engineering, Feng Chia University, No.100, Wenhua Rd., Xitun Dist., Taichung City 40724, Taiwan (ROC)
[2]Ph.D. Program of Mechanical and Aeronautical Engineering, Feng Chia University, No.100, Wenhua Rd., Xitun Dist., Taichung City 40724, Taiwan (ROC)

Abstract. Instant noodles have become a very common food in recent years, and because of its convenience and affordability, it is now one of the most profitable products in the food industry. While it is a very convenient food product to carry around, its packaging tends to be inconvenient and not eco-friendly. The packaging of instant noodles can be categorized into two different types: The first is small, and very light, packed in plastic bags, requiring that customers supply their own bowl in order to enjoy the food; the other is packed in a paper, or Styrofoam bowl, usually with additional space in the bowl for customers to add water. The dimensions of the bowl make it less portable, and the construction materials are not eco-friendly. We build a model design that combines the benefits of each packaging type (Fig.2). The size remains as small as the plastic bag type, while simultaneously having the expandable space of the bowl type. The proposed model is made compact during shipping, but its foldable design allows the customer to readily reconstruct it into a bowl for use. Using recycled paper and PE film for materials, we have ensured that, with the use of special recycling technology, the packaging is 100% recyclable. The result is an eco-friendly, light, and purposeful design.

1 Introduction

Today, technological advancements in transportation have come to provide individuals with countless opportunities for both domestic and international travel. For those who travel often it can be easy to miss the comforts of home. For these flight crewmembers, or jet setters, the convenience and familiar comfort of instant noodles make them a popular choice. The traditional instant noodle however, lacks convenience due to its packaging.

Instant noodles were invented by Ando Momofuku, a Taiwanese immigrant living in Japan. The inspiration for inventing instant noodles came when Momofuku observed that the queues for ramen restaurants were incredibly long. The traditional instant noodle packaging can be divided into two kinds: bowl and bag. Each packaging type has its pros and cons. We developed a design that combines the strengths of each type into one—a light

[*]Corresponding author: chenname0317@gmail.com

packaging containing a foldable bowl structure. This packaging is also made of recycled paper and PE film to achieve the goal of eco-friendly manufacturing.

1.1 Traditional instant noodle packaging analysis

1.1.1 Bag-packed instant noodles

The first generation of instant noodles was packed in small bags. The resulting product is very small, light, and easy to carry around. This method, however, requires the consumer to inconveniently travel with a bowl in which to assemble the ingredients to make the food.

1.1.2 Bowl-packed instant noodles

The bowl-packed instant noodles were also invented by Ando Momofuku in 1791. The reason for Momofuku inventing the bowl-packed instant noodles was to attract Americans, who, at that time, would commonly only have plates in their home. For them, making this style of Asian cuisine could be very inconvenient. The first generation of bowl packaging was made with polystyrene, but its high cost made the retail price 3 times higher than that of the bag-style instant noodles. People refused to pay that much, encouraging Momofuku to make slight changes to the ingredients, and the tableware, to make it popular.

1.2 Recycled paper with PE film

The recycled paper procedure and technology have now well developed. In recycled paper tableware, for example, the technology today allows for 60-70% reuse of the recycled paper. This means that 60-70% of recycled paper can be remade into eco-friendly products.

By using this recycled material with PE films, we are now able to make a new and eco-friendly bowl. After the bowl has been used, it is able to be recycled repeatedly, until it is impossible to be reused. This concept allows people to enjoy the convenience with little harm to the planet.

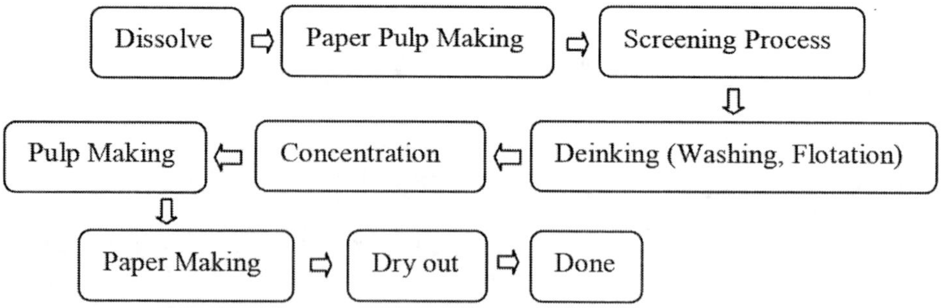

Fig. 1. Procedure for Recycling and Reproducing Paper Tableware.

1.3 Bio-degradable packaging

Our target is to make the whole product a degradable or recyclable item, so the materials we choose must satisfy this condition. Today, many bio-degradable materials are available, and because we are using the material in contact with food, we require one that does not contain harmful chemicals. For the wrapping, we chose a bio-degradable plastic made of Starch, Polylactic Acid and fiber protein. The use and composition of this material differs very little from traditional plastic, except in the ability to be degraded by Aerobic microorganisms, water, and oxygen in the soil to form carbon dioxide and water. This crucial difference ensures that we achieve the target of being eco-friendly, while simultaneously protecting the food from bacteria.

2 Design and methods

We have completely redesigned the traditional instant noodle bowl. As previously discussed, the recycled paper material chosen for the bowl is perfectly suited to a folding structure. The folding structure could be very difficult if we did not calculate the size right. After lots of hard work, we came up with a foldable structure that we think is the best result.

2.1 Size

Our goal is to make this product very easy to carry. Therefore, keeping the size and appearance minimal is a fundamental characteristic of the design. According to our research, similar products on the market waste too much space on the width. This design can efficiently reduce the length and give the noodles room to stretch. Still, it occupies too much space in our luggage. To rectify this, we minimize the size of the paper bowl down to a size that is almost equal to that of the noodles: The common size of instant noodles is 15 cm x 15 cm x 1.5 cm, and we made our design 15.5 cm x 15.5 cm x 2 cm.

2.2 The individual packaging of the noodle

The packaging of instant noodles can be divided into two categories: those with individual packaging and those without. The reason for some companies choosing to give the noodles additional individual packaging is to guard against bacteria, or other potential health risks. We think this step in making a food product is very important. Therefore, we chose a bio-degradable material for the packaging of the noodles. This not only achieves our goal of using eco-friendly materials, but also gives customers an added degree of safety.

2.3 Eco-friendly tableware

Our design background of this product is to let people carry it easily and the concept of easy and convenient is that there is no need for extra preparations. Therefore, it is necessary that tableware is included in the product. Generally, when people eat instant noodles, they will use chopsticks, and a spoon, but the existing disposable tableware of these two items are neither eco-friendly, nor convenient.

A spork which combines the functionality of the spoon and fork, is the most practical tool for this situation – allowing the customer to enjoy the noodles without any inconvenience.

To achieve the condition of eco-friendly, we chose to make the spork out of the same recycled paper and PE film as the bowl. By doubling the layer of paper, the spork was made stronger and easier to hold.

2.4 Folding structure

The most unique quality of this product is its folding structure. These folding structures allow the instant noodle box to remain very small. The fold design can be separated into three parts.

2.4.1 Overall structure

The design can be described as a waterproof, eco-friendly paper divided into the folding structure. Every face has a layer on the upper, and lower side. Each upper layer is connected by the connector, and the lower layer has seams to form a water proof surface after putting PE film on.

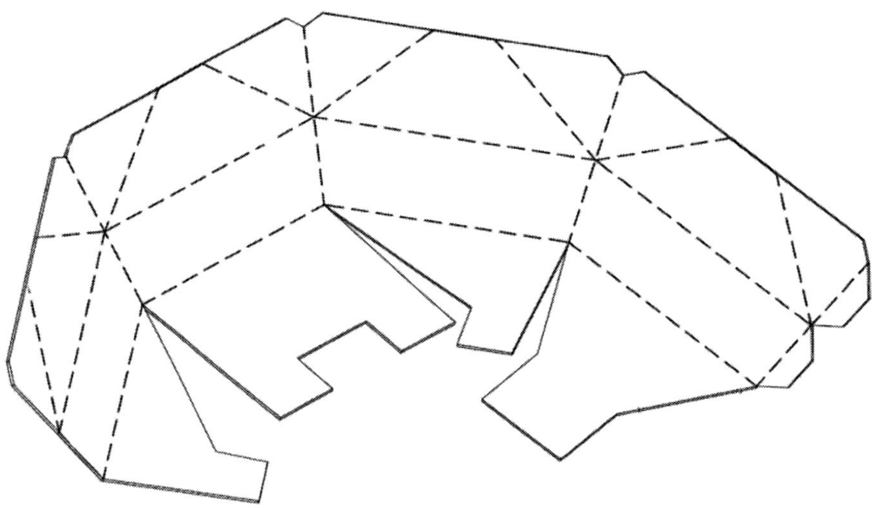

Fig. 2. The overall structure of the foldable instant noodle bowl.

2.4.2 Unfolded state

When the bowl is in the unfolded state, which means functioning state it provides customers with a large, wide space in which to pour hot water, and gives the noodles room to expand. The unfolded state can be seen as a wide, hollow square, or pillar shape. After unfolding the upper layer, the box can be extended to almost double along its length. Each face is connected to each adjacent side to prevent hot water from spilling out.

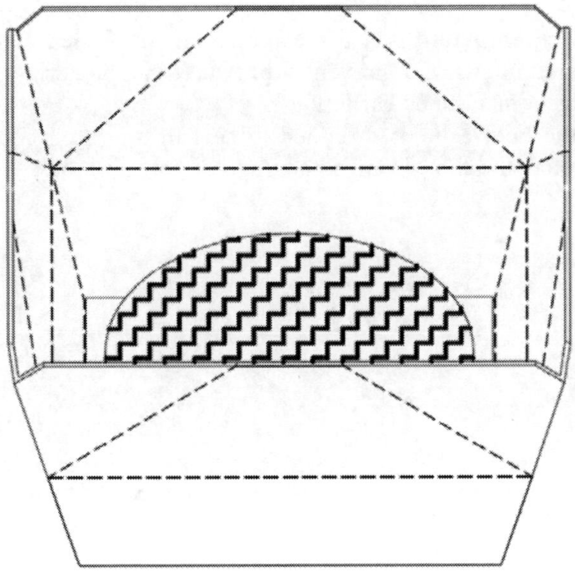

Fig. 3. Unfolded state of the instant noodle bowl.

2.4.3 Unfolding (folding) state

To unfold the bowl, a customer simply needs to lift two sides of the layers. Due to the design of the connectors, each side will be easily lifted at once. Additionally, when the bowl is being folded in the factory, the supply chain only needs to lightly press the side of the connectors to fold the box.

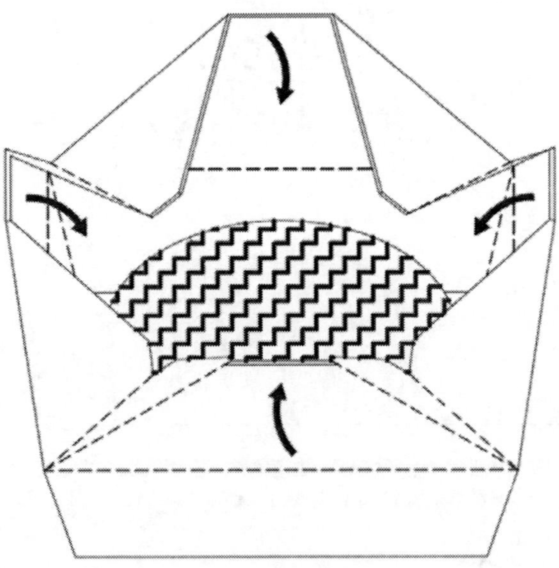

Fig.4. Unfolding (Folding) state of the instant noodle bowl.

2.4.4 Folded state

Because of the folding structure, when the box is in the folded state it is impossible to completely seal the contents. To prevent from dirt and bacteria, the noodle itself has individual packaging as mentioned earlier.
On the surface of the box, a lid is designed to seal up the box. This lid is composed of aluminum foil to protect the food from exposure to direct sunlight.

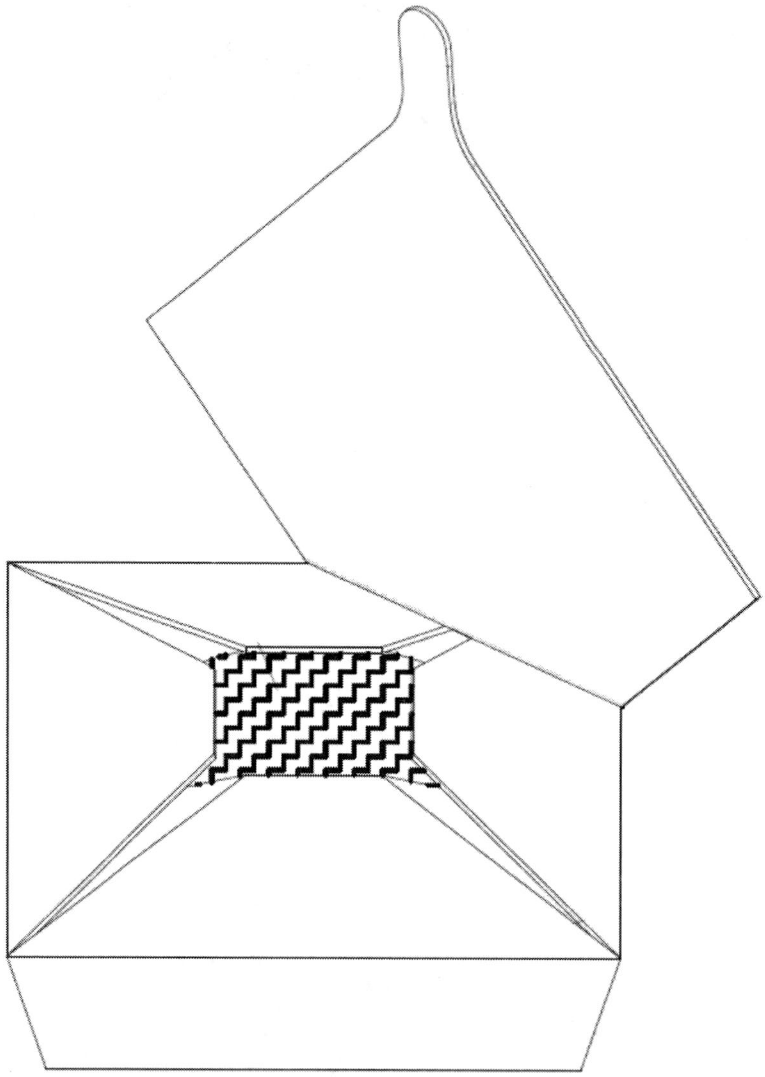

Fig.5. Folded state of the instant noodle bowl.

2.4.5 Side view of the box

By looking at the side view of the box, we can readily observe the folded state. The connectors will lie on top of the noodles, and it is demonstrated that the size of the box is almost as the same size as the noodles.

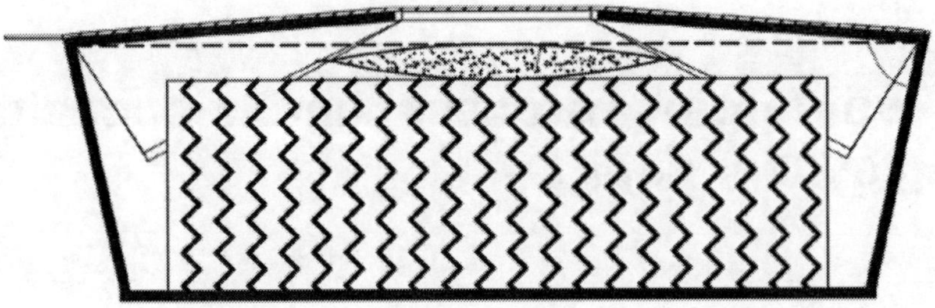

Fig.6. Side view of the folded instant noodle bowl.

3 Conclusions

No one wants to travel with heavy luggage. Commuters of all kinds are seeking a light package, or material, for their goods and food items. Reducing the size of the instant noodle bowl not only provides a lighter choice to customers, but also results in better profit margins for the manufacturer. The compact size of the product allows manufacturers to ship larger quantities, resulting in increased sales, and reduced carbon emissions. Additionally, the small size of the packaging can also be used to store other foods, not just noodles. Therefore, transporting food to the third world, or people trapped by natural disasters can be lot easier too.

All eco-friendly materials also reduce the harm to our injured planet, helping the next generation realize a better future.

References

1. J.C. Lee, Taiwan instant noodle expo (2008)
2. C.H. Chen, Eat everywhere: instant noodle, body and diet space, **27** (2010)
3. K. Ambrozic, M.E. Dempsy, Reuse, recycle … reduce: a greener library with print management **88** (2012)
4. S. Xiangke, Design and engineering of biobased materials -- process engineering & thermal recycling of poly(lactic acid), and studies in functional silane and siloxanes. **92** (2014)
5. Greentumble editorial team, How is paper recycled: step by step, https://greentumble.com/how-is-paper-recycled-step-by-step/ (2015)
6. Recyclebank, Because you asked: what kinds of disposable plates can I recycle? , https://livegreen.recyclebank.com/because-you-asked-what-kinds-of-disposable-plates-can-i-recycle (2015)
7. H.Y Chen, The influence of environmental awareness on green products purchase intention: the mediation effect of environmental behavior, **21** (2016)

Study on forging process and die design of parking sensor shell

Tung-Sheng Yang* and *Jhong -Yuan* Li

Department of Mechanical Manufacturing Engineering, National Formosa University,

64 Wunhau Road, Huwei, Yunlin, Taiwan 632, ROC

Abstract. The process of precision forging has been developed recently because of its advantages of giving high production rates and improved strength. For complete filling up, predicting the power requirement and final shape are important features of the forging process. A finite element method is used to investigate the forging force, the final shape and the stress distribution of the parking sensor shell forging. The stress-strain curve of AL-6082 is obtained by the computerized screw universal testing machine. The friction factor between AL-6082 alloy and die material (SKD11) are determined by using ring compression test. Stress-strain curve and fiction factor are then applied to the finite element analysis of the parking sensor shell forging. Maximum forging load, effective stress distribution and shape dimensions are determined of the parking sensor shell forging, using the finite element analysis. Then the parking sensor shells are formed by the forging machine. Finally, the experimental data are compared with the results of the current simulation for the forging force and shape dimensions of the parking sensor shell.

1 Introduction

Parking sensors are proximity sensors for road vehicles designed to alert the driver to obstacles while parking. These systems use ultrasonic sensors to proximity detectors to measure the distances to nearby objects via sensors located in the front or rear bumper fascia or visually minimized within adjacent grills or recesses. Manufacturing 3C products by applying cold bulk forming processes is a common production technique owing to its well-known advantages, such as reduced material and time costs and the resulting increased strength of the products. Moreover, the applications of cold forged aluminum alloy parts have been increasing because of the important needs for weight reduction of automobile. However the associated process planning and tool design are complex tasks.

The finite element method (FEM) is one of the most powerful tools of computer aided engineering (CAE) to solve various problems in metal forming processes. Commercial CAE software finds use in the areas of process verification and modification as well as for design improvement [1, 2, 3]. Aluminum alloys have been widely used to manufacture lightweight automotive bodies and parts. Jensrud et al. [4] proposed a new thermo-mechanical process for aluminum forging. They showed that it is possible to reduce the

Corresponding author: tsyang@nfu.edu.tw

load without reduction of the hardness under the proposed optimum processing conditions. Khaleed et al. [5] performed three-dimensional analysis including thermal effects for a flash-less cold forging product of aluminum alloy. They obtained a good agreement with experimental results and optimized the initial shape of the workpiece. The ring compression test has proven to be one of the most popular and commonly used methods for quantitative evaluation of friction conditions in bulk forming operations. The ring compression test was firstly used by Kunogi [6] as a qualitative method of comparing lubricants for cold extrusion. Lee and Altan [7] used an upper-bound velocity field that considers bulging has been applied to cylinder and ring upsetting. According to the relationship between inner diameter changes and height reduction, the calibration curve of constant shear friction factor for a 6:3:2 rings have been predicted.

This study development the aluminum alloy forging process and die design of parking sensor shell. The stress-strain curve of AL-6082 is obtained by the universal testing machine. The friction factor between AL-6082 and die material are determined by using ring compression test. Stress-strain curve and fiction factor are then applied to the finite element analysis of the parking sensor shell forging to predict the forging force and the shape of parking sensor. The experimental data are compared with the results of the current finite element simulation

2 Flow stress and friction factor of 6082 aluminum alloy

The computerized screw universal testing machine is used to determine the mechanical properties. A tensile specimen is a standardized sample cross-section. It has two shoulders and a gauge section in between. The shoulders are large so they can be readily gripped, whereas the gauge section has a smaller cross-section so that the deformation and failure can occur in this area. Figure 1(a) shows the dimensions of the tensile specimen. The tested tensile specimen is AL 6082 alloy. The test process involves placing the test specimen in the testing machine (Fig. 1(b)) and slowly extending it until the test specimen fractures (Fig.2(a)). During this process, the elongation of the gauge section is recorded against the applied force. The data is manipulated so that it is not specific to the geometry of the test sample. Thus the stress-stain curves and fracture values of AL 6082 alloy are obtained. Figure 2(b) presents the stress-stain curves of AL 6082 alloy at room temperature. To obtain the shear friction factor, the specimens are produced outer diameter, inner diameter and height as 18mm, 9mm and 6mm (6: 3: 2), respectively. The ring specimens are compressed with different deformation and lubricants. Fig. 3(a) shows the results of ring compression test. Both the reduction ratio of height and reduction ratio of inner diameter are calculated. Fig. 3(b) presents the friction factors of five specimens are between 0.10-0.15, and the average value of friction factor is about 0.13.

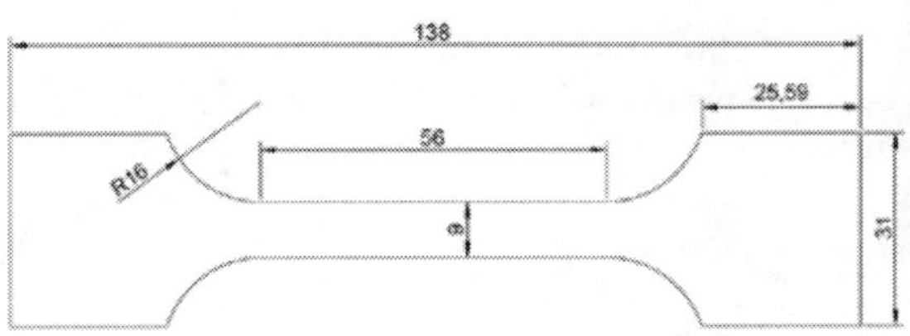

(a) Dimensions of the tensile specimen

(b) Tested tensile specimen

Fig. 1. Dimensions of the tensile specimen and tested tensile specimen.

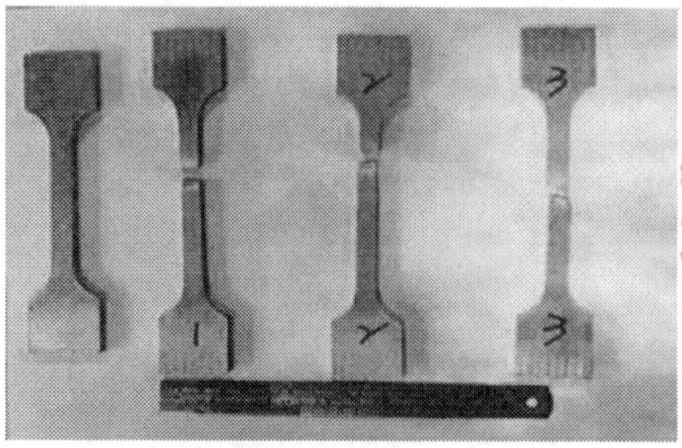

(a) Fractured specimen

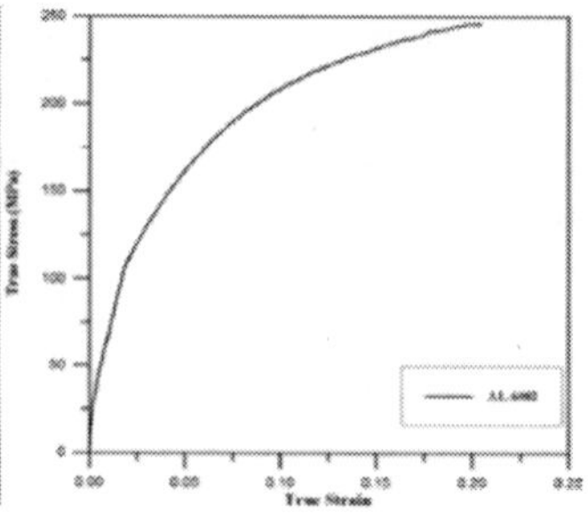

(b) True stress-stain curve

Fig. 2. fractured specimen and true stress-stain curve.

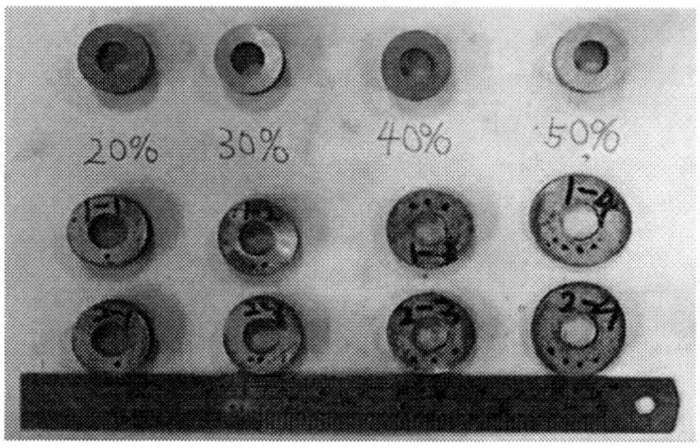

(a) Results of ring compression test

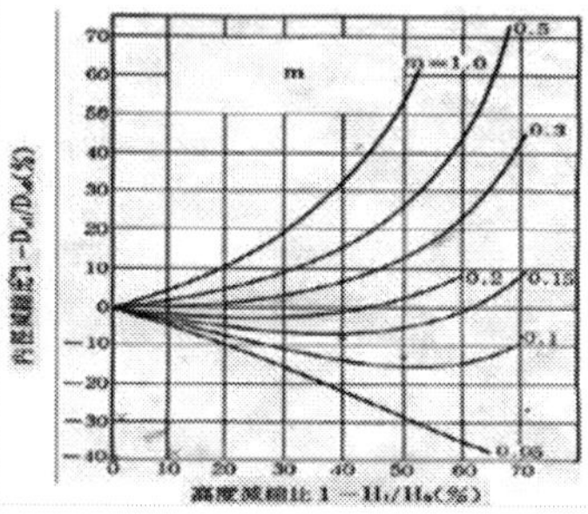

(b) Friction factors of the specimens

Fig. 3. Results of ring compression test and friction factors of the specimens.

3 Finite element analysis of forging of parking shell

3.1 Production and die design

The parking sensor shell is mainly assembled with the mating sensor. Thus the shape of the sensor is used to determine the inner groove of the shell. Figure 4 shows the parking sensor assembly which include the sensor and shell. The top of the shell must to be fairly flat so that the ultrasonic transmission can accurately receive. In addition, due to the requirement of the parallelism, verticality and concentricity the further machining is needed after forging. Therefore the design of forged shell must take into account of machining allowance (about 0.3 mm) from the product design of shell. Figure 5 shows the product design of parking sensor shell. Forging dies include punch, die and other parts as shown in Fig. 6. The ejector rod is used to eject the forged piece from the die. Figure 7 shows the schematic diagram of the punch and die.

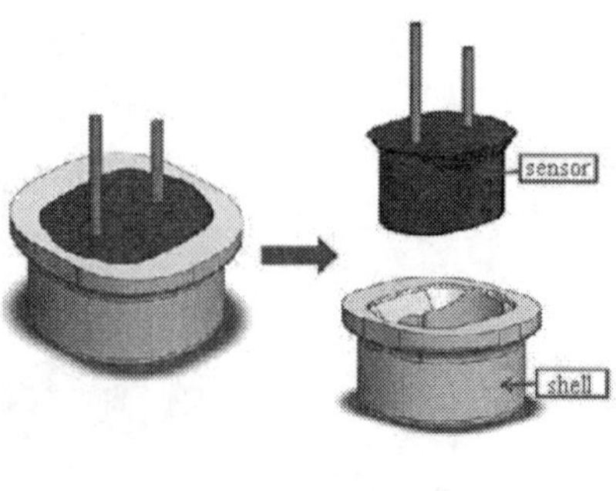

Fig. 4. Parking sensor assembly diagram (including sensor and shell).

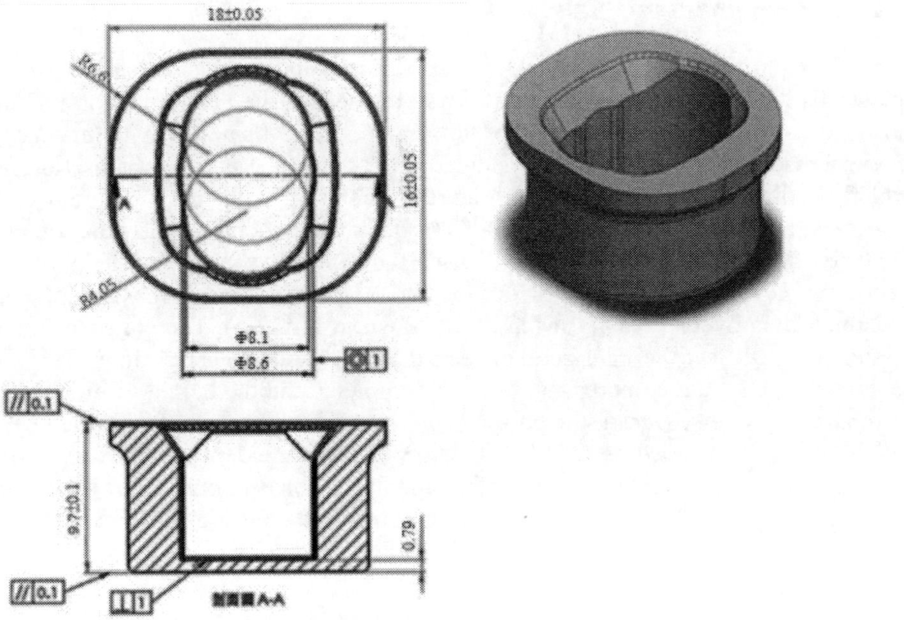

Fig. 5. Diagram of product design of parking sensor shell.

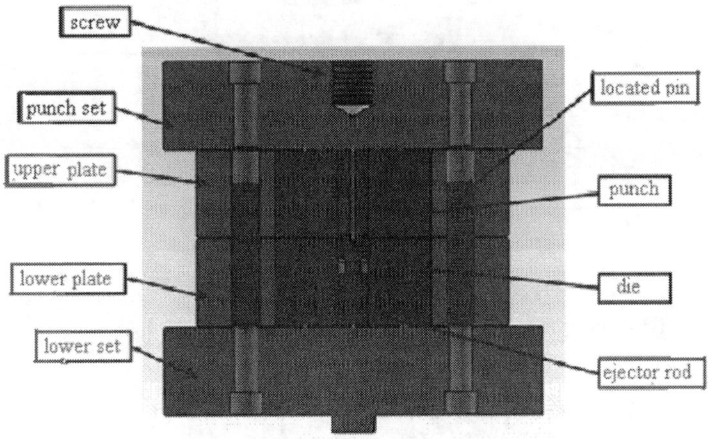

Fig. 6. Forging dies.

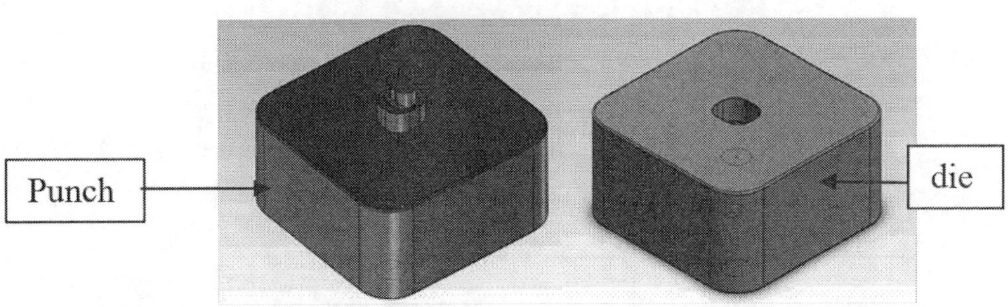

Fig. 7. Schematic diagram of punch and die.

3.2 Finite element analysis.

The FEM software DEFORM-3D [6] has been applied to simulate the plastic flow of materials during the forging process of parking sensor shell. The flow stress curves are the input data during the simulation of forging process. In addition to material properties, preprocessor also require input of detailed process parameters such as friction factor, punch velocity, dimensions of billet, etc. Figure 8 shows the schematic diagram of the parking sensor shell forging. The material is formed by the punch and the consequence the billet material flows into the shape of die. During the analyses, the die and the punch are assumed to be rigid, whereas the billet is assumed to be plastic. The flow stress of AL6082 billet obtained by universal testing machine is shown in Fig. 2(b). The constant friction factor is assumed as 0.13 at the interface between the billet and die. The value of friction factor is obtained by the ring compression test. The diameter and the height of billet are 15 mm and 6 mm, respectively. During the parking sensor shell forging simulation the element number of the billet is assumed as 80000-180000. When the element numbers exceed 140000 the forging force approach to a steady value about 190 tons (Fig. 9). Thus the element number is assumed as 140000. The details of simulation parameters are shown in Table 1.

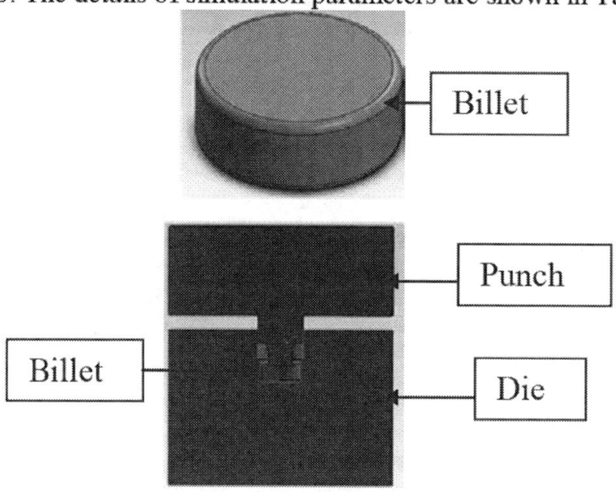

Fig. 8. Schematic diagram of parking sensor shell forging.

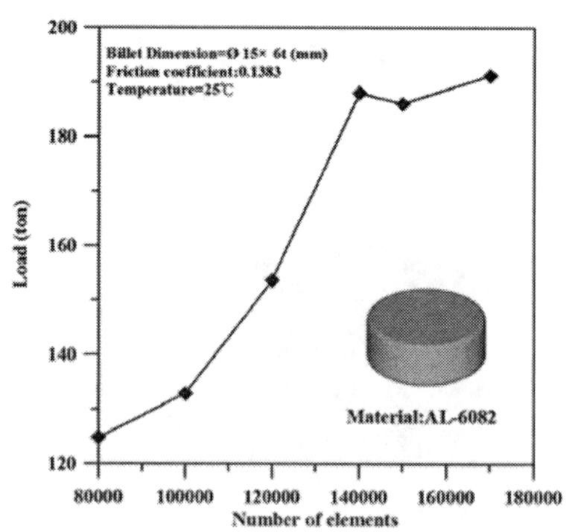

Fig. 9. Effect of number of elements on forging load.

 https://doi.org/10.1051/mateconf/201818500020

Table 1. Details of simulation parameters.

friction factor	0.13
temperature (°C)	25
punch speed (mm/min)	10
billet dimension (mm)	Ø15 x 6 t
max. forging force (ton)	188

Figure 10 shows the variation of the punch loads with punch stroke. The punch load increases as the punch stroke increases. When the material flows into the die cavity completely, the punch load increases rapidly, the maximum forging load is about 188 tons. Figure 11 shows the effective stress distribution of billet when the material flows into the die cavity completely. The maximum value of effective stress occurs at the fillet of die. Figure 12 shows the simulated shape of parking sensor shell.

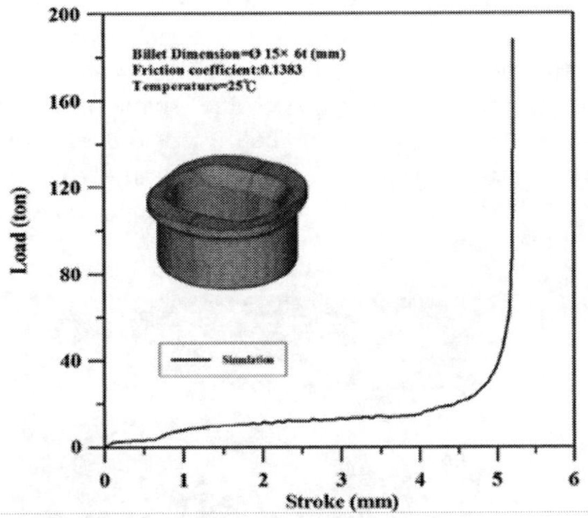

Fig. 10. Variation of the punch loads with punch stroke.

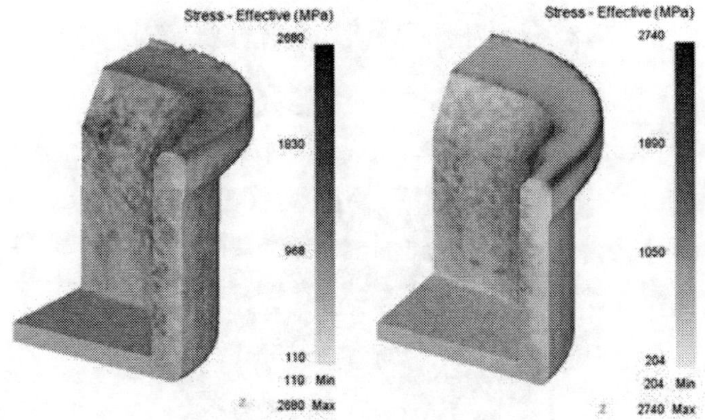

Fig. 11. Effective stress distribution of billet.

Fig. 12. Simulated shape of parking sensor shell.

4 Experimental results

Figure 13 shows the device of experiment including the forging machine, punch, die and guide pins. Guide pins are used to locate the movement of the upper and lower die. The test machine is 200 tons hydraulic test equipment, the forming speed is 10mm/min. The billet placed between punch and die. The material of billet is AL6082, and the material of die is SKD11. The diameter of billet is 15 mm, and the height of billet is 6 mm. The lubricant used in current experiment between die and billet is graphite, which is also used in the ring compression test. The die is stationary and the punch moved toward to the die until the material flow into the die cavity completely. Figure 14 shows the experimental result of forging the parking sensor shell. The experimental shape of parking sensor shell is near the simulated result which shown in Fig. 12. The dimensions of forged part are larger than the designed dimensions as shown in Fig. 5 about 0.30 mm, which is machining allowance of the product design of shell. The experimental punch load (about 191tons) agrees well with the simulation result (188 tons).

Fig. 13. Device of experiment (punch, die and guide pins).

Fig. 14. Experimental result of parking sensor shell.

5 Conclusions

The stress-strain curves of the AL6082 are obtained by the universal testing machine. The friction factor between the die and billet is obtained by ring compression test with different deformation and lubricant. The lubricant is graphite and the experimental friction factor is 0.13. Therefore the stress-strain curves and experimental friction factor are as the input parameters during the finite element analysis of parking sensor shell forging process. Maximum forging load, effective stress and shape dimensions are determined of the parking sensor shell forging, using the finite element analysis. Finally, the parking sensor shell parts are formed by the forging machine under the conditions using finite element analysis. The punch load is close to the simulation result. The dimensions of the experimental forged part agree well with the simulation result and initial design of forged part.

References

1. M. Markus, K. Markusr, W. Knu, A. Taylan. J. Mater. Proc. Tech. **33** 75 (1992)
2. T. Ishikawa, T. Ishiguro, N. Yukawa, T. Goto. CIRP Annals–Manu.Tech. **63** 289 (2014)
3. Z. Wuhao, H. Xinghui, H. Lin. Proc. Eng. **207** 442 (2017)
4. O. Jensrud, K. Pedersen. J. Mater. Proc. Tech. **80**, 156 (1998)
5. H.M.T. Khaleed, Z. Samad, A.R. Othman, M.A. Mujeebu, A.B. Abdullah, M. M. Zihad. J. Manu. Proc. **13**, 41 (2011)
6. M. Kunogi. J. Sci. Res. Inst. **50**, 215 (1956)
7. C.H. Lee, T. Altan. J. Eng. Indu. **16**, 77 (1972)

Finite element simulation and analysis of saw cutting

Dyi-Cheng Chen*, *You-Hua* Chen, *Ci-Syong* You, and *Shin-Han* Huang

Department of Industrial Education and Technology, National Changhua University of Education, Changhua 500, Taiwan

Abstract. In this paper, Finite Element Analysis (FEA) used DEFORM^{TM} 3D to construct 6061 aluminum sawing simulation. Before sawing, we pushed the speed and cutting speed under pressure as a sawing process parameters. Then there are three groups of combination conditions in the study. Analysis and discussion based on loading, effective stress, cutting temperature and blade wear. According to the impact parameters the structural design of the saw blade is improved by taking full account of the mutual influence of sawing parameters.

1 Introduction

In order to improve the processing efficiency and product surface quality, especially to meet the strict requirements of all types of parts, sawing is an important metal processing method. Finite Element Analysis (FEA) is an effective auxiliary design tool, which can represent the material properties of workpieces including the function of temperature, strain and strain rate. It fully considered the interaction between cutting parameters of sawing turning tools. According to the relationship between parameters improved the impact of structural design. Research on sawing machining includes: Toru et al. [1] heat-processes the cutting tool, transfers the heat from the heated cutting tool, and simulates the workpiece. Cutting experiments prove that the tool heating mode improves the machinability. Stepan et al. [2] verification is used to calculate the stability of fixed milling and to measure general numerical algorithms that are validated in the work environment. It is able to predict the stability conditions of the cylindrical milling cutter geometry during the milling process. Jaroslava and Zdenk [3] studied the tool life of cutting edge radii, which machining the steam turbine shell. Umbrello [4] used DEFORM^{TM} 2D software to simulate the orthogonal cutting of titanium alloy, which changed the optimum cutting parameters of titanium alloy including the tool angle, temperature, velocity and feed rate. Diniz and Micaroni [5] compared the surface roughness with different feed rates, cutting speeds, dry cutting and cutting fluid analysis. Totis et al. [6] using a dynamic mathematical model through finite element analysis for cutting test to improved tool geometry optimization. The experimental results proved the excellent characteristics of the new device. Wang [7] compared with SCMT and CNMM positive angle turning tool for different feed rates, cutting depth during

*Corresponding author:dcchen@cc.ncue.edu.tw

the workpiece force analysis. CAKIR and ISIK [8] used finite elements to analyze the vibration frequency and the buckling strength during cutting. Mitrofanov et al. [9] used finite element analysis to analyse the temperature distribution of the tool and workpiece during orthogonal cutting. Ceretti et al. [10] used DEFORM^{TM} 2D to simulate the effects of parameters on the turning process. Attanasioa et al. [11] used experimental and simulation results showing a positive correlation between wear and cutting speed or feed rate.

2 Related theories

2.1 Finite element simulation analysis

DEFORM^{TM} 3D software is based on the design system and machining process, through the two-dimensional or three-dimensional flow deformation of the analytical model to simulate the metal material on the mold forming, ductile damage values, deformation temperature, plastic flow velocity, stress and strain distribution situation. This software can greatly save the time and money of mold-testing. And it can provide important information, such as the impact of different materials and heat conduction on the finished product.

During the forming process, due to the impact of plastic deformation, the processing of defects are mostly caused by ductile material damage. The brittle fracture of metal material is based on the microscopic phenomenon that the crystal cracks and the atoms are mutually combined and destroyed that is the "Cleavage fracture" . The amount of strain generated prior to material failure is minimal or no plastic deformation and no early warning of damage. Chen et al. [12] revised the "Cockcroft & Latham" guidelines to formulate a non-dimensional decision rule called "Normalized C & L criterion" :

$$\int_0^{\bar{\varepsilon}_f} \frac{\sigma_{\max}}{\bar{\sigma}} d\bar{\varepsilon} = C \tag{1}$$

where $\sigma_{\max}$ is the maximum tensile stress, $\bar{\varepsilon}$ is the effective strain, $\bar{\sigma}$ is the effective stress, $\bar{\varepsilon}_f$ is the effective strain at failure, and C is the damage value.

Mold wear calculation model: To analyze the mold wear which selected the wear calculation model, DEFORM analysis software gives two calculation models. Here choose Archard wear mode, the formula as follows:

$$Z_{AB} = \int K \frac{p^a v^b}{H^c} dt \tag{2}$$

In the formula, Z_{AB} is worn, p is the contact pressure, v is the relative velocity, H is the hardness of the mold, dt is the increment time, a and b are experimentally coefficients; a= 1, b = 1, c = 2 and K = 0.000002.

2.2 Saw blade cutting

Figure 1 shows sawing schematic diagram of 6061 aluminum alloy using finite element analysis. It is interface shear friction as 0.12 and interface heat transfer coefficient as 0.004. The hardness of 6061 aluminum is 54 and initial temperature as 68°F.It can be seen that the Fig.1 (a) is before processing, Fig.1(b) is during processing, Fig.1(c) is after processing. The tool material is a rigid material and the displacement in the x, y, and z directions of the workpiece boundary condition is completely fixed. It can be seen in Figure 2-Figure 4.

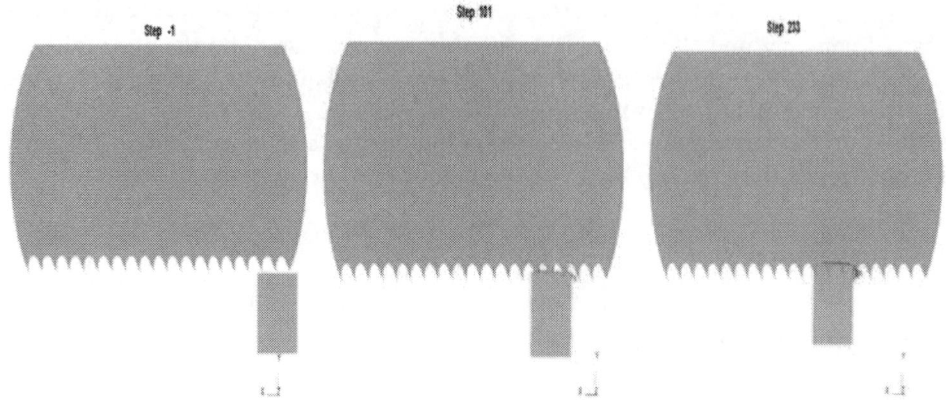

(a) Before processing (b) During processing (c) After processing

Fig. 1. 6061 aluminum alloy sawing schematic diagram.

The workpiece material selects 6061 aluminum alloy (6061), and the tool material uses high speed steel. The displacement of the workpiece in the x, y and z directions are completely fixed. Figure 2 is the half-section of before sawing. Figure 3 shows the half-section of during sawing. Figure 4 shows the half-section of after sawing.

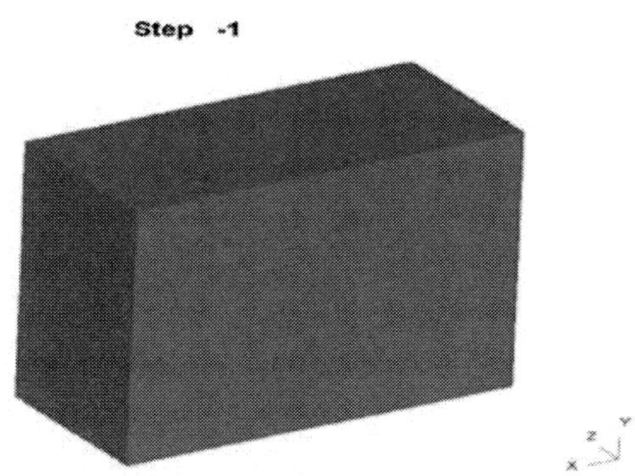

Fig. 2. Before sawing for 6061 aluminum alloy.

Fig. 3. During sawing for 6061 aluminum alloy.

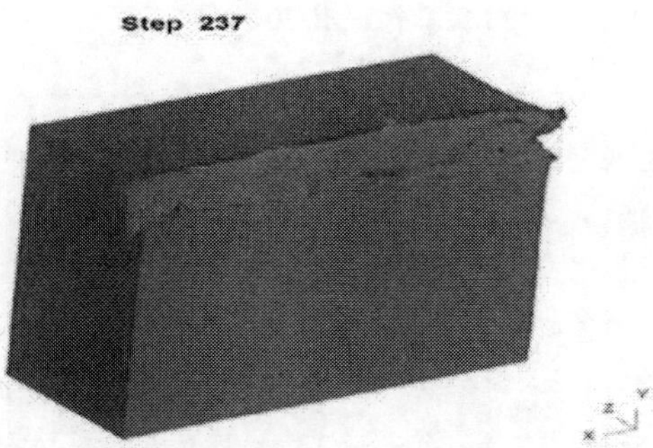

Fig. 4. After sawing for 6061 aluminum alloy.

3 Results and discussions

Table 1 presents a simulation parameter table for the sawing speed difference, which setting the simulation analysis of the change parameter, forward speed of 2, 4, 6 mm/sec, and the pressing speed of 0.2, 0.4, 0.6 mm/sec.

Table 1. Simulation parameters table.

Numbering	X-axis forward speed (mm / sec)	Y-axis downward pressure speed (mm / sec)
1	2	0.2
2	4	0.4
3	6	0.6

Figure 5 shows no. 1 saw blade of the X Load (N) analysis. It can be seen that the load increases with the increasing time, which indicates that the bigger thrust can saw the material. Figure 6 presents no. 1 saw blade of the Y Load (N) analysis. It can be seen that the load increases with the increasing time. It indicated that the using greater downward pressure to saw the material. Figure 7 presents no. 1 saw blade of the Z Load (N) analysis. It shows the maximum load of 20.8N, and large difference between X load and Y load. The line shape produced up and down shock. It can be seen that the sawing Z-axis has no effect. Figure 8 compared with X Load (N) between no. 1, no. 2 and no. 3 at the same sawing depth that the each number's load is about 200N. It indicates there is no significant relationship between the sawing speed and X load.

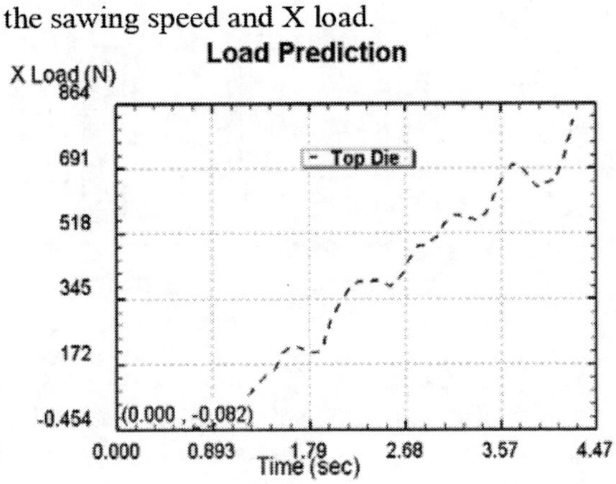

Fig. 5. X Load (N) analysis for No. 1 saw blade.

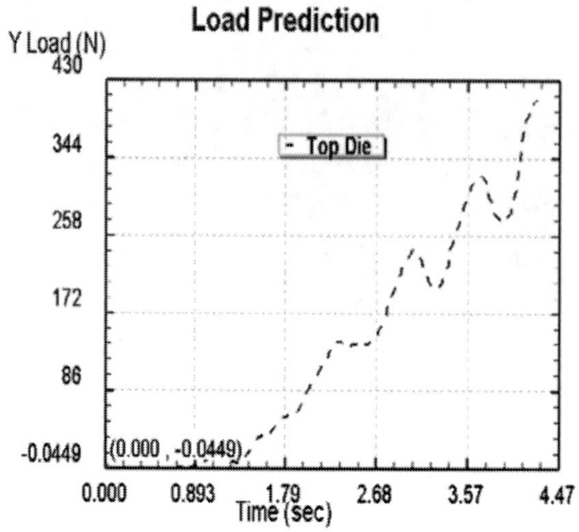

Fig. 6. Y Load (N) load analysis for No. 1 saw blade.

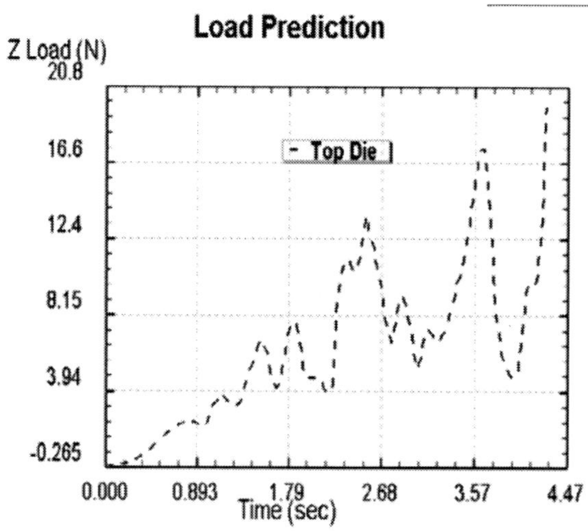

Fig. 7. Z Load (N) load analysis for No. 1 saw.

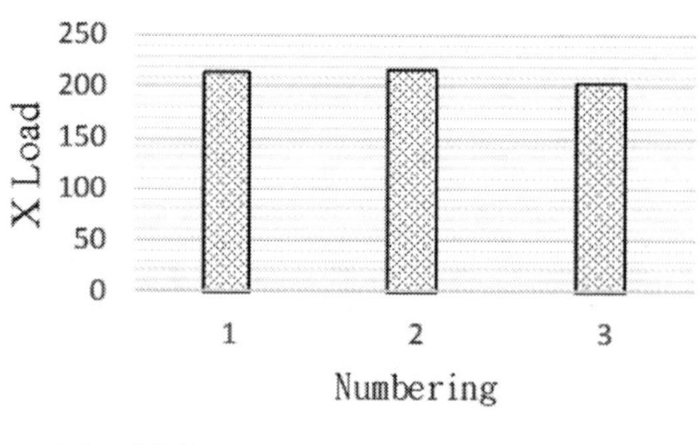

Fig. 8. Comparison chart on X load (N).

Figure 9 compares the temperatures with each number of saw blades. It can be seen that the higher sawing speed causing the lower temperature, but less significant. Figure 10 shows the temperature analysis of aluminum alloy, when the sawing time increases, the material temperature is increased. Figure 11 shows the temperature simulation of 6061 aluminum alloy. It can be seen that the temperature is evenly distributed in the upper half of the sawing material.

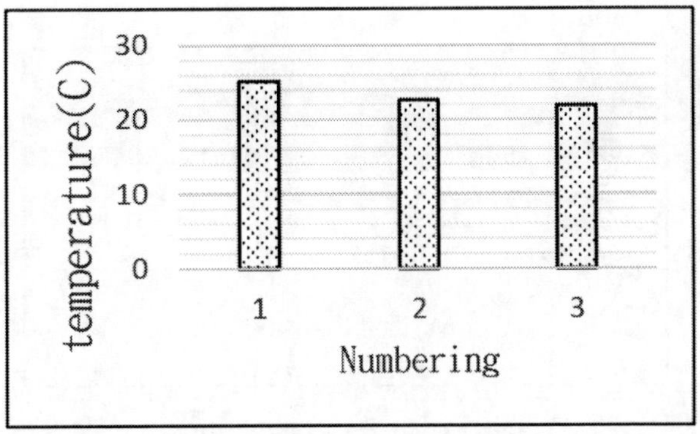

Fig. 9. Comparison chart of saw blade temperature for each number.

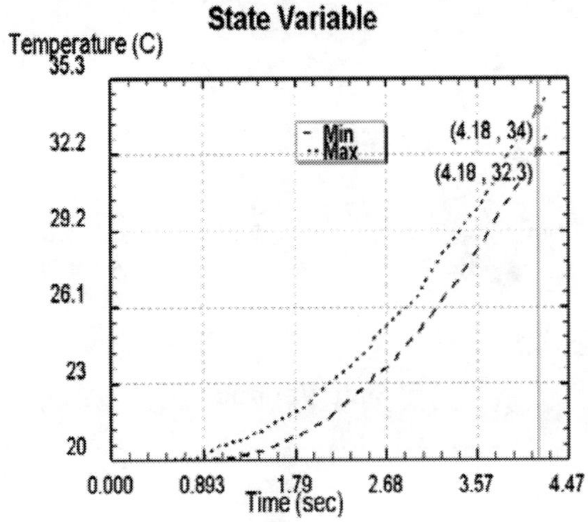

Fig. 10. Temperature analysis of 6061 aluminum alloy.

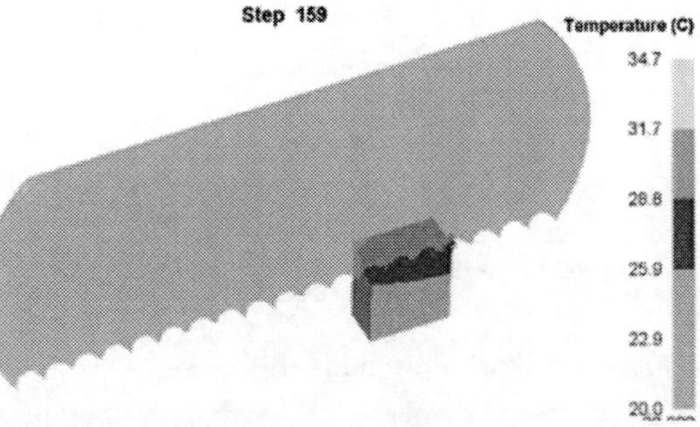

Fig. 11. Temperature simulation of 6061 aluminum alloy.

Figure 12 shows the comparison of effective stress for 6061 aluminum alloy. It can be seen that sawing speed and effective stress no significant relationship. Figure 13 shows the effective stress simulation for parameter no. 1. The maximum of effective stress concentrates on the contact area between the serration tip and the workpiece. The contact between saw teeth and chip produces stress to cause bending. Figure 14 shows the effective stress analysis. The effective stress shows stable trend after 1 second. Figure 15 simulation the wear of saw blade. The rightmost saw tooth wear is minimum and the leftmost relative maximum wear, because the rightmost teeth contact material less time, and the left tooth contact material longer time. The largest wear part of saw tooth is occurred on the saw tooth tip.

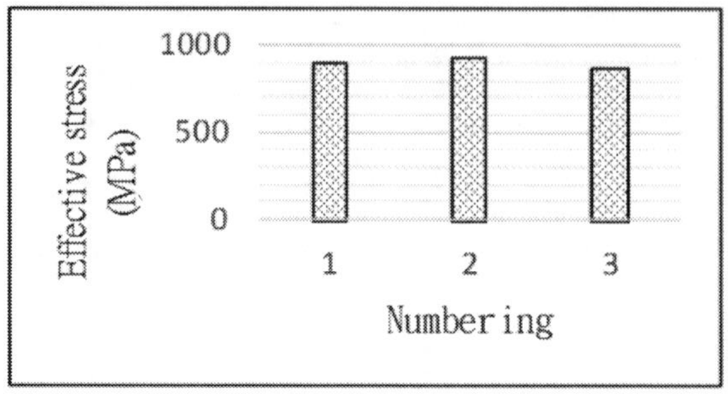

Fig. 12. Effective stress comparison chart of 6061 aluminum alloy.

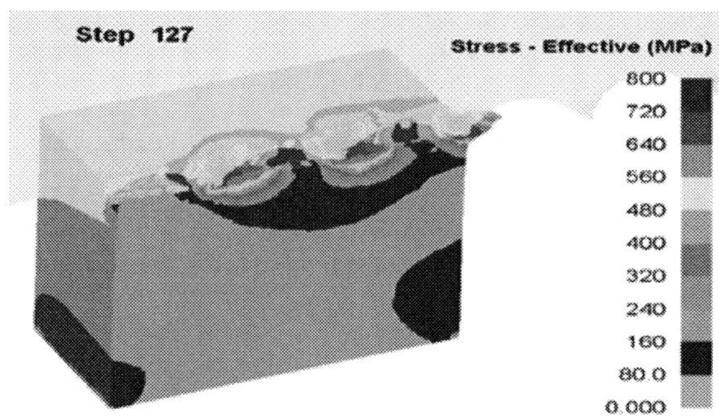

Fig. 13. Effective stress simulation of 6061 aluminum alloy.

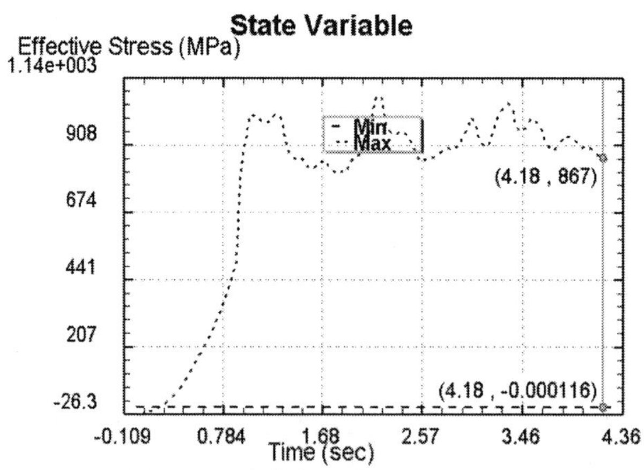

Fig. 14. Effective stress analysis for No.1 aluminum alloy.

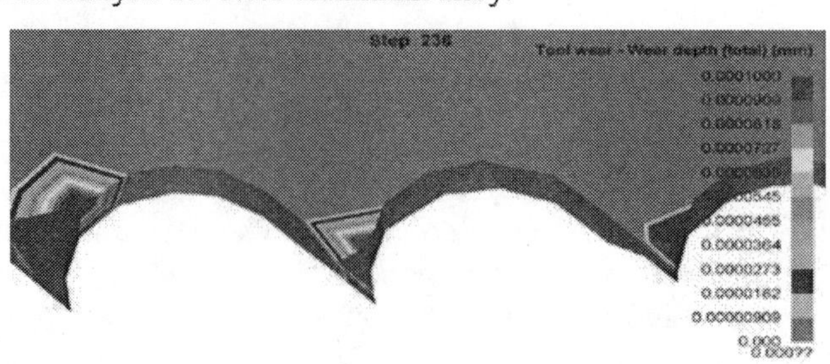

Fig. 15. Saw blade wear simulation.

4 Conclusions

In this paper, Finite Element Analysis (FEA) used DEFORM^{TM} 3D to construct 6061 aluminum sawing simulation. The results simulated are as follows:
There is no guide angle to allow the cutting edge to smoothly cut in when the first tooth cuts into the material. It cause the aluminum alloy produced a greater stress. A stable cutting process is occurred after the second tooth entered. The maximum stress from cutting is occurred the site where the point and the tip touches. Effective strain increases with an increase of blade feed, and the effective strain is distributed higher at the end of the cutting process. Chip is accumulated in the end and has a greater effective strain. Initial contact to the material produces an instantaneous high temperature. The temperature increases with an increase of cutting edge feed, and the highest chip temperature has just formed. The largest part of the wear is the sharp edge of the tool, and the maximum depth of simulated wear is 0.0001mm.

References

1. K. Toru, H. Kanako and M. Mamoru, CIRP Annals - Manufacturing Technology, **36**, 105-108 (2014)
2. G. Stepan, J. Munoa, T. Insperger and M. Surico, CIRP Annals - Manufacturing Technology, **63**, 385-388 (2014)
3. F. Jaroslava and J. Zdenk., Procedia Engineering, **69**, 565-573 (2014)
4. Umbrello, Journal of Materials Pressing Technology, **196**, 79-87 (2008)
5. A.E. Diniz and R. Micaroni, International Journal of Machine Tools & Manufacture, **42** , 899-904 (2002)
6. G. Totis, O. Adams, M. Sortino, D. Veselovac and F. Klocke, Measurement, **49**, 164-181 (2014)
7. J. Wang, Journal of Materials Processing Technology, **97**,114-119 (2000)
8. M. Cemal CAKIR and Y. ISIK, Materials and Design, **26**, 105-112 (2005)
9. A.V. Mitrofanov, V.I. Babitsky and V. V. Silberschmidt, Journal of Materials Processing Technology, **153-154**, 233-239 (2004)
10. E. Ceretti, C. Lazzaroni, L. Menegardo, T. Altan, Journal of Materials Processing Technology, **98**, 99-103 (2000)
11. A. Attanasio, E. Ceretti, A. Fiorentino, C. Cappellini, C. Giardini. Wear, **269**, 344-350 (2010)
12. C.C. Chen, S.I. Oh and S. Kobayashi, Ductile Fracture in Axisymmetric Extrusion and Drawing, J Eng. Ind., **101-102**, 36 (1979)

Understanding the organizational critical activities of manufacturers in case studies

Wei-Hsi Hung[1,*], *Tsung-Yueh* Lu[2], *Mei-Fang* Wu[2], and *Yun-Chen* Lin[3]

[1]Department of Management Information Systems, National Chengchi University, No.64, Sec.2, ZhiNan Rd., Wenshan District, Taipei City 11605, Taiwan, R.O.C.

[2]Department of Industrial Engineering & Systems Management, Feng Chia University, 100 Wenhwa Road, Taichung 40724, Taiwan, R.O.C.

[3]Department of Information Management, National Chung Cheng University, 168 University Road, Min-Hsiung, Chia-Yi 62102, Taiwan, R.O.C.

Abstract. Manufacturers tend to carry out certain important and continuous activities with respect to internal operations which could contribute to their success. These activities are termed organization critical activities (OCAs). The main objective of this study is to elicit the OCAs in the manufacturing industry to understand what can cause them to be successful. Through the case study approach including the interviews with the senior managers from three manufacturing companies in Taiwan and a series of prioritizing activities, 12 OCAs have been identified. This study found that Taiwanese manufacturers placed more importance on the activities of quality and cost control while placed relatively less importance on the marketing related activities. More findings and future research suggestions are also provided.

1 Introduction

Manufacturers tend to carry out certain important and continuous activities with respect to internal operations which could contribute to their success. These activities, are termed organization critical activities [4], being granted significant resources and supervision from the senior management. Organization Critical Activities (OCAs) was introduced by Hung [4], and is defined as: the activities constantly conducted by the organization for the success in the industry. They are regarded as an essential to the short-term, mid-term, and long-term success of the organization. Thus they receive a huge amount of resources and frequent supervised and directed by the senior management.

However, there is the lack of research studying what organization critical activities are conducted by manufacturers in order to ensure their success. The main objective of this study is to elicit the OCAs in the manufacturing industry to understand what can cause them to be successful.

[*]Corresponding author: fhung@nccu.edu.tw

2 Organization critical activities

Activities are the implementations of works within the organization [10]. They are established in strategies, tactics, and various organizational levels [5, 6, 8]. Activities are to be implemented during works, and they are the minimum unit of work in the organization which will support the works of different departments and levels. Different departments and levels will support different activities, yet they will all result in the same result, which is the enhancement of performance and efficiency of work due to support by activities.

It was defined in research of Hung [4] that organization critical activities are the activities to be constantly implemented by an organization for success of industry. Organization Critical Activities (OCA) are regarded as an essential element for the short-term, mid-term, and long-term success of industry. Thus they will receive a huge amount of resources and frequent supervision and instruction from high level management.

Hung [4] summarized the definitions of process flow by various scholars [1, 2, 3, 7, 11, 12], and proposed that processes are mutually correlated activities among various levels. This kind of activity is embedded in a set of rules and resources. It can enforce and grant social activities as shown in Table 1 below.

Table 1. The inclusion relation among process flow, task, activity, and OCA [4].

Name	Content	Range
Process	Activity, entity, rule, resource	1、2、3、4
Task	Resource and activity	2、3、4
Activity	Critical and non-critical	3、4
OCA	Critical activity for business success	4

The comparison of activity, task, and process involving OCA is as shown in the figure above. The activity is an essential core of a given process of task. An activity contains less processes and tasks. Organization critical activity is the activity constantly carried out by organization for achieving success, which must be greatly supported by high level supervisors with a great deal of resources. They are the activities to be constantly carried out by an organization for achieving success. They are the activities and essential element for an industry to achieve short-term, mid-term, and long-term success. It is only focused on whether or not the activity itself can achieve the organizational success.

In project management, critical path is a method for determining the starting and ending date of a project. The result of this method is to find a critical path, or to form a shortest chain of activities from start to finish. The critical path of activities is called critical activities [9], and the project management process flow can contribute to identification of task items of critical activities. Even though the term of "Critical Activities" in project management is very similar to the term of "Organization Critical Activities", critical activities actually refer to the short-term timeframe dedicated to completing a given task in project management, rather than a long-term activity for organizational success.

3 Methodology

In this study we use the manufacturing industry in Taiwan as an example to investigate what OCAs are conducting in order to ensure manufacturing companies' success. The data collection and analysis methods were case study and thematic coding. A total of 15 supervisors from 14 domestic manufacturing companies were participated in the interviews, and all interviews were voice recorded and analyzed. The basic company information, company background, and personal information of respondent are as shown in Table 2.

Table 2. Basic information of respondent.

Respondents	Years of work	Job/Duty	Industry
I1	7	Chief Officer of Safety, Health, and Environmental Protection	Electronic components industry
I2	10	Chief Officer of IT Department	Automobile/Motorcycle parts retailers
I3	16	R&D Manager	Automobile/Motorcycle parts retailers
I4	9	IT Chief Leader	Storage/Transportation industry
I5	12	Procurement Manager	Machinery and hand tool industry
I6	13	Business Manager	Chemical raw material manufacturing industry
I7	10	Sales Manager	Electronic components industry
I8	8	Head of Production	Electronic components industry
I9	12	Deputy R&D Director	Chemical raw material manufacturing industry
I10	15	Business Manager	Electrical machine manufacturing industry
I11	6	Chief of General Affairs and Procurement	Storage/Transportation industry
I12	8	Deputy Business Manager	Chemical raw material manufacturing industry
I13	9	Sales Manager	Chemical raw material manufacturing industry
I14	9.5	Chief of Production	Electronic components industry
I15	9.5	Chief of Factory Affairs	Chemical raw material manufacturing industry

In this study, the one-on-one interview was adopted with the semi-structural interview approach. Before each interview, the research objectives were introduced to the interviewee. During the interview, the interviewer asked the respondent how his/her company defined success without providing any information related organization critical activities, and then gradually guide the interviewee to describe the content and measure of organization critical activities of the company. After receiving the answer, the interviewer provided the list of organization critical activities to be confirmed and corrected by the interviewee in order to update the current OCA list. After the interviews, two experts made the suggestions and corrections to ensure that we obtained the most appropriate list of OCAs in the manufacturing industry.

4 Results and discussions

The results of interviews at initial stage of this study have revealed a total of 12 OCAs in the manufacturing industry. The first OCA is the development of new products. The second OCA is the development of new technology. The third OCA is marketing research. The fourth OCA is the expansion of domestic and foreign customer bases. The fifth OCA is sustaining existing customers. The sixth OCA is providing specialized services. The seventh OCA is ensuring product quality. The eighth OCA is ensuring the quality of raw material. The ninth OCA is enhancing production efficiency. The tenth OCA is providing personnel training. The eleventh OCA is increasing the efficiency of internal staff. The twelfth OCA is cost control.

After eliciting a list of OCAs, we then interviewed the senior managers from the other three manufacturing companies for prioritizing the 12 OCAs. Relevant company document was also analyzed to triangulate the given priorities. Results of the priorities given by the three case companies are shown in Table 3. The average priority of each OCAs was calculated for generating the final list of OCAs for the three manufacturing companies.

Table 3. Priorities of OCAs given by the three case companies.

Priority	OCAs	Priorities Given by the Case Companies			
		Case A	Case B	Case C	Average
1	Ensuring product quality	1	4	4	3
2	Ensuring the quality of raw material	4	5	3	4
3	Cost control	2	8	5	5
4	Development of new product	7	9	1	5. 7
5	Development of new technology	5	10	2	5. 7
6	Sustaining the existing customers	11	2	6	6.3
7	Increasing the efficiency of internal staff	9	1	10	6.7

7	Enhancing production efficiency	3	6	11	6.7
9	Marketing research	6	7	8	7
10	Expansion of domestic and foreign customer bases	12	3	7	7.3
11	Providing personnel training	8	11	9	9.3
12	Providing specialized services	10	12	12	11.3

According to Table 3, the top three OCAs are ensuring product quality, ensuring the quality of raw material, and cost control, which means that quality and cost are two major activities to achieve success for the manufacturers. The next two are development of new product and development of new technology, which means research and development related activities are important to the companies' success. Interestingly, market research and expansion are relatively in the lower level of priority.

5 Conclusions

Very few studies in the past have focused on eliciting the OCAs in the manufacturing industry. This study aims to investigate what critical activities are carrying out by the manufacturers in Taiwan. Through the case study approach including the interviews with the senior managers from three manufacturing companies and a series of prioritizing activities, 12 OCAs have been identified. This study also found that Taiwanese manufacturers placed more importance on the activities of quality and cost control while placed relatively less importance on the marketing related activities. Yet, they believed that quality and cost control, manufacturing efficiency, marketing, training, and services are all important to their success. From the government perspective, this study allows the government to better understand the current operation situations of Taiwanese manufacturing industry, and thus to help them develop appropriate policies for them to achieve their success. From the enterprise perspective, the results of this study can provide reference for manufacturers in other countries to elicit their OCAs and to better manage their activities for success. Future research can apply the methods used in this study in other countries or industries for better understanding of achieving organizational success.

References

1. S. Biazzo, Bus. Process Manage. J, **6**, 99-112 (2000)
2. A. Clark, A. Baxter, Manage. Account, **70**, 54-55(1992)
3. S. Convey, *CMA Magazine*, **65**, 20-24(1994)
4. W.H. Hung, PhD Thesis(2006)
5. A. Howard, A. Kochhar, J. Dilworth, Int. J. Oper. Prod. Manage, **22**, 7-29 (2002)
6. D. Leonard, R. McAdam., J. Eur. Ind. Tr, **26**, 4-13(2002)

7. R.G. Lee, B.G. Dale, Bus. Process Manage. J, **4**, 1463-7154(1998)
8. G. Svensson, Manage. Decis, **40**, 574-583(2002)
9. J.H. Son, M.H. Kim, J. Syst. Sofwr, **58**, 211-219(2001)
10. P.B.B. Turney, *Strategic Finance*, **73**, 20(1994)
11. P.P. Tallon, K.L. Kraemer, *Americas Conference on Information Systems* (1998).
12. M. Zairi, Bus. Process Manage. J, **3**, 64-80(1997)

Multi-tasking multi-machine scheduling system for multi-stage multi-criteria production

Shao-Chen Liu[1], *Yu-Ren* Chen[1], *Cheng-Ju* Kuo[1], *Tzu-Yu* Lin[1], *Kuo-Cheng* Ting[1], *Don-Lin* Yang[1,*], *Hsi-Min* Chen[1], and *Yi-Chung* Chen[2]

[1]Department of Information Engineering and Computer Science, Feng Chia University, Taichung, Taiwan
[2]Department of Industrial Engineering and Management, National Yunlin University of Science and Technology, Yunlin, Taiwan

Abstract. In recent years, many major manufacturers have been incorporating Industry 4.0 technologies such as preventive fault detection, automated scheduling algorithms, and component management to increase productivity and reduce production costs. Achieving this objective requires a substantial amount of working capital to acquire large quantities of new machinery, equipment to extract data from the machinery, and high-priced big data analysis software. However, most factories in the world are small- or medium-sized companies and have not enough capital to replace their machinery or purchase big data analysis software. It is therefore almost impossible for these factories to reach the goal of Industry 4.0. Furthermore, most of the conventional automated production scheduling methods only consider a single criterion in scheduling, which is not applicable for actual situations. This study therefore proposed a multi-tasking multi-machine scheduling system for multi-stage multi-criteria production to address various shortcomings in existing methods. To achieve this goal, we proposed a novel concept based on skyline queries to assist in the scheduling process. Also, a data structure of "heap" is applied in this work to accelerate the scheduling process. The experimental results demonstrated the validity of the proposed approach.

1 Introduction

Put simply, Industry 4.0 involves the use of large quantities of machinery data, sensors, the Internet of Things, big data analysis, and collaboration between man and machine to increase the overall productivity of a factory. To achieve this objective, factories must spend a substantial amount of working capital to install new machinery, extract data from the machinery, and use high-priced big data analysis software for scheduling and production. However, most factories in Taiwan are small- or medium-sized companies and have not enough capital to replace their machinery or purchase big data analysis software. It is therefore almost impossible for these factories to reach the goal of Industry 4.0.

In consideration of this, many researchers have been developing inexpensive ways to help these manufacturers achieve Industry 4.0. For instance, Kuo *et al.* [4][9] proposed

[*]Corresponding author:dlyang.tw@gmail.com

methods that combine machine maintenance personnel experience, add-on triaxial accelerometers, and artificial neural networks for machinery fault detection and successfully applied these methods to spring factories. Kuo *et al.* [5]successfully developed an RFID-based component management system that records what components a machine contains and what processing task it is performing, thereby greatly increasing component management efficiency. However, a comprehensive view of all relevant research revealed that only these two types of cost reduction techniques were discussed. No discussion has been made on automated scheduling under the Industry 4.0 model, so some discussion is warranted to address this issue.

Generally speaking, optimizing the schedules of work orders is one of the most effective ways to increase a factory's production efficiency. A good schedule can efficiently allocate various resources, increase factory productivity, and enable machinery to work continuously in the absence of any major malfunctions. It is because of this advantage that researchers have actively conducted studies on scheduling in the past. Common scheduling methods can be divided into single-machine scheduling and multi-machine scheduling depending on the number of machines. [3][6][8] are examples of single-machine scheduling methods that can design schedules for the shortest work time or shortest lead time. In contrast, multi-machine scheduling methods can make schedules based on two concepts: job shops and flow shops. However, we have found that most of the above methods only considered a single criterion to select suitable scheduling results. In other words, they mixed multiple criteria into a single score and then chose the schedule with the highest score. The major issue is that the results will not be able to adapt to various situations in a factory. Suppose we only consider total work time and delivery delay time, and each of the criteria contributes to 50% of the scheduling score. The total work time of Schedule A is 50 hours (contributing 0.5 points to the scheduling score), and the delivery delay time is 1 hour (contributing 0.5 points to the scheduling score). In contrast, the total work time of Schedule B is 40 hours (contributing 0.7 points to the scheduling score), and the delivery delay time is 5 hours (contributing 0.35 points to the scheduling score). A conventional scheduling algorithm would give Schedule B as the solution because it has a higher total score than Schedule A. However, we must understand that such a solution may be reasonable in some circumstances but unreasonable in others. Even in the same factory, there may be different considerations at different times. If the customer is more easygoing, and a delay of 1 hour or 5 hours makes no difference, then the factory may choose Schedule B because it requires less work time. In contrast, if the customer places a lot of emphasis on the delivery time, then the factory will have to choose Schedule A over Schedule B because they can deliver the order faster. This example makes it clear that even in the same factory, Schedules A and B are both possible choices, so both should be presented to factory personnel for consideration rather than just providing Schedule B. Since the conventional scheduling method has a serious flaw, new methods are needed to address it.

The existing method that is closest to the scheduling approach which we propose is the multi-tasking multi-machine scheduling algorithm for single-stage multi-criteria production presented by Kuo *et al.* [9].The method that they proposed can efficiently assist factories in scheduling single-stage work orders and is particularly helpful for processes with only a single stage, such as the manufacturing of springs or screws. However, it cannot be applied to factories with multi-stage manufacturing processes, such as household appliance factories or semiconductor fabs. This study therefore developed a multi-tasking multi-machine scheduling algorithm for multi-stage multi-criteria production to resolve this issue.

When developing the target algorithm, we also found some parts of conventional scheduling methods that were unsuitable for Industry 4.0 and needed to be overcome. During computation, most conventional scheduling algorithms assume that the execution time for each work order is precise and known. For instance, the execution time can be obtained from the SOP operation manual of each component or the past experience of the scheduling personnel. However, these SOP operation manuals or experiences came from the Industry 3.0 era or older, when all components followed standard specifications and could only be used in mass production. The Industry 4.0 era emphasizes component specialization and small-scale production, so many components are only manufactured in several batches. We therefore cannot accurately estimate the production time of each component based on previous statistics. In response to this, we used add-on triaxial accelerometers to understand machine conditions at each time point and then determine the amount of production time that is needed for each component based on said conditions. Ultimately, we can optimize scheduling based on the time. Experiments demonstrated the validity of the proposed approach.

The remainder of this paper is arranged as follows. Section 2 introduces related works, Section 3 presents the methodology used in this study, Section 4 explains the experiment results, and Section 5 contains the conclusions of this study and suggestions for future research.

2 Related works

2.1 Sensors commonly used in machinery

Previous research [9] shows that the most common physical features in machinery are motion, vibration, noise, and temperature. To detect these features, we can use triaxial accelerometers (for motion and vibration), microphones (for noise), and thermocouples (for temperature). To simplify the problem, we used only triaxial accelerometers to detect the features in machinery, but the approach that we applied to triaxial accelerometers can be easily applied to other types of sensors.

2.2 Scheduling algorithms

A good scheduling algorithm can efficiently allocate resources, arrange work orders, and increase the production efficiency of a factory. Previous research [9] mentioned that scheduling algorithms can be viewed from three aspects. First, in terms of the objective, scheduling algorithms can be divided into process types and production types. Second, in terms of product structure, they can be divided into one-piece products or assembly products. Finally, in terms of the number of machines, they can be divided into single-machine operations or multi-machine operations [9].

2.3 Skyline queries

In recent years, skyline queries [1][2][7] have become a part of decision-making processes in database systems. Aside from identifying data points with better performance in only a few dimensions out of many, skyline queries can also make queries based on the needs and preferences of different users. Say for example that a user wants to take a trip to the beach and wants to stay at a hotel that is closer to the beach or has lower room rates. The available hotel information is as shown in Table 1. Skyline queries can be used to find the hotels that best fit the user's needs and preferences. Let us map the data in Table 1 onto a graph with the distance from the beach as the X axis and the room rate as the Y axis, as shown in Fig. 1. If we compare Hotel A and Hotel B, the former is closer to the beach and less expensive than the latter (which we simply put as the former dominating the latter), so the user will choose Hotel A rather than Hotel B. In this case, the information of Hotel B

Table 1. An example of hotel database.

Name	Cost	Distance
A	10	1
B	12	1.7
C	10	2.5
D	4	3.5
E	8	3.2
F	6	2
G	14	3

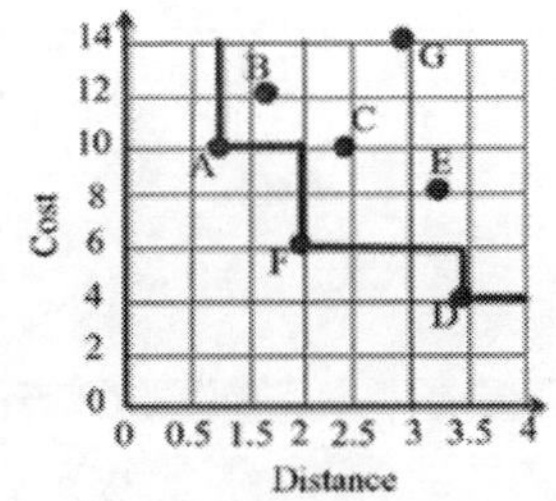

Fig. 1. A graph representation of Table 1.

does not need to be provided to the user as an option. In contrast, if we compare Hotel A and Hotel D, we will find that the two are incomparable; Hotel A is closer to the beach, but Hotel D has a lower room rate. Either is a possible choice for the user, so both should be recommended to the user. The key of skyline queries is to find data points that are not dominated by any other data points in the dataset. In Fig. 1, for example Hotels A, D, and F are the skyline query results because they are not dominated by any other points.

3 Methodology

This section introduces the methodology of this study, including data collection, time analysis, and the scheduling approach.

3.1 Data collection

This study mainly used triaxial accelerometers. During the study, we collected acceleration values along the X, Y, and Z axes and used them to measure the motions and vibrations of the machinery. Figure 2 presents an actual example of the collected data along the X axis. The horizontal axis presents the number of data items, while the vertical axis presents acceleration in m/s^2. Note that the values may contain positive as well as negative values because the sensors have fronts and backs; accelerations in different directions are opposite to each other in terms of positivity and negativity. We can also see from Fig. 2 that aside from the motion information of the machine, the data collected from a triaxial accelerometer also contains a lot of noise. We therefore employed a band pass filter to filter out the high-frequency and low-frequency noise.

3.2 Time analysis

After collecting acceleration signals from the machines, we can then use the signals to determine the precise amount of time that the machine needs to manufacture a piece of each type of component. The operation method is divided into two parts. The first part is windowing the acceleration signals and extracting the feature values of each windowed signal. The second part is finding the cycle of the feature values of the windowed signals and using the cycle to calculate the precise amount of time needed for each workpiece. We explain these two parts in more detail below.

The signal windowing approach in the first part is slightly different from conventional methods, which basically overlap signals in different windows before analysis. In contrast, the focus of this paper is to conduct cycle analysis, so the signals in different windows do not have to overlap. Figure 3, for example, displays the windowed results of the X axis in Fig. 2. Once the signals have been windowed and segmented, we extract five feature values from each windowed signal: (1) maximum value: the maximum value of all of the data

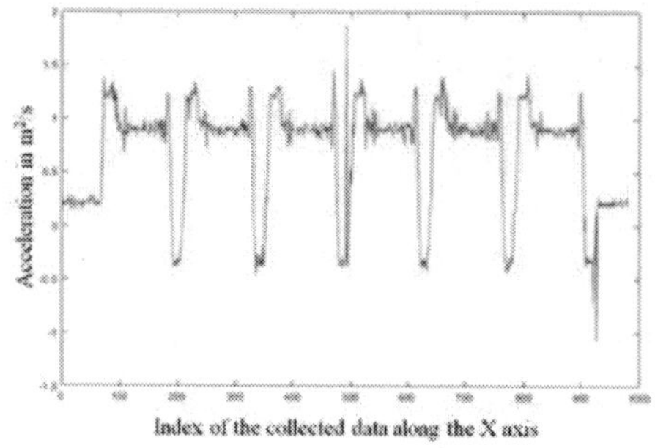

Fig. 2. The collected data along the X axis.

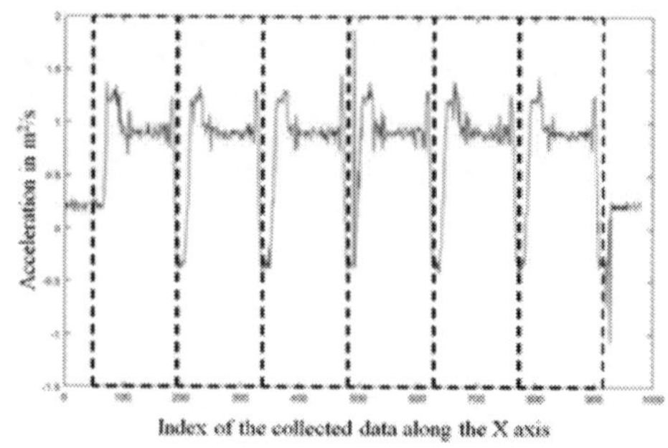

Fig. 3. The windowed result of the collected data along the X axis.

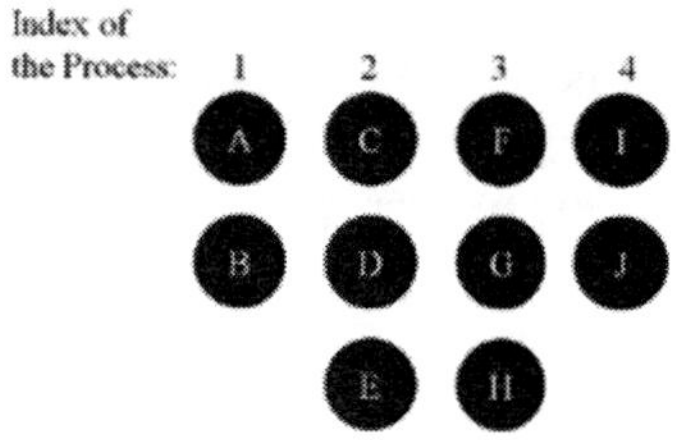

Fig. 4. An example of manufacturing processes.

points in each window, (2) minimum value: the minimum value of all of the data points in each window, (3) mean value: the mean value of all of the data points in each window, (4) energy value: the sum of squares of all of the data in each window divided by the length of the window, (5) standard deviation: the standard deviation of all of the data points in each window.

In the second part, we examine the cycle of maximum and minimum values of these five feature values in the series of windowed signals and derive five sets of cycle data. Ultimately, the average of the 30 sets of cycle data serves as the precise amount of time that the machine needs to finish manufacturing a certain component and is used to estimate the amount of time needed to complete work orders for the same type of component but in a different quantity.

3.3 Target scheduling algorithm

The multi-tasking multi-machine scheduling algorithm for multi-stage multi-criteria production developed in this study was designed based on the concept of skyline queries. It mainly assumes that each work order requires multiple stages of manufacturing and that each stage of manufacturing may be performed by multiple parallel machines. However, the machines may vary in efficiency or have different limitations. Take Fig. 4 as an example. This factory has four manufacturing processes, the first of which is performed by two parallel machines. It takes these two machines 1 hour and 2 hours respectively to complete the first manufacturing process (derived from the time analysis in the previous section), and the yield score of both machines is 2. The second manufacturing process is performed by three parallel machines, and it takes them 3 hours, 4 hours, and 1 hour respectively to complete the manufacturing process (derived from the time analysis in the previous section). Their yield scores are 1, 2, and 1, and so on and so forth for the third and fourth manufacturing processes. To not lose generality, we specifically assumed that once a machine begins manufacturing, it does not stop until one type of component is completed. To insert an additional work order, the machine must complete the current component before the proposed algorithm can be applied for rescheduling.

Below we explained the process of the algorithm by using Fig. 5. When starting up the algorithm, we first initialize a heap and a list. The former caches all of the current scheduling results of the algorithm, while the latter is used to find the query results of the currently known multiple criteria. The first step of the algorithm is to calculate all of the possible schedules when only the first work order is considered. Suppose that there are four work orders that need to be scheduled: a, b, c, and d. The results of this step are <a>, <b>, <c>, and <d>, where <•> represents a schedule. The second step is to place the results of the first step into the heap. The schedules are arranged in ascending order based on the total sum of all of their dimensions. The conditions for each dimension are customized by the factory, such as total work hours and delay time. In the third step, the first schedule (i.e., the schedule with the smallest total of all the dimensions) is extracted from the heap and checked. The check is divided into two parts: the first determining whether the scheduling combinations are completed (i.e., all work orders have been scheduled) and the second determining whether the schedule is dominated by known results in the list. Based on the results of the two parts, we can derive the four possible outcomes below.

Case 1: the schedule is not complete and is not dominated by a result in the list. In this case, we cannot be sure whether expansions of the schedule will include other multi-criteria schedule results, so we expand the schedule, place it in the heap, and rearrange the order for the next round of checks. For example, with work orders a, b, c, and d to be scheduled, suppose that Schedule <c, d> is not dominated by any result in the list. Schedule <c, d> is first expanded to <c, d, a> and <c, d, b>, and the expanded results are rearranged into the heap based on the total of the dimension values to await the next round of checks.

Case 2: the schedule is not complete and is dominated by a result in the list. In this case, the schedule in question will inevitably be dominated by a result in the list no matter how the schedule combination is expanded. This is because the conditions of the schedule will only increase with the number of work orders, so it cannot be the final multi-criteria scheduling result. We can therefore directly remove this schedule combination from the heap. The expanded schedule combinations will not be checked, either, which accelerates the algorithm. For example, suppose that with work orders a, b, c, and d to be scheduled, Schedule <a, b> is found to be dominated by a result in the list. Schedule <a, b> is then directly eliminated from the heap, and <a, b, c, d> and <a, b, d, c> will not be checked.

Case 3: the schedule is complete and is not dominated by a result in the list. In this case, the schedule is a result of the target query, so it is directly added to the list.

Case 4: the schedule is complete and is dominated by a result in the list. In this case, the schedule cannot be the result, so it is directly removed from the heap.

The steps above are repeated until there are no more schedules in the heap.

4 Experimental results of our simulation

In the simulation experiment of this study we used synthetic datasets because we wanted to verify whether the proposed approach can be applied to different factories for any given number of manufacturing stages and any given number of machines. To do this, we can only conduct simulations with synthetic data. Table 2 presents the data that we generated, where P1 to P5 represent five processes in a factory. Numbers 1 to 35 under each of P1 to P5 mean the indexes of machines in different process. For example, the machine 5 under P1 means the fifth machine in the first process. Also, each machine has two scores with regard to a single process. To not lose generality, the calculations of the two scores can be set by the user, such as total work time or delay time. In this experiment, we compared the result of the proposed method with that of a brute force approach, which involved finding all of

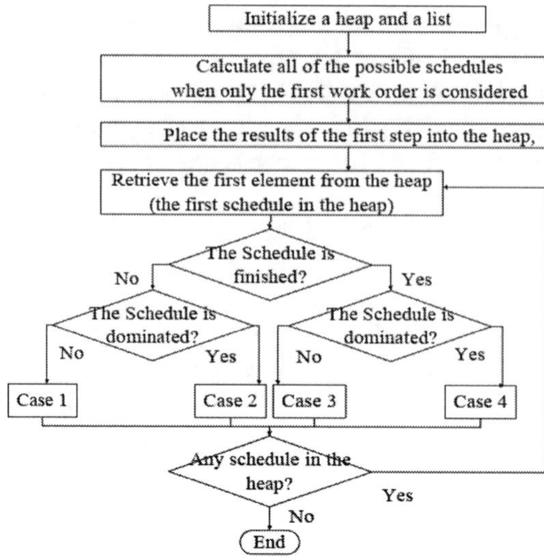

Fig. 5. The flow chart of the proposed scheduling algorithm.

Table 2. The factory data we used in this work, where 'Pi' represents a process and 'Sj' means a score.

# of machine in P1	S1 of machines in P1	S2 of machines in P1	# of machine in P2	S1 of machines in P2	S2 of machines in P2
1	7	4	8	5	4
2	9	5	9	9	1
3	6	8	10	1	5
4	2	3	11	8	9
5	3	9	12	2	8
6	2	7	13	6	3
7	7	5	14	3	7

# of machine in P3	S1 of machines in P3	S2 of machines in P3	# of machine in P4	S1 of machines in P4	S2 of machines in P4	# of machine in P5	S1 of machines in P5	S2 of machines in P5
15	5	8	22	9	1	29	9	3
16	3	4	23	2	6	30	3	6
17	1	6	24	7	5	31	4	4
18	6	9	25	4	7	32	5	8
19	9	2	26	8	4	33	8	1
20	1	6	27	1	8	34	2	9
21	6	9	28	6	2	35	6	7

the schedule combinations and then applying a skyline query to the schedule combination using a BNL algorithm [2].The experiment was conducted in Windows 10 64-bit with Intel i7-6500U and 16G RAM, and the program was written using Python.

In Figure 6 we compare the results of our proposed approach and the brute force approach. The process of the brute force algorithm is finding all possible schedule combinations in the first step and then using the BNL algorithms to find the skyline results. As can be seen, no matter what parameter combination was used, the brute force approach (top four red lines) required more time than the proposed approach (blue lines at bottom). This is because the proposed approach can effectively reduce the number of schedules that need to be checked, whereas the brute force approach cannot do so. Consequently, the brute force approach required significantly more time than the proposed approach. Furthermore, we can see that the computation time increased with the number of manufacturing stages and the number of machines in both the proposed approach and the brute force method.

This is because increasing any such variable will increase the number of schedule combinations, so the algorithms need more time to finish checking them. Of particular note, the amount of time that the brute force method required increased exponentially because the computation time of the BNL algorithm used in the brute force method increases exponentially with the quantity of data that needs to be checked. In contrast, the amount of time required by the proposed approach only presented linear increases because the number of schedule combinations that need to be checked does not increase too much even if the numbers of manufacturing stages and machines increase.

5 Conclusion and future research

In recent years, many industries have begun to incorporate Industry 4.0 technologies to increase industry value. However, doing so requires a substantial amount of working capital, which is difficult for many small factories. Especially in Taiwan, most of the manufacturers are small- or medium-sized companies and have not enough capital to incorporate Industry 4.0 technologies. In view of this, our study proposed a novel approach to help these manufacturers achieve automated scheduling at lower costs. Furthermore, the production concept of small quantities and great variety in Industry 4.0 is different from that of large quantities and small variety in Industry 3.0. Since existing scheduling methods may no longer be applicable, we also proposed a new approach in this regard. Accordingly, we developed a multi-tasking multi-machine scheduling system for multi-stage multi-criteria production. In future research, we will further develop a scheduling algorithm that allows for the insertion of additional work orders or impromptu maintenance to better satisfy the actual needs of factories.

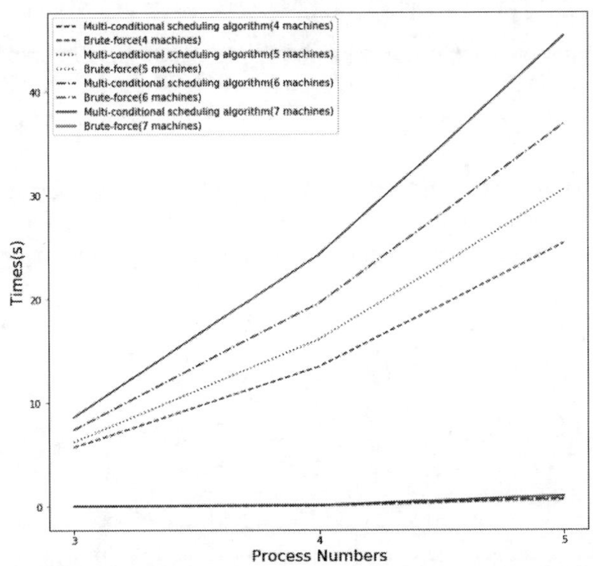

Fig. 6. The comparisons between the proposed method and other methods.

Acknowledgment

This work was supported in part by the Ministry of Science and Technology of Taiwan, under contracts MOST 106-2119-M-224-003 and 106-2218-E-126-001.

References

1. I. Bartolini, P. Ciaccia, and M. Patella, SaLSa: Computing the skyline without scanning the whole sky, *Proceedings of Conference on Information and Knowledge Management*, pp. 405-414, 2006
2. S. Borzsonyi, D. Kossmann, and K. Stocker, The skyline operator, *Proceedings of International Conference on Data Engineering*, pp. 235-254, 2001
3. D. Ivanov, A. Dolgui, B. Sokolov, F. Werner and M. Ivanova, A dynamic model and an algorithm for short-term supply chain scheduling in the smart factory industry 4.0, *International Journal of Production Research*, vol. 54, no. 2, pp. 386-402, 2016
4. C.J. Kuo, K.C. Ting, and Y.C. Chen, State of product detection method applicable to Industry 4.0 manufacturing models with small quantities and great variety: An example with springs, *Proceedings of IEEE International Conference on Applied System Innovation*, pp. 1650-1653, 2017
5. C.J. Kuo, K.C. Ting, J.C. Ying, S.K. Chen, and Y. C. Chen, The RFID-based real-time monitoring system and the management algorithm of RFIDs, *Proceedings of International Conference on Machine Learning and Cybernetics*, pp. 499-503, 2017
6. G.J. Kyparisis and C.P. Koulamas, Flexible Flow Shop Scheduling with Uniform Parallel Machineries, *European Journal of Operational Research*, vol. 168, pp. 985–997, 2006
7. D. Papadias, Y. Tao, G. Fu, and B. Seeger, An optimal and progressive algorithm for skyline queries, *Proceedings of 2003 ACM SIGMOD International Conference on Management of Data*, pp. 467-478, 2003
8. D.N. Tahar, F. Yalaoui, C. Chu, and L. Amodeo, A Linear Programming Approach for Identical Parallel Machinery Scheduling with Job Splitting and Sequence-dependent Setup Times, *International Journal of Production Economics*, vol. 99, no. 1-2, pp. 63-73, 2006
9. C.J. Kuo, K.C. Ting, Y.C. Chen, D.L. Yang, H. M. Chen, Automatic machine status prediction in the era of Industry 4.0: Case study of machines in a spring factory, Journal of Systems Architecture: Embedded Software Design, vol. 81, pp. 44-53, 2017

Computer aided design and analysis on distributors in DAC columns

Chen-Jiann Lin[1,*], *Tseng-Hsiang* Tse[1] , *Liu- Cheng* Che[1], and *Liang-Ming* Tsai[2]

[1]Department of Mechanical and Automation Engineering, I-Shou University, 84001 Kaohsiung City, Taiwan
[2]Department of Chemical Engineering, I-Shou University, 84001 Kaohsiung City, Taiwan

Abstract. Dynamic axial compression (DAC) columns are key elements in simulated moving bed, which is a chromatography process in drug industry and chemical engineering. In this study, rules for designing distributors are proposed based on mass conservation and validated by experiments, the computer aided design (CAD) and the computational fluid dynamics (CFD). Experimental works are conducted to choose feasible numerical parameters for simulations. In CFD, the transient laminar flow fields are governed by the momentum and species transport equations with Darcy's law to model the porous zone in the packed bed. Results show that CFD combined with CAD solid modelling is a good approach to explore detailed flow fields in DAC columns and carry out parameter analysis for innovative designs. For further testing and evaluation, a new model of compound distributor is designed, 3D printed and processed in factory for practical applications in preparative chromatography.

1 Introduction

Simulated moving bed (SMB) chromatography process, which is a multicolumn chromatography process, has been applied to various industrial fields, such as foods, petrochemicals, fine chemicals, and pharmaceuticals [1]. The counter current movement of the adsorbent bed is simulated by periodically shifting the inlet and outlet nodes to the direction of the flow during each switching period. Therefore, the SMB process can achieve high productivity and low solvent consumption [2] and allows significantly higher yield and purity compared to a batch chromatography [3, 4].

Generally, the performance of chromatography depends on the combined effects of DAC column dynamics, the pre and post column geometry [5]. In the beginning of the column, an even fluid distribution must be achieved cross-sectionally through careful design in order to give satisfactory separation efficiency [6, 7]. Therefore, the distributor at the column inlet will be vital to provide plug flow, hence improve the column efficiency. At the same time appropriate packing materials have to be chosen and the column has to be carefully packed to maintain the flow pattern as close as possible to a plug flow [8, 9]. The above mentioned factors are especially important to columns with large diameter. If these factors were not considered, band distortion and unsymmetrical peaks would be observed because flow misdistribution can result from faulty header design [7, 10]. Therefore, DAC

[*]Corresponding author: james88@isu.edu.tw

columns tend to be of low length over diameter (L/D) for high-productivity considerations, which results in poor flow distributions and reduces the column efficiency. In DAC columns, the distributor (collector) plays a key role to make the plug flow in columns. Manufacturers in this field have been devoted to improving the distributor's performance to raise the efficiency of their products from DAC columns [11, 12].

CFD has well been known for more than 20 years as a pragmatic and reliable tool to describe local hydrodynamics in various separation processes [13]. Its modelling has been applied to investigate the transport phenomena in the preparative chromatography column [14-16]. Therefore, it is proposed to characterize in detailed hydrodynamics of DAC columns based on the recent advances of flow characterisation by CFD, and then to deduce numerical treatments that can be beneficial to simulations. In the present study, CFD is used to access the internal information inside the DAC column and the flow is considered to be laminar due to low flow speed in the micro-particle packed column [9, 17].

For the present research, we present the governing equations for the physical problem and introduce CAD and CFD techniques to simulate flow fields in the DAC column in section 2. Rules to design distributors at the header of DAC columns are also proposed in this section. Results and discussions are stated in section 3, which includes experiments, numerical treatments for CFD, CFD code validations and parameter analysis to investigate flow phenomena. Furthermore, a model of compound distributor is displayed and the prototype is 3D-printed, which is referred to develop DAC columns with low L/D for future practical applications. Remarks are concluded in the final section, which includes the present contributions and future works.

2 Methodology

The present simulation is simplified as an incompressible laminar flow of a single species that does not generate heat. Inside the DAC column, non-slip boundary conditions are used for the walls and the packed bed is viewed as a porous medium in which the flow of a Newtonian fluid is known to follow Darcy's law at a low Reynolds number. The flowchart for the design of DAC columns in this study is illustrated in Fig. 1.

2.1 Governing equations

2.1.1 Equations in physical domain

The transient conservation equation for transport of a scalar quantity ϕ is demonstrated by the following equation written in integral form for an arbitrary control volume *V* as follows [18]:

$$\frac{\mathrm{d}}{\mathrm{dt}}\oint_{\mathrm{V}} \rho\phi \,\mathrm{dV} + \oint \rho\phi\, \vec{\mathrm{v}}\cdot \mathrm{d}\vec{\mathrm{A}} = \oint \Gamma\, \nabla\phi \cdot \mathrm{d}\vec{\mathrm{A}} + \int_{\mathrm{V}} \mathrm{S}_{\phi}\mathrm{dV} \tag{1}$$

where ρ indicates density; $\vec{\mathrm{v}}$ is velocity vector; $\vec{\mathrm{A}}$ is surface area vector; V is volume; Γ is the diffusion coefficient for ϕ and S_ϕ is source term of ϕ per unit volume. The symbol ϕ can be replaced by 1 for the continuity equation, and u, v and w are the momentum equations in the X, Y and Z directions, respectively.

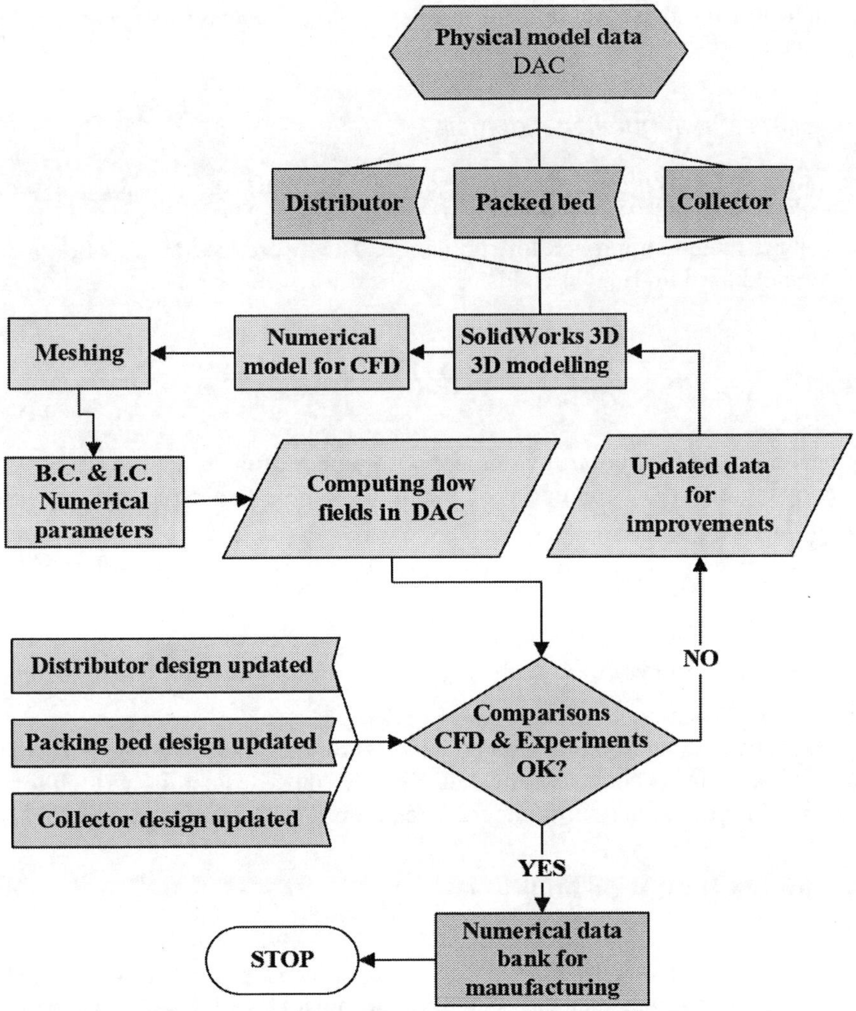

Fig. 1. The flowchart for the design of DAC columns.

Inside the DAC column, the packed bed can be simulated by porous medium, which is expressed as extra source terms including viscous loss (Darcy's law) and inertial loss. The added source terms to model the porous effect is expressed as

$$S_j = -\left(\frac{\mu}{\alpha} v_j + C' \frac{1}{2} \rho |v| v_j\right) \tag{2}$$

where α is permeability ($1/\alpha$ is viscous resistance); C' indicate the inertial resistance factor and $|v|$ is the magnitude of the velocity. If the fluid flows very slowly in porous zone for some cases, the inertial loss can be ignored to save the computing time.

In the present work the pulse input was applied and the injected sample was dilute solution. No adsorption between the solute and the stationary phase takes place. Therefore the mass loss between the liquid phase and solid phase could be neglected concerning momentum balance. The species concentration conservation used in present calculation is

$$\frac{d}{dt} \oint_V \rho C_i \, dV + \oint \rho C_i \vec{v} \cdot d\vec{A} = \oint D_i \nabla C_i \cdot d\vec{A} + \int_V S_C \, dV \tag{3}$$

where C_i indicates the mass fraction for species i; D_i is the mass diffusion coefficient for species i in the mixture.

2.1.2 Equations in computational domain

Since the fluid used in this study is IPA (Isopropyl alcohol), the density is assumed to be constant in the flow field. Equation (1) can be expressed in generalized coordinates using the finite volume method approach for simulation. The discretized formulation for each cell (control volume) based on Eq. (1) yields

$$\sum_{f}^{N_f} v_f \phi_f A_f = \sum_{f}^{N_f} \Gamma_\phi (\nabla \phi_f)_n A_f + S_\phi V \tag{4}$$

where f denotes face of the control volume; N_f is number of faces enclosing cell.

After combing boundary conditions, equation (4) can be linearized and expressed in the computational domain as:

$$a_p \phi = \sum_{nb} a_{nb} \phi_{nb} + b \tag{5}$$

where the subscript nb refers to neighbour cells; a_p and a_{nb} are the linearized coefficients for ϕ and ϕ_{nb}; b is the source term including Eq. (2).

The number of neighbours for each cell depends on the grid topology, but will typically equal the number of faces enclosing the cell. Similar equations can be written for each cell in the grid and this results in a set of algebraic equations to be solved using CFD techniques.

2.2 Rules for distributor design

The modified distributor was designed according to some rules proposed by this study. The rules are on the assumption that the plug flow is inviscid without circulation existing in channels of the distributor. The rules are made by satisfying the law of mass conservation and stated as follows:

1. Draw a radial line from the centre of the distributor surface to get a radius and evenly divide the radius to N segments (Method 1), or divide the radius to make rings, including the inner circle, with equal area (Method 2).
2. Make N-1 concentric circles on the surface to form N-1 rings and an inner circle in which the feed tube (inlet) is located. Calculate the areas and denote them S_1 to S_N from inner to outer region on the distributor surface.
3. Determine Mn (n=1 to N-1) uniformly spaced distribution openings surrounding each ring. Denote the opening area as An (No restrictions for the shape of the openings and an being different from ring to ring for Method 2).
4. Make sure opening-area density, $(\sum_{1}^{Mn} A_n)/S_n$ on each ring (n), including the inner circle, have the same value; it will keep the same total flow rate for each ring and the inner circle.
5. Denote the segment length of each channel (including its branch) from the inlet to the opening as L_j; then make sure total length $(\sum_{1}^{J} L_j)$ is the same for each channel on the

surface. Besides, the cross-sectional area from the inlet to the opening along the channel is kept invariant to make flow rate conservation.

2.3 Computer aided design and analysis

2.3.1 Solid modelling and mesh

For most preparative and process scale chromatographic columns, DAC column ends include distributors (or collectors) and frits to distribute (or collect) fluid over the cross section of the column. Figure 2 displays illustrations of the DAC column with collector excluded. The engineering sketch, numerical model by SoldWorks 3D CAD and a mesh system for CFD computations are displayed in Fig. 2(a), 2(b) and 2(c), respectively [19]. The mesh is unstructured grid system with mixed hexahedra and tetrahedral grids. To effectively compute the present flow fields, the flow domain was split into smaller subdomains. We employed the mixed grid types to build the mesh around the distributor and frits. In the packed bed zone, structured hexahedral grids were used to fit the space inside the column. Moreover, finer grids were imposed along boundary walls to resolve the laminar boundary layer. Grid number for our parameter study was determined to be 0.8 million from grid independent tests and validated with experiments. The regions near wall, inlet and outlet were allocated a slightly higher mesh density as a velocity gradient was present. The desktop PC, 32G RAM with 4 cores 2.7GHz CPU, performed computation with iteration no. around 3000 and 3hrs computation for each case.

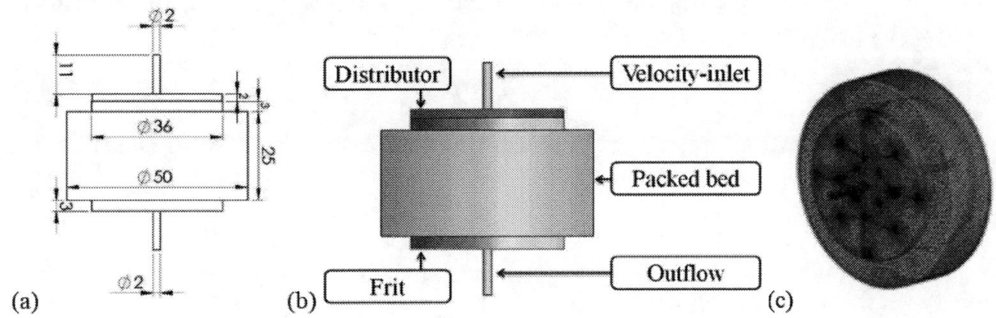

Fig. 2. Illustrations of the DAC column. (a) Engineering sketch. (b) Numerical solid modelling by CAD. (c) Mesh system for CFD computations.

2.3.2 CFD for detailed flow fields

The equations governing the present incompressible flow fields have no link between the continuity and momentum equations due to lack of the equation of state. To solve the set of algebraic equations derived from the governing equations, the SIMPLE algorithm was employed to carry out the calculations [19, 20]. SIMPLE is an acronym for Semi-Implicit Method for Pressure Linked Equations. SIMPLE actually solves the pressure correction equation, which is derived from combining the discrete continuity and momentum equations. Therefore, the set of momentum and continuity equations are coupled and nonlinear so we solve the equations iteratively. The momentum equations are solved for the velocities; however, the newly obtained velocities don't satisfy continuity. Corrections to velocities and pressure are proposed to satisfy the discrete continuity equation in SIMPLE algorithm. After iterations for the above correction computations, the final solution can be obtained if the estimated residuals approach within required tolerances.

In this study, the 3D numerical geometry model for the DAC column is created based from the engineering drawing as shown in Fig. 2(a)(b). Then we use CFD pre-processor, ANYSYS MESHING to transform the geometry model to the mesh system as shown in Fig. 2(c). After the mesh system is imported to the CFD FLUENT solver, numerical parameters, such as fluid properties, initial and boundary conditions have to be input. Besides, parameters for computation iterations and post processing for contour plotting need to be provided. According to the present numerical model, the velocity at inlet of the DAC column is set to be a constant value, which comes from experiments. The porous medium includes two zones, the frit part and the packed bed part as shown in Fig. 2(b). In fact, the collector at the rear portion of the DAC column is the same as the distributor in geometry. For the sake of limited computing resource, half of the DAC column can be adopted to study the header only and the outflow boundary condition are provided at outlet. The stationary solution of the momentum conservation was calculated first. After checking the convergent flow field, the concentration balances were switched on to make the transient simulation.

3 Results and discussions

Following the rules as stated previously, several new designs for distributor have been developed as shown in Fig. 3. Figure 3(a) illustrates the distributor with 3X4 circle openings located in a circle and 2 rings with equal area (Method 2). Figures 3(b) and 3(c) illustrate distributors with 3X4 circle openings and cross openings located in 3 circles, respectively, with same line segment radially (Method 1). In Fig. 3(c), N is three; M_n is four; J is two with open-area density equal to 8.83%.

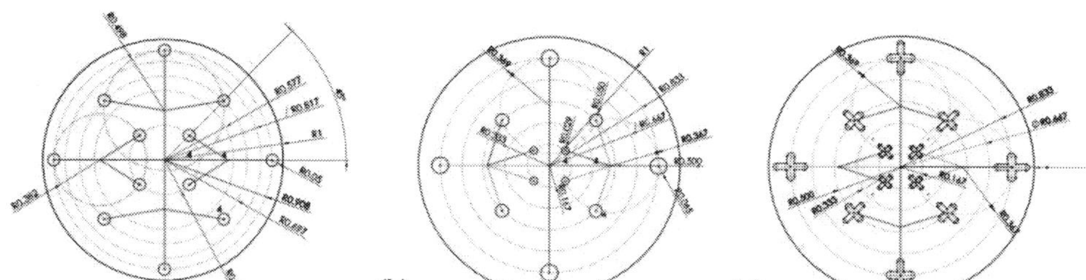

(a) (b) (c)

Fig. 3. Example models developed for the distributor by proposed rules (a) 3X4 openings by Method 2. (b) 3X4 circle openings by Method 1. (c) 3X4 cross openings by Method 1.

3.1 Validation experiment

To show the feasibility of our proposed rules for distributor design, two models were used for experiments. The original distributor (commercial distributor) and the new designed one for experiments, including the solid modelling and the fabricated workpiese, are displayed in Figs. 4(a) and 4(b) and Figs. 5(a) and (b), respectively. The experiments were conducted at the Centre for Advanced Chromatographic Processing of I-Shou University. The working fluid is the solvent, ethanol, and the solute is 20mL caffeine. The packed bed inside the column is full of SiO_2 with 0.37 porosity and acts as the porous zone in computations. The UV detector detected signals with wavelength at 254 nm for caffeine. The experimental data were collected as an electronic signal, all values of which were divided by the area under the eluted pulse curve. Therefore, the obtained data can be regarded as the normalized concentration. Parameters for experiments are thereafter used for boundary conditions in CFD simulations.

To validate the flow profile improvement from the new distributor in DAC column, the number of theoretical units (NTU) is adopted. The concept of the number of theoretical units in chromatography has been developed to described and quantify the differences in the separating efficiency based on the eluted peak shape [5]. The efficiency of a DAC column is a statistical calculation of the standard deviation of the peak in unit time. It is measured in NTU or the height equivalent to a theoretical plate (HETP), which are defined as follows:

$$\text{HETP} = \text{L} / \text{NTU} \tag{6}$$

where L is the length of the packed bed and NTU stands for the number of theoretical units.

HETP is the preferred description of column efficiency as columns of any geometry can be compared as their efficiency per unit length. A higher NTU or a lower HETP indicates better column efficiency and it can be clearly demonstrated in Fig. 6, in which HETP from the new designed distributor are higher than that of the commercial one for two flow rates. However, experiments in Figures 6(a) and 6(b) show that better performances are obviously presented for longer length columns (high L/D) and poor performances for shorter length ones (low L/D) because there is no time for flow to distribute uniformly in a short column and the good distributor helps to overcome this drawback.

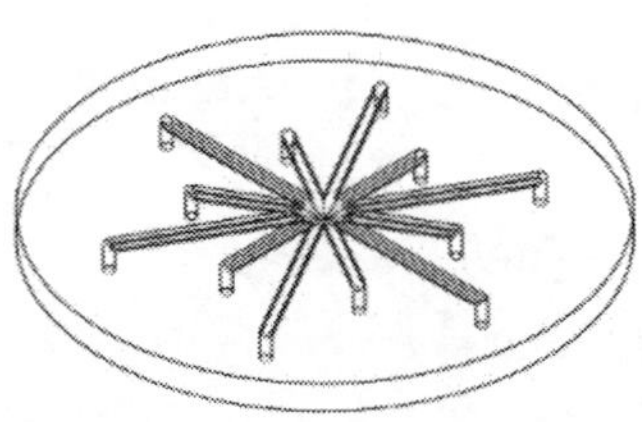

(a)

(b)

Fig. 4. The original distributor for experiments. (a) Solid modelling. (b) Fabricated workpiece.

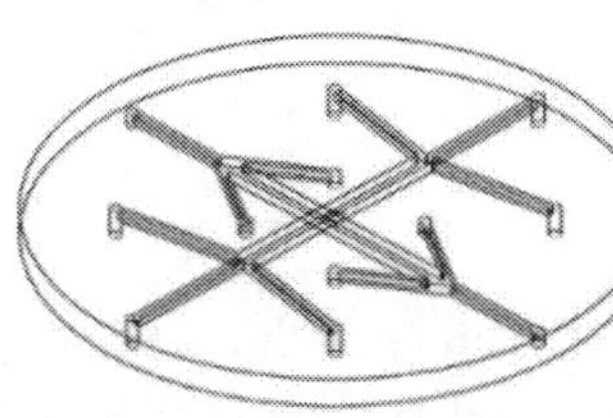

(a)

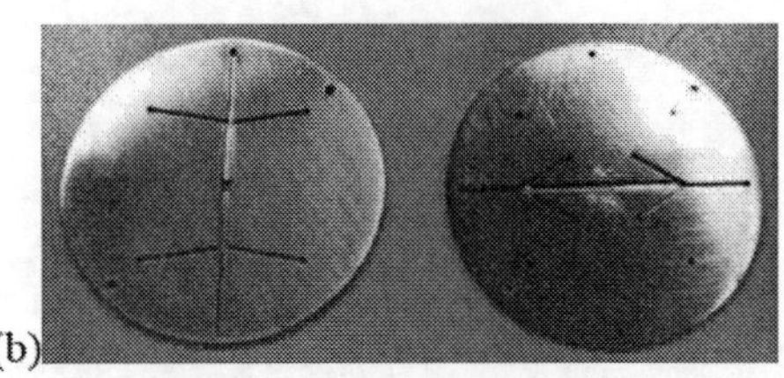

(b)

Fig. 5. The new designed distributor for experiments. (a) Solid modelling. (b) Fabricated workpiese.

3.2 Numerical treatments for CFD

The transient flow models were simulated based on the Finite Volume Method in ANSYS Fluent. The DAC column was initially filled with the solvent (ethanol) which remained in steady state. In addition, there was no free solute in ether the mobile phase or the stationary phase. After some elapsed time, the free solute mixture (caffeine) was introduced into the column using pulse injection. Furthermore, zero-gradient concentration was imposed on the wall and no-slip condition was applied for the eluent flow. Porous media are modelled by the addition of a source term to the standard momentum equations. This momentum source term contributes to the pressure gradient in the porous cell, creating a pressure drop that is proportional to the fluid velocity in the cell. In present numerical calculations, the rate of convergence slowed if the pressure drop was relatively large in the flow direction in a porous region. This slow convergence can occur because the momentum source term yields a loss of diagonal dominance in the matrix of discretized equations (Eq. (4)) solved. The best remedy for poor convergence of a problem involving a porous medium is to supply a good initial guess for the properties across the medium. Another possible way to deal with poor convergence in computation is to disable the porous media model temporarily and obtain an initial flow field. With the porous media model turned off, FLUENT will treat the porous zone as a fluid zone and calculate the flow field accordingly. Once an initial solution is obtained, or the calculation is proceeding steadily to convergence, we thus enable the porous media model and continue the calculation with the porous region included.

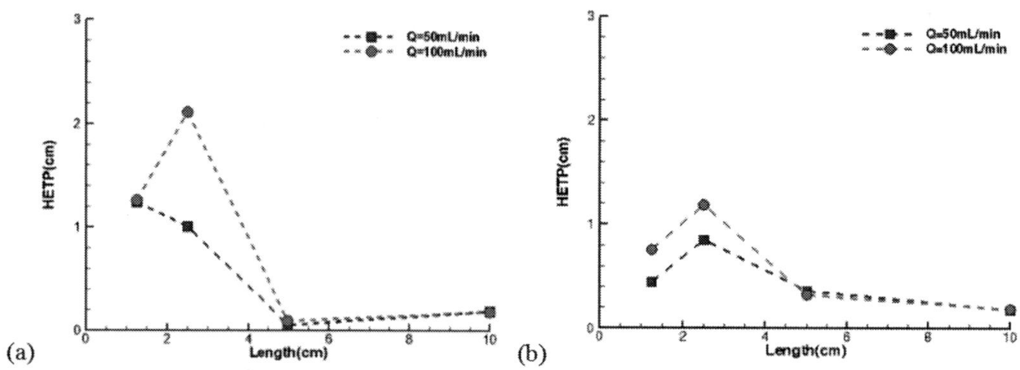

Fig. 6. Comparisons of HETP versus the column length for two flow rates. (a) The commercial distributor. (b) The new designed distributor.

Based on the Kozeny-Carman correlation in Eq. (7) and following the solution strategies as mentioned, we can obtain the convergent calculations and confirm numerical parameters for further simulation. The Kozeny-Carman correlation is represented by the following equation [9]:

$$\alpha = \frac{d_p^2 \cdot \varepsilon^2}{150(1-\varepsilon)^2} \tag{7}$$

where α is permeability; ε is porosity and d_p is diameter of packing particles. The porosity is the volume fraction of fluid within the porous region (i.e., the open volume fraction of the medium). The porosity is set to be 1.0 (or permeability $\alpha \to \infty$; viscous resistance $1/\alpha \to 0$) if the medium is treated as completely open. According to data from the manufacturer of the DAC column; $d_p = 50\times10^{-6}$m; $\varepsilon = 0.37$; $\alpha = 5.75\times10^{-12}$m^2. Therefore, viscous resistance is evaluated to be 1.74×10^{11}m^{-2}. Thus at the beginning of calculations, this value is a trial to seek the appropriate one which meets the experimental data within the required tolerance. In Fig. 7, the experimental datum was searched out via the CFD by the trial value as an initial input and the process is a kind of reverse engineering to obtain the values we expect. Figure 7 illustrates the relationship of pressure drop versus inlet flowrate for the DAC column in experiments and CFD calculations. Experimental and CFD data agree well and the curve reflects their linear property. Thus a linear function about the relationship of pressure drop versus inlet flow rate for the present DAC column can be proposed. In fact, in laminar flows through porous media, the pressure drop is typically proportional to velocity, which has been well known and described by the Darcy's law [7].

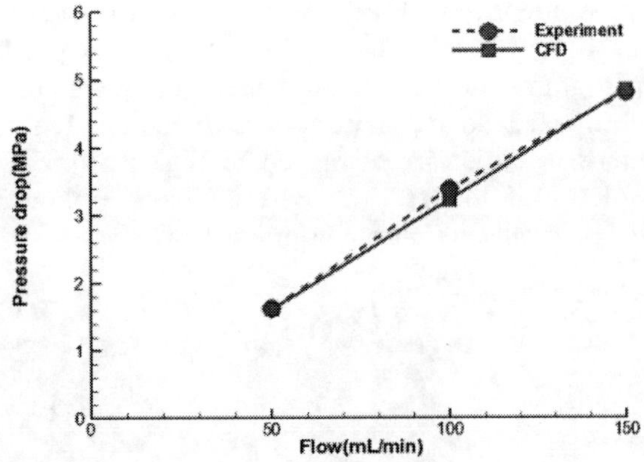

Fig. 7. Pressure drop versus inlet flow rate for the DAC column.

3.3 Parameter analysis

CFD is a good tool to describe the transient detailed flows by the time marching technique as shown in Fig. 8 (column length L=25mm; flow rate Q=100mL/min). Figure 8 shows the mass fraction (concentration) contour evolution of the caffeine injected by a step function pattern after a period of time of achieving the steady-state flow field. The diameter of the tube connected to the column is much smaller than the column diameter. Without the distributor in the region near the inlet, the fluid particle near the axis will flow faster than that near the sidewall, which results in the curvature of the concentration profile. The wing of the curved profile may explain the peak tailing of the eluted peak because it will be eluted after the bulk part of the profile. Apparently, the distributor helps to forming plug flow at inlet (Figs. 8(a) and 8(b)) because of mass conservation and CFD helps to providing the visualization of concentration fields to see the distributor's contribution at various elapsed times (60 sec., 65 sec. and 70 sec.).

For comparing with the experimental data, the simulated chromatogram is the eluted curve of the area-weighted mass fraction average at outlet versus time. Besides, all simulated area-weighted values were normalized with dividing by the area, which was the integral of the eluted pulse curve. Figure 9 shows comparisons of experimental and simulated chromatograms for the original distributor in Fig. 9(a) and the new designed one in Fig. 9(b). Apparently, the experimental chromatogram of the new designed distributor appear to have higher peak and less trailing as compared with the original one, and the new distributor can provide better chromatography performance as shown in Fig. 6. However, simulated chromatograms are only minor distinct as compared by both distributors. It is possible that solid-phase properties were not realistically modelled and simulated. The flow and concentration profiles after the distributors were diffused in the same mechanism across the packed bed, and resulted in very similar chromatogram.

Before going further to study the pattern effect of the input solute on chromatograms, the different viscous resistances of the packed bed were simulated. The chromatogram with the homogeneous viscous resistance in porous zone is illustrated as the red line for comparisons in Fig. 10. In heterogeneous cases, the chromatogram of the green line for the small resistance in the radial direction shows the shorter retention time and the highest peak in Fig. 10(b). On the contrary, the chromatogram of the blue line for the small resistance in the axial direction shows the longer retention time and the lowest peak height. Viscous resistances are relevant to the flow speed and the ability to diffuse across the packed bed. Figure 11 shows the effect on the chromatogram for various injected patterns with the same time period. The eluted curves are similar in shape in Fig. 11(b) except the peak heights due to mass conservation of solute in Fig. 11(a). Besides, patterns of the injected solutes at the same time period seem to be less relevant to patterns of chromatograms. Figure 12 shows comparisons of chromatograms for the same quantity of solutes but different time periods. Four time periods of injected solute patterns are demonstrated in Fig. 12(a) and the respective chromatograms at outlet are displayed in Fig. 12(b). For the same quantity of inputs, the injected solutes of longer time period result in lower peak height, longer retention time and wide peak pulse in chromatograms.

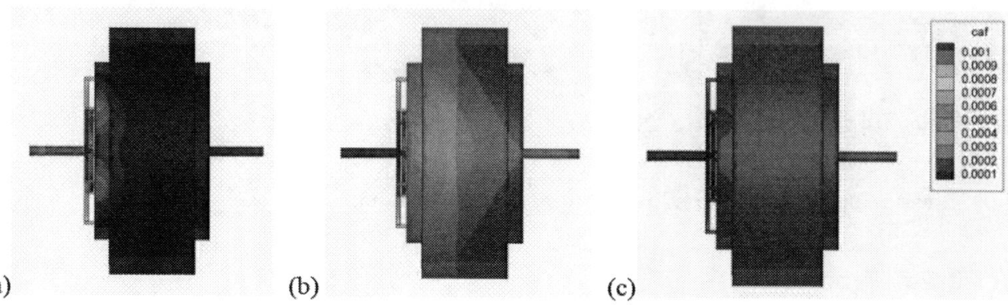

Fig. 8. Concentration contours of caffeine at various times (a) 60 sec. (b) 65 sec. (c) 70sec.

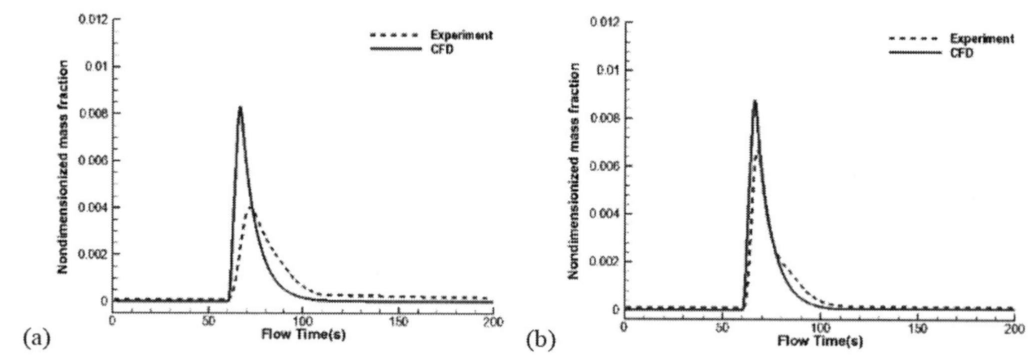

Fig. 9. Comparisons of experimental and simulated chromatograms at outlet. (a) For the original distributor. (b) For the new designed distributor.

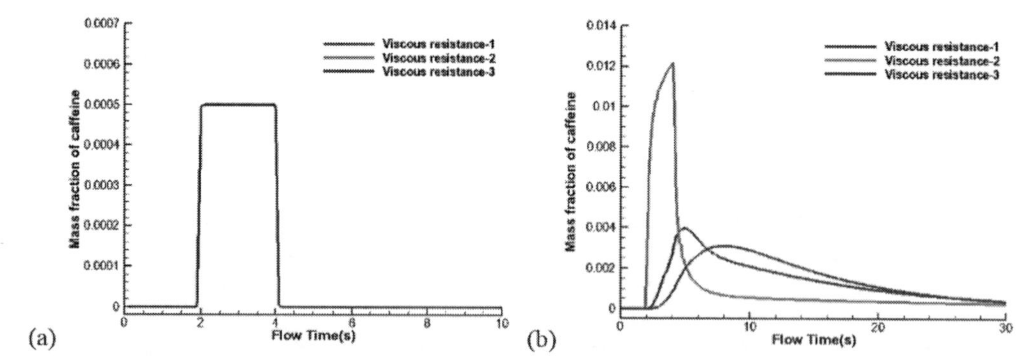

Fig. 10. Comparisons of chromatograms for different viscous resistances. (a) Three solutes of same quantity and patterns. (b) The chromatograms at outlet.

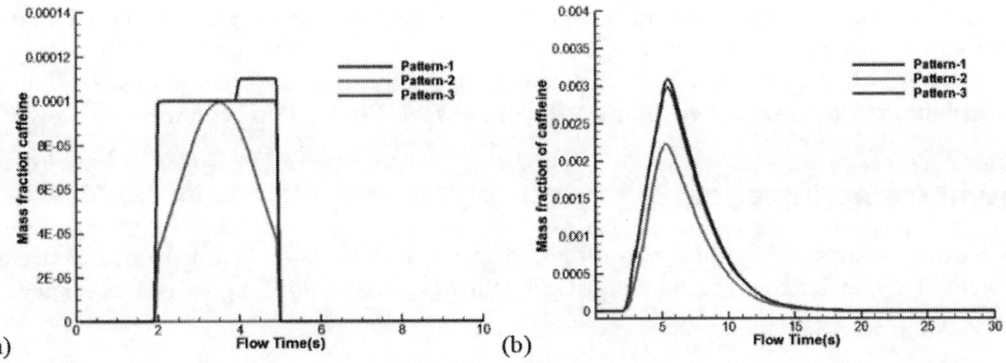

(a) (b)

Fig. 11. Comparisons of chromatograms for different injected solute patterns. (a) Three injected solute patterns. (b) The chromatograms at outlet.

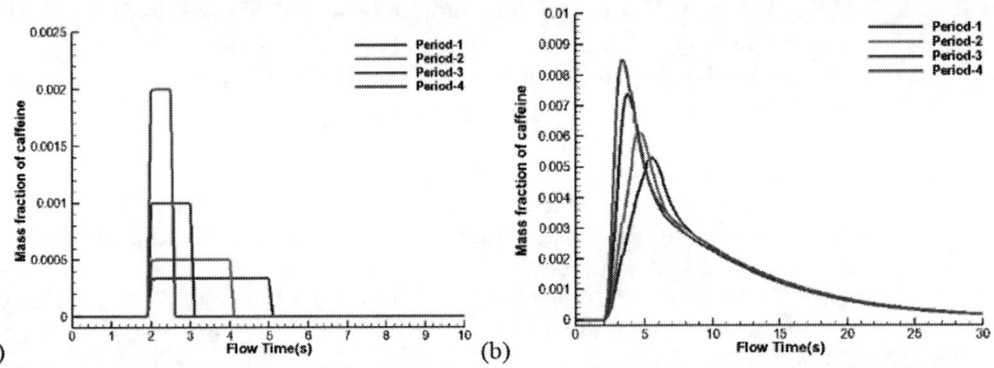

(a) (b)

Fig. 12. Comparisons of chromatograms for same quantity of solutes but different time periods. (a) Different injected patterns of solutes. (b) The chromatograms at outlet.

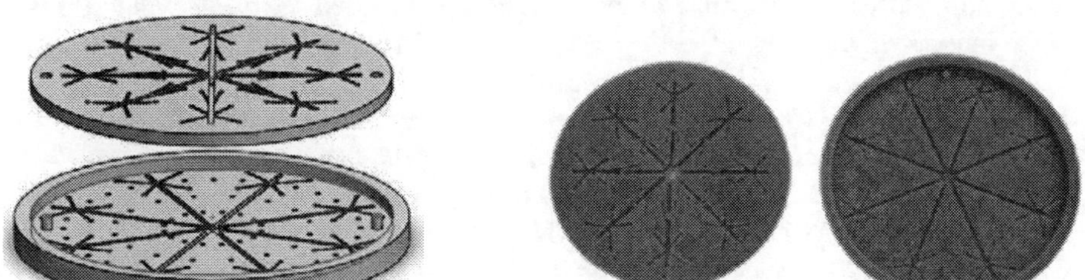

(a) (b)

Fig. 13. The new compound distributor. (a) Solid modelling. (b) Prototypes by 3D printing.

4 Concluding remarks

The rules for designing distributors are proposed for DAC columns in this study. The new distributor was developed based on the proposed rules and validated by experiments, which shows the new distributor performs better than the original one. In addition, the CAD and CFD techniques were used to build solid modelling and thereafter conduct numerical analysis, which can provide detailed numerical flows and visualizations. CFD analysis show that there exist the longer retention time and the lowest peak height in chromatograms for small viscous resistance in the axial direction in heterogeneous beds. Moreover, patterns of the injected solutes at the same time period seem to be less relevant to patterns of chromatograms due to highly diffusive property in packed beds. For the same quantity of inputs, injected solutes of longer time period result in lower peak height, longer retention time and wide peak pulse in chromatograms. Based on present computer aided techniques, the new compound distributor has been designed, 3D printed and processed in factory for testing, evaluation and practical chromatographic applications. It is anticipated that this distributor can perform better column efficiency especially for DAC columns with low L/D.

Acknowledgements

Financial Assistance for this research provided by the Taiwanese Ministry of Science and Technology, under contract number MOST 106-2218-E-214-001, is greatly appreciated.

References

1. G. Guiochon, J. Chromatography. A **965**, 129 (2002)
2. J.Y. Song, K.M. Kim, C.H. Lee, J. Chromatography. A **1471**, 102 (2016)
3. M.T. Liang, R.C. Liang, J. Chromatography. A **1229**, 107 (2012)
4. M.T. Liang, et al., The Journal of Supercritical Fluids. **79**, 324 (2013)
5. S. Tan, B.C. Khoo, MIT Libraries. https://dspace.mit.edu/handle/1721.1/3700 (2003)
6. Q.S. Yuan, et al., J. Chromatography. A **831**, 149 (1999)
7. Y. Wu, C. Ching, Chromatographia. **57**, 329 (2003)
8. J.C. Park, K. Raghavan, S.J. Gibbs, J. Chromatography. A **945**, 65 (2002)
9. Y. Li, C.W. Park, Industrial & engineering chemistry research. **37**, 2005 (1998)
10. F.G. Lode, et al., J. Chromatography. A **796**, 3 (1998)
11. C.C. Allen, N.C. Durham, *Fluid distributor*. (United States Patent, US 4537217, 1985)
12. K. Gebauer, J. Kranse, M. Lundkvist, *Fluid distributor unit*. (United States Patent, US 2012/0074051, 2012)
13. W. Shyy, *Computational modeling for fluid flow and interfacial transport*. (Courier Corporation, 2014)
14. W. Choopakdee, et al., in *Computer Aided Chemical Engineering*. edited by P.S. Varbanov, J. J. Klemeš, L. P. Yen, **33**, 1393 (2014)
15. L.F. Gomes, et al., *Chemical Engineering Science*. **132**, 46 (2015)
16. L.F. Gomes, et al., *Chemical Engineering Science*. **153**, 188 (2016)
17. Y. Cohen, A. Metzner, AIChE Journal. **27**, 705 (1981)
18. Y.N. Jeng, J.L. Chen, Numerical Heat Transfer, Part B: Fundamentals. **22**, 211 (1992)
19. *ANSYS FLUENT 13 Theory Guide*. (ANSYS, Inc. USA, 2010)
20. S.V. Patankar, *Numerical heat transfer and fluid flow*. (Hemisphere Pub. Co New York, 1980)

Forming parameters optimizations with drawing force and die stress in wire rod drawing using rotating die under Coulomb friction

Ing-Kiat Tiong, *Un-Chin* Chai, and *Gow-Yi* Tzou*

Department of Mechanical and Automation Engineering, Chung Chou University of Science and Technology, Changhua, 510, Taiwan

Abstract. An optimization research is performed on the related forming parameters of wire rod drawing through a rotating die under Coulomb friction. The optimization research is conducted through finite element method (FEM) simulation combined with Taguchi method. There are two drawing characteristic optimizations have been carried out. They are the optimizations with drawing force and die stress. The forming parameters considered in this study are half die angle, frictional coefficient, die fillet, and rotating angular velocity of the rotating die. The same procedure is carried out in both optimizations. The geometrical models of the wire rod, top die and rotating die are constructed firstly in SolidWorks and imported into the FEM simulation software named DEFORM 3D. With the aid of Taguchi method, the simulation experiments are carried out. The results such as drawing force, die stress, and the corresponding signal-to-noise (S/N) ratio are obtained and compared. The influence rank of the forming parameters and the optimal combination of parameters are obtained through the response table for both optimizations. The results such as effective stress, effective strain, velocity field, drawing force, and die stress are studied. The results show that the minimizations of drawing force and die stress are successfully achieved.

1 Introduction

Metal rod drawing is a well-known manufacturing process which applies a tensile force to stretch a metal rod into a desired cross-section or shape. This process is carried out at room temperature and it is considered as a cold working process. Sas-Boca et al. [1] carried out a FEM simulation on wire drawing process including the analysis of stress distribution and determined the optimum process parameters for the desired wire diameter. Chevalier [2] studied about the influence of geometrical parameters and friction on the quality of final wire through FEM simulation, and the result showed the friction ratio has no effect on the stress distribution, but it affects the global drawing force. Tang et al. [3] investigated the influence of drawing parameters on the forming of the micro copper tube with straight grooves and the results showed the reduction and die angle affect the drawing force while

*Corresponding author: gowyitzou@gm.ccut.edu.tw

the die angle affects the stress in the multi-pass drawing process. Haddi et al. [4] performed a study on the relationship between drawing parameters and they developed a modified model to obtain the minimum drawing force and stress through the optimal combination of parameters obtained. El-Dominaty et al. [5] studied the thermal aspects of the wire-drawing process and the results obtained can be used for selecting the process parameters so that the minimization of temperature rise can be achieved. Kleinermann et al. [6] applied FEM simulation to solve a more complex problem named as inverse problem including parameters identification, initial geometry, and tool shape design. Wang et al. [7] successfully performed die design and modification by using advanced process simulation software and proved that it is better than old empirical method. Beland et al. [8] performed an optimization research on cold drawing process of 6063 aluminium tubes for reducing the drawing steps and drawing stress in the drawn tube. Tiernan et al. [9, 10] developed a machine for carrying out wire drawing without the use of conventional dies and they also presented a mathematical model of die-less drawing to represent the process of wire deformation. Walter et al. [11] performed process simulation on cold forming process and demonstrated die stress analysis on industrial problems. The results showed that the three-dimensional simulation is very important for analyzing the die stress accurately. The reference mentioned above is related to the traditional drawing with a fixed die or without the use of a die. Tzou et al. [12] has demonstrated the FEM simulation using DEFORM 3D on rotating drawing with constant shear friction and they successfully explored the effect of forming parameters on the drawing characteristics. They also proved that the use of rotating die is able to reduce the drawing force and increase the material flow.

The aim of this study is to study the relationship between drawing characteristics and drawing parameters by performing the optimization research on the related forming parameters of metal rod drawing through a rotating die under Coulomb friction. In this study, the influence rank of forming parameters and the optimal combination of parameters are obtained.

2 Pre-processing of drawing characteristics optimization

Fig. 1 shows the schematic diagram of metal rod drawing with a front die and a rotating die. The material of the metal rod is set as SWRCH22AB. The flow rate of the metal rod is σ=832.233$\varepsilon^{0.382}$ MPa and this indicates that it is an elastic-plastic material. Coulomb friction is assumed along the interface between dies and work-piece. In this study, the metal rod is pulled by the top die with a fixed velocity, V_0. As shown in Fig. 1, V_0 is 33.5 mm/s and the rotating die rotates with an angular velocity, ω (rad/s) in an anticlockwise direction. The forming parameters related to drawing process are shown in Fig. 2; they are half die angle (α^o), frictional efficient (μ), die fillet (R, mm), and angular velocity of the rotating die (ω, rad/s). Rigid-plastic FEM simulations are carried out through DEFORM 3D. With the aid of Taguchi method, the process of drawing characteristic optimizations is performed. The drawing parameters mentioned earlier are considered as the control factors for this study. Each control factor has three different levels. The overall simulation conditions are summarized in Table 1.

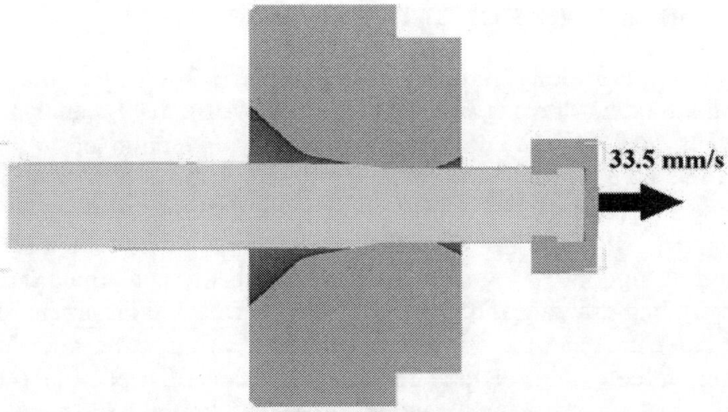

Fig. 1. Schematic diagram of metal rod drawing with rotating die and front die.

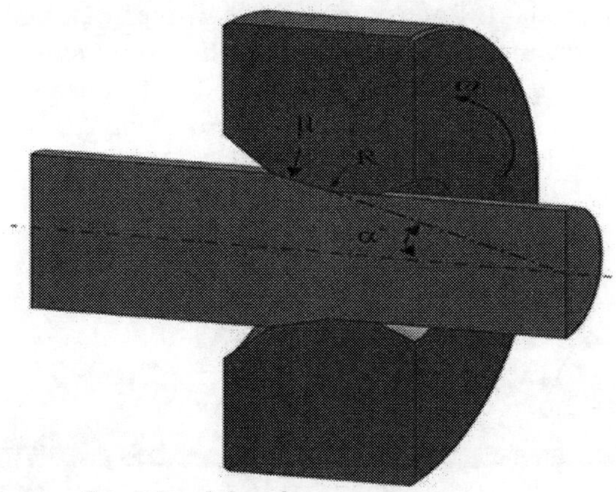

Fig. 2. Forming parameters of metal rod drawing.

Table 1. FEM simulation conditions.

Metal Rod Conditions	
Material	SWRCH22AB
Object type	Plastic
Young's Modulus, E [GPa]	210
Poisson's Ratio, υ	0.3
Power law [MPa]	$\sigma = 832.233\varepsilon^{0.382}$
Number of Elements	60000
Diameter of Metal Rod, D_0 [mm]	6.4
Velocity of Metal Rod, V_o [mm/s]	33.5
Die Conditions	
Diameter of Die, D_d [mm]	5.6
Half Die Angle, α [°]	10, 14, 18
Frictional Coefficient, μ	0.05, 0.10, 0.15
Die Fillet, R [mm]	3, 5, 7
Rotating Angular Velocity, ω [rad/s]	2.5, 4.5, 6.5

3 Optimization analysis of drawing force

The objective of performing drawing force optimization is to obtain the minimum drawing force in a rotating drawing process. Fig. 3 shows the fishbone diagram for drawing force optimization involving the control factors and levels mentioned in the previous part. Table 2 shows the $L_9(3^4)$ orthogonal table for drawing force optimization. Referring to Table 2, nine sets of simulation experiments with different combinations of drawing parameters are carried out. The values of the parameters are determined by referring to the fishbone diagram in Fig. 3. FEM simulation experiments are performed through DEFORM 3D. The corresponding drawing force and S/N ratio for each experiment are obtained and summarized in Table 2. According to the results from Table 2, experiment 1 has the lowest value of drawing force (4.33 kN) and the highest value of S/N ratio (-12.73), whereas experiment 9 has the largest value of drawing force (6.35 kN) and the smallest value of S/N ratio (-16.06). Since minimum drawing force is required, experiment 1 is determined as the best result for this study and experiment 9 is considered as the worst result. Table 3 is the response table of Taguchi method. Referring to Table 3, the frictional efficient (μ) has the highest influence on the drawing force following by half die angle (α), die fillet (R) and angular velocity of the rotating die (ω). In Table 3, the level of each control factor with the highest value of S/N ratio is selected and combined to form the optimal combination of parameters. The optimal combination of parameters is $A_1B_1C_3D_2$ (α=10°, μ=0.05, R=7 mm, ω=4.5 rad/s).

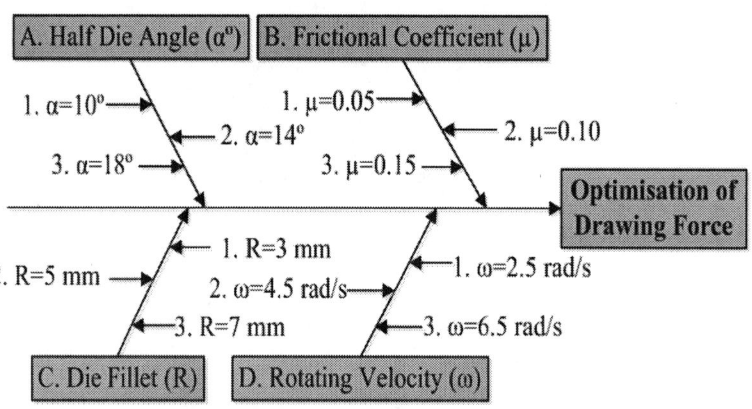

Fig. 3. Fishbone diagram for optimization of drawing force.

Table 2. $L_9(3^4)$ orthogonal table for drawing force optimization and simulation results.

Experiment	A(α)	B(μ)	C(R)	D(ω)	Drawing Force (kN)	S/N Ratio
1	1	1	1	1	4.33091	-12.73
2	1	2	2	2	4.97430	-13.93
3	1	3	3	3	5.53731	-14.87
4	2	1	2	3	5.36975	-14.60
5	2	2	3	1	5.44790	-14.72
6	2	3	1	2	6.27132	-15.95
7	3	1	3	2	4.61509	-13.28
8	3	2	1	3	5.72152	-15.15
9	3	3	2	1	6.35088	-16.06

Table 3. Response table of S/N ratio for drawing force optimization.

S/N Ratio	A (α)	B (μ)	C (R)	D (ω)
Level 1	-13.8441	-13.5381	-14.6097	-14.5043
Level 2	-15.0903	-14.6031	-14.8635	-14.3885
Level 3	-14.8302	-15.6233	-14.2914	-14.8718
E(1->2)	-1.2462	-1.0651	-0.2538	0.1158
E(2->3)	0.2601	-1.0201	0.5721	-0.4833
Effect	1.25	2.09	0.57	0.48
Rank	2	1	3	4

Fig. 4 shows the diagram of effective stress, effective strain, and velocity field for experiment 9 which is the worst result. In Fig. 4, the maximum effective stress of experiment 9 is 734 MPa and its maximum effective strain is 0.720 mm/mm. The velocity experienced by the metal rod is set to be the same value for all the experiment simulations. However, the range of the velocity field for each experiment simulation occurs to be slightly different. The range of velocity field for experiment 9 is between 23.7 mm/s and 33.5 mm/s.

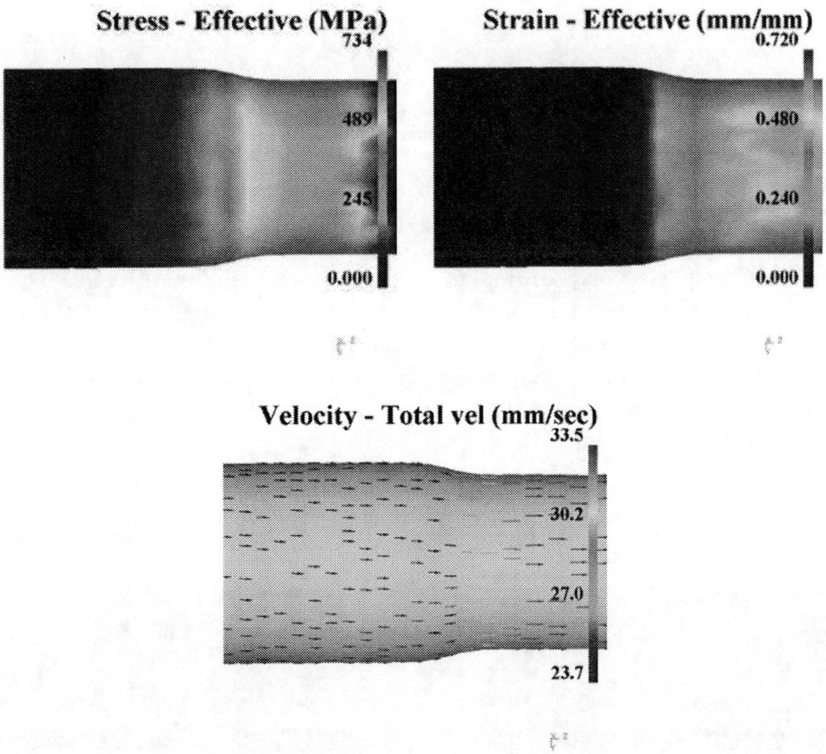

Fig. 4. Diagram of effective stress, effective strain, and velocity field for experiment 9 (worst result).

Fig. 5 shows the diagram of effective stress, effective strain, and velocity field for experiment 1 which is the best result. Referring to Fig. 5, the maximum effective stress for experiment 1 is 644 MPa and its maximum effective strain is 0.511 mm/mm. Both of the maximum effective stress and strain for experiment 9 is higher than experiment 1. The range of velocity field for experiment 1 is ranged from 25.4 mm/s to 33.5 mm/s which are overall higher than experiment 9.

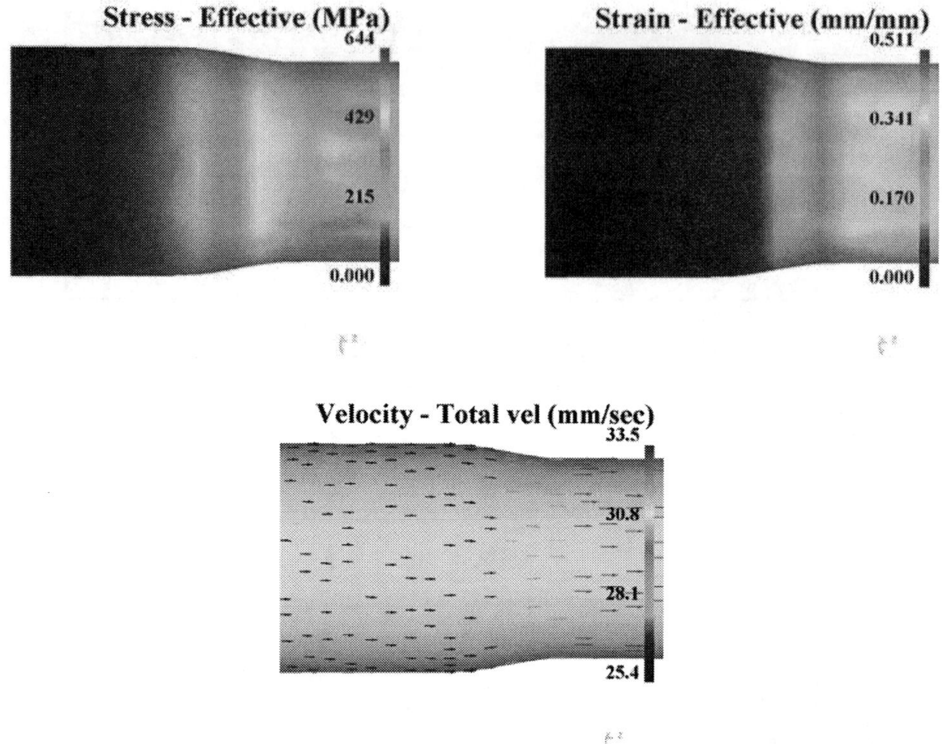

Fig. 5. Diagram of effective stress, effective strain, and velocity field for experiment 1 (best result).

Fig. 6 shows the diagram of effective stress, effective strain and velocity field for the optimal result ($A_1B_1C_3D_2$). According to Fig. 6, the maximum effective stress for the optimal result is 614 MPa and its maximum effective strain is 0.452 mm/mm. The maximum effective stress and strain are lower than experiment 1 and 9.The range of velocity field for the optimal result is ranged from 25.5 mm/s to 33.5 mm/s which are higher than the best and worst simulation results. The minimization of metal rod drawing force is successfully carried out.

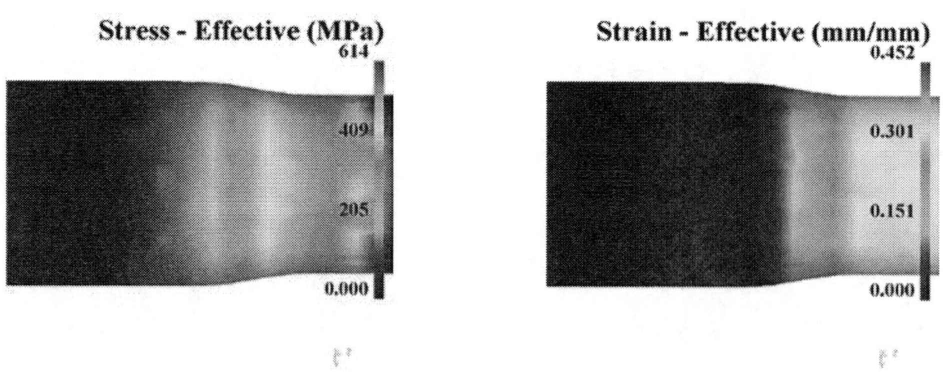

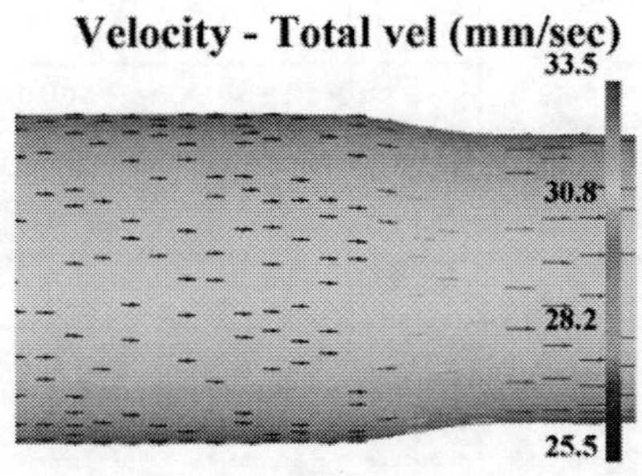

Fig. 6. Diagram of effective stress, effective strain, and velocity field for the optimal result.

4 Optimization analysis of die stress

Fig. 7 shows the fishbone diagram for the optimization of die stress involving the control factors and the corresponding levels. Table 4 shows the $L_9(3^4)$ orthogonal table for die stress optimization. According to Table 4, there are nine simulation experiments with different combinations of parameters carried out for performing the die stress optimization. The combinations of parameters are similar for the optimizations of drawing force and die stress because the same orthogonal table is used for both optimizations. The value of levels for each control factors are chosen by referring to Fig. 7. FEM simulations are carried out by using DEFORM 3D. The die stress for each simulation experiment and its corresponding S/N ratio are obtained and recorded in Table 4. From Table 4, the results show that experiment 3 has the lowest die stress (770.94MPa) and the highest value of S/N ratio (-57.74) while experiment 1 has the largest die stress (1330.77MPa) and the smallest S/N ratio (-62.38). The optimization of die stress aims to minimize the die stress. Since the minimum die stress is required, experiment 9 with the lowest die stress is considered as the best result and experiment 8 is determined as the worst result. Table 5 is the response table for the optimization of die stress. In Table 5, the die fillet (R) has the most significant influence on the die stress following by half die angle (α), frictional efficient (μ), and angular velocity of the rotating die (ω). Referring to Table 5, the level of each control factor with the highest S/N ratio is chosen and combined to form the optimal combination of parameters. The optimal combination of parameters is determined as $A_1B_3C_3D_2$ (α=10°, μ=0.15, R=7 mm, ω=4.5 rad/s).

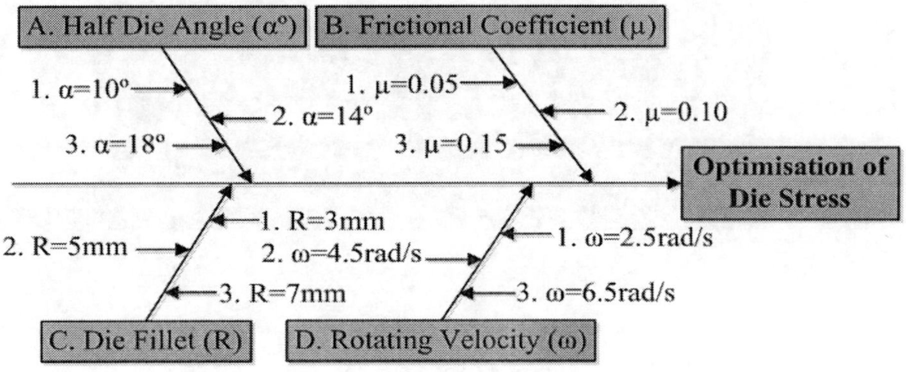

Fig. 7. Fishbone diagram for optimization of die stress.

Table 4. $L_9(3^4)$ orthogonal table for die stress optimization and simulation results.

Experiment	A (α)	B (μ)	C (R)	D (ω)	Die Stress (MPa)	S/N Ratio
1	1	1	1	1	1099.00	-60.82
2	1	2	2	2	913.23	-59.21
3	1	3	3	3	770.94	-57.74
4	2	1	2	3	1061.46	-60.52
5	2	2	3	1	941.71	-59.48
6	2	3	1	2	1191.46	-61.52
7	3	1	3	2	904.09	-59.12
8	3	2	1	3	[redacted]	[redacted]
9	3	3	2	1	978.26	-59.81

Table 5. Response table of S/N ratio for die stress optimization.

S/N Ratio	A (α)	B (μ)	C (R)	D (ω)
Level 1	-59.2573	-60.1541	-61.6079	-60.0358
Level 2	-60.5060	-60.3907	-59.8462	-59.9525
Level 3	-60.4718	-59.6904	-58.7810	-60.2469
E(1->2)	-1.2487	-0.2366	1.7616	0.0833
E(2->3)	0.0342	0.7003	1.0652	-0.2944
Effect	1.25	0.70	2.83	0.29
Rank	2	3	1	4

Fig. 8 shows the diagram of die stress for experiment 8 which is the worst result of all the experiment simulations. Referring to Table 4, the average die stress of experiment 8 is 1330.77MPa.

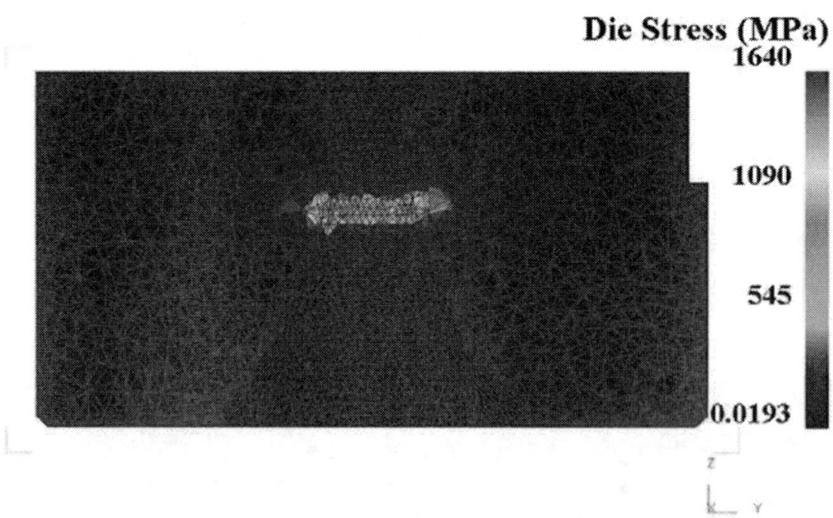

Fig. 8. Diagram of die stress for experiment 8 (worst result).

Fig. 9 shows the diagram of die stress for experiment 3 which is the best result among all the experiment simulations. According to Table 4, the average die stress of experiment 3 is 770.94MPa and it is lower than the die stress of experiment 8.

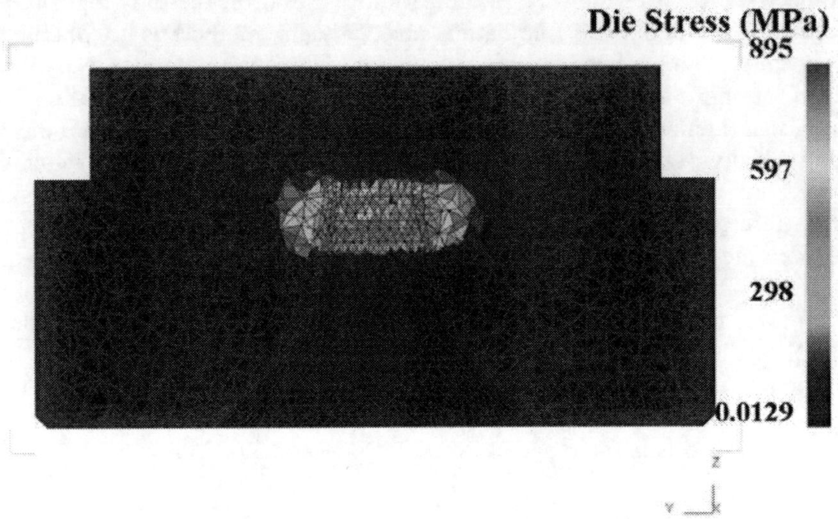

Fig. 9. Diagram of die stress for experiment 3 (best result).

Fig. 10 shows the diagram of die stress for the optimal result ($A_1B_3C_3D_2$). The average die stress for the optimal result is 750.89MPa. The average die stress of the optimal result is lower than both of the best and worst simulation results. The minimization of die stress for metal rod drawing process is successfully performed.

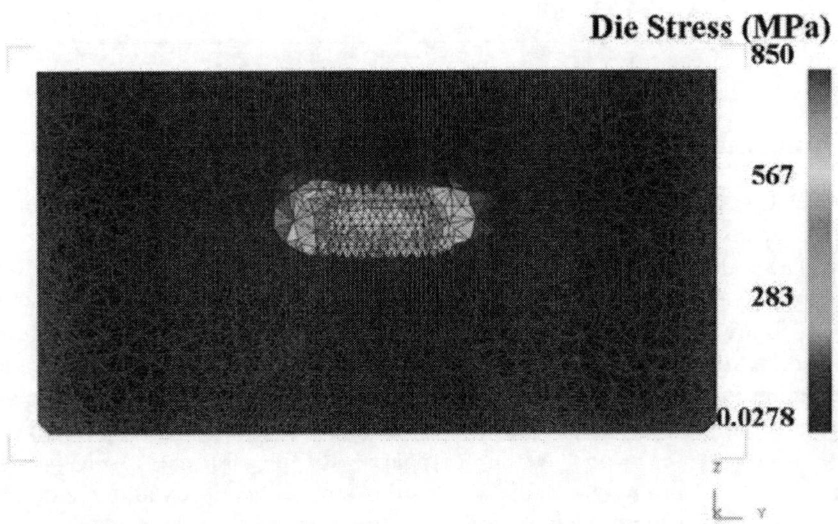

Fig. 10. Diagram of die stress for the optimal result.

5 Conclusions

5.1 Optimization of drawing force

The objective of performing the optimization research on the drawing force for the metal rod drawing through FEM simulation and Taguchi method is to obtain the minimum drawing force. The influence rank of the drawing parameters to the drawing force is obtained through the response table of Taguchi method. The frictional coefficient (μ) has the largest influence to drawing force following by half die angle (α), die fillet (R), and angular velocity of the rotating die (ω). The optimal combination of drawing parameters for drawing force optimization is $A_1B_1C_3D_2$ (α=10°, μ=0.05, R=7 mm, ω=4.5 rad/s). The drawing force, effective stress, effective strain, and velocity field for each experiment simulations are obtained and compared. Table 6 shows the improvement rate of the best and the worst results compared to the optimal result for the drawing force optimization. The drawing force of the optimal result is obtained to be 4.17 kN which is smaller than the best (4.33 kN) and the worst results (6.35 kN). The drawing force is improved by 34.40% and 3.81% compared to the best and the worst results respectively. The improvement rate of the effective stress is in the range between 4.66% and 16.35% while the improvement rate of the effective strain is in the range between 11.55% and 37.22%.

Table 6. Improvement rate (%) of the best and worst results compared to the optimal result for the drawing force optimization.

	Best Result (BR)	Worst Result (WR)	Optimal Result (OR)	Improvement (OR-BR)/BR	Improvement (OR-WR)/WR
Effective Stress (MPa)	644	734	614	4.66 %	16.35 %
Effective Strain (mm/mm)	0.511	0.720	0.452	11.55 %	37.22 %
Drawing Force (kN)	4.33	6.35	4.17	3.81 %	34.40 %

5.2 Optimization of die stress

The objective of performing the optimization research on the die stress for the metal rod drawing by applying FEM simulation and Taguchi method is to obtain the minimum die stress. The influence rank of the drawing parameters to the die stress is obtained through the response table. Die fillet (R) has the highest influence on the die stress for metal rod drawing following by half die angle (α), frictional coefficient (μ) and angular velocity of the rotating die (ω). The optimal combination of drawing parameters for die stress optimization is $A_1B_3C_3D_2$ (α=10°, μ=0.15, R=7 mm, ω=4.5 rad/s). The die stress for each experiment simulations is summarized and compared. Table 7 shows the improvement rate of the best and the worst results compared to the optimal result for the die stress optimization. According to Table 7, the die stress is improved by 2.60% and 43.58% compared to the best and the worst results respectively.

Table 7. Improvement rate (%) of the best and worst results compared to the optimal result for the die stress optimization.

	Best Result (BR)	Worst Result (WR)	Optimum Result (OR)	Improvement (OR-BR)/BR	Improvement (OR-WR)/WR
Die Stress (MPa)	770.94	1330.77	750.89	2.60%	43.58%

Thanks for the financial support from Ministry of Science and Technology (MOST 104-2221-E-235-002) to carry out this research successfully.

References

1. I.M. Sac-Boca, M. Tintelecan, M. Pop, D.A. Ilutiu-Varvara, A.M. Mihu, Pro. E. **181**, 187-192 (2017)
2. L. Chevalier, J. Mat. P. Tech, **32**, 145-153 (1992)
3. Y. Tang, D.S. Ou, Z.P. Wan, L.S. Lu, B. Lian, T. N. M. S. China, **21**, 2264-2269 (2011)
4. A. Haddi, A. Imad, G. Vega, Mat. D, **32**, 4310-4315 (2011)
5. A. El-Domiaty, S.Z. Kassab, J. Mat. P. Tach, **83**, 72-83 (1998)
6. J.P. Kleinermann, J.P. Ponthot, J. Mat. P. Tech, **139**, 521-526 (2003)
7. K.L. Wang, V. Argyropoulos, J. Mat. P. Tech, **166**, 345-358 (2005)
8. J.F. Beland, M. Farard, A. Rahem, G, D'Amours, T. Cote, App. Math. M., **35**, 5302-5313 (2011)
9. P. Tiernan, M.T. Hillery, J. Mat. P. Tech., **155-156**, 1178-1183 (2004)
10. P. Tiernan, M.T. Hillery, J. Mat. P. Tech., **10**, 12-20 (2008)
11. J. Walter, S. Kurtz, W.T. Wu, J. Tang, J. Mat. P. Tech., **71**, 64-70 (1997)
12. G.Y. Tzou, D.C. Chen, S.H. Lin, Key. E. Mat, **716**, 63-67 (2016)

Design of deep learning on intelligent levelling system for industry 4.0 technology

Sung-Yu Tsai and *Jen-Yuan* (*James*) Chang*

Department of Power Mechanical Engineering, National Tsing Hua University, Hsinchu 30013, Taiwan

Abstract. Sheet metal is widely used in the industry for metal forming purposes, such as metal stamping and laser cutting as shown. It is often winded and stored in a coil form in order for better transportation. In the recent years, industry 4.0 has been a widely discussed topic in terms of industry manufacturing solutions, the manufacturing is required to be more flexible, efficient and also require more customization. In conventional coil levelling system, the machine settings are often tuned by the experienced technicians with many years of experiences. However, as industry 4.0 focused on information process through real objects, it is required to digitize the experience through deep learning method. Therefore, it is required to be adapted through data information transfer between real world and machines, or even machines to machines. In addition, the data information is often processed and analysed through computers which are often desired to mimic the operations of the experienced machine technicians through machine learning or deep learning methods. This paper is aimed to describe and develop the deep learning algorithm with application based on coil levelling system. Finally, through this paper, design of the deep learning algorithm with application based on coil levelling system is verified.

1 Introduction

The metal forming industry is considered as a main part in the mass production of automotive, electronics and communication or even coils service centres. The conventional metal forming production line can be seen in Fig. 1, where it consists of a coil line and a press machine as the two main components of the production line. The coil lines provide the overall handling of the production such as levelling the coil sheet and sheet transportation to the press machine. The press machine provides the forming process of the production.

The schematics of the coil levelling machine can be illustrated in Fig. 2. The main components of a conventional coil line are the decoiler and the levelling machine consists of several rollers. The coil is stored on the decoiler, which is used to unwind coils of sheet metals and pass it through leveller to level the sheet metal using rollers. In addition, it also feed to the press machine for metal forming process. Once the material is threaded inside

*Corresponding author: jychang@pme.nthu.edu.tw

the leveller, the material is then going through alternating bending of stretching and compression levelling process by rollers, from which process the ultimate goal is to remove shape defects and to flatten sheet metals for other applications [1].

Once the material is levelled, it is then transported to the press machine for metal forming process. The levelling part of the process is considered as the most important part of the coil handling system as many of the blanking processes require the sheet to be flat for the press to form the important shapes. In addition, the coil service centres require the sheet to be flat to stack up the sheets for packing purpose.

In the past few years, smart manufacturing or industry 4.0 has been a widely discussed topic in many different manufacturing industries across the world. Smart manufacturing is focused on cyber physical system (CPS) [2], with the operations of the physical world through information processing via information networks such as IoT. Therefore, intelligent manufacturing is required to be adapted through data information transfer between real world and machines, or even machines to machines. In addition, the information is often analysed through computers which are often desired to mimic the operations of the experienced machine technicians. Through machine learnings or deep learning methods to catch up the intelligence from experienced technician can certainly fill the existing gap and needs in levelling process to reach the goal of the industry 4.0.

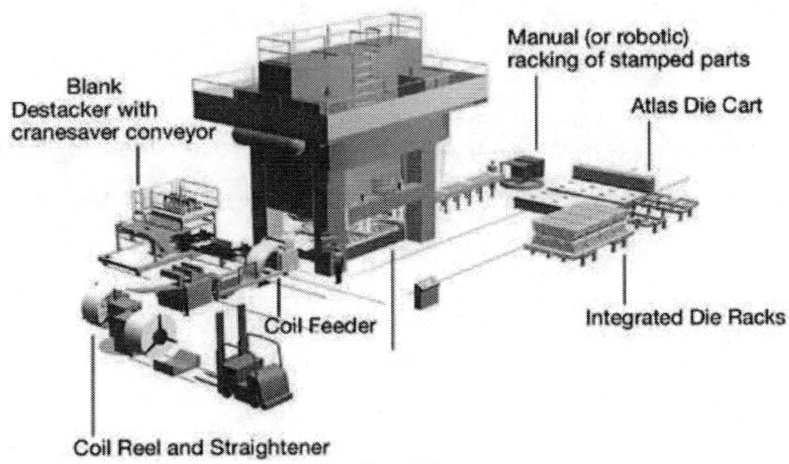

Fig. 1. Conventional metal forming production line [3].

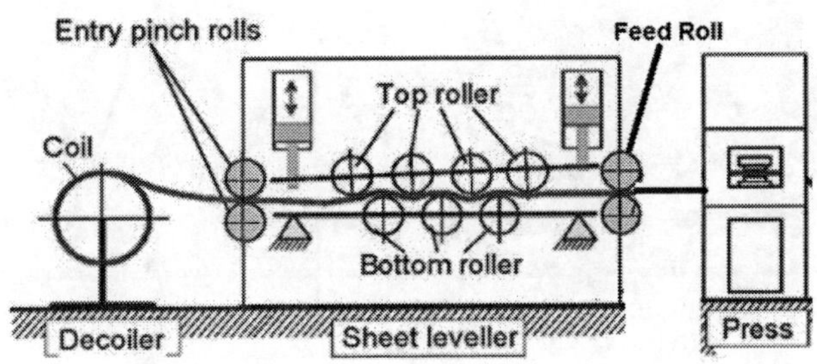

Fig. 2. Schematics of a coil levelling system.

2 Background theory

1.2 Roller levelling

The principle of roller levelling is the use of non-flat metal materials undergoing series of elastic bendings by rollers. The elastic bending process consistes of alternating positive and negative bending stresses of the metal material to overcom the material's pre-existing residual stress that caused the material's original non-flatness. Through series of such bending process, the goal is to reduce or even remove the material's residual stress so as to traighten and flatten the metal material. This process can be considered and modelled as an elastic beam of plate going through series of elasitc bending. To simply the idea, an Euler-Bernulli beam as shown in Fig. 3 is adopted to describe the elastic behaviour of the sheet metal in the levelling process. When a sheet metal beam processes through roller levellers, as shwon in Fig. 3 and measured against its neural axis, each section of the sheet metal beam will have one portion with compressive load and the other holds extensive load. Accoring to Euler-Bernulli beam model [4], for a prismatic beam having thickiness h and width b, its maximum normal stress caused by bening moment M occurs at farest distance from its nueral axis.

$$\sigma_{max} = \frac{M}{I} \cdot \frac{h}{2} \qquad (1)$$

where I is the second moment of area about neutral axis. The given moment is the result of its internal normal stress σ which can be described by:

$$M = \int_{-\frac{h}{2}}^{+\frac{h}{2}} \sigma b y \, dy \qquad (2)$$

Here y is the distance measure from the beam's neural axis. Assume the stress-strain relation is

$$\sigma = Ee + Fe^n \qquad (3)$$

in which strain can be written as $e = y/R$, the intermal bening moment can then be written as:

$$M = \int_{-\frac{h}{2}}^{+\frac{h}{2}} \left(\frac{Eby^2}{R} + \frac{Fby^{n+1}}{R^n} \right) dy$$

$$= \frac{E}{R} I_1 + \frac{F}{R^n} I_n \qquad (4)$$

where $I_1 = \int_{-\frac{h}{2}}^{+\frac{h}{2}} by^2$ and $I_n = \int_{-\frac{h}{2}}^{+\frac{h}{2}} by^{n+1} dy$

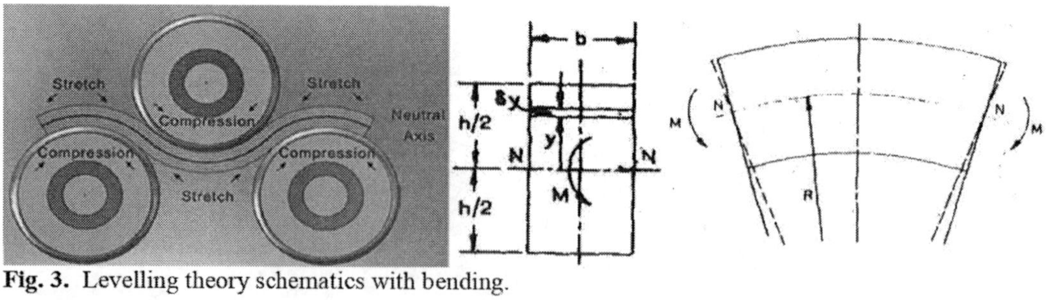

Fig. 3. Levelling theory schematics with bending.

2.2 Deep learning

There are three main categories of deep learning methods, such as convolutional neural networks (CNN), deep belief networks (DBN) and stacked auto-encoder (SAE) [5]- [8].

Since this application is more suitable to an unsupervised learning, a stacked auto-encoder neural network is applied. Stacked auto-encoder is a type of neural networks which contains multiple layers of sparse auto-encoders in which the output of each layer is connected to the inputs of the sequential layers [9]. The schematic plot of auto-encoder is shown in Fig. 4. The features of the stacked auto-encoder can be used in different dimension. In addition, one of the advantages of this algorithm is that it does not share the problem of gradient diffusion.

Supposed that there are only a set of unlabeled training examples that is $\{x(1),x(2),x(3),\ldots\}$ where xi ∈ Rn, and y is the predicting target value. An auto-encoder network is an unsupervised learning algorithm which applies backpropagation and setting the target values to be equal to the inputs. That is $y(i)=x(i)$.

An auto-encoder takes an input with transformation process from the input layer to the hidden layer. Such process is called encoding. On the other hand, the transformation process from the hidden layer to the output layer is called decoding. Therefore, the encoding and decoding process is presented by

Auto-encoder

$$d = S(W_1x+b_1) \quad (1)$$

$$y = S(W_2x+b2) \quad (2)$$

Where S represents sigmoid function, d represent the hidden layer and W_1 represents the weights matrix between input layer and hidden layer. On the other hand, W_2 represents the weights matrix between the hidden layer and the output layer and b_i represents the bias. As a result, the auto-encoder learns a function of $h(w,\ b)\ (x)=x$. In addition, the auto-encoder is also trained with back-propagation algorithm used for optimization such as gradient decent. Supposes there is m number of training samples, the cost function for one sample is:

$$J(W,b;x,y)=\frac{1}{2}\left\|h_{w,b}(x)-y\right\|^2 \quad (3)$$

$$J(W,b)=\left[\frac{1}{m}\sum_{i=1}^{m}\left(\frac{1}{2}\left\|h_{w,b}\left(x^{(i)}-y^{(i)}\right)\right\|^2\right)\right]$$

$$+\frac{\lambda}{2}\sum_{l=1}^{n_1-1}\sum_{i=1}^{s_l}\sum_{j=1}^{s_l-1}(W_{ji}{}^{(l)})^2 \quad (4)$$

where λ represents the weight decay parameter, n1 represents the number of layers n the network, s1 represents the number of units of layer l, also called a weight decay term, which tends to decrease the magnitude of the weights and helps to prevent overfitting.

In addition, by adding sparsity constraints, it will help to find the useful features, which constrains the activation of hidden units to small value, and make the overall cost function to be

$$Jsparse(W,b)=J(W,b)+\beta\sum_{j=1}^{s}KL(\rho\parallel\hat{\rho}_j) \quad (5)$$

where

$$\mathrm{KL}(\rho\parallel\hat{\rho}_j)=\rho\log\frac{\rho}{\rho_j}+(1-\rho)\log\frac{1-\rho}{1-\hat{\rho}_j} \quad (6)$$

where:

$KL(\rho \parallel \hat{\rho}_j)$	sparsity penalty term
β	controls the weight of the sparsity penalty term
$\hat{\rho}_j$	average activation of hidden unit j
ρ	sparsity parameter
s	number of hidden units

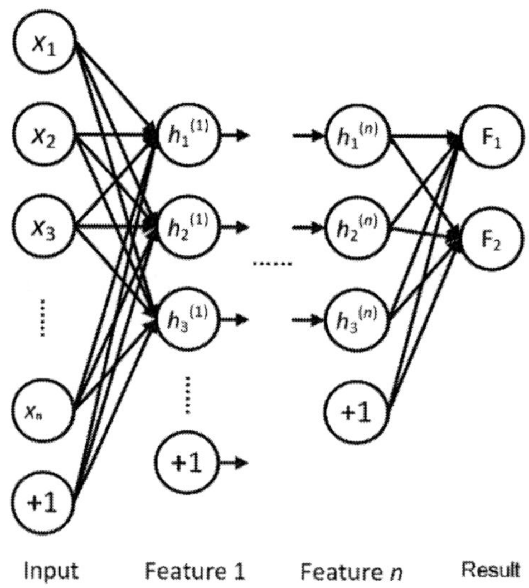

Fig. 4. Schematics of auto-encoders.

3 Methods

In this research, it is required to analyse the coil set residual stress as it is often found to be the most common defects amount coils. In addition, the amount of residual stress is different from the outer coil to the inner coil. In order to simulate such circumferences, the experiment is carried out by feeding different sheet metal material with different pre-rolled circumference to simulate the coilset residual stress from outer circumference of the coil to inner circumference of the coil. The sheet metal that is used in this experiment is the most commonly used SPCC cold rolled steel in the industry. Each coil has outer diameter and inner diameter, the inner diameter is set as 508 mm as it is an international standard for every coil. The common outer diameter is often set as 1200 mm. Therefore, to simulate such coil, set relationship, the 3 different circumferences have chosen to simulate the coils from 1200 mm outer diameter to 1000 mm and to 800 mm as shown in Fig. 5. This will identify the residual stress amount the coil. Two different thickness 0.6 and 0.8 mm, width and pre-rolled circumference have used as the input factors for this experiment. Each sheet metal is fed into the leveler; it is going through 9 rolls with alternate bend which breaks the residual stress caused by coil set and level the sheet metal material. After each of the sheet metal is leveled, it is moved to the measurement table as shown in Fig. 6. The flatness of each sheet metal sample is measured. The inlet and outlet roll indentation is adjusted to find the best position for fitting such curvatures.

Fig. 5. Coil with different section of circumference.

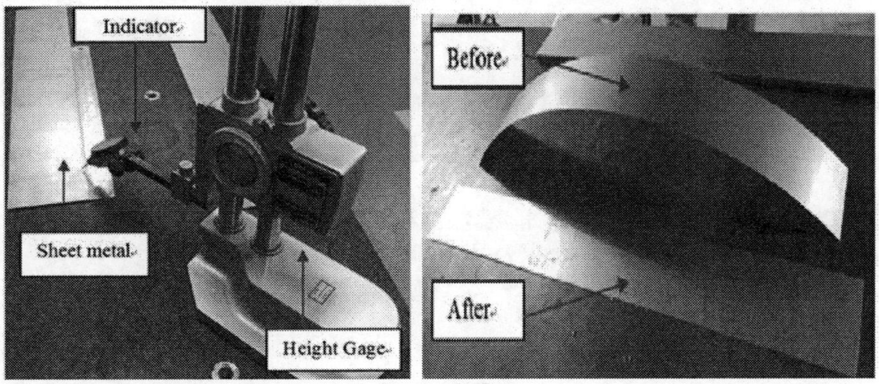

Fig. 6. (a) Experimental setup with (b) Before and after.

4 Results and Discussions

The data are simulated and examined between the simulated value and the real experimental value. The results are shown in both Fig.7 and Fig.8. The success rate of the algorithm increases as the number of sample increases. It is divided into 4 different circumference categories which represent the coil usage from outer diameter to inner diameter. It is shown that the algorithm does not function well within the previous 200 samples. However, this algorithm is reaching to steady state after 600 samples to achieve successful rate around 90%. In general, it is observed that the success rate of the simulation will be more if the parameters are chosen properly. It is concluded that SAE algorithm is shown to have effect on such application. In the future work, other deep learning methods can be used to compare such application and improve on current solutions.

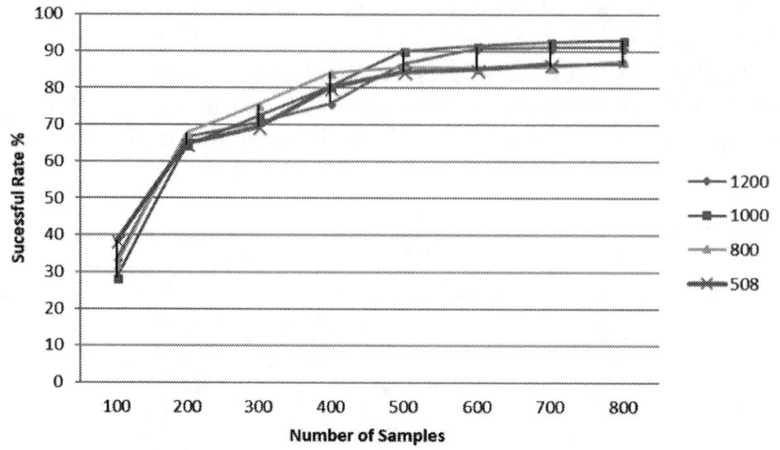

Fig. 7. Successful rate vs number of samples with 100s of samples at 0.6 mm.

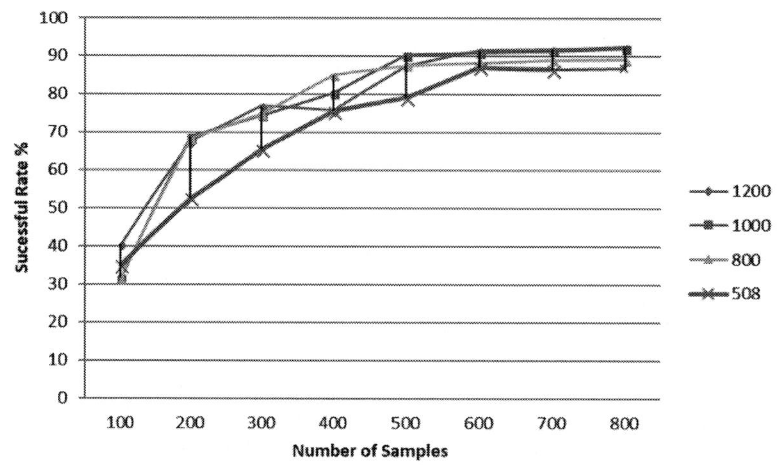

Fig. 8. Successful rate vs number of samples with 100s of samples at 0.8 mm.

5 Conclusions

The aim of this research is to design an intelligent levelling system for industry 4.0 technology. In this paper, conventional mechanics theory is discussed with supporting theory. In addition, deep learning theory is applied with supportive mechanics theory. The main achievement of this research is to digitalize the past experiences from a senior machine technician. In the research, the relationship between coil set residual stress and the roll indentation were conduction and compared with the operator experiences. Finally, the simulated data have shown to have over 90% of the success rate with sufficient number of samples.

References

1. P.B. Mellor, Engineering Plasticity, London: Van Nostrand Reinhold Company, (1973)
2. B. Sniderman, M. Mahto, and M. Cotteleer, Industry 4.0 and manufacturing ecosystems, Deloitte University Press, 2016. [Online]. Available: http://dupress.deloitte.com/dup-us-en/focus/industry-4-0/manufacturing-ecosystems-exploring-world-connected-enterprises. (Accessed 9 10 2016).
3. G.S. Vasilash, Pressworking: Cross-Bar Press Considerations, Automotive, 15 4 2000. [Online]. Available: http://www.adandp.media/articles/pressworking-cross-bar-press-considerations. [Accessed 20 5 2015].F. De Lillo, F. Cecconi, G. Lacorata, A. Vulpiani, EPL, **84** (2008)
4. T. Shogo, F. Yoneaki, Recent trend and problems of straightening technologies of thick plates, Journal of the Japan Society for Technology of Plasticity, pp. 408-11, (1999)
5. B. Wu, L. Chan, T. Lee, and L. Ao, A study on the precision modeling of the bars produced in two cross-rolls straightening, in J. Mater. Proc, Tech, (2000)
6. J. Xue, L. Cui, X. Hu, and X. Liu, Analysis of contact angle and curvature during roller levelling process, China Metal, vol. 19(2), pp. 23-6, (2009)
7. H. Huh, J. Heo, and H. Lee, Optimization of a roller levelling process for Al7001T9 pipes with finite element analysis and Taguchi model, Int. J. Mach. Tool. Manuf., vol. 43, pp. 345-50, (2003)
8. E. Doege, R. Menz, and S. Huinink, Analysis of the levelling process based upon an analytic forming model", CIRP Annals - Manufacturing Technology, p. 191–194, (2002)
9. E. Theis, Everything you need to know about flatteners and levelers for coil processing, THE FABRICATOR, 2 (2003)

Embedded smart box for legacy machines to approach to I 4.0 in smart manufacturing

Yi-Chih Fan and *Jen-Yuan* (*James*) Chang*

Department of Power Mechanical Engineering, National Tsing Hua University, Hsinchu 30013, Taiwan

Abstract. This paper introduces the design of a Linux-based embedded controller which includes machine state detection application for legacy machines and manufacturing line. For Industrie 4.0 (I4.0), it is important to acquire, manipulate, and transmit machine operating states or physical data to form useful information. However, many existing legacy machines lack of controller or sensor(s) to response to their operating status. Some machine controllers cannot be connected to provide internal parameter(s) by means of communication. Gathering machine operating state should be the first priority to approach to I4.0. This paper adopts widely used Raspberry PI as the core platform to build Embedded Smart Box (ESB). It uses external sensors to detect the machine operating status to compute the machine's availability (one of Overall Equipment Efficiency factors) and measures current to calculate the power consumption. In this research, the combination of embedded system and sensors can be a smart box for legacy machines. Such cost-effective design would help users to take the useful data from the machines and construct the base of I4.0 system even without the existence of the controller. This embedded-based design methodology has great potential implications that might fundamentally change the legacy factories into I4.0 smart one.

1 Introduction

Manufacturing process information sent to notify user anywhere and anytime in wired and wireless environment to enable the controlling and monitoring between machines and mobile device is the trend in industry [1]. Gathering related machine data, processing, storing and sending or notifying user is the basis of I4.0. I4.0 is the mega-trend in manufacturing. Its key components includes Internet of Things (IoT), Machine to Machine (M2M), and integration of vertical and horizontal production device. Thoben et al. provided an overview of I4.0-related initiatives and some selected application examples [2]. The basic design for upgrading machine into I4.0 is building the data acquisition (DAQ) system to collect and store machine-related information. Then creating the online monitoring systems to display and notify user when specific situation occurs in manufacturing process. High cost and newer machine could embrace Industrial Personal Computer (IPC), Database, and Web architecture to build such DAQ and pertinent Web user interface for machine.

*Corresponding author: jychang@pme.nthu.edu.tw

Since complex monitoring design and relative high costs make those systems not practical for legacy processing machine. An open source and low-cost Embedded system, data acquisition, database and web system using Raspberry Pi (RPi), applied to target equipment could be a suitable and cost-effective solution [3]. Many machines in manufacturing sites with average lifetimes exceeding twenty years and serviceable are called "legacy" equipment. They are incompatible with networks connection and are still being used. How to improve such legacy machines to make them approaching IoT and being compatible presents big problems. In these legacy machines, sensing operating status and sending information outside should be achieved to form IoT capability. The signal tower in legacy machines indicates operation status, however, it is hard to identify whether the machine is in actual operating or just in idle state. For many different legacy machines, various operation statuses detecting method by current signal is proposed to identify the actual operating condition [4]. For no sensor-embedded machine, Cheng-Ju Kuo et al. [5] used add-on simple sensors to obtain various feature values from that machine, and designed procedures to predict status of machines to help factories with legacy machines to overcome barriers to build Industry 4.0 function. For match the main trend to build the high degree of automation, the economic efficiency of highly productive machine is determined by the manufacturing equipment availability. Many methods are investigated to increase the availability [6]. For machine availability monitoring and process planning, [7] develop an Internet- and Web-based service-oriented system to enable real-time monitoring during working process. It introduced tiers of architecture to implement the connection between shop flow machines and web-based monitoring system. For building status sensing and web-connecting ability in legacy machine, this paper focus on building cost-effective Web-based monitoring embedded system to improve legacy machine digitalization and internet connection function. Also help legacy machine approach to IoT

2 Material and methods

Many small- and medium-sized factories have a large amount of legacy machines, these factories cannot afford advance communication device to implement DAQ and internet-connection in those machines. The ESB architecture this paper proposed is as Fig.1.

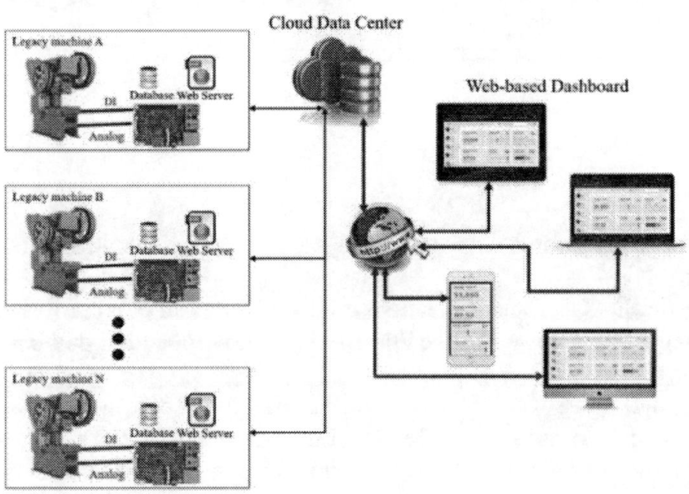

Fig. 1. System architecture.

RPi with open source embedded Linux operating system (Raspbian) is used to connect legacy machine, gathers information about machine operating status (Digital Input, DI) and current/voltage (Analog Input, AI) value. ESB proceeds some calculation and comparison, then stores these data into on-board MySQL database server. ESB also installed Apache web-server on board, user(s) can connect to ESB web page to browse the current monitoring data or historical trend plot. For notifying user the machine status on time, ESB adopted the Websocket data exchange technology to build a message tunnel with Browser HTML. If any of emergency status occurs in monitoring process, ESB will send message to user's web page on time. In local Ethernet network environment users can connect to ESB and browse the data or interact with system directly. Considering the future system expansion, Cloud Data Centre (CDC) could be built to store ESBs data and achieve remote monitoring feature.

Legacy machine's controller used to control machine action and display the machine status by light tower or 7-segment LED, many of them doesn't provide output connection pin for post-processing, also lack for sensor to sense physical state. For monitoring the machine power consumption and calculating the availability of machine, ESB measured current, voltage, and operating status. Fig.2 depicts the design of signal connection to implement these requirements.

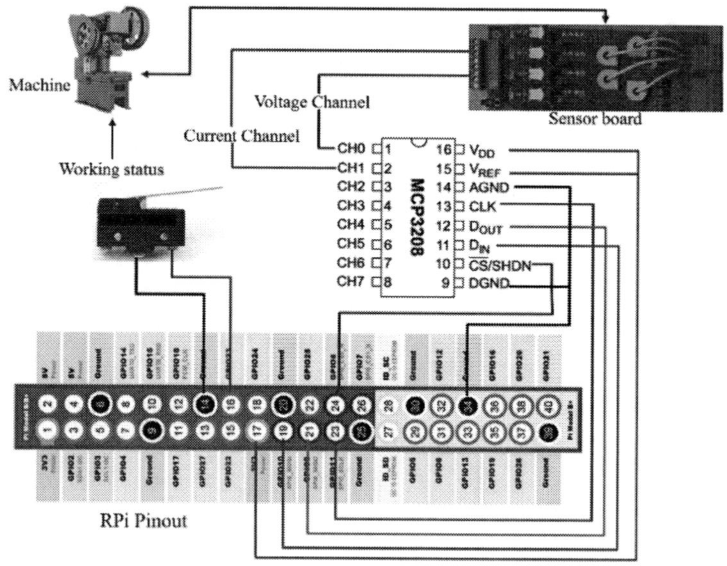

Fig. 2. Pinout and signal connection layout.

2.1 Software process and modules

For detecting the working state, voltage and current, the system software includes three executing parts:

- Run main program to handle the DAQ process. Read setup content and check recording time is up or not. If the recording time is up, write the current data into embedded database server.
- Invoke the embedded server, including web-related and database for providing current data display and storing recorded data. Based-on embedded Linux system, MySQL server stores all the data manipulated by system. Apache web server brings the web pages to interact with user for system setting, data querying and displaying. Web-Socket server implements the direct and fast data exchange method between user and system through web pages.

- Lunch crontab for routine checking for hourly, daily, and monthly calculation of monitored data. Some advance data processing, including hourly/daily/monthly data summarizing, are separated from main program to reduce the flow complexity and increase the system flexibility. Data produced by these processes are also stored into embedded MySQL.

The whole system software flow chart shown in Fig. 3.

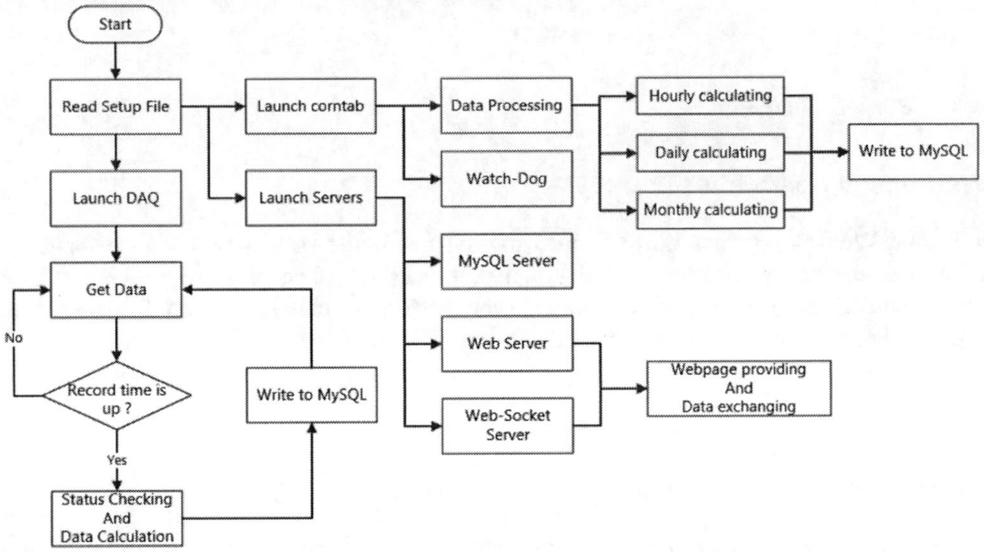

Fig. 3. System software flow chart.

As mentioned above, power consumption and availability are key parameters. Main program acquires machine's voltage and current to calculate the power consumption. The energy E in kilowatt-hours (kWh) per day is equal to the power P (V*I) in watts (W) multiply the number of usage hours per day t and divided by 1000 watts per kilowatt as following:

$$E_{(kWh/day)} = (V \cdot I)_{(W)} \cdot t_{(h/day)}/1000_{(W/kW)} \tag{1}$$

Availability, one of the three OEE (Overall Equipment Effectiveness) factors shown in Fig. 4, is calculated as the ratio of Run Time to Available Time [8, 9]:

Availability = Run Time / Available Time

Available Time is shift time subtracts break time. Run Time is simply Available Time less Stop Time, where Stop Time includes unplanned Stops (e.g. Breakdowns) or planned Stops (e.g. Transitions).

Run Time = Available Time − Stop Time

Assume one day, the Shift Time is 8 hours (480 minutes), break time is 30 minutes. Breakdowns and transitions time is 50 minutes. Then we have

Available Time = 480 – 30 = 450

Run Time = 450 – 50 = 400

Availability = 400/450 = 0.889 = 88.9 %

From detecting the working status signal, system can verify whether machine is working or not. If DI (depicted in Fig.2) is high/low, then represents the machine is working/down. Calculate the summation of working and down time can obtain machine availability.

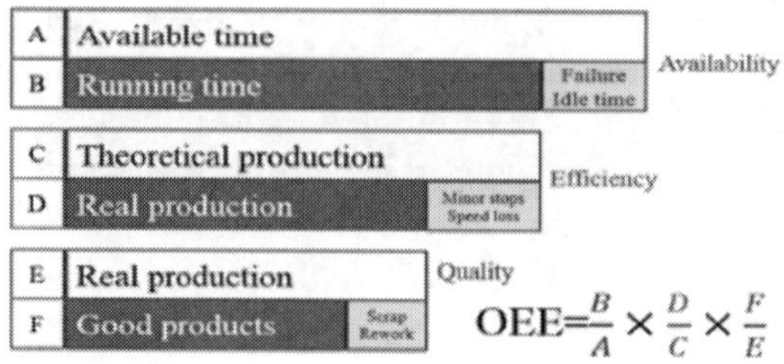

Fig. 4. Three factors of OEE.

2.2 Web-based operation

Internet- and Web-based for remote monitoring system is the main trend in I4.0 for smart manufacturing to bring convenient and consistent user interface anytime and anywhere. Based on embedded Linux system and web server; ESB combines HTML5, CSS, Javascript and web socket technologies to build the web pages as interface for user to interact with. No matter what device users run the Http Browser, users will get the same display content sent by ESB.

3 Experiment and results

To detect the machine working status, one can connect reciprocal part to Limit Switch component. Analog transducers can be used to measure machine's voltage and current. ESB acquires these three signals and calculates the power consumption and availability. For continuous monitoring situation, machine state will not change rapidly, and in general one can set the time interval in storing data to save system resources and storage space. Once time interval is up; then stores data to embedded MySQL database. Fig. 5 shows typical one day machine data which is under perfect working situation. The monitored values show working state, break time and lunch time. Machine starts at 08:00 and shutdown at 17:00, lunch time is 60 minutes and break time summation is 30 minutes. If unplanned down time appears, assumed it is x minutes, then availability will be

$$\frac{60\cdot 9-60-30-x}{60\cdot 9-60-30}\cdot 100\% \qquad (2)$$

Assume the unplanned down time is 50 minutes, e.g. there is 20 minutes down time for tuning machine and 30 minutes for fixing problem, the availability will be $\frac{540-60-30-50}{540-60-30}\cdot 100\% = 88.9\%$.

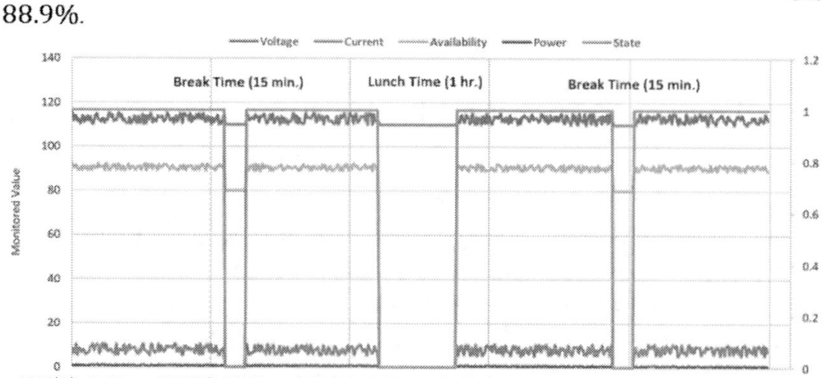

Fig.5. One day well-working machine data.

The actual machine availability shown in Fig. 6 varied from 70% to 93.5% in one month. Actually, no one machine is perfect and provides 100% availability. From the availability trend plot, one can understand the lower productivity timestamp or the trend, and make some plan to promote machine performance.

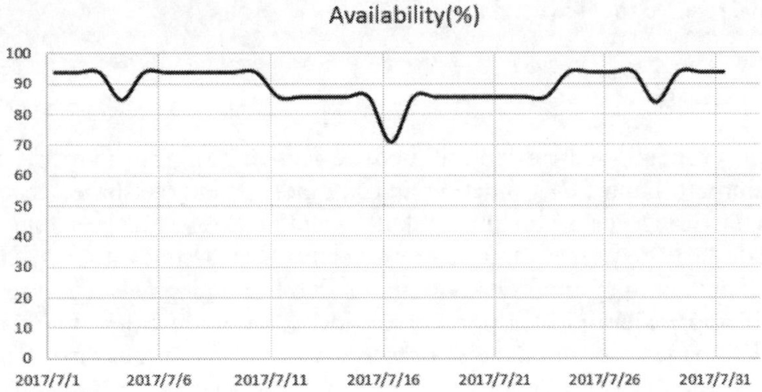

Fig.6. One month availability.

For being suitable on multi-user's different device display screen and providing the consistent view, web-based development tools are adopted. A lot of clients connect to server might slow down the data exchange speed due to the complex connection process between client and server. ESB adopted web socket to create the message exchanging tunnel to omit the re-connecting process. Server sent current monitored data to all clients at the same time. The desktop and mobile web pages show the current status as Fig. 7.

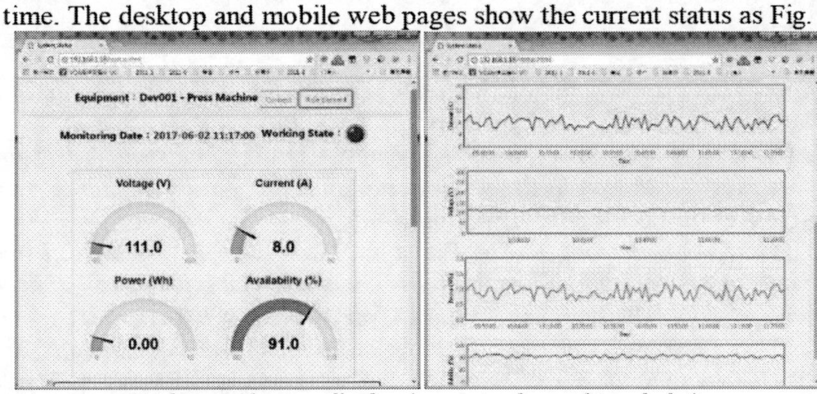

Desktop web page display (current value and trend plot)

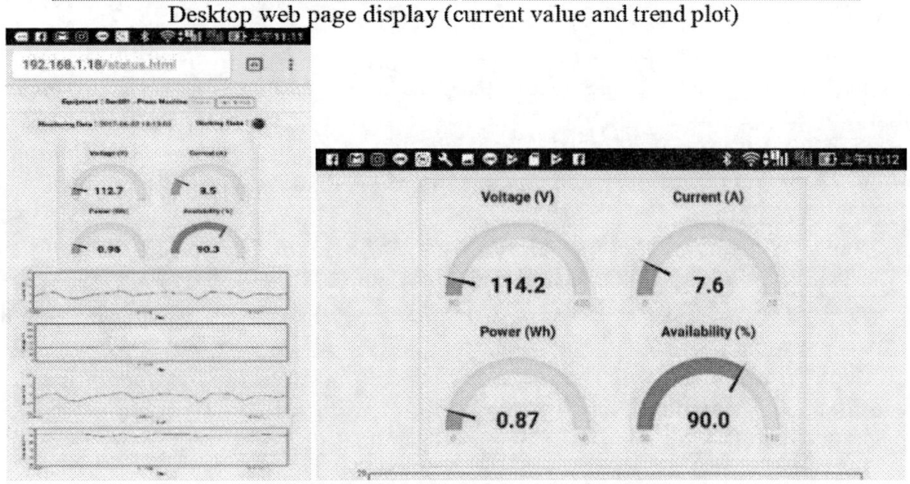

Mobile display (portrait and landscape)

Fig.7. Desktop and mobile current status web page display.

Current data shown in Fig.7 include working state (LED on/off), monitored parameters gage and trend plot. No matter what or how many devices used to request the monitoring page, all of them will obtain the same data on time. If user's Browser window keeps alive, ESB will send data automatically via web socket tunnel when system produces new data. Users always receive the newest data.

4 Conclusions

This paper proposed and carried out a cost-effective design to retrofit a large amount of legacy machine to build DAQ function and intranet- or internet-based feature to approach smart manufacturing for I4.0. Simple but essential DAQ function applied to take the machine state provide user on time data to verify the machine situation, and derive other important manufacturing coefficients (e.g. availability and power consumption). Intranet- or internet-based feature creates the world-wide and remote connection ability for legacy machine. Web pages user interface design lead users observe real-time data and trend plot to understand the current and tendency of monitored machine parameters. The proposed benefits include scalable design from single machine to cluster, cost-effective to improve the many of existence legacy machines, and Web- and internet-based to meet the I4.0 trend. To retrofit and promote the legacy machine toward I4.0 in mart manufacturing and extract the availability and power consumption for user reference and manufacturing planning, this paper introduces compact and effective approach to implement.

References

1. D.H Kim, J.Y. Song, *Mobile and remote operation for M2M application in upcoming u-manufacturing.* JMST, Vol. **22**(1): pp. 12-24, (2008)
2. K.D. Thoben, S. Wiesner, and T. Wuest, IJAT, *Industrie 4.0" and Smart Manufacturing – A Review of Research Issues and Application Examples.* International Journal of Automation Technology, Vol. **11**(1), pp. 4-16, (2017)
3. R.I.S. Pereira, I.M. Dupont, P.C.M Carvalho,S.C.S Jucá, *IoT embedded linux system based on Raspberry Pi applied to real-time cloud monitoring of a decentralized photovoltaic plant.* Measurement, Vol. **114,** pp. 286-297, (2018)
4. M. Maeda, Y. Sakurai, T Tamaki, Y Nonaka, *Method for Automatically Recognizing Various Operation Statuses of Legacy Machines.* Procedia CIRP, Vol. **63**, pp. 418-423, (2017)
5. C.J. Kuo, K.C. Ting, Y.C. Checn, D.L. Yang, H.M Chen, *Automatic machine status prediction in the era of Industry 4.0: Case study of machines in a spring factory.* JSA,Vol. **81**, pp. 44-53, (2017)
6. B. Denkena, P. Bluemet, S. Kroening, J. Roebbing, *Condition based maintenance planning of highly productive machine tools.* Production Engineering, Vol. **6**(3), pp. 277-285, (2011)
7. L. Wang, *Machine availability monitoring and machining process planning towards Cloud manufacturing.* CIRP-JMST, Vol. **6**(4), pp. 263-273, (2013)
8. Vorne, Itasca IL, USA. *Calculate OEE.* [Online]. Available: https://www.oee.com/calculating-oee.html, Accessed on: Dec. 20, 2017
9. P.J. Zepf, *How to Calculate Overall Equipment Effectiveness: A Practical Guide.* 2013, [Online]. Available:: https://www.automationworld.com/article/topics/oee/how-calculate-overall-equipment-effectiveness-practical-guide, Accessed on: Dec. 22, 2017

Application of hot forming quenching on patchwork blank using two-stage refilled friction stir spot welding

D.S. Jo [1], *J.H.* Kim [1], *D.C.* Ko [2], and *B.M.* Kim [3,*]

[1]Division of Precision and Manufacturing System, Pusan Nat'l Univ., Geumjeong-gu, Busan, South Korea, 609-735
[2]Department of Convergence Science, Pusan Nat'l Univ., Geumgeong-gu, Busan, South Korea, 609-735
[3]School of Mechanical Engineering, Pusan Nat'l Univ., Geumgeong-gu, Busan, South Korea, 609-735

Abstract. The purpose of this study is to apply Hot Forming Quenching (HFQ) on Patchwork Blank of AA6061 using Two-stage refilled Friction Stir Spot Welding (TFSSW). TFSSW is a developed joining method to improve joint strength of conventional Friction Stir Spot Welding (FSSW) and it consists of two stages. The first stage is a conventional FSSW process and second stage is a refilling process for refilling keyhole. The Design of Experiment (DOE) was used to optimize the TFSSW process parameters. A hat shape forming test was performed using a patchwork blank manufactured with the optimal process parameters to investigate validity of applying HFQ. Formed hat shape part was sequentially heat treated with artificial aging (T6) condition. The hardness of the weld zone was measured to confirm a drop of mechanical property in comparison with the conventional cold forming, which shows the validity of HFQ application to patchwork blank using TFSSW.

1 Introduction

In the recent years, aluminum alloy has been widely used in automotive industries to achieve higher fuel efficiency and improve crashworthiness. Also, Patchwork blank is adopted to manufacture automotive parts to maximize weight lightening: (1)However, there are several difficulties to improve patchwork blanks particularly for aluminum alloys. Because of their less familiar weldability requirements associated with conventional welding methods such as laser welding and resistance welding; (2)Therefore, in this study, Two-stage refilled Friction Stir Spot Welding(TFSSW) is used instead of the conventional welding method. Fig.1 shows the process flow of FSSW and TFSSW. TFSSW is a developed joining method to improve joint strength of conventional FSSW process and it consists of two stages; (3)TFSSW and FSSW is a solid state welding process that is

*Corresponding author: bmkim@pusan.ac.kr

convenient for welding of aluminium alloys. FSSW has various advantages of low cost and low energy compared with the conventional welding method. Furthermore, recent studies on FSSW have been actively conducted. Effect of welding parameters on mechanical performance and microstructure of welded specimens(4-6), optimization of FSSW process parameters(7,8) and effect of tool geometry on joint strength(9-11) were investigated.

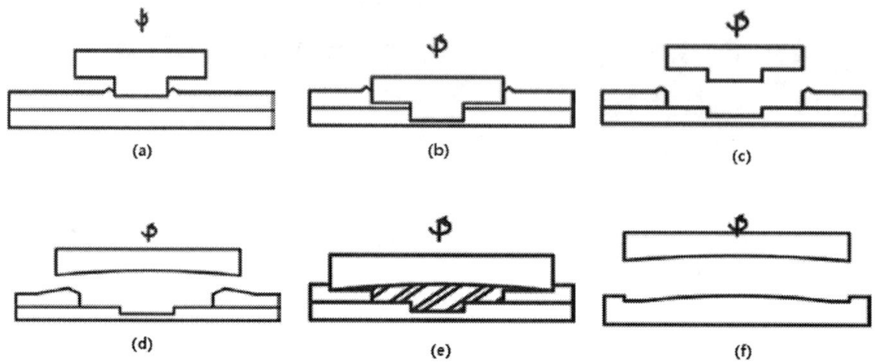

Fig. 1. Two-stage refilled Friction Stir Spot Welding process (a)-(c) first stage(conventional FSSW process), (d)-(f) second stage(TFSSW refilling process).

However, there is no study on forming of patchwork blank applying FSSW and TFSSW yet. Therefore, this study applied the forming of patchwork blank with Hot Forming Quenching(HFQ) using FSSW and TFSSW. The process flow is a step of forming the welded patchwork blank by heating it in a furnace, followed by press forming. HFQ is a process of hot forming a plate heated to a solution treatment temperature quenching it by a cooling channel provided in the die, and simultaneously performing hot forming and quenching(12-13).

In this study, a hat shape forming test was performed using a patchwork blank manufactured with the optimal TFSSW process parameters obtained by Design Of Experiment(DOE) to investigate validity of applying HFQ. Formed hat shape part was sequentially heat treated with artificial aging (T6) condition. The hardness of weld zone was measured to confirm drop of mechanical property in comparison with the conventional forming method, which shows the validity of HFQ application to patchwork blank using TFSSW.

2 Optimization of TFSSW process parameters

In order to manufacture patchwork blank using TFSSW, TFSSW process parameters and tool were selected. Many studies conducted on the optimization of FSSW process parameters. Based on previous studies, the TFSSW process parameters were selected as the tool rotational speed, plunge depth and dwell time, which have the greatest influence on joint strength. According to previous study (3), the tool of TFSSW 1stage(conventional FSSW) tool has pin diameter of 3mm, height of 1mm and shoulder diameter of 10mm, concaveness of 3°, 2stage(refilling process)tool has no pin, 15mm of shoulder diameter and concaveness of 5° as shown in Fig.2.

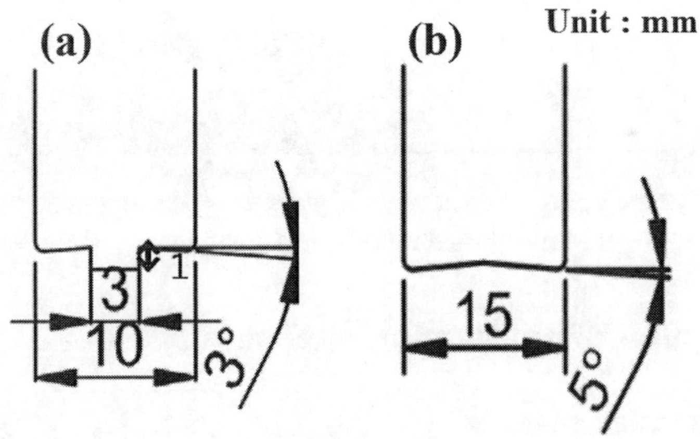

Fig. 2. (a) TFSSW 1stage tool (b) TFSSW 2stage tool.

The Patchwork blank was manufactured with AA6061-T6 (1.2t, 1.5t) and the process parameters of TFSSW and FSSW were optimized to best joint strength using the Taguchi method, one of the DOE. All experiments were carried out using a Friction Stir Process Machine Seeing Fig.3.

Fig. 3. Friction Stir Process Machine.

All specimens were welded by friction stir welding using fixtures. Lap shear test was carried out to evaluate the joint strength and it performed on a MTS machine. Fig.4 shows the appearance of a typical lap shear test specimen after TFSSW process. In order to perform TFSSW process parameters optimization, the L9 orthogonal array table was prepared and lap shear test was performed accordingly. The influence of each process parameter on the maximum shear force was evaluated using the Taguchi method, one of the Design Of Experiment, and the average values at each level of process parameters are the main effect plot. In first stage, Taguchi method results showed that tool rotational speed is the most significant factor with a percentage contribution of 44.57% followed by plunge depth and dwell time with a contribution of 13.88% and 9.04% respectively. In second stage, Taguchi method results showed that plunge depth is the most significant factor with a percentage contribution of 44.57% followed by tool rotational speed and dwell time with a contribution of 13.88% and 9.04% respectively. In addition, the optimal levels of the rotational speed, dwell time and plunge depth were found to be 1400rpm, 8seconds and 2.6mm in first stage and 1400rpm, 4seconds and 0mm in second stage.

Fig. 4. appearance of a typical lap shear test specimen.

3 Application of hot forming quenching

3.1 Experimental procedure

A hat shape forming test was performed using a patchwork blank manufactured with the optimal process parameters. Fig.5 showed a patchwork blank with a 300mm×260mm AA6061-T6 1.5mm thickness and a 150mm×120mm AA6061-T6 1.2mm thickness with 9 spot weld points. The tools consisted of punch, pad and die. A total of 21 cooling channels were used to overcome the non-uniformly distributed temperature of the tools and cooling water constantly flowed through the channels to cool the tools during hot forming quenching. Next, the patchwork blank heated at the solution temperature for 15minutes is transferred to the hat type die and press formed. During Forming, the punch speed was fixed at 20mm/s and the holding force were 48000N, respectively. In addition, due to the cooling channel in the die, quenching was performed simultaneously with forming and for sufficient quenching time, a quenching time of 10seconds was given after forming. Subsequently artificial aging (T6 condition) was performed. Fig. 6 showed the state of patchwork blank after hat shape forming.

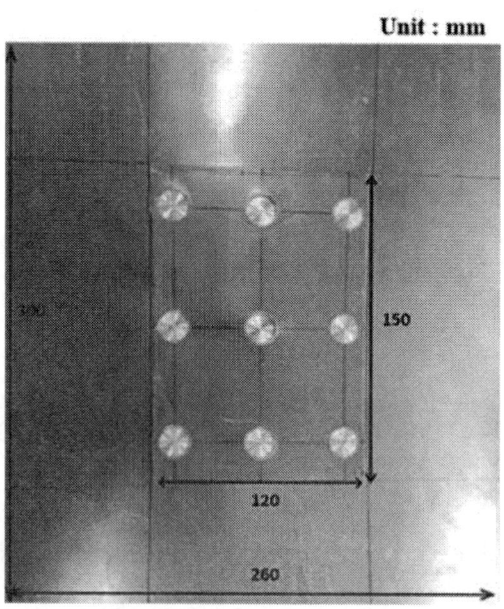

Fig. 5. Appearance of patchwork blank using TFSSW.

Fig. 6. Appearance of formed hat shape.

3.2 Measurement of weld zone hardness

The hardness of the weld zone before and after the forming was measured to confirm that a drop of the hardness of the weld zone, which is one of the disadvantages of the TFSSW. Formed hat shape part was heat treated with artificial aging (T6) condition. Sequentially after cutting the hat shaped part, the hardness from the center of the weld zone to 12mm was measured at intervals of 2mm. The results of Vicker's hardness measurements are shown in Fig.8 as the average hardness of the left, center, and right. It can be confirmed that the hardness is restored when hot forming quenching is applied.

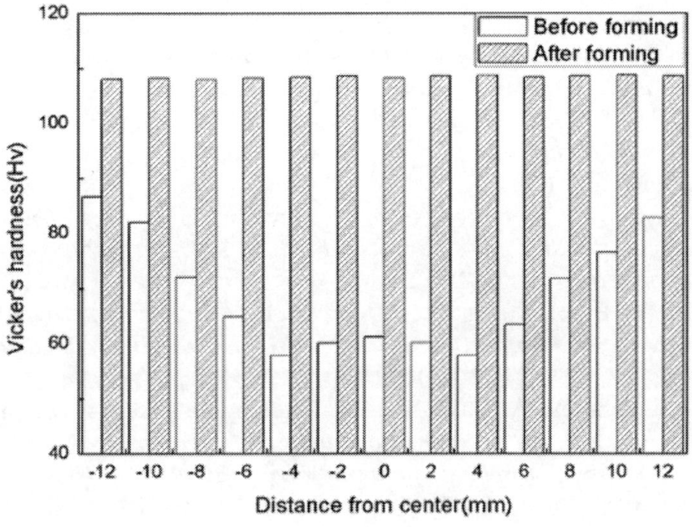

Fig. 8. Result of hardness measurement.

4 Conclusions

The objective of this study is to apply hot forming quenching on patchwork blank using TFSSW. Using the Taguchi method, one of the design of method, the optimal process parameters with the best joint strength were selected. Rotational speed, dwell time and plunge depth were found to be 1400rpm, 8seconds and 2.6mm in first stage(conventional FSSW) and 1400rpm, 4seconds and 0mm in second stage(refilling process). Also, a hat shape forming test was performed using a patchwork blank manufactured with the optimal process parameters. As a result of forming, forming was performed without defect. Formed hat shape part was heat treated with artificial aging(T6) condition. The hardness of weld zone before hot forming quenching is average 70Hv and after hot forming quenching is average 110Hv, respectively. It can be seen that the hardness of the material recovered.

Acknowledgment

This work was supported by the National Research Foundation of Korea(NRF) grant funded by the Korea government(MSIP) (No. 2012R1A5A1048294).

References

1. M. Merklein, M. Johannes, M. Lechner and A. Kuppert, J. Mater. Process. Technol., *A Review on Tailored Blanks-Production, Applications and Evaluation*, **214**, 151-164 (2014)
2. M. C. Stasik, R.H. Wagoner, Int. J. Form. Process. *Forming of tailor-welded aluminum blanks*, **1**, 9-34 (1998)
3. M. Sajed, J. Manuf., Processes, *Parametric study of two-stage refilled friction stir spot welding*, **24**, 307-317 (2016)
4. Z. Shen, X.Yang, S.Yang, Z. Zhang, Y.Y. In, Mater Des, *Microstructure and mechanical properties of friction spot welded 6061-T4 aluminum alloy*, **54**, 766-778 (2014)
5. Z. Zhang, X. Yang, J. Zhang, G. Zhou, X. Xu, B. Zou, Mater Des, *Effect of welding parameters on microstructure and mechanical properties of friction stir spot welded 5052 aluminum alloy*, **32**, 4461-4470 (2011)
6. H. Bisadi, M. Sajed, S. Sattari, The 5th National Conference of Metals and Materials Forming, *Effect of welding parameters on mechanical properties of friction stir spot welded aluminum alloys* (2011)
7. M. Tutar, H. Aydin, C. Yuce, N. Yavuz, A. Bayram, Mater Des, *The optimization of process parameters for friction stir spot-welded AA3003-H12 aluminum alloy using a Taguchi orthogonal array,* 789-797 (2014)
8. Y. Bozkurt, M. K. Bilici, Mater Des, *Application of Taguchi approach to optimize of FSSW parameters on joint properties of dissimilar AA2024-T3 and AA5754-H22 aluminum alloys,* **51**, 513-521 (2013)
9. M. Sajed, M. Sedighpour, A. Rasouli, S. H. Daneshpour, The 1st National Conference of Applied Mechanics, *An experimental study on effect of tool probe geometry on strength of friction stir spot welded aluminum sheets* (2014)
10. H. Badarinarayan, Y. Shi, X. Li, K. Okamoto, Int J Mach Tools Manuf, *Effect of tool geometry on hook formation and static strength of friction stir spot welded aluminum 5754-O Sheets* **49**, 814-823 (2009)
11. Y. Tozaki, Y. Uematsu, K. Tokaji, Int J Mach Tools Manuf, *Effect of tool geometry on microstructure and static strength in friction stir spot welded aluminum alloys*, **47**,

2230-2236 (2007)

12. X.B. Fan, Z.B. He, S.J. Yuan, P. Lin, Mater. Sci. Eng., *Investigation on Strengthening of 6A02 Aluminum Alloy Sheet in Hot Forming-quenching Integrated Process with Warm Forming-dies*, **587**, 221-227 (2013)
13. D.H. Ko, J.H. Kim, C.J. Lee, D.C. Ko, B.M. Kim, Trans. Kor. Mech. Eng, *Evaluation of Formability and Mechanical Characteristic for Hot Forming Quenching in Sheet Forming of Al6061 Alloy*, **37**, 483-490 (2013)
14. B.C. Shang, Z.M. Yin, G. Wang, B. Liu, Z.Q. Huang, Mater. Des., *Investigation of Quench Sensitivity and Transformation Kinetics during Isothermal Treatment in 6082 Aluminum Alloy*, **32**, 3818-3822 (2011)

Self-loosening of 3D printed bolted joints for engineering applications

Jun-Hee Wi, *Kwang-Hee* Lee, and *Chul-Hee* Lee*

Department Mechanical Engineering, Inha University, 100 inha-ro, Nam-gu, Incheon, 22212, South Korea

Abstract. A bolt joint is a simple element that joins mechanical components. Self-loosening of bolted joints occurs due to the vibrations caused by motors and engines, resulting in the breakage of machines, and potentially serious safety problems. Recently, developments in 3D printing technologies have enabled the fabrication of detailed components. These technologies can be used for producing fasteners using 3D printed bolts. Many researchers have proposed a theoretical model for self-loosening of the bolt, and experimental studies on the self-loosening phenomenon have been advanced. However, studies on the self-loosening of 3D printed bolts have not been conducted. Therefore, it aims to confirm the self-loosening phenomenon and the safety of 3D printed bolts through experiments and finite element simulation. A lateral vibration test system is constructed and self-loosening of the bolt is evaluated by observing the axial force according to the vibration cycle by using a strain gauge. This study compared the self-loosening of 3D printed bolts and steel bolts by changing the preload and amplitude. In addition, the experimental results are verified through finite element simulation. Through this study, it is expected that 3D printed bolts will be used more frequently in situations where specially shaped bolts are needed.

1 Introduction

Bolts are the most efficient and simplest parts used to combine mechanical components. The most serious problem associated with the tightening of bolts is self-loosening. Experimental observations indicate that the self-loosening of a bolt begins with the relaxation of the initial preload. The self-loosening in such a scenario is called the first stage of self-loosening of the bolt. After the clamping force drops below a certain value, the nut begins to unwind [1]. This is the second stage of self-loosening of the bolt. If the bolt loosens, it causes a serious safety problem. Many researchers have studied the relevant principles and performed experiments to predict the circumstances under which the bolt self-loosens. Junker suggests that the self-loosening of the bolt has a greater effect when a vibration occurs in the shear direction of the bolt than when a vertical vibration occurs [2]. Therefore, a lateral load test based on the Junker tester has been carried out.

*Corresponding author: chulhee@inha.ac.kr

Recently, a number of small and complicated parts are being produced by 3D printing methods using various materials because of the advantages of employing 3D technologies in manufacturing. This improvement in 3D printing technologies will enable the use of 3D printed bolts and nuts as fasteners. However, no research has been conducted on the self-loosening and safety of 3D printed bolts and nuts. Therefore, the self-loosening characteristic of the 3D printed bolt is evaluated in this study and the results are compared with that of the steel bolt. The simulation and experiment are conducted to predict the self-loosening characteristics of steel and 3D printed bolts under various preload and relative displacements of jointed part conditions. The first stage of self-loosening of the bolt not only leads to the second stage of self-loosening but can also lead to a serious accident due to a decrease in clamping force. Therefore, the first stage of self-loosening is only focused in this study. It is assumed that the results obtained from this study can be used as a reference to design structures that use 3D printed fasteners.

2 Experiments

2.1 3D printed specimen

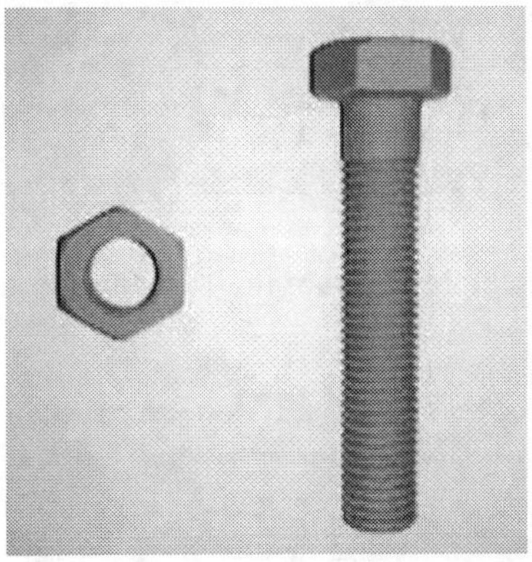

Fig. 1. Specimen of 3D printed bolt and nut.

3D printing technologies have been widely used in various applications where complicated parts are required, because the technologies have the ability to produce parts without additional processes, compared to other manufacturing technologies. The specimen of bolt and nut are 3D printed using Stereolithography Apparatus (SLA) printer (XFAB) as shown in Fig. 1 to evaluate self-loosening characteristic under various operating conditions. The bolts and nuts were printed in the same size as the M12X70 steel bolt.

A tensile test, according to ISO 527, for SLA material (PRECISA 779) was conducted to obtain the material properties of the 3D printed bolt and nut. The tensile tests were performed with the universal testing machine (UTM). The tensile test resulted on a yield strength of 26 MPa. The obtained material properties are used to calculate the appropriate preload of the 3D printed bolt during the self-loosening tests and the finite element simulations.

2.2 Test setup for self-loosening of bolted joint

Experimental equipment is specifically designed to evaluate the self-loosening of bolted joints as shown in Fig. 2. A holder with an eccentricity of 0.25 mm, 0.5 mm, and 0.75 mm is connected to the motor (1,500 W, BLDC220 V, 2,000 RPM) shaft and the linear motion (LM) guide. Therefore, the horizontal movement (perpendicular to the axis of bolt) can be provided as a transverse load to the clamped parts. A load cell (BONGSHIN, DBBP-2 t) is placed be-tween the LM guide and the upper plate of the clamped parts to directly measure the force exerted on the bolt. Several pins (1 mm diameter) are inserted between the two plates so that the relative movement of the plate is more likely to occur. A strain gauge is also attached to the shaft of the bolt to monitor the change in the clamping force in real time. The size of the steel bolt is M12 and its strength grade is 4.6. A rotary type encoder is attached to the end of the bolt and the nut is fixed to the bolt using an adhesive (Epons AR-16) to purely monitor the rotation of the bolt in the first stage of self-loosening.

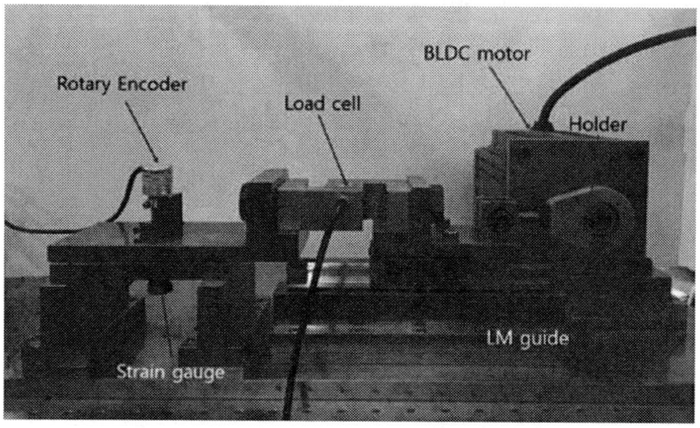

Fig. 2. Experimental equipment for self-loosening of bolted joints under vibration condition.

The test is carried out to observe the self-loosening of the bolt by providing a relative movement between the lower and the upper plates with a constant operating frequency. According to Bickford and Eccles, the appropriate level of preload corresponds to an axial stress is 50% to 75% of the nominal yield strength of the bolt [1, 6]. Based on this, the steel bolt and 3D printed bolt were tightened with preloads ranging from 9.5 kN and from 1.01 kN, respectively.

The relative displacement condition is altered to monitor how it affects the self-loosening characteristic for the steel bolt specimen and the 3D printed bolt.

3 Finite element simulations

A simulation model of bolted joint was configured to evaluate the first stage of self-loosening under vibration conditions. It consists of a bolt, a nut, and two plates. The bolt is modelled using the dimension of M12X70 hex bolt. The 3D finite element model is prepared using Hypermesh as shown in Fig. 3 [7]. This study focuses on observing the first

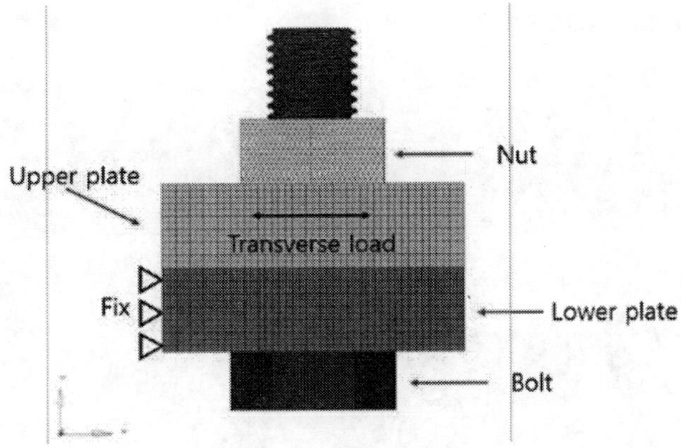

Fig. 3. 3D finite element mesh model.

stage of self-loosening for 3D printed bolt. Therefore, elastic deformations are only considered for the materials used in the simulation. The elastic modulus of the 3D printing material is 3.9 GPa and Poisson's ratio is 0.3.

Another important parameter for the analysis is the coefficient of friction between the contact surfaces. The threading friction between the bolt and the nut has been experimentally determined in the previous study and is 0.09 [8]. The coefficient of friction between the upper plate and the lower plate is defined as 0.16. For other contact interfaces (lower plate–bolt, upper plate–nut), the coefficient of friction is defined as 0.1.

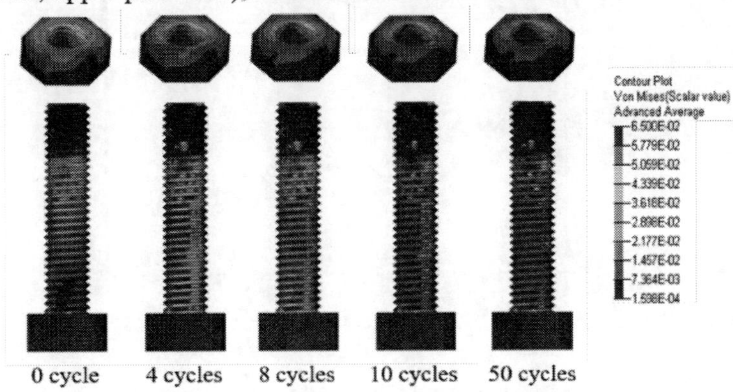

Fig. 4. Stress changes in the 3D printed bolt and nut model during the cycles of transverse load. (unit GPA)

The calculated preload is defined by applying the rotating motion of the nut in the jointed parts. The tightening torque is applied by the rotated nut and the required preload is obtained. After that, the simulation of the self-loosening phenomenon is conducted while the relative motion between the fixed lower plate and the upper plate is being applied. The maximum relative displacement is set to 0.25 mm. Each loading case proceeded in 50 cycles. Fig. 4 shows the stress change of the 3D printed bolt during the simulation. During the self-loosening, the stress is observed on the contact surface between the bolt and plates. The stress is reduced when the first stage of self-loosening is finished (after 10 cycles). Therefore, it is assumed that the clamping force is reduced to lower stress in the 3D printed bolt.

4 Results and discussions

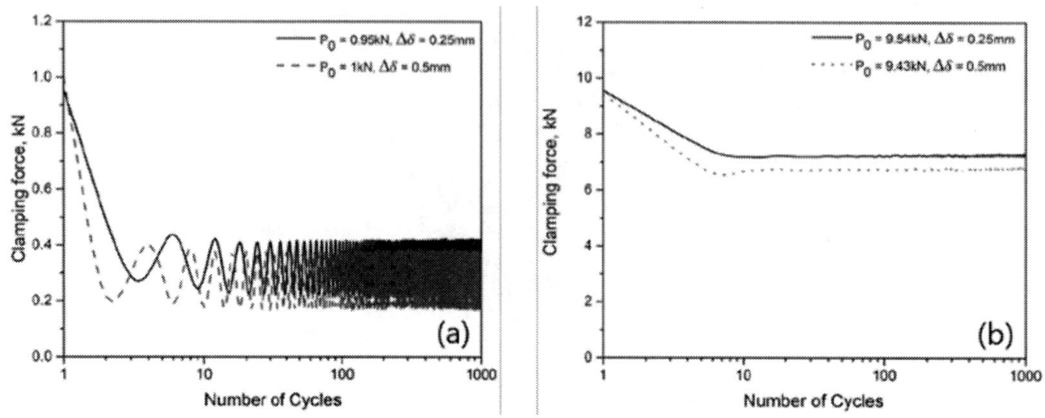

(a) 3D printed bolts (b) Steel bolts

Fig. 5. Reduction in clamping force of bolts according to the number of cycles.

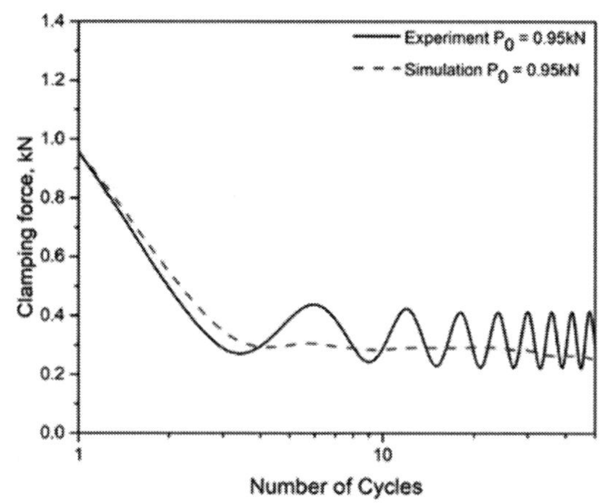

Fig. 6. Comparison of clamping force reduction in simulation and experiment.

Fig. 5 shows the experimental result for the reduction in the clamping force of the 3D printed bolt and steel bolt according to the number of cycles with changes in initial preload and relative displacement. The initial values of preload are set to the calculated values as close as possible and are 50% of the nominal yield strength. In the Fig. 5(a), the initial preload of the 3D printed bolt is 1.01 kN. In Fig. 5(b), the loss of the clamping force is observed when the relative displacements are 0.25 mm and 0.5 mm while keeping the initial preload set to 9.5 kN. It is assumed that the first stage of self-loosening ends when the clamping force becomes stable. As the amplitude of the relative displacement increases, the slope of the loss of clamping force becomes larger and the first stage of self-loosening occurs earlier. In general, the clamping force slowly decreases after the first stage of self-loosening. However, the clamping force remains constant after the first stage because the nut is fixed to the bolt, preventing the rotating motion of the nut. The self-loosening of the 3D printed bolt begins earlier than that of the steel bolt.

A graph comparing the self-loosening phenomena of 3D printed bolts with the self-loosening phenomena using finite element (FE) simulation and experiment is shown in Fig. 6. The preloads for the 3D printed bolt are set at 0.95 kN. The relative displacement of the bolted plates is 0.25 mm. In the FE simulation results, the clamping force decreases and the self-loosening tendency is in good agreement with the experimental results. The same preload for steel bolt cannot be applied to the 3D printed bolt owing to the lower strength of 3D printing material. However, when specially designed bolts are to be used, they will be more advantageous than steel bolts.

Whether the first stage or second stage has a significant influence on self-loosening process depends on the material of the bolt. For low-class bolts, the first stage of self-loosening becomes dominant. More than 30% of the clamping force can be reduced in the first stage of self-loosening [9].

In this study, only the first stage of self-loosening, where the rotation of the nut does not occur, is observed. In most previous studies, the second stage of self-loosening of the bolt is considered to be the cause of the self-loosening, and the second stage of self-loosening is considered to be more dangerous than the first stage of self-loosening. However, since the first stage of self-loosening of the bolt leads to the second stage of self-loosening and can cause big accidents, the first stage of self-loosening should also be studied carefully.

5 Conclusions

1. The first stage of self-loosening is affected by the material of the bolt as well as the relative displacement of the plate. 3D printed bolts exhibit a clamping force loss of more than 30% due to the nature of the material.

2. The FE simulation can be used to identify the first step of the bolt self-loosening. This can be used to remove elements that can cause the second stage of self-loosening before a critical issue occurs.

3. The 3D printed bolt loosens before the steel bolt, but it is more advantageous than the steel bolt in the fastening situation where the special shape is more important. Using materials similar to steel rather than conventional 3D printing materials would be more advantageous than steel bolts in terms of safety and specific design.

References

1. J.H. Bickford, *An Introduction to the Design and Behavior of Bolted Joint* (Marcel Dekker, New York, 1995)
2. G.H. Junker, SAE International, SAE Paper 690055 314-335 (1969)
3. R.F. Finkelston, Machine Design, **44** 122-125 (1972)
4. T. Sakai, Bulletin of JSME, **21** 333-340 (1978)
5. T. Sakai, bulletin of JSME, **21** 1385-1390 (1978)
6. W. Eccles, SAE International, SAE Paper 930578 (1993)
7. Hypermesh, On-line Help and Technical Support, Altair Engineering Inc.
8. Y. Jiang, M. Zhang, T.W. Park, C.H. Lee, ASME Pressure Vessel and Piping Conference, 1083 (2002)
9. Y. Jiang, M. Zhang, T.W. Park, C.H. Lee, ASME Journal of Mechanical Design, **125** 518-526 (2003)

Fastener products lightweight design and forming process simulation

Jinn-Jong Sheu*, *Chien-Jen* Ho, *Cheng-Hsien* Yu, and *Kuo-Ting* Wu

Department of Mold and Dies Engineering, National Kaohsiung University of Science and Technology, No. 415, Jiangong Road, Sanmin District, Kaohsiung City 80778, Taiwan

Abstract. In this research, an integrated design system was established to design the product of nuts with flange and generate the lightweight geometry of product. The multi-stage forming process was evaluated using the CAE simulations. The topology optimization method was used to achieve the lightweight design, that included keeping necessary geometrical features and remove the excess volumes. The topological discrete model had been remodelled into a meaningful geometry which is able to satisfy the requirement of proof load of fastener specification. The final design of the lightweight geometry was adopted to test the capability of carrying proof load required using CAE simulations with the boundary conditions of the related ASTM standard. In the evaluation stage, the finite element method was used to do the topology optimization, the proof load evaluation, the forging process and the die stress analysis. The simulation results showed the lightweight design was able to reduce the weight of product and maintain enough mechanical strength. The proposed process and die designs were able to obtain the lightweight product without defects.

1 Introduction

The light weight design is the main stream of vehicle components in order to decrease both of the material utilization and the energy consumption. Fasteners are the basic and the most important components to assemble the neighbouring components of a vehicle to hold the car body structures. The light weight design of fasteners is the easiest way to cut the weight of a car without modifying the original component designs. To change the material of fasteners from steel to titanium alloy or high strength aluminium alloy is another option of lightweight design but the higher materials cost was required. New geometry design of fasteners is able to cut the weight of fasteners using the original material and the same forming process.

The new product design takes into consideration of the product strength requirement and the light weight capability. Unseld et al. [1] proposed the product development procedure using the CAE technology to evaluate both of the mechanical strength and the forming process. The results were able to achieve weight reduction up to 25% and the cost reduction potential up to 10%. Alves et al. [2] proposed the tube nut and screw forming

*Corresponding author: jjsheu@nkust.kuas.edu.tw

process design using the combination of the extrusion, the ironing, and the expanding methods. The hollow types of screws and nuts are very efficient for the lightweight designs. The topology analysis method can cut the unnecessary materials of a component under the specified loading and the boundary conditions. Emmelmann et al. [3] used topology optimization technique to design a lightweight laser-remote-scanner components successfully. Lightweight of structure is crucial to design the dynamically moving system. Yüksel et al. [4] proposed two different contact models of the linear guide of a five-axis milling machine to present the effects of the stiffness of the contact elements. The entire machine assembly depends on the least stiff components of the linear guide. It is crucial to deal of the contact conditions for the topology optimization process. The surface contact model is able to obtain the best results that the stiffness is increased by 20% and the structure weight is decreased by 10%. Goss et al. [5] had studied the high strength fastener cold forging process with the work hardening materials using the flownets analysis of the CAE and the experimental results, respectively. The benefit of using work hardening material is the saving of heat treatment, yet the forging process is more difficult and the tool life may be poor.

The testing methods of the fasteners for different conditions are proposed to mimic the real applications. Andrew et al. [6] proposed a new testing method to evaluate the failure of the high strength nut made from nickel-base super alloy which is used on the A-10 aircraft. A combination of shanking and pre-load and push the nuts to extreme limits was tested to evaluate the subpar capability. The proposed testing method is able to provide quantitative data to evaluate the impact of the manufacturing process on the performance of the bolt and nut combination. Hagiwara et al. [7] developed a fastener testing machine in accordance with the ISO 16047 torque / clamp force testing standard. The strain gauges of the force and the torque measurement are designed to test the clamp force and the torque applied on the testing bolt. Izumi et al. [8] studied the tightening and loosing mechanism of the fasteners using the FEM method. The shear stress of the threaded fastener predicted with the conventional mechanics is over estimated. The tightened torque is lower than the prediction and results in the loose of fastener earlier than the expectation period. While the FEM results are in good agreement with the experiment tests. The sliding phenomena of the contact face indicated the loosing condition is occurred. Castelluccio et al. [9] adopted FEM simulation to study influences of the geometry, the material, and the friction on the load-displacement of a fastener. A three-dimensional model is capable of describing the load-displacement of the elastic and the plastic deformation of the threaded fastener. ASTM-F606/F606M-16[10] are the standard mechanical strength testing methods for the threaded bolt and nut. This paper also used this standard to evaluate the mechanical capability of the designed fastener.

The demonstrated product of lightweight fastener is a hexagonal nut with flange. The geometry of the fastener was obtained using the topology optimization to obtain the feasible lightweight geometry patterns. Design of Experiment (DOE) method was adopted to obtain the exact geometry parameters of the fastener. CAE simulations were performed to evaluate the performance of the lightweight design. The proposed design flow of a lightweight fastener was shown in Fig. 1.

2 Material and geometry of the high strength fastener

The high strength fastener property class (P.C.) 9.8 M6 nut was adopted for the demonstration of the lightweight design. The material of the nut is 10B21 and the main dimensions of the nut were designed according to the ISO/DIS 4161 standard [11]. The dimensions of nut and the flow stress model of the material were shown were shown in Fig. 2a and Fig. 2b, respectively. The flow stress of 10B21 was obtained by cylinder upsetting experiment at strain rate 0.1, room temperature (20°C). The volume of the nut is 470.77mm^3.

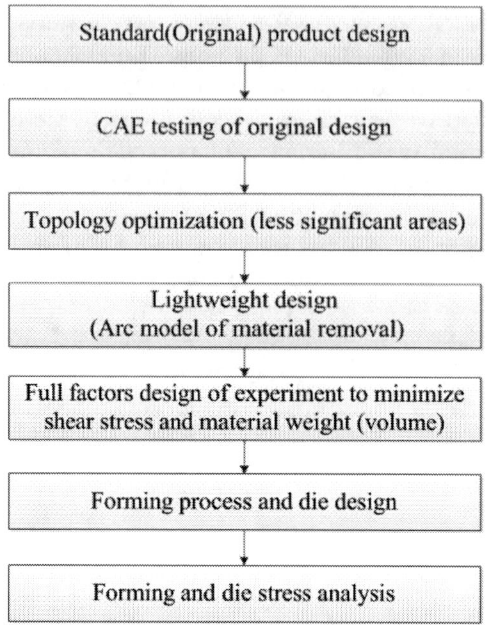

Fig. 1 Flowchart of the lightweight design of fastener.

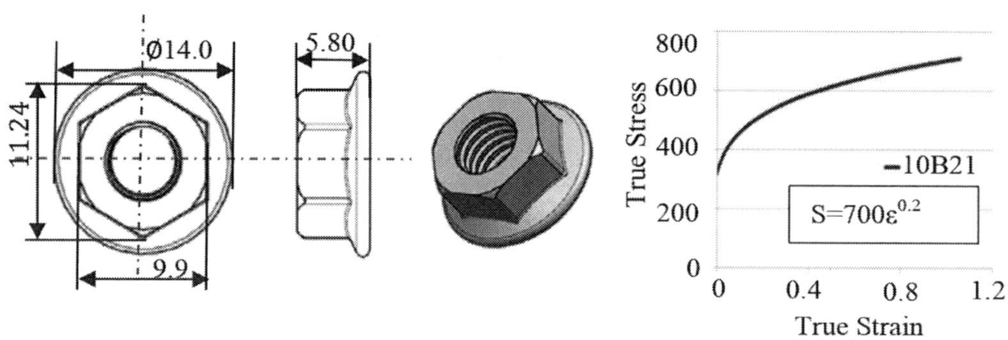

(a) Dimensions and geometry (b) Flow stress curve

Fig. 2 (a) Dimensions of the fastener and (b)the flow stress of billet material.

The mechanical property testing method of the designed fastener was in accordance with the ASTM standard [10]. The CAE model for the proof load test was shown in Fig. 3. The material model of the workpiece was assumed to be plastic and presented using power law model. The mandrel and the bottom die were assumed to be rigid. The billet was modelled using126,738 tetrahedral volume meshes and 25,822 nodes with the minimum mesh size of 0.125mm. The meshes outer area of the billet were refined with the density ratio 0.1(one-tenth of the original meshes). The mandrel was moved downwardly with the velocity of 25mm/min. The Columb friction coefficient was set to 0.07 for steel die. The stroke of the mandrel was stopped while the poof load of nut reaching the limit value 13,100 N [12].

3 Topology optimization and DOE lightweight design optimization

Topology optimization was carried out to find the less significant areas to cut the materials of the lightweight design. The quantative amounts of the removed materials were determined using the DOE optimization. The CAE model of the mechanical testing, the topology optimization of the material removal, and the suggested geometrical parameters for DOE optimzation were shown in Fig. 4. The topology optimzation results had revealed the areas of the removable materials were distributed on the top and the side faces of the nut. A half arc model defined by the radius *R* and the center distance *L* with respect to the nut center was proposed to cut the material at the less significant areas of the topology analysis results. The design factors *R* and *L* were normalized, the radius of the arc was divided by the side length of nut L_{side}, and the arc center was divided by the parallel side distance $L_{parallel}$, to obtain better DOE results. The design factors and the levels of factor were shown in Table 1.

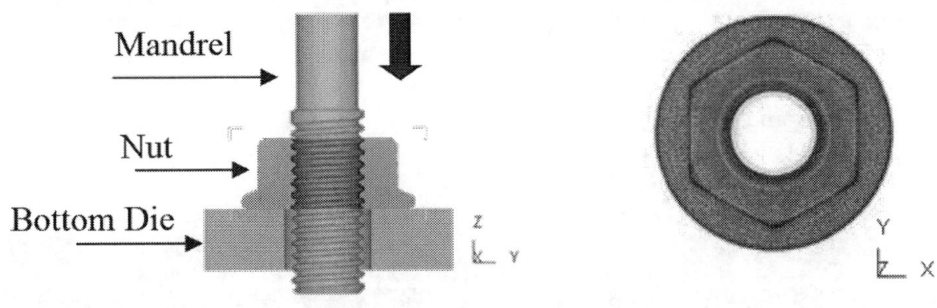

Fig. 3 Testing method and the CAE model of the designed nut.

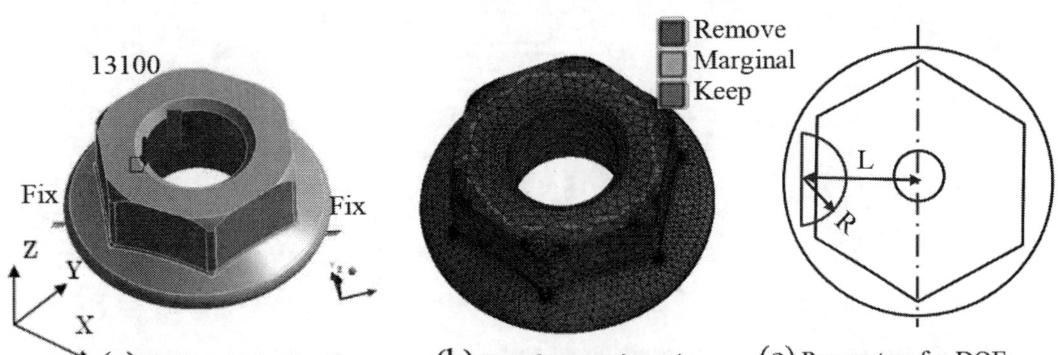

(a) CAE model of testing (b) Topology optimzation (c) Parameters for DOE

Fig. 4 Topology analysis model and Parameter Definition.

The goal of the DOE optimization is to minimize both of the maximum shear stress in the Z-R direction(τ_{zr}) and the Volume(V) of remained materials. The weight facotrs of 60% and 40% were proposed to calculate the cost function of the ith case (CF_i) of DOE in Eq.1 considering the effects of the stress with more sigificance and the volume with less significance in this paper. Penalty value 1 was given in the case of the maximum shear stress higher than the allowance of shear 160 Mpa of the nut material to prevent the designs of product wihtout enough mechanical strength. The signal to noise ratio (S/N) for *n* times repeated tests with corresponding results y_i of the smaller-the-better optimization is shown in Eq.2.

Table 1. Design factors and corresponding levels.

Factors \ Levels	1	2	3
$L/L_{parallel}$	1	1.1	1.2
R/L_{side}	0. 8	1	1.2

$$CF_i = \left(\frac{\tau_i - \tau_{min}}{\tau_{max} - \tau_{min}} * 0.6 + \frac{V_i - V_{min}}{V_{max} - V_{min}} * 0.4\right) + Penalty\ value \quad (1)$$

$$S/N = -10log\left(\frac{1}{n}\sum_{i=1}^{n} y_i^2\right) \quad (2)$$

4 Forming process and die design

The proposed forming operation processs including the billet cutting (OP1), the heading (OP2), the upsetting (OP3), the hexagonal forging (OP4), the flange forming (OP5), and the finishing (OP6) operations. The selected billet size was 9 mm in diameter and 6.95 mm in height according to the required product volume (442.28mm^3) of lightweight design. The forming process and the corresponding die designs were shown in Fig. 5 for the production of the optimal designed fastener.

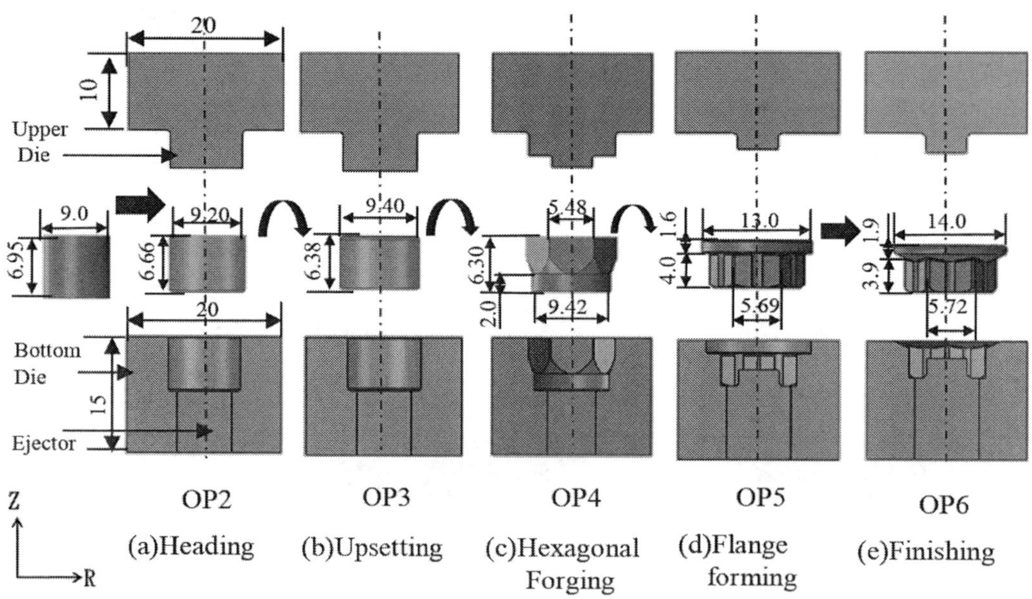

Fig. 4 Forming process and die design for nut forging.

5 Forming process and die stress analysis

The CAE models of the forming process and the die stress analyses were shown in Fig. 6. The billet material is boron steel 10B21 with the Young's module 210 GPa, Poisson ratio 0.3, yielding stress 320MPa, tensile strength 510MPa, and elongation 15%-30%. For the forming process analysis, the tools were assumed to be rigid. The billet was modelled with tetrahedral volume mesh using minimum size 0.131mm as shown in Fig. 6 (a). The tool-billet friction was modelled using the constant shear friction factor of 0.12. For die stress analysis, all of the tools were meshed and assumed to be elastic. The top surface of punch and the bottom surface of the die were assumed to be fixed, the die surface loads were mapped from the forming process simulation results. The CAE models and the boundary conditions of the die stress analysis for finishing forming were shown in Fig. 6(b) and Fig. 6(c). The die material adopted was Tungstun carbide (WC) with the Young's module 650 GPa, Poisson ratio 0.24, compressive strength 3130MPa, and tensile strength 1660MPa.

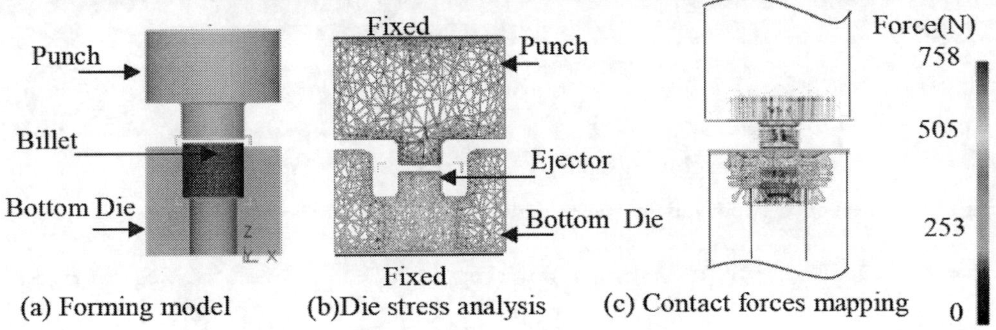

Fig. 5 CAE modellings of forming and die stress analyses.

6 Results and discussions

6.1 Mechanical Strength analysis of lightweight designs

The full factor DOE design layout and the mechanical testing analysis results were shown in Table 2. The response chart of the analysis results was shown in Fig.7(a) which gives the optimum combination of the normalized center distance ($L/L_{parallel}$) and the normalized radius (R/L_{side}) of the lightweight cutting arc were 1.2 and 0.8, respectively. The optimum lightweight design of the product was shown in Fig.7(b) which showed the product volume is 442.28mm^3 and achieved 6.05% weight reduction in comparison with the original product design (volume 470.77mm^3).

The mechanical strength of the origianl and the lightweight designs were evaluated uisng the proposed strength testing simulaiton method. The smaller of the maximum shear stress in the ZR-direction is the better deisgn. The maximum shear stress in the ZR-direction of the standard and the lightweight designs shown in Fig.8 were 128MPa and 131MPa , respectively. Both of the maximum shear stress of these two designs are lower than the allowance of shear stress(160 MPa). That means both of the product designs were qualified.

Table 2. Trial runs of the full factor DOE and theanalysis data.

Trials	$L/L_{parallel}$	R/L_{side}	τ_{zr}	Volume(mm^3)	Light weight (%)	CF	S/N
1	1	0. 8	151	401.13	14.79	0.524	5.613
2	1	1	166	377.09	19.9	1.646	-4.329
3	1	1.2	171	348.96	25.87	1.6	-4.082
4	1.1	0. 8	139	424.02	9.93	0.442	7.092
5	1.1	1	147	405.47	13.87	0.482	6.339
6	1.1	1.2	155	381.72	18.92	0.5	6.021
7	1.2	0. 8	131	442.28	6.05	0.4	7.959
8	1.2	1	136	431.11	8.42	0.427	7.391
9	1.2	1.2	143	413.72	12.12	0.458	6.783

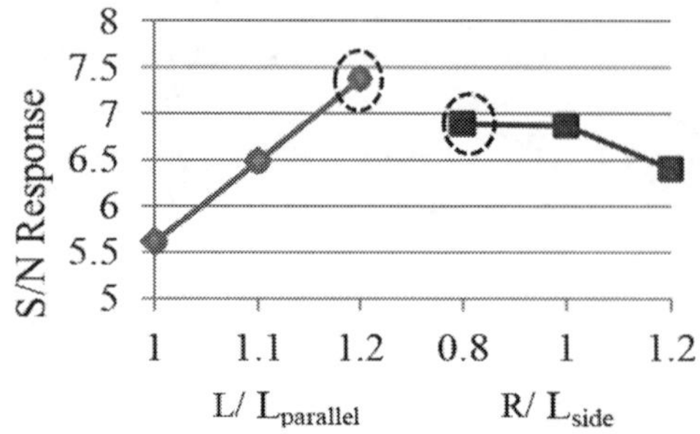

(a) S/N Response graph

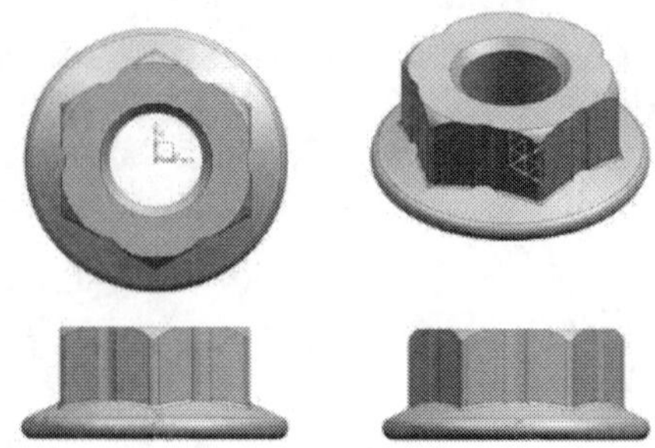

(b) Lightweight nut

Fig. 7 S/N Response graph and the corresponding optimum geometry desing of nut.

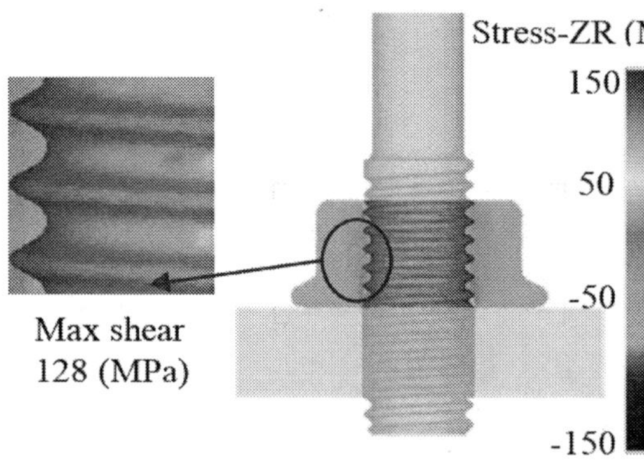

(a) Shear stress of standard design

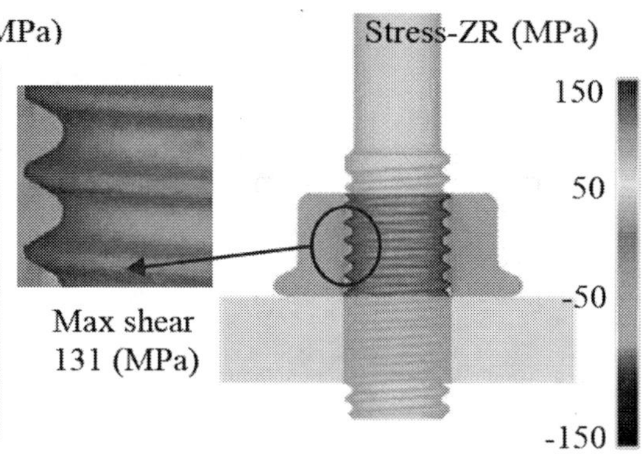

(b) Shear stress of lightweight design

Fig. 8 Mechanical strength testing results of the original(left) and the lightweight design(right).

6.2 Forminmg process simulation

The froming process simulation was carried out to verify the feasibility of manufacturing and the final property class of the product. The main dimesions of the forged prat were shown in Fig. 9(a). Dimensions of the flange diameter, 3.98, the parallel side distance, 9.87, and the height, 5.80, were all fullfiilled with the specification of the tolerance. The maximum effective stress of the product was higher than 880 MPa as shown in Fig. 9(b). The simulation results had shown the mechanical strength was increased and fullfill with the property class 9.8 requiremnt of the ISO 898-1:2013 standard (Ys:720MPa, Ts:900MPa).

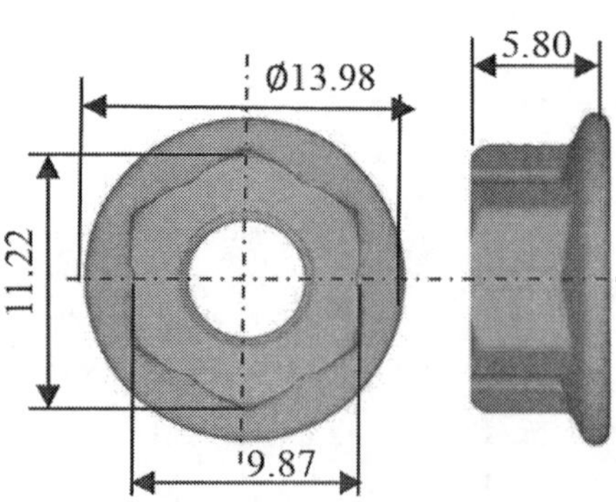

(a) Geometry of product

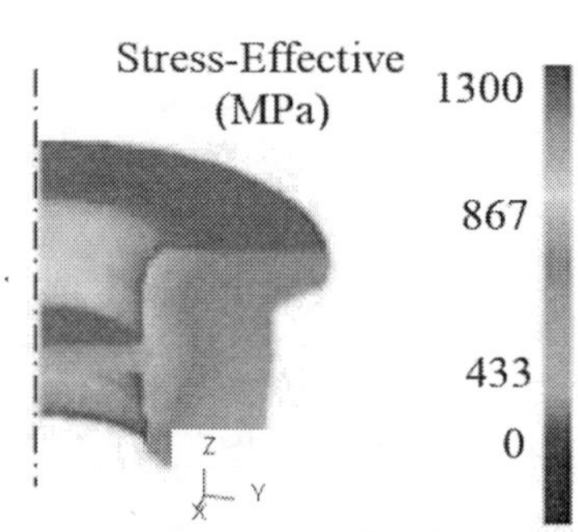

(b) Effective stress distribution

Fig. 6 Geometry and effective stress distribution of the lightweight product.

6.3 Die stress analysis

Distributions of the die stress in the circumferential (theta)direction for the forming operations 2 to 6 were shown in Fig. 10. The die srtresses in operations 5 and 6 are too high for the die material. It shows the one-piece die design for operation 5 and 6 is not feassible. Split die insert and the compressive ring designs are required tot solve the die failure problem.

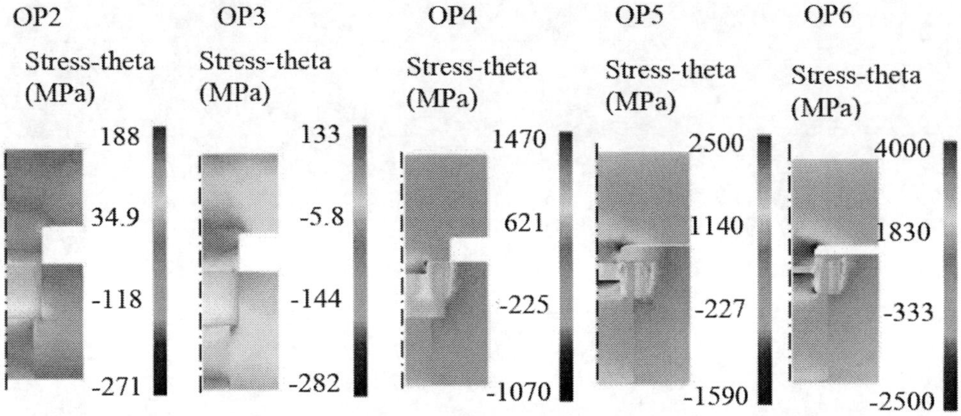

Fig. 7 Die stress distribution in the circumferential (theta) direction for the nut forming process.

7 Summary

This paper presented the lightweight design method via CAE simulations of the mechanical property testing, forming process simulation, and the die stress analysis. The results of this study showed

(1) The CAE technologies are capable of creating lightweight design feasibly.
(2) Integration of the topology optimization and theDOE technique is possible to design the lightweight products with qualified mechanical property.
(3) The high stress of dies had revealed the one-piece die design is not feasible for mass production of nuts. The compressing ring designs were required to protect the die insert and give enough die life for the massive production of nut.

References

1. P. Unseld, L. Kertesz, S. Hasler , ICFC (2015)
2. L.M.Alves, C.M.A.Silva, P.A.F.Martins, J CLEAN PROD, **135**, 1673-1680 (2016)
3. C.Emmelmann, M.Kirchhoff, F.Beckmann, Phys Procedia,**12**, 459-464 (2011)
4. E.Yüksel, E.Budak, A. S.Ertürk, Procedia CIRP, **58**, 487-492 (2017)
5. D.C. Goss, J MATER PROCESS TECH, **98**, 135-142 (2000)
6. D.L. Andrew, S.S. Carlson, J. H. Macha, R. T. Pilarczyk, ENG FAIL ANAL, **74**, 35-53 (2017)
7. M. Hagiwara, Y. Mano, M. Hamada, K. Nunogami, H. Ozeki, S. Ito , 7, J ADV MECH DES SYST, **3**, 324-332 (2009)
8. S. Izumi, T. Yokoyama, A.Iwasaki, S. Sakai , ENG FAIL ANAL, **12**, 604-615 (2005)
9. G.M.Castelluccio, M.R.W.Brake, Comput Struct, **186**, 1-10(2017)
10. ASTM-F606/F606M-16, (2016)

11. ISO 4161, (2012)
12. ISO 898-1, (2013)

Design and analysis for manipulator fixture applications

Ming-Chu Hsieh[1,*] and *Zhen-Hong* Khong[2]

[1]Department of Mechanical Engineering, National Chin-Yi University of Technology, Taiwan

[2]Department of Mechanical Engineering, National Chin-Yi University of Technology, Taiwan

Abstract. The manipulators arms are the most important components in automatic machinery and equipment. These arms must conform to product requirements and match important accessory devices such as clamping jaw so that the entire process operates properly. Therefore the most important issue is that the manipulator arm has only one clamping jaw, followed by other issues such as welding and related features. In this research the manipulator arm of automated machinery with attached equipment was optimized, and the design serves as an object of study for multifunctional applications. This research emphasizes on customized manipulator design to create products with distinguished styles and characteristics, and serve as a basis for an improvement on innovative design. Manipulator arms from different manufacturers with various head-disc size and screw-hole locations were investigated, and concept of human palm was incorporated in the design to create a mechanical fixture that can be fitted into the arm. Overall design including main body of the manipulator, pneumatic cylinder set, gear and connecting rod combination, planet gear set, and linkage rods and fixture. These components were combined to offer the functions of the manipulator. Commercial software package, SolidWorks, was used to construct the model for the manipulator arm, and CAE analysis was implemented to identify the stress on the structure and possible interferences of the mobile components. Stress analysis was also performed on the gear set under pneumatic loading to ensure that the gear mechanism has a sufficient strength. The results of this study showed that an arc shaped manipulator body is the most easily handled during installation and most effectively controlled during operation. The simple profiles of the manipulator also lower the space required for storage. The design of the gear set and integrated application of the shaft helps prevent loosening of gear and shaft during operation. In this study, a manipulator was designed with three different functions, each fixture is powered by an isolated pneumatic cylinder, and this mechanism can be manipulated to form 0° to 180° clamping action in accordance with requirements of the task. The methods and results of this study serve as a reference for machinery industry to achieve high quality product, and also as a basis for innovative design.

* Corresponding author: smj@ncut.edu.tw

1 Introduction

With the progress of time, the uses of robots for various occasions with diverse applications are increasing in the market. Presently, the definition of a robot is very inclusive; i.e., an automation system combined with a machine structure is called a robot. The objective behind the development of robots is to simplify life, substitute humans in harsh work environments, or execute difficult actions. Firstly, the platform for robotic development is anthropomorphosis, i.e., a robot imitates the human vision of judging objects and the actions of human hand joints and fingers. Therefore, mechanical arms refer to automatic control equipment that imitates the functions of the human arm to complete various operations.

With the increase in the demand of manufacturers for automatic production, the development of the mechanical arm industry has accelerated. Generally, mechanical arms are substitutes for humans to execute dangerous, massive, and dull work. Currently, the applications of mechanical arms are automatic gripping jaw, welding, paint spraying, assembling, etc. The technologies related to the mechanical arm involve subjects of machinery, electronics, automatic control, artificial intelligence (AI), kinematics, biomechanics, sensor technology, information processing, bionics, neural network, material science, etc. Thus, the application of mechanical arms to the manufacturing industry is accompanied with cost reduction, various processing modules, and labor-saving performance, and is an important topic worthy of discussion.

The application of mechanical arms in high-tech automated production lines and in handling or assembling has been popularized, such as spraying, welding, and assembling the car body in the automobile industry. The geometric shapes and functions of mechanical arms in various fields are varied owing to different usage objectives and environments. Mechanical arms are automatic control equipment that completes various operations by imitating the functions of the human arm; they possess multi-joint connections, and can act and move linearly in plane and three-dimensional spaces. A mechanical arm is composed of a mechanical main body, controller, servo mechanism, and sensor, and its actions are set by the program input by a computer according to the operation demands.

The design of a low degree of freedom system enables robots to be lightweight and simple [1], and further assembles the system itself during the simulation and controller design. Additionally, it helps in going beyond the theoretical level to study the practical applications of the system. The methods of track production proposed in *Robotics* may be generally divided into several methods such as trajectory control, speed control, acceleration control, and torque control.

In medical application, the Da Vinci robotic surgery system assists doctors in performing surgeries, where the 3D-HD image technology and simulated wrist surgical instruments are jointly used to form smaller and more accurate operative lacerations. The Keio University in Japan designed a five-fingered robot hand [2] equipped with capabilities for performing space missions, solving Rubik's Cube problems, classifying medicines, and even cooking every day dishes such as pancakes. However, it is rather impossible for a mechanical arm to complete certain complex manual operative tasks, such as flexible rolling, rotating, bending, and friction-sensing, because it consists of multiple parts and components, including crank, connecting rod, joint, slider, gear, belt, and bearing [3].

Renault et al. from France employed virtual technology to conduct a simulation study on an artificial hand and arm in order to realize the action of grabbing [4]. Ma Ruqi et al. from the Harbin Institute of Technology [5] applied virtual technology to the joint simulation of a mechanical arm and used PD control to effectively realize the tracking movement of the mechanical arm. There was a location sensor on the hand joint to sense the angle of the joint and a force sensor on the tendon to measure the stretch and flexion conditions of the fingers. This information was utilized to further control the movement of the joint and fingers in order to realize the actions of grabbing objects, etc. [6].

The clamping jaw on the mechanical arm is designed for specific tasks and has specific clamped parts. As the conception of clamp fixture, the clamping jaw is the tested institution of the axial bending moment clamp fixture [7] formulated by ASTM, and this institution uses the upper and lower brackets to complete the counterpoint of supporting structures consisting of a 1/16 inch pressure head in the center and three 1/8 inch balls on the lower circle; its test method corresponds to the pressure head pressing test coupons from above. Marin and Ferreira [8] proposed an optimized fixture design for a certain workpiece and proposed optimized fixture location and form under conditions ensuring that the workpiece would not move. According to the literature review, the continuous improvement of the mechanical arm has assisted people. Further, the mechanical arm has adopted more optimized functions compared to the traditional mechanical arms, making the fixture automation device more practical.

Research on practical robots started in the 1950s and it shifted toward practical robotic applications from theoretical academic based research, and thus revolutionized industrial development [9] in several fields. In 1956, the first industrial mechanical arm was created by Joseph F. Engelberger, known as the father of the mechanical arm. In 1962, the first industrial robot in the world was installed in the production line of the GM Automobile Company in the United States [10].

2 Design and research method for manipulator fixture

The major contents of this design include the design of a manipulator and three connecting rods and fixtures. The pneumatic cylinder power operation is used to drive a gear set via a belt power source pulling connecting rod and the fixture is used to achieve telescopic actions. The main advantage of the pneumatic cylinder is that it can reduce the weight of the manipulator and it can further prevent excessively fast operations by serving as a buffer during the process of stretching out and retracting. Another advantage of the pneumatic cylinder is that it can prevent excess indoor operation temperature by replacing the application of the battery valve with a traditional manipulator. The pneumatic cylinder adopts pneumatic control to drive the gear and belt pulling the connecting rod and fixture institutions. The fixture is effective for clamping industrial parts and different parts in life. It can meet working requirements by the simultaneous operation of one to three connecting rods and fixtures to complete the application and dimension design of the whole manipulator. The expanded diagram of the fixture application design of the manipulator is shown below is shown in Figure 1.

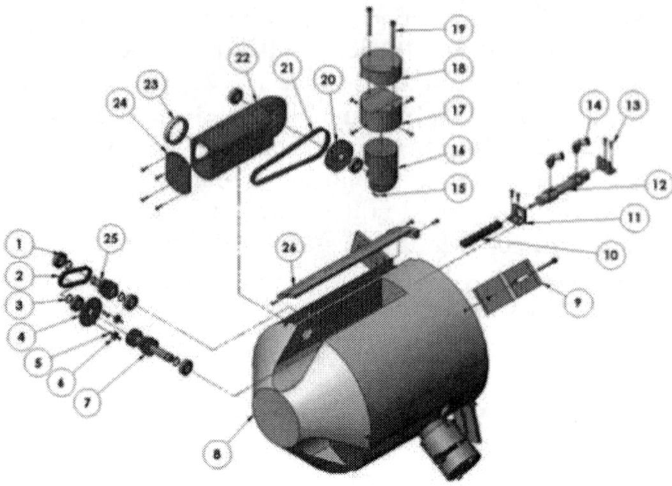

Fig. 1. The expanded diagram of the fixture application design of the manipulator.

The advantages and disadvantages of traditional robot and homemade robot Description: Generally, traditional manipulators are fixed and combined with the mechanical arm to perform a single clamping operation. In the production line or automobile industry, several mechanical arms are applied to execute the entire work flow. However, if one manipulator breaks down in the production line, the production line has to be stopped in order to replace the failed one and the repair process affects the working efficiency; moreover, small and medium enterprises usually do not have sufficient mechanical arms for replacement. Therefore, this study aims to design an optimized multi-functional manipulator that can operate like a human palm to solve these problems. It has three connecting rod functions and fixtures, and the workpiece can stretch out to operate whenever necessary. Moreover, failed clamping jaw or electro welding of the workpiece can be replaced directly, which will reduce the consumption and expenditure of mechanical arms for small and medium enterprises are shown in Table 1.

Table 1. The advantages and disadvantages of traditional and self-made hand grips.

Project	Traditional manipulator fixtures	Homemade manipulator fixtures
Special Requirement	Single Function	Multi-function
Environmental Awareness	Single-grip fixture application	Three-link function manipulator fixture application
Form Usage	Single with robotic arm execution	Three connecting rod functions and fixtures can move between 0° to 180°; mechanical arm execution
Workpiece Replacement	Manipulator needs to be replaced entirely	Fixture on the connecting rod of manipulator can be separated and changed directly.
Production Efficiency	Complete parts independently	Can complete a number of parts
Production Line Application	Completed by multiple mechanical arms and manipulators	A small amount of mechanical arm and manipulator to complete
Environmental Application	Suitable for high efficiency industrial field	Suitable for Economical industrial field

3 Application control and stress analysis of manipulator fixture

In this study, SolidWorks simulation was used to conduct finite element analysis and von Mises stress analysis was employed to reduce excessive driving force between gears within the manipulator to avoid the deformation of parts. Repeated analyses and modifications were carried out to obtain the optimum gear application. AISI 1045 steel was selected as the material, with properties of Young's modulus E = 204 GPa, Poisson's ratio $\upsilon = 0.29$, and density $\rho = 7850$ kg/m^3 are shown in Table 2. Model fixity offers the follower axial restraint geometry, and limits the force received by the gear contact angle on the surface of follower to apply restraint.

Table 2. Material properties.

name	AISI 1045 steel
Yield Strength	530MPa
Tensile Strength	625MPa
Elasticity Modulus	204GPa
Poisson's ratio	0.29
Material density	7.85g/cm^3
Shear Modulus	80GPa

Gear force analysis was performed on key parts and a fixed hinge of the gear center confined the rotation of the gear center and axle. Model loading analysis was conducted by applying a force of 47 N exerted by pneumatic cylinder at the top of the followeris shown Figure 2-13. The same force of 47 N was applied according to the rotating order of the gear and the force effect was observed to ensure sufficient stress strength for applications in part and material designs are shown in Table 3.

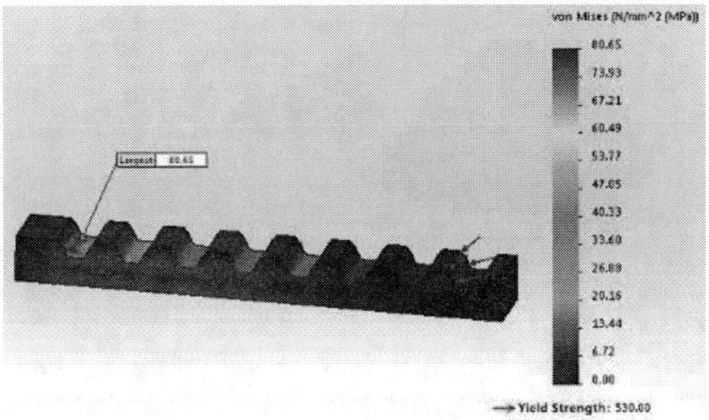

Fig. 2. Rack (8 teeth) von Mises stress diagram.

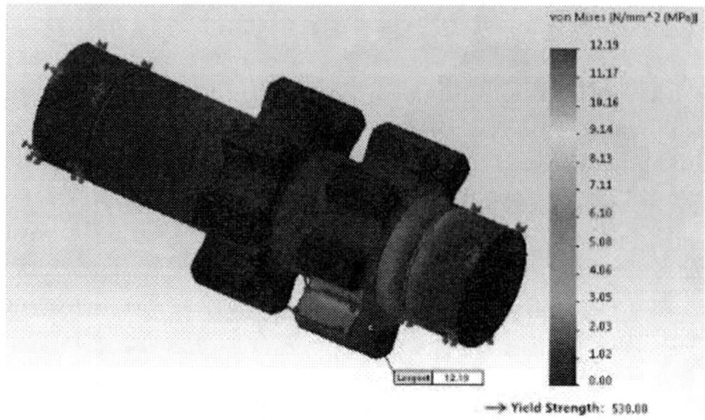

Fig. 3. Gear shaft combination -1 (two gears gear 8 teeth) von Mises stress diagram.

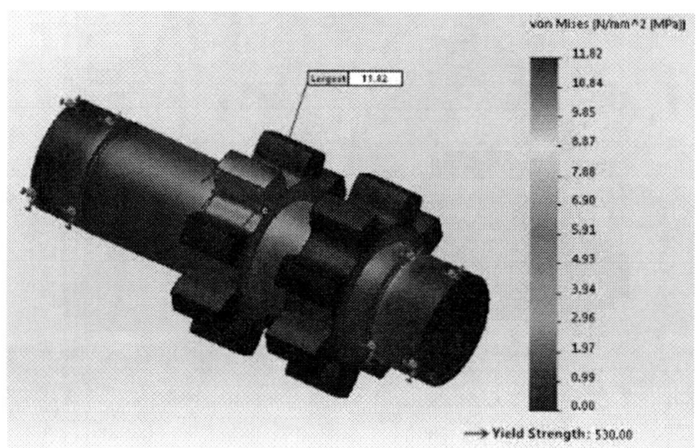

Fig. 4. Gear Shaft Combination - 2 (two gears are 8 teeth) von Mises stress diagram.

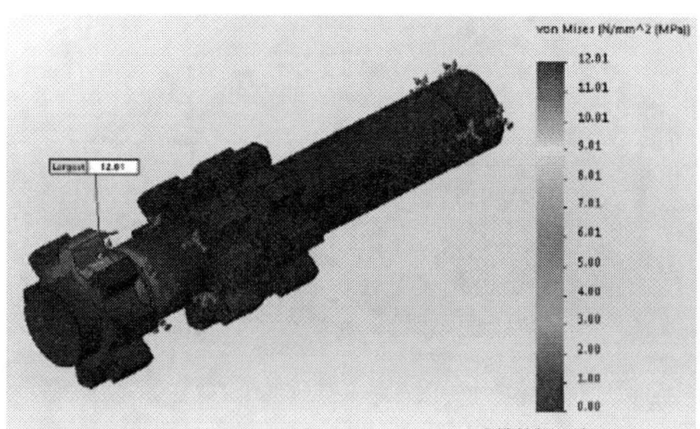

Fig. 5. Gear shaft combination -1 (three gear 8 teeth) von Mises stress diagram.

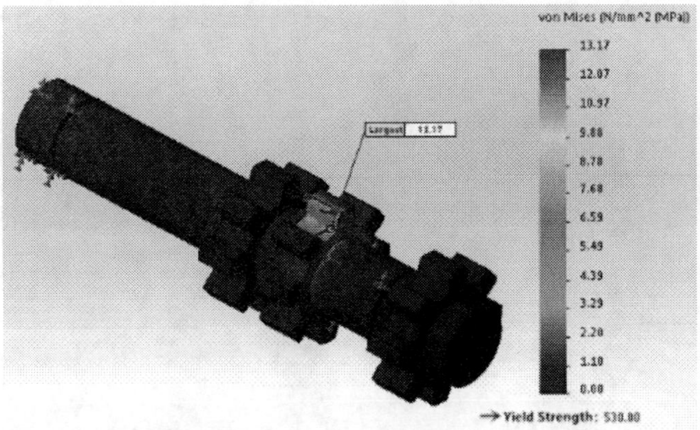

Fig. 6. Gear shaft combination -2 (three gear 8 teeth) von Mises stress diagram.

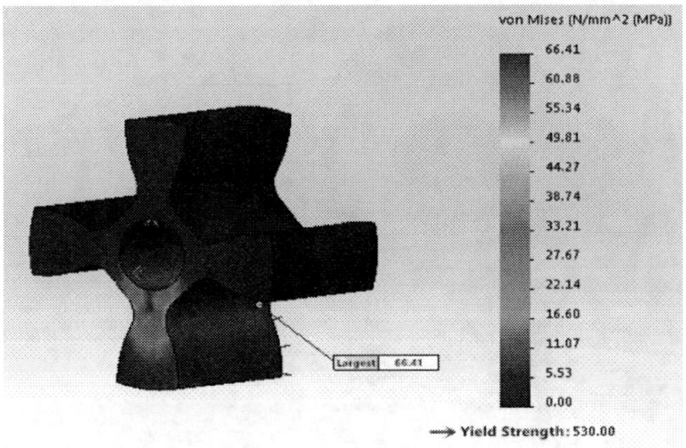

Fig. 7. Below the planet gear (gear 4 teeth) von Mises stress map.

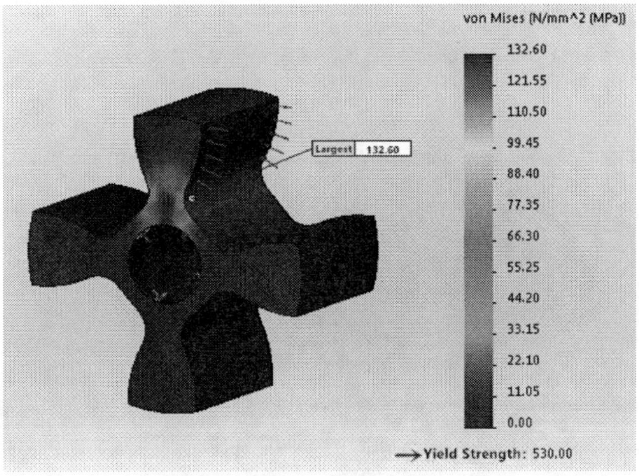

Fig. 8. Planetary gear above (gear 4 teeth) von Mises stress diagram.

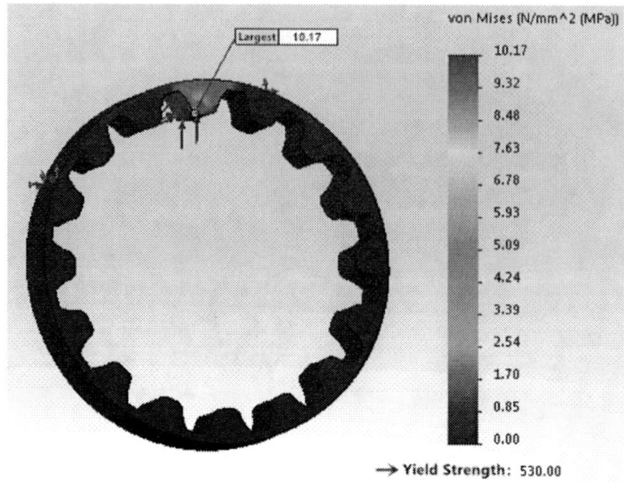

Fig. 9. Planetary gear (gear 16 teeth) von Mises stress diagram.

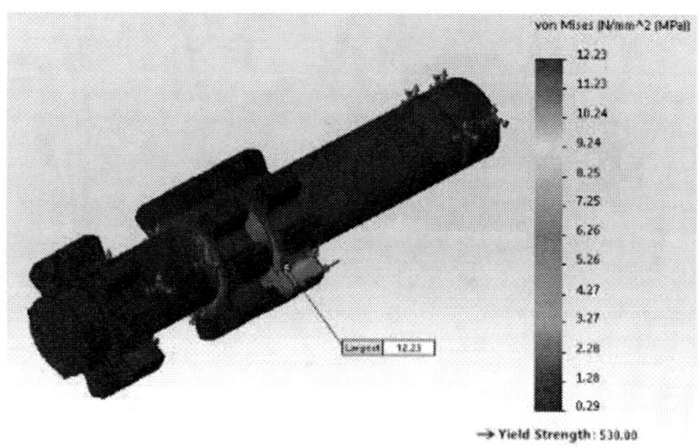

Fig. 10. Gear shaft combination -3 (three gear 8 teeth) von Mises stress diagram.

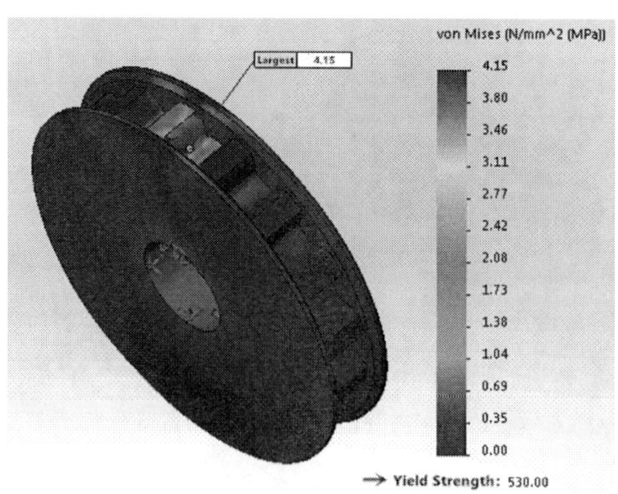

Fig. 11. Gear (16 teeth) von Mises stress diagram von Mises stress diagram.

Table 3. von Mises stress under gear stress condition.

Type of Gears	Gear weight (g)	von Mises stress (MPa)
Rack (8 teeth)	19.17	80.65
Gear shaft combination -1 (two gears gear 8 teeth)	33.63	12.19
Gear shaft combination -2 (two gears gear 8 teeth)	33.63	11.82
Gear shaft combination -1 (three gear 8 teeth)	54.81	12.01
Gear shaft combination -2 (three gear 8 teeth)	54.81	13.17
Below the planet gear (gear 4 teeth)	1.05	66.41
Planetary gear above (gear 4 teeth) von Mises stress diagram	1.05	132.60
Planetary gear (gear 16 teeth)	8.45	10.17
Gear shaft combination -3 (three gear 8 teeth)	54.81	12.23
Gear (16 teeth) von Mises stress diagram	49.89	4.15

von Mises stress was applied to the manipulator fixture; planetary external gear supported the interaction of the connecting rod and fixture, and equivalent stress of this structure was calculated under a yield strength of 530 MPa to obtain the maximum load above the fixture of 680.68 N. Structure damage, due to excessive stress when clamping industrial accessories, was also avoided, as shown in Fig. 12.

When the load above the fixture is 680.68 N, the maximum stress received by connecting rod head is 27 MPa. Such conditions will not exceed 530 MPa yield strength to cause excessive deformation and damage , as shown in Fig. 13.

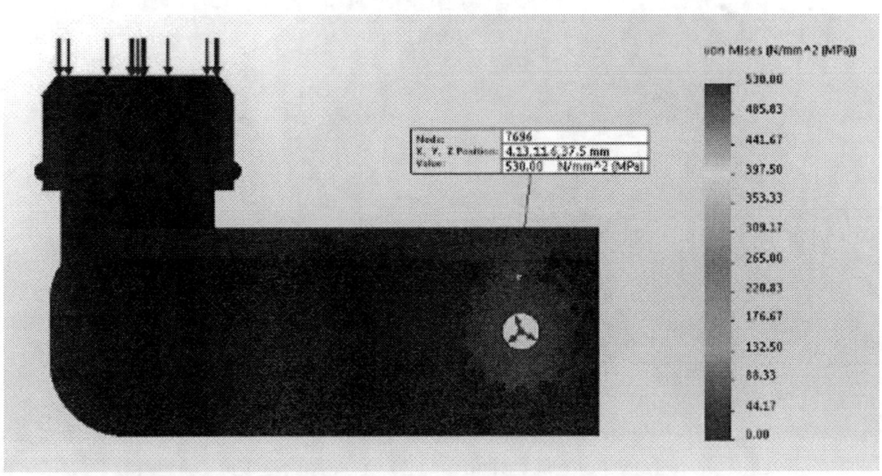

Fig. 12. Drawing of maximum load of planetary external gear under yield strength.

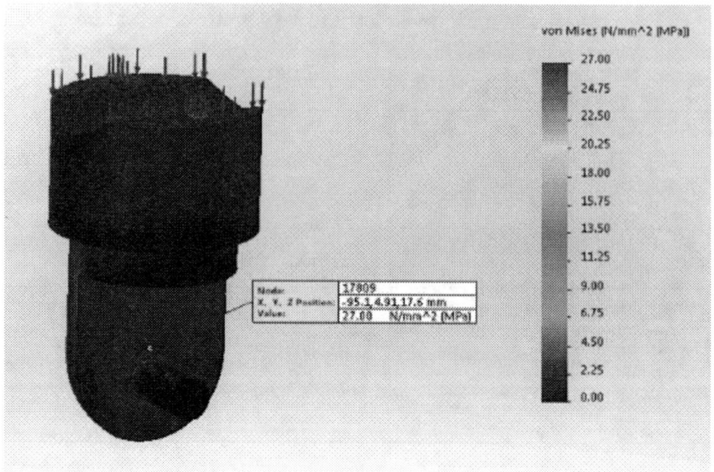

Fig. 13. Drawing of von Mises stress of fixture and connecting rod head.

The design of the manipulator fixture in this study emphasized a flexibility and action pattern that is similar to a human palm, and the simplified institution design and freedom in the degree of motion control made it possible to perform the required actions is shown in Fig. 14.

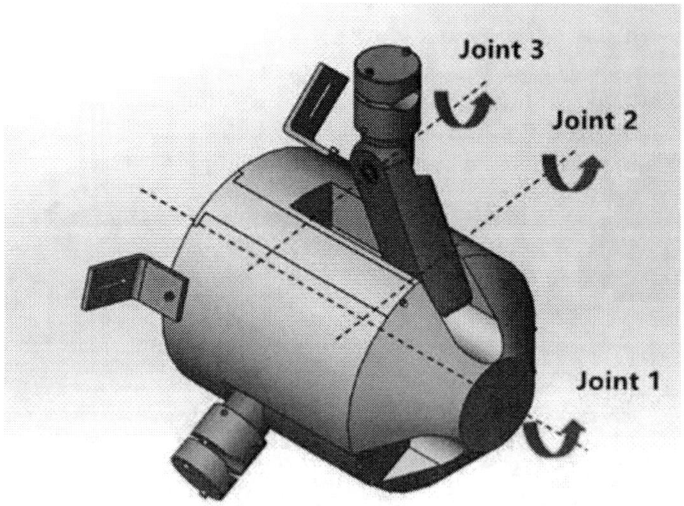

Fig. 14. Robot fixture design freedom of use map.

Manipulator fixture application joint (1) can be fixed with the mechanical arm and rotated from 0° to 360°; then, internal pneumatic cylinder of the manipulator drives the gear to rotate the internal planetary gear set of joint (2) by rotating the connecting rod from 0° to 180°; finally, joint (2) connecting rod is transmitted to joint (3) for uniform rotational degree of freedom from 0° to 180° are shown in Table 4.

Table 4. Robotic fixture application design freedom table.

Joint 1	0° to 360°
Joint 2	0° to 180°
Joint 3	0° to 180°

Firstly, section A is the main power source to match racks for the distance control pushed out by the pneumatic cylinder. The pushed distance rack will enable the first gear of the two gear sets to rotate reversely, and the coaxial second gear will rotate in the reverse direction to pull the belt and rotate the first gear of another three-gear set; then, the pulled first gear and second gear will rotate coaxially to drive the whole planetary gear set in section B; the planetary external gear has to be welded and fixed with the connecting rod to rotate the connecting rod from 0° to 180° due the rotation of the pinion. Finally, in section D, the third gear of three-gear set rotates coaxially to pull the belt and combines the connecting rod head and fixture to jointly rotate from 0° to 180°as shown in Fig. 15-16.

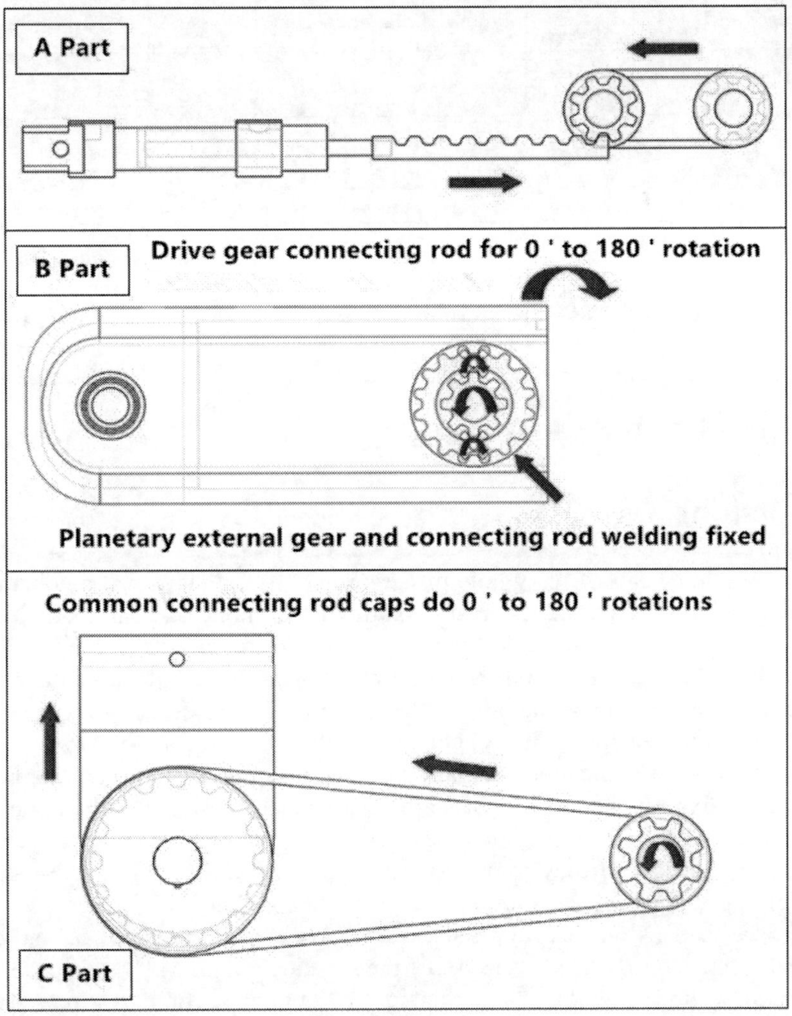

Fig. 15. Manipulator application motion control simulation.

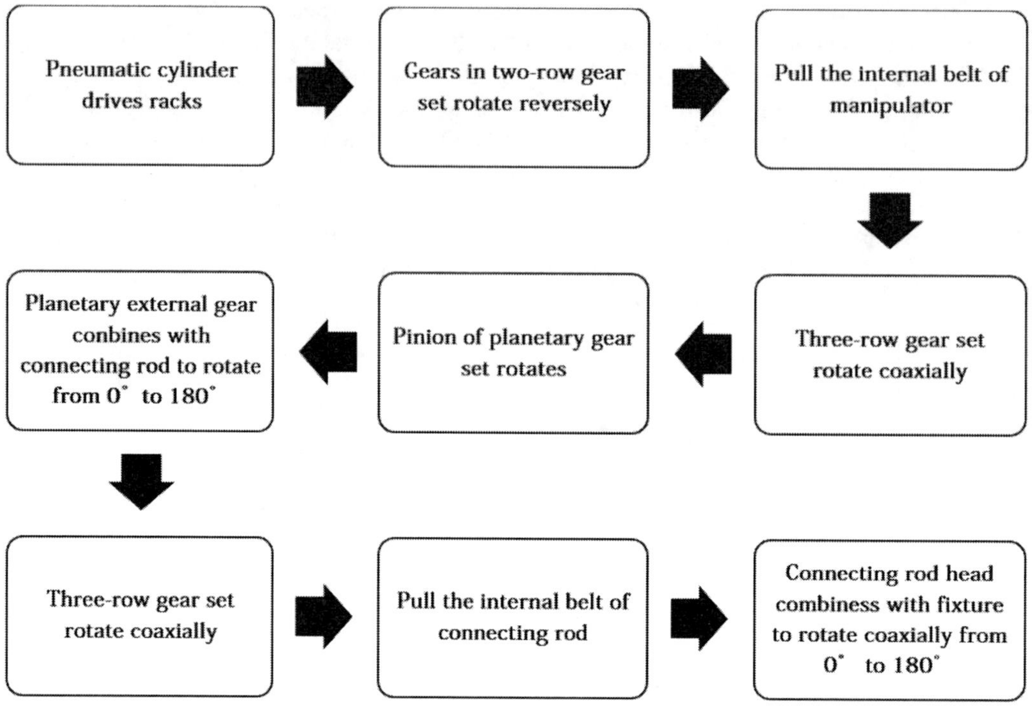

Fig. 16. Flow Diagram of Transmission Operation Mode.

4. Conclusions

In this study, the design of the whole process refers to the industrial mechanical arm for coordinating the design size of the manipulator, the number of gear teeth in the overall application and gear appearance design that conforms to the overall application of the manipulator in consideration of movement, and the ideal angle of the connecting rod and fixture operation that has to be achieved; the transmission of every gear must match with one-piece to complete the task. The conclusions of this study are as follows:

1. The concept of a simulated human palm is used to design the application of a manipulator fixture; sizes of all parts are designed as stipulated in the main body of the manipulator; three connecting rods are applied to functions for achieving stretching out and drawing back, and they jointly rotate with the fixture; a fixture is used to clamp three kinds of industrial accessories or household appliances.
2. In response to the gear institution stress analysis, design of the gear, and selection of materials, the stress effect caused by the application of 47 N force exerted by a pneumatic cylinder will be calculated to obtain sufficient stress strength for applications. Then, the maximum limit of stress for every gear institution is calculated and the results show that the maximum limit is 308.85 N in order to avoid the deformation and damage of gears due to excessive stress limits. Finally, the maximum load above the fixture is 680.68 N, and the connecting rod head will not be damaged when the stress is less than 27 MPa.
3. After setting the electrical control circuit, the motion control pattern provides signals to the air source of the electromagnetic valve with five connecters and two positions. The buffer enables the pneumatic cylinder to combine the actions of stretching out and drawing back to operate the whole transmission system architecture. Finally, the connecting rod performs the action of stretching out and drawing back to achieve

rotations and applications between 0° to 180° with the fixture.

4. In compliance with users' demands, the application of the manipulator fixture to motion simulation will be applicable for one to three connecting rods and fixtures operations. One connecting rod and fixture can achieve separate applications of stretching out and drawing back; two connecting rods and fixtures can achieve double-applications of stretching out and drawing back; finally, three connecting rods and fixtures can be used if the users have extensive demands, and the fixtures can be adopted to clamp effective industrial equipment applications and meanwhile achieve improved production speed.

References

1. R.J. Full and D.E. Koditschek, Templates and Anchors: Neuromechanical Hypotheses of Legged Locomotion on Land, Journal of Experimental Biology, 1999. 202(23): pp. 3325-3332
2. I. Yamano and T. Maeno, Five-finger Robot Hand using Ultrasonic Motors and Elastic Elements, in Proceeding of the 2005 IEEE International Conference on Robotics and Automation, pp. 2673-2678, 2005
3. M. Saitohand and K. Furuta, Generalized minimum variance control of slider-crank mechanism, in SICE Annual Conference, pp. 2205-2209, 2007
4. M. Renault and F.B. Ouezdou, Dynamic simulation of hand-forearm system, IEEE International Workshop on Robot and Human Interactive Communication, pp. 20-25, 2001
5. R.Q. Ma, S.H. Hao, W.F. Zheng, M. H. Hao, B.Y. Song, Research on coordinated simulation of robot arm based on MATLAB and ADAMS. Machinery Design & Manufacture, pp. 93-95, 2010
6. P. Michelman, Precision Object Manipulation with a Multi-fingered Robot Hand, IEEE Transactions on Robotics and Automation, Vol. 14, No. 1, pp. 105-113, 1998
7. ASTM, Standard test method for biaxial flexure strength (modulus of rupture) of ceramic substrates, ASTM F394-78. American Society for Testing and Materials, Philadelphia, PA. 1996
8. R.A. Marinand and P. M.Ferreira, Optimal placement of fixture clamps: Minimizing the maximum clamping forces, Journal of Manufacturing Science and Engineering, Transactions of the ASME, Vol. 124, No. 3, pp. 686-694, 2002
9. J.Q. Gan, E. Oyama, E. M. Rosales, H. Hu, A Complete Analytical Solution to the Inverse Kinematics of the Pioneer2 Robotic Arm, J. of Robotica, vol. 23, pp. 123-129, 2005
10. H. P. Huang, Y.T. Chen, R.J. Wang, M.K. Chi, Development of the multi-axis control platform for robot arm, IEEE, Proceedings of Annual Conference, pp. 2326-2331, 2010

https://doi.org/10.1051/matecconf/201818500032

A study of microstructure and tensile property with casting S45C welding

Chin-Guo Kuo[1], *Lih-Ren* Hwang[2,*], and *Yen-Tien* Kang[3]

[1]Department of industrial education, National Taiwan Normal University, Taipei City 10610, Taiwan
[2]Department of Mechanical and Automation Engineering, Chung Chou University of Science and Technology, Yuanlin City, Changhua County 51022, Taiwan

Abstract. For energy saving and longer life time of a large casting iron may use of weld repairs or make up to more save manufacture time and energy than re-casting. Such as breakage, crack and wear in large casting iron castings used weld to repair casting for eco-energy.
The purpose of this study is to investigate the dissimilar joint property of S45C carbon steel with the multi-layer method and the multi-pass welding method. Four weld rods, experimentally examined the influence on microstructures, which were marked as A, B, C and D, respectively. The microstructures of joint bead were observed by the optical microscope. The specimens of tensile test were made according to ASTM E8M. The experimental results were shown as following:

1. The pearlite structures and the bainite structures were formed in welded zone by the weld rod of middle carbon steels. The yield strength and ultimate strength of the specimens were 360MPa and 460MPa.
2. The feature of welded specimen was close to stainless steel due to the stainless steel rod included some Cr and Ni. In addition, a ferritic-austenitic two-phase stainless steel was observed in optical microscope.

1 Introduction

Generally, the big constructional machines are used to manufacturing by foundry (as Fig.1). In solidification, shrinkage accompany with many defects in working. Those defects like crack, wear or shrinkage are often fixed by welding and therefore weld traces are found in such machine as engines or cases. The joint consequence not only related to the skill but also need to take care the mechanical properties. Thus, it must base on material science knowledge to prevent the damages, not experience or favor.

Fig.1. Example of constructional S45C steel.

The weld mechanical property concerns the microstructures and the composition. To investigate the joint property, the method of multi-layer and multi-pass welding[1-4] is one way often used. Multiple complicated phase transformation happens under heat treatment[5,6]. Because of the different cool rate and dissimilar materials, the weld bead involves several particular microstructures that affect mechanical property[7-10]. In addition, the joint property affected a lot by the rod composition[11-13] and the microstructure analysis plays important role in understanding of joint strength. Therefore, the present research aimed at the microstructure analysis of casting S45C welding and the tensile property.

*Corresponding author: lrhwang@dragon.ccut.edu.tw

2 Experiments

The base metals (BM) in the present study are S45C carbon steel. Its dimension is length 100mm x width 50mm x thickness 16mm, respectively. In order to avoid temperature affect in deformation, keeping cool with water is necessary. Every metal bulk has 30 degree in bevel angle[14], 2 mm in depth and 2 mm in space. Prior to experiment, a backing strip is weld spotted at the bottom which is a carbon steel plate and whose volume is 100×25×6mm, as shown in Fig.2.

The acetone is utilized cleaning particles on the weld zone. Experimental process shows in Fig.3. Experiment equipments include: Olympus BX51 used for OM microstructure, SMAC ARC-300 used for flat-weld, AG-10TE used for tensile test, JEOL-JSM 6360 used for SEM picture and OXFORD INCA Energy 300 used for EDS analysis.

Fig. 2. V-butt of S45C base metal.

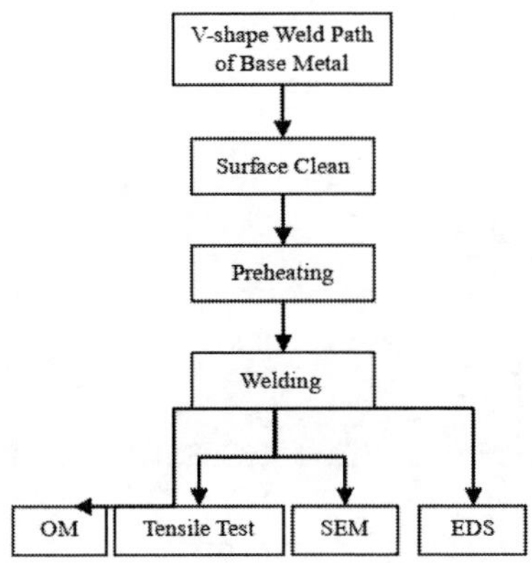

Fig. 3. Experimental flow chart.

There are four kinds of rods those are popular in industry. Every one of them is from separate company and specification is for CNS E4303 carbon steel rod, marked as A, B, C and D, respectively. Their individual parameters show in Table 1 below. Before welding, residual steam must be removed carefully as following：

1. These rods keep dry in 300℃ to 500℃for 60 minutes until welding. When weld beginning, rods keep in a dry container between 100℃ to 150℃, whose quantities is going to be used out in half day.
2. To avoid porosity formed, the arc must be moved back and forth.
3. A uniform weave beading, not over three times the electrode diameter, will aid greatly in preventing undercut in butt welds.
4. Based on different thickness, the metals preheated in 80℃ for a period of time.
5. Corners held with fixtures in order to avoid distortion.

Table 1. Welding parameters of every weld.

Filler	A	B	C	D
Diameter (mm)	3.2	3.2	3.2	3.2
AC (A)	100~140	100~130	80~110	80~120
Actual AC (A)	130	130	110	120
Method	Flat weld	Flat weld	Flat weld	Flat weld

The method of multi-layer and multi-pass welding is according to the functional conditions of SMWA rectifier. Temperature keeps in interpass-temperature while welding and weld slag and oxide impurity are removed with brush. The aim at this study is to analyze the microstructure and mechanical property of joint in between base metal and rod.

Selected microstructure zones of weld were observed by optical microscope and which is showed in Fig.4 : Base Metal(BM)、Hot Affect Zone (HAZ)、Fusion Line、Cover Pass、Filler Pass and Root Pass. In addition, according to ASTM E8M, all welded bead are deformed into tensile specimens and which are show in Fig.5 and Table 2. After tensile fracture, different cross section features from base metal are observed by JEOL-JSM 6360 Scanning Electronic Microscope. Then, OXFORD-INCA Energy 300 is used for qualitative analysis.

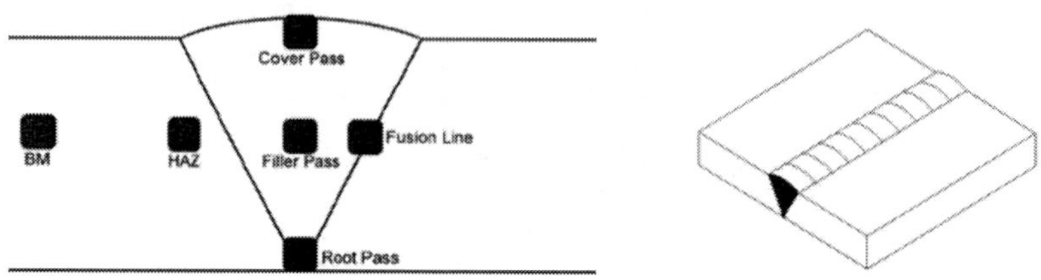

Fig. 4. Selected weld position for microstructure.

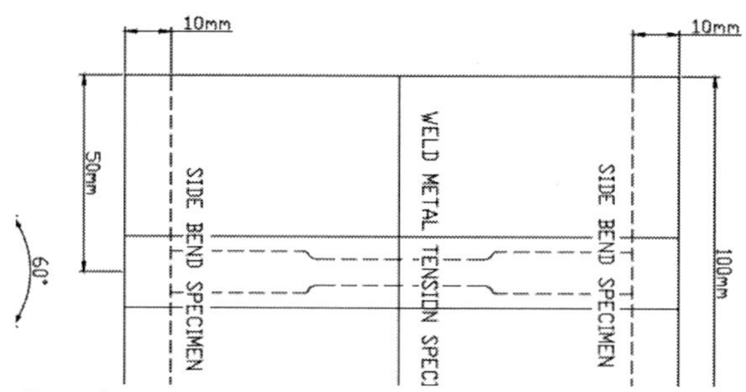

Fig.5. Sample position of tensile specimen.

Table 2. ASTM E8M Rule for tensile test.

	Standard Specimen				
	mm	mm	mm	mm	mm
G	62.5±0.1	45.0±0.1	30.0±0.1	20.0±0.1	12.5±0.1
D	12.5±0.2	9.0±0.1	6.0±0.1	4.0±0.1	2.5±0.1
R	10	8	6	4	2
A	75	54	36	24	20

G---Length D---radius R---minimum radius
A---the scale of reduction of the minimum length

3 Experimental results

3.1 Microstructure observation

Fusion Zone is divided into three parts to discuss, which are cover pass, filler pass and root pass. Rod contents of those, marked as A, B and C, have few impurities as Si, Mn, P and S. All of their compositions and microstructures are similar. Pearlite exit in filler pass and root pass. Because root pass reaches low temperature faster than that filler pass, so that pearlite in root is fine. The Bainite microstructure, which the upper Bainite appeared more than the lower Bainite, was observed at the cover pass. The weld microstructures showed in Fig.6 to Fig.8.

As shown in Fig. 9, Dendritic ferrite in filler pass and dual phase in cover pass were found obviously. Owing to the rod D comprised Cr and Ni, so that Austenitic-Ferritic Stainless Steel (a mixture of austenite and ferrite) formed during heat treatment.

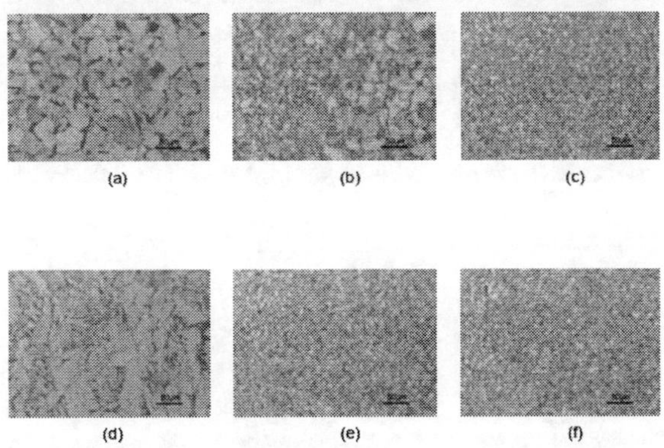

Fig. 6. Microstructure of the weld bead A (a) BM (b) HAZ (c) Fusion Line (d) Cover Pass (e) Filler Pass (f) Root Pass.

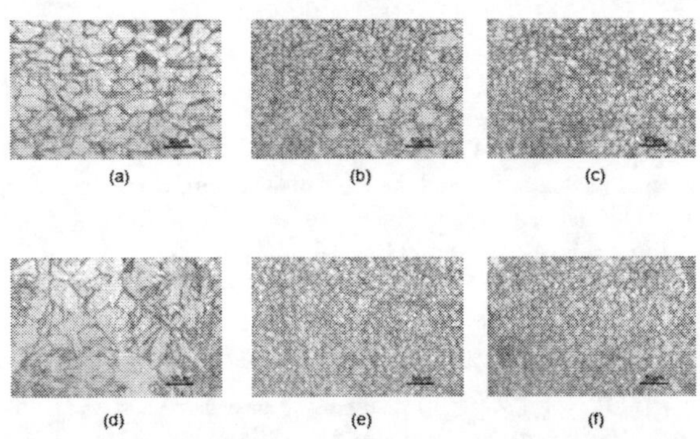

Fig. 7. Microstructure of the weld bead B (a) BM (b) HAZ (c) Fusion Line (d) Cover Pass (e) Filler Pass (f) Root Pass.

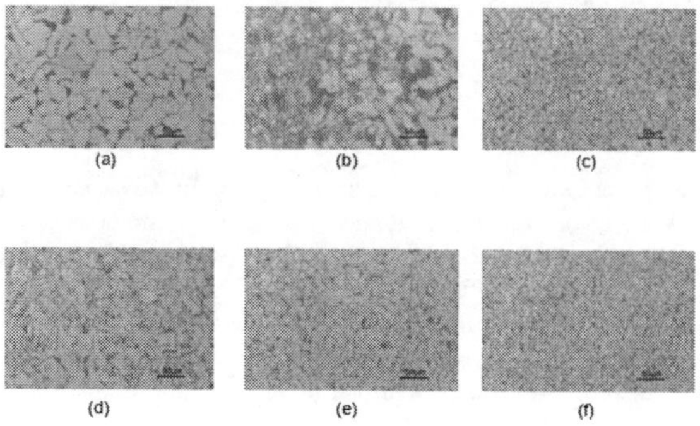

Fig. 8. Microstructure of the weld bead C (a) BM (b) HAZ (c) Fusion Line (d) Cover Pass (e) Filler Pass (f) Root Pass.

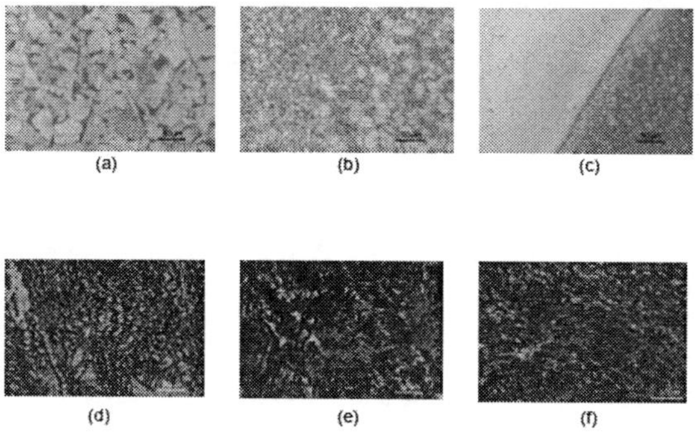

Fig. 9. Microstructure of the weld bead D (a) BM (b) HAZ (c) Fusion Line (d) Cover Pass (e) Filler Pass (f) Root Pass.

3.2 Tensile experiment

Specimens were according to ASTM standard and made of full weld bead. Table 3 showed the ultimate strength, the yield strength, the elongation and the reduction of area (R. A.). The ultimate strength of BM and D are 664 MPa and 702 MPa, respectively. The yield strength of BM and D are 475Mpa and 486MPa.

All of the heat treated specimens, the elongation and the R.A. of BM were worse than others. That indicated the ductile of the welding specimens enhanced by the heat treatment. However, the heat treatment only improved slightly the yield strength and ultimate strength of the rod D, as showed in Fig. 10.

Table 3. Tensile test data.

Specimen	Average Elongation (%)	Average Yield Strength (MPa)	Average Ultimate Strength (MPa)	Average R.A. (%)
A1	36	367	456	63
A2				
A3				
B1	30	359	477	68
B2				
B3				
C1	36	358	460	66
C2				
C3				
D1	41	486	702	68
D2				
D3				
BM1	18	475	664	53
BM2				
BM3				

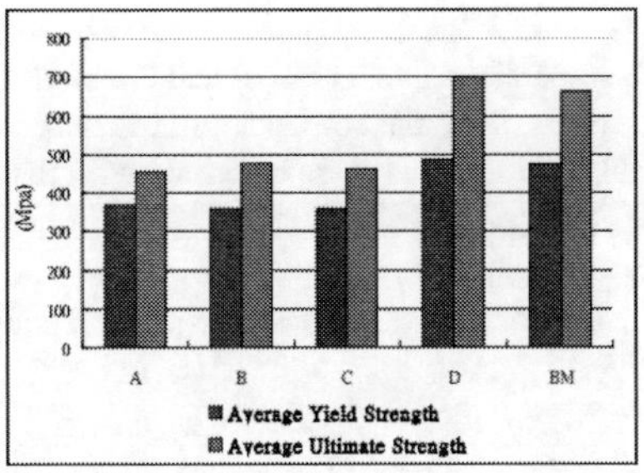

Fig. 10. Bar chart of tensile test.

3.3 SEM and EDS analysis

Different from cleavage of BM, all specimens represent vary ductile fracture about dimple(as Fig.11). Some carbide-like particles are inside the dimple. Either material void caused by particles inside dimple or by interface between particle and matrix. Ductile fracture is importantly based on the volume fraction of carbide, distribution and appearance. While the method of multi-layer and multi-pass welding induced phase transformation, the particles and distribution affected the ductile property in heat treatment.

When tensile test started, initial dimples exited some brittle particles which concentrate many dislocations. Micro-voids happened in the interface between brittle particle and matrix. Dimples were elongated by tensile stress and they laterally connected to each other with internal stress. Finally, micro-cracks formed a lot of dimples on ductile fracture surface.

As shown in Fig.12, specimens of D are with the finest dimple in fracture surfaces. It means that the micro-voids are many and concentrated. Overall, fracture surfaces are totally matched to their mechanical property.

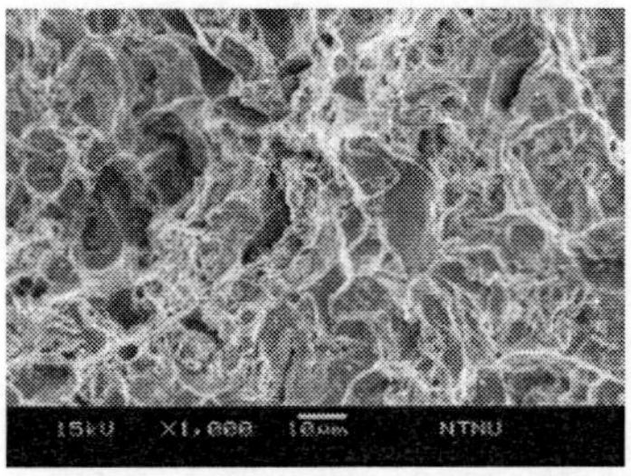

Fig.11. Fracture surface of base metal.

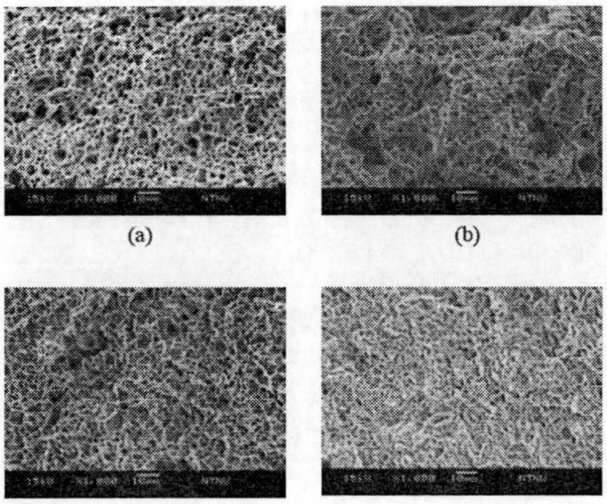

(a) (b)

(c) (d)

Fig.12. Fracture surface of A, B, C and D weld.

3.4 EDS analysis

The compositions of all specimens are presented in Table 3. On one hand marked as A, B and C, they are close to normal carbon steel. Besides Fe and C, there are some elements Si, Mn, P and S exited. Their compositions are similar, but each weight percent is kind of different. On the other hand, the compositions of D are with 27.3% Cr, 0.66% Co and 9.17% Ni. It is similar to the dual phase stainless steel, whose features are better strength than stainless steel 304 and good resistance to corrosion. The EDS analysis shows in Table 4.

Table 4. EDS Analysis of weld bead.

wt% / specimen	Fe	C	Si	Mn	P	S
A	99.11	0.42	0.08	0.29	0.05	0.05
B	98.63	0.50	0.25	0.53	0.04	0.05
C	98.99	0.55	0.21	0.17	0.03	0.05
D	60.98	0.01	1.39	0.49	27.30	0.66
BM	98.59	0.48	0.30	0.57	0.02	0.04

4 Conclusions

The composition of the rod D is close to stainless steel. It causes austenitic-ferritic stainless steel or dual phase stainless steel in cover pass as shown in Fig. 9. On the other hand, the compositions of the others are close to middle carbon steel so pearlite and bainite exit from root pass to cover pass as shown in Fig.6 to Fig.8.. The yield strength and the ultimate strength of them are smaller than base metal.

Every cross section appears ductile failure. Of all, dimple is finer in specimen D and it connects to better ductile property in mechanical test. In order to application in casting S45C material, the research is a reference for choosing weld rod kind.

Acknowledgments

This work was supported by National Taiwan Normal University under contract No. 98031017.

References

1. K. Ogawa, H. Yamaguchi, S. Kaga and K. Sakaguchi, Optimization of Friction Welding Condition for S45C Steel Using a Statistical Technique, Vol. 24, No. 2, pp. 47-53 (1993)
2. R. Kacar and O. Baylan, An Investigation of Microstructure Property Relationships in Dissimilar Welds Between Martensitic and Austenitic Stainless Steels,Vol. 25, pp. 317-329 (2004)
3. S.D. Meshram, T. Mohandas and G. Madhusudhan Reddy, Friction Welding of Dissimilar Pure Metals, Vol. 184 , pp. 330-337 (2007)
4. A. Ozekcin, H. W. Jin, J.Y. Koo, N.V. Bangaru, R. Ayer, G. Vaughn, R. Steel and S. Packer, A Microstructural Study of Friction Stir Welded Joints of Carbon Steels, Vol. 14, No. 4, pp. 1053-5381 (2004)
5. S. Murugan, D.K. Rai, P.V. Kumar, T. Jayakumar, B. Raj and M.S.C. Bose, Temperature Distribution and Residual Stress Due to Multipass Welding in Type 304 Stainless Steel and Low Carbon Steel Weld Pads,Vol. 78, pp. 307-317 (2001)
6. Y. Ueda, E. Takahashi, K. Fukuda and K. Nakacho, Transient and Residual Stress in Multi-Pass Welds, *Transactions of JWRI*, Vol. 3, No.1, pp. 59-67 (1974)
7. M. Sahin, Joining with friction welding of high-speed steel and medium-carbon steel, Vol. 168, pp. 202-210 (2005)
8. A.P. Reynolds, W. Tang, T. Gnaupel-Herold and H. Prask, Structure, properties, and residual stress of 304L stainless steel friction stir weld, Vol. 48, pp. 1289-1294 (2003)
9. D.J. Lee, J.C. Byun, J.H. Sung and H.W. Lee, The dependence of crack properties on the Cr/Ni equivalent ratio in AISI 304Laustenitic stainless steel weld metals, Vol. 513-514 , pp.154-159 (2009)
10. A. Celik and A. Alsaran, Mechanical and Structural Properties of Similar and Dissimilar Steel Joints, *Materials Characterization* Vol. 43, pp.311-318 (1999)
11. R.H. Koztowski, Composite of austenitic-ferritic stainless steel, Vol. 53, pp. 239-246 (1995)
12. H. Fujii, L. Cui, N. Tsuji, M. Maeda, K. Nakata and K. Nogi, Friction stir welding of carbon steels, Vol. 429, pp.50-57 (2006)
13. Ling Cui, Hidetoshi Fujii, Nobuhiro Tsuji and Kiyoshi Nogi, Friction stir welding of a high carbon steel, Vol. 56, pp. 637-640 (2007)
14. P. Wongpanya, Welding Residual Stresses in Two Competing Single V-Butt Joints, Vol. 19, No. 1, pp. 67-75 (2009)

Bee colony optimization with Taguchi method for solving the dynamic economic dispatch

Chia-Sheng Tu[1], *Hsi-Shan* Huang[1,*], *Ming-Tang* Tsai[2], and *Fu-Sheng* Cheng[2]

[1]College of Intelligence Robot, Fuzhou Polytechnic, Fujian, China
[2]Department of Electrical Engineering, Cheng-Shiu University, Kaohsiung, Taiwan

Abstract. Dynamic economic dispatch is to minimize the cost of power production of all the participating generators over a time horizon of 24 hours in one day. The dynamic economic dispatch with non-smooth cost functions, for which is formulated the optimal dispatch model of generations by considering the ramp up/down scheduling of power. This paper presents a Bee Colony Optimization (BCO) that applies the Taguchi Method (TM) to solve the Dynamic Economic Dispatch problem. The Taguchi method that involves the use of orthogonal arrays in estimating of the non-smooth cost function and Bee Colony Optimization is used to find the objective function under the operational of system constraints. The Taguchi method can global optimization for fast local convergence by minimizing the cost function in a few iterations. The effectiveness and efficiency of the TM-BCO is demonstrated by using a 10 unit of IEEE case with non-smooth fuel cost functions and is more effective than other previously developed algorithms. Moreover, the proposed approach presents significant computational benefits than traditional random search method especially for multi-unit systems with larger numbers of non-smooth cost functions and more complicated dynamic economic dispatch.

1 Introduction

The Dynamic Economic Dispatch (DED) problem is to determine the optimal power outputs of all generating units to minimize total fuel cost while satisfying the operational constraints and load demand. DED is the complexity of search space, which the probable large number of variables in scheduling process and complicated constraints. The DED of non-smooth cost functions is a complex and high-dimensional problem with multiple constraints. A number of approaches have been developed for solving the DED problem. The two-step method is proposed to solve the DED with valve-point effect in [1] and the harmony search algorithm employed to solve the DED problem for Micro-grid in [2]. Recently, some artificial intelligent techniques have been presented to solve the DED problem and have shown their effectiveness in [3-4]. The Bee Colony Optimization (BCO) employs employed bees, onlooker bees and scout bees moving patterns to research the feasible solution in [5].

*Corresponding author:hw5400@hotmail.com

The BCO has many of the advantages of biological intelligence in searching and it has stability in higher-dimensional search. The Taguchi Method (TM) of experimental design has been widely used in industry and the objective function will be best obtained by changes in these primary factors in the direction indicated. The [6-7] employ the Taguchi Method that involves the use of orthogonal arrays to solve the economic dispatch problem with non-smooth cost functions. The present paper applies the Bee Colony Optimization with Taguchi method to solve DED problem with non-smooth fuel cost functions. In the BCO-TM procedure, the Taguchi Method is embedded in the BCO in order to improve the behaviour patterns of each bee swarm and increase its search efficiency and accuracy in high dimensions. The TM-BCO for dynamic economic dispatch problems to increase the searching efficiency and the proposed method is feasible, robust, and more effective than previously algorithms.

2 The problem formulation

The dynamic economic dispatch planning of ancillary service shall satisfy the electric energy demand, and must consider whether the generators can ramp up or down the power in a short period of time. The mathematical model of power dispatch has multiple variables, which are mostly confined to the constraints of generator characteristics. The dynamic economic dispatch problem is to minimize the total operating cost. The total operating cost includes fuel cost and start cost of units which is formulated as (1).

$$\begin{aligned} Cost &= \sum_{t=1}^{h}\sum_{i=1}^{n} F_i^t(P_i^t) \\ &= \sum_{t=1}^{h}\sum_{i=1}^{n} a_i P_i^{t2} + b_i P_i^t + c_i + \left| e_i \times \sin(f_i) \times (P_{i,\min} - P_i^t) \right| \end{aligned} \tag{1}$$

where h is the scheduling hour and n is the total number of generators. P_i^t is the power output and F_i^t is cost function of a committed unit i at hour t. The a_i, b_i, c_i, e_i and f_i are the coefficients of unit i with non-smooth cost function.

The constraints of dynamic economic dispatch problem are described as follows.

1. The load balance

$$Load^t = \sum_{i=1}^{n} P_i^t \tag{2}$$

The sum of all P_i^t should be equal to the total load demand $Load^t$ at hour t.

2. The spinning reserve

$$Load^t + SR^t = \sum_{i=1}^{n} P_{i,\max}^t \tag{3}$$

where SR^t is the spinning reserve and set to 10% of $Load^t$.

3. The ramp up and ramp down constraints of generator

$$P_{i,ur}^t = P_i^{t-1} + UR_i \tag{4}$$

$$P_{i,\max}^t = \min[P_{i,\max}, P_{i,ur}^t] \tag{5}$$

$$P_{i,dr}^t = P_i^{t-1} - DR_i \tag{6}$$

$$P_{i,\min}^t = \max[P_{i,\min}, P_{i,dr}^t] \tag{7}$$

UR_i and DR_i are the upper ramp up and lower ramp down which $P_{i,ur}^t$ and $P_{i,dr}^t$ constraints of generator of unit i at hour $t-1$. The $P_{i,\max}^t$ and $P_{i,\min}^t$ are selected for $\min[P_{i,\max}, P_{i,ur}^t]$ and $\max[P_{i,\min}, P_{i,dr}^t]$.

4. Power generation limitation

$$P_{i,\min}^t \le P_i^t \le P_{i,\max}^t \tag{8}$$

$P_{i,\max}^t$ and $P_{i,\min}^t$ are the maximum and minimum generation limits of unit i at hour t.

3 The proposed methodology

In this paper, the Taguchi Method (TM) integrated with Bee Colony Optimization (BCO) to enhance the performance of the optimization approach. This BCO mimics the food foraging behaviour of honey bees and the swarm also consists of three categories, scout bees, employed bees, and onlooker bees. The Taguchi method of experimental design is a suitable method for parameter optimization problems with a given computable and this research is to minimize a cost function.

3.1 Bee colony optimization

The BCO includes the employed bees, onlooker bees phases, and scout bees phases and can be described as follows.

3.1.1 Employed bees

The better part of the employed bees fly considering the social and cognitive information achieved by the swarm. Each bee knows its current optimal position of personal experiences and also knows the current global optimal position among all bees in the population. The employed bees try to reach compatibility between local search and global search. Each employed bee makes a change on the position of food sources to generate a new food sources in the neighbourhood and as (9).

$$x_{eb}^{T+1} = x_{eb}^T + [c_1 \cdot rand \cdot (Pbest_{eb} - x_{eb}^T) + c_2 \cdot rand \cdot (Gbest - x_{eb}^T)] \quad eb = 1, 2, \cdots\cdots, N_1 \tag{9}$$

where $rand$ is the random numbers between 0 and 1; c_1 and c_2 are the acceleration constants in this paper set to 2.05 and 2.05; T is the current iteration and $T_{\max}$ is the maximal iteration set to 200; $Pbest_{eb}$ is of the population best solution and $Gbest$ is global optimal solution of employed bees at iteration T; eb is the number of employed bee population and N_1 is the total number of employed bee.

3.1.2 Onlooker bees

The onlooker bees in the bee colony optimization will follow the employed bees to obtain nectar information and modified by using the probabilistic selection method as (10) to follow the employed bees as (11).

$$prob(x_{eb}^{T}) = \frac{1/Cost(x_{eb}^{T})}{\sum_{ob=1}^{N_1} 1/Cost(x_{eb}^{T})} \tag{10}$$

$$x_{ob}^{T+1} = x_{ob}^{T} + rand(c_1 \cdot rand \cdot x_{ob}^{T} - x_{eb}^{T}) \qquad ob = 1, 2, \cdots\cdots, N_2 \tag{11}$$

where $prob(x_{ob}^{T}) = [x_1^T, x_2^T, \ldots, x_{N_2}^T]$ is the better $Cost$ value at the food source; ob is the number of onlooker bee population and N_2 is the total number of onlooker bee.

3.1.3 Scout bees

In the BCO, the scout bees working model was modified to the average value of the global optimum solution and all solution locations as (12) and (13).

$$x_{sb}^{T+1} = x_{sb}^{T} + [rand \cdot (Gbest - L \cdot mean^{T})] \qquad sb = 1, 2, \cdots\cdots, N_3 \tag{12}$$

$$L = round(1 + rand) \tag{13}$$

where L is the average of all variable solutions in the iteration T; sb is the number of scout bee population and N_2 is the total number of scout bee. The population size for employed bees, onlooker bees, and scout bees are total population 40%, 40%, and 20%, respectively and total population size is set to 100.

3.2 Taguchi method

For example, the unit in Taguchi Method of non-smooth cost function. The $F_i(P_i)$ is a cost function of several variables as Figure 1. and that is IEEE 3 units system for economic dispatch with valve point effects.

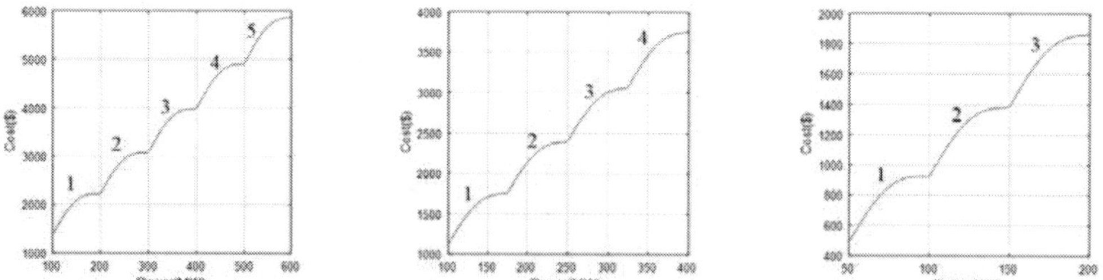

Fig. 1. Taguchi Method of non-smooth cost function.

In Taguchi method, a subset of the possible value combinations is used and to reduce the time consumed in conducting experiments. The Taguchi Method based on orthogonal arrays is introduced and chosen to provide sufficient information to determine the effects of each factor. Table 1 shows the orthogonal array shown in Taguchi Method each unit factor has different levels. In the present example, unit 1 is denoted by $P_{1,1}, P_{1,2}, \cdots\cdots, P_{1,k} = [1, 2, \cdots\cdots, 5]$ for smooth cost function, the unit i can be expressed as $P_{i,1}, P_{i,2}, \cdots\cdots, P_{i,k} = [1, 2, \cdots\cdots, k]$. The Taguchi Method array in Table II is in the form of $L_M(q^m)$, where q is the number of levels each factor has, m is the maximum number of factors the table can handle, M is this table total tests for the required using, the orthogonal arrays shows in Table 1.

Table 1. $L_{12}(5^3)$ Orthogonal arrays.

Exp.	1	2	3
1	1	1	1
2	1	2	2
3	1	3	3
4	2	1	2
5	2	2	3
6	3	1	3
7	3	4	1
8	4	3	1
9	4	4	2
10	5	2	1
11	5	3	2
12	5	4	3

3.3 Bee colony optimization with Taguchi method

Bee colony optimization with Taguchi method for dynamic economic dispatch approach proceeds as shown in Figure 2.

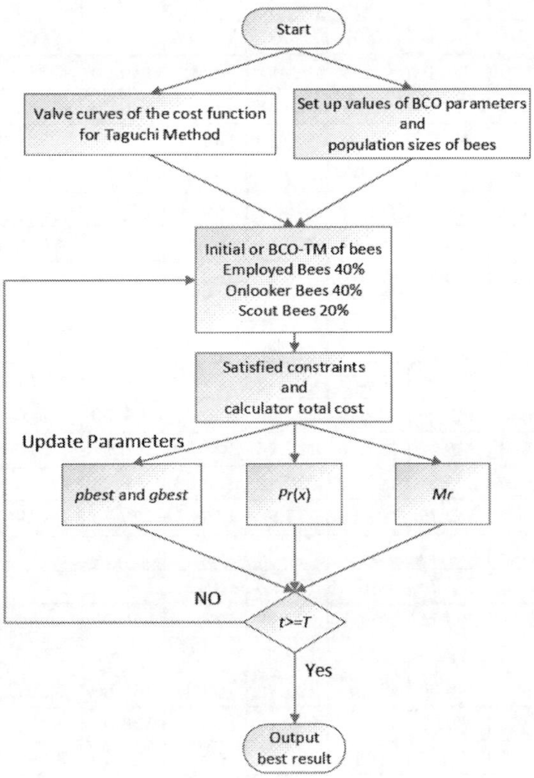

Fig. 2. The flowchart of TM-BCO.

4 Simulation results

This case consists of 10 generator units considering the valve-point effects for dynamic economic dispatch. Exactly the same data of all units as given in [8] will be used in this example and the load curve of 24 hours in one day as shown in figure 3. The algorithms were implemented in the programming language Matlab 2014b with Intel Core i7 3.2GHz computer was used. Table 1 shows the results of TM-BCO programming during a total scheduling of 24 hours. The TM-BCO simulation result is given in Table 2. From Tables 2 and 3 show that the proposed algorithm did have a more robust structure and a better capability to avoid local optimum and explore new solution territories.

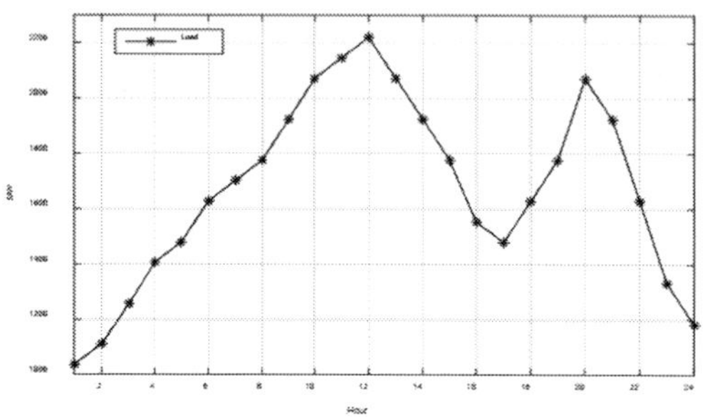

Fig. 3. The load of 24 hours in one day.

Table 2. TM-BCO simulation result for dynamic economic dispatch.

hour	Unit										Load
	1	2	3	4	5	6	7	8	9	10	
1	150.00	305.01	73.00	60.00	73.00	124.16	128.19	47.65	20.00	55.00	1036.00
2	230.00	309.39	73.00	61.95	73.00	74.37	130.00	77.65	25.64	55.00	1110.00
3	299.31	303.76	73.00	60.00	120.88	117.64	130.00	77.40	21.01	55.00	1258.00
4	379.31	298.98	86.06	60.00	170.88	160.00	128.37	47.40	20.00	55.00	1406.00
5	299.31	323.49	166.06	60.00	220.88	132.67	130.00	72.46	20.14	55.00	1480.00
6	379.31	385.95	220.15	60.00	170.88	160.00	128.64	47.28	20.79	55.00	1628.00
7	394.44	305.95	300.15	68.40	220.71	160.00	130.00	47.00	20.35	55.00	1702.00
8	385.05	385.95	339.64	68.67	170.71	160.00	130.00	60.98	20.00	55.00	1776.00
9	465.05	460.00	303.61	110.77	217.06	115.51	130.00	47.00	20.00	55.00	1924.00
10	470.00	460.00	312.91	160.02	243.00	144.07	130.00	77.00	20.00	55.00	2072.00
11	470.00	460.00	340.00	202.32	222.75	160.00	130.00	85.51	20.41	55.00	2146.00
12	470.00	460.00	340.00	242.10	227.98	159.41	130.00	115.51	20.00	55.00	2220.00
13	456.52	460.00	340.00	192.10	177.98	153.80	130.00	86.19	20.41	55.00	2072.00
14	376.52	395.64	295.09	142.10	227.98	160.00	130.00	116.19	25.48	55.00	1924.00
15	405.55	315.64	339.79	123.80	177.98	116.82	129.76	86.19	25.46	55.00	1776.00
16	328.79	235.64	293.48	118.70	127.98	160.00	130.00	84.41	20.00	55.00	1554.00
17	302.06	309.03	213.76	68.70	177.98	148.21	130.00	54.87	20.40	55.00	1480.00
18	377.50	389.03	293.76	60.00	127.98	127.73	130.00	47.00	20.00	55.00	1628.00
19	457.50	393.76	269.95	60.00	177.98	153.79	130.00	58.02	20.00	55.00	1776.00
20	470.00	460.00	340.00	110.00	227.98	160.00	130.00	88.02	31.00	55.00	2072.00
21	464.67	396.27	340.00	123.42	177.98	158.63	130.00	58.02	20.00	55.00	1924.00
22	384.67	316.27	314.05	73.42	127.98	160.00	129.60	47.00	20.00	55.00	1628.00
23	304.67	236.27	240.96	60.62	78.46	159.02	130.00	47.00	20.00	55.00	1332.00
24	225.21	225.12	177.64	60.00	73.74	160.00	130.00	47.19	30.10	55.00	1184.00

Table 3. Comparison results with methods.

Methods	Maximal Converged Cost (NT$)	Minimal Converged Cost (NT$)	Average Converged Cost (NT$)	Average number of generations to converge	Average Execution time (sec.)
PSO	1053155.41	1036159.77	1048154.39	183	1.11
BCO	1048365.21	1035777.05	1043381.41	191	1.27
TM-PSO	1042451.73	1032186.89	1036819.28	175	1.67
TM-BCO	1036487.25	1031492.40	1033374.74	181	1.72

Table 3 shows the maximal, minimal and average converged cost, average number of generations to converge and average execution time achieved by various methods over 100 runs. This paper uses TM-BCO to calculate the minimal converged cost is 1031492.40(NT$) and average execution time is 1.72(sec.). Table 4 shows the PSO, BCO, TM-PSO and TM-BCO of 24 hours minimal cost for dynamic economic dispatch. The convergence characteristics for PSO, BCO, TM-PSO and TM-BCO as shown in Figure 4.

Table 4. Simulation methods.

hour	Cost (NT$)			
	PSO	BCO	TM-PSO	TM-BCO
1	29114.02	28407.65	28346.71	28366.00
2	30633.91	30527.17	30626.08	30733.43
3	33808.95	33600.48	33468.29	33650.64
4	37110.60	37378.18	37021.51	36840.56
5	38299.45	38553.47	38818.59	38805.25
6	41813.25	41583.53	42197.28	41703.79
7	43333.25	43536.87	43633.23	43225.63
8	45074.65	45238.67	44650.71	45374.64
9	48865.07	48719.45	48384.47	48474.59
10	52392.50	52590.43	52455.52	52583.82
11	54428.73	54539.42	54312.35	53970.65
12	56369.91	55847.45	56046.31	55817.13
13	52326.79	52379.08	52509.16	52137.67
14	48977.08	49014.90	48628.64	48634.18
15	45977.13	45608.79	45598.94	45569.79
16	41200.17	40462.33	40469.53	40537.55
17	38765.64	38728.77	38838.81	38663.78
18	42413.83	41593.75	41980.78	41281.98
19	44949.14	45443.06	45004.72	45029.77
20	52378.54	52535.34	52130.21	52472.83
21	48520.42	48807.13	48685.71	48514.70
22	41846.32	42082.17	41536.21	41714.50
23	35504.70	35902.82	34976.91	35437.72
24	32055.71	32696.14	31866.21	31951.79
Total	1036159.77	1035777.05	1032186.89	1031492.40

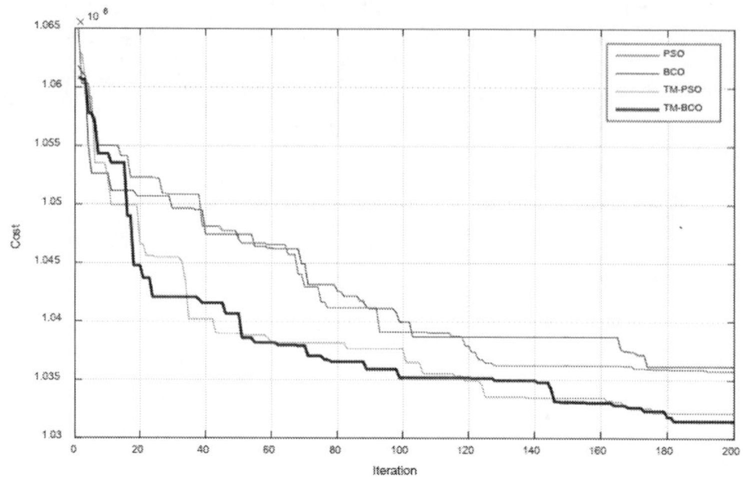

Fig. 4. The convergence characteristics for method.

5 Conclusions

The Bee Colony Optimization with Taguchi Method is capable of solving the constrained dynamic economic dispatch problem. The TM-BCO was incorporated the orthogonal arrays to select a better unit factor, consequently enhancing BCO. The TM-BCO can provide a powerful global exploration capability of IEEE 10 units. From test result proved the effectiveness of the proposed algorithm in solving dynamic economic dispatch problems under a reasonable execution time. The TM-BCO has great potential for further applications for solving the economic dispatch problems in real-world applications.

References

1. Pengpeng Yang, Chunping Zhu, Long Zhao, Mingqiang Wang, Xingyao Ning, Yuan Liu, 2016 IEEE Region 10 Conference, (2016)
2. Ritesh Jha, Nand K. Meena, Anil Swarnkar, Nikhil Gupta, K. R. Niazi, 2015 Annual IEEE India Conference, (2015)
3. Taher Niknam, Farank Golestane, Bahman Bahmanifirouzi, 2011 IEEE Power Engineering and Automation Conference, **1** (2011)
4. A. Bakirtzis, V. Petridis, S. Kazarlis, IEE Proceedings-Generation, Transmission and Distribution, **141** (1994)
5. D. Karaboga, B. Akay, Journal of Applied Mathematic Computation 2009, (2014)
6. Ying Cai, Derong Liu, Proceedings of the 44th IEEE Conference on Decision and Control, (2005)
7. Derong Liu, Ying Cai, IEEE Transactions on Power Systems, **4** (2005)
8. Yan Chen, Jinyu Wen, Lin Jiang, Shijie Cheng, IET Generation, Transmission & Distribution, **7** (2013)

The implementation of an interactive gaming machine of “Mafia Wars”

Jsung-Ta Tsai[1], *Yen-Ming* Tseng[1,*], and *Andrian Muzakki* Firmansyah[2]

[1]College of Intelligence Robot, Fuzhou Polytechnic, Fuzhou, Fujian, China
[2]Department of Digital Entertainment and Game Design, Taiwan Shoufu University, Taiwan

Abstract. This article describes the implementation of a gaming machine of “Mafia Wars” using Arduino UNO board and some other electrical components for the construction. The gaming machine is designed by the following items, such as game story setting, interactive elements for the game play, game art, state machine for the game process, coding, electrical components, and hardware construction, etc. There are INTRO, ROULETTE, SLAP, WAR and End Round states in this game. Some important interactive mechanisms with Arduino board for the gaming machine are designed and illustrated in this study. The hardware of the gaming machine consists of an Arduino UNO, a LCD screen, a photo resistor, several resistors, LEDs, buttons, plastic coins, game art decoration and the wood shielding box. Finally, the resultant functions of the gaming machine are tested by playing and the testing results demonstrate the good working performance.

1 Introduction

“Mafia Wars” is a multiplayer social network game developed by Zynga company. “Mafia Wars” worked on Facebook and it was a popular game in which players start a mob family with their friends and run crime actions. The developers often added new weapons and features for the game to successfully aggregate more players. Aki Järvinen [1] has ever invastigated the game design of “Mafia Wars” game for its good social interactive properties and Samuel Greengard [2] stduied the influence of the popularity of virtual goods and currencies to the behavior of the players. Three years later, the sequel “Mafia Wars 2” was released in October 2011 but shut down in December 2012 by sequence. However, “Mafia Wars” has become a historic game. It is significant to develop a similar game machine with different interactive ways to play. In this study, an interactive game machine is designed and implemented. The name of the game machine is also called “Mafia Wars”, because a similar game story to the former is adopted. The interactive ways of playing “Mafia Wars” gaming machine are developed for two players to fight with each other at the same time.

* Corresponding author: swk1200@qq.com

2 Game design

2.1 Concept of gaming machine (Gameplay) [3]

1. The game story is based on the "Mafia Wars" for players to act as a gangster leader.
2. Goal: Keep alive and kill all the opposite members.
3. Rules: A game round proceeds by "Roulette", "Slap" and "War" state cycle. At the beginning, "Roulette" state randomly produce a bingo condition for players to react, then "Slap" deals with the response from players to determin a winner, who can get a gun or cannon. Finally, the winner in "War" state has ability to shoot up, the defender should dodge well to avoid his gang member killed.
4. Victory: When a player has finally no gang member alive, the game ends and he lose the game. The other one with gang members alive will be the winner.
5. The roulette panel is formed by 6 LEDs arranged in a circle, it indicates the bingo conditions.
6. The gaming machine consists of Arduino UNO, sensors, elecrical components, LCD display, roulette, push buttons and a buzzer to interact with players.
7. The LCD display would show the corresponding information for each state.The gaming machine is constructed by a handmade wood box.
8. The apperance of 3D design model for the gaming machine is shown as follows (Fig.1):

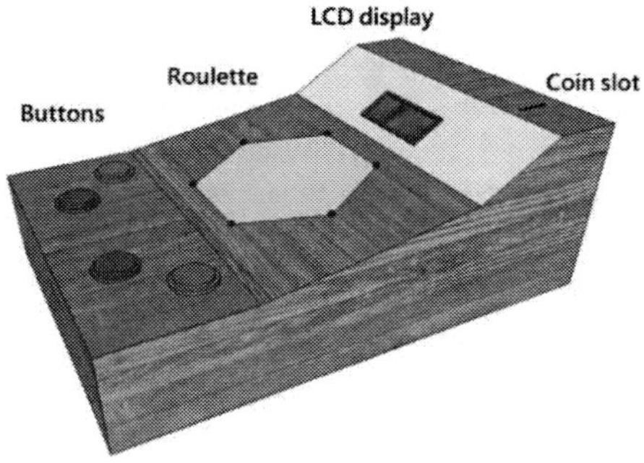

Fig. 1. The 3D design model of "Mafia Wars" game machine.

2.2 Game story

There was a mafia gang living in Madou town for many years. One day there came another mafia gang from a neighboring city. They often fight hardly when they encountered in the street. After some negotiations failure, the mafia gang bosses declared war to defeat the other gang. Only the winner can rule the town. Now you become the leader of a mafia gang and get 5 elite members. Your job is to eliminate other mafia gang by killing all of their elite gang members. But don't forget to escape from Town Sheriff (Police Car), and protect your member from Mad Bull.

2.3 Game Art

The characters and tools of Town Sheriff, Police Car, Mad Bull, Gun and Cannon are designed and sketched in Table 1.

Table 1. Game art of "Mafia Wars" gaming machine.

Town Sheriff	Police Car	Mad Bull	Gun	Cannon

2.4 Mechanism of "Mafia Wars" gaming machine

2.4.1 Flow chart of gaming mechanism

A simple flow chart design for "Mafia Wars" gaming mechanism is illustrated as shown in Fig. 2. The game begins with Intro (intoduction), Roulette, Slap, War to End Round.

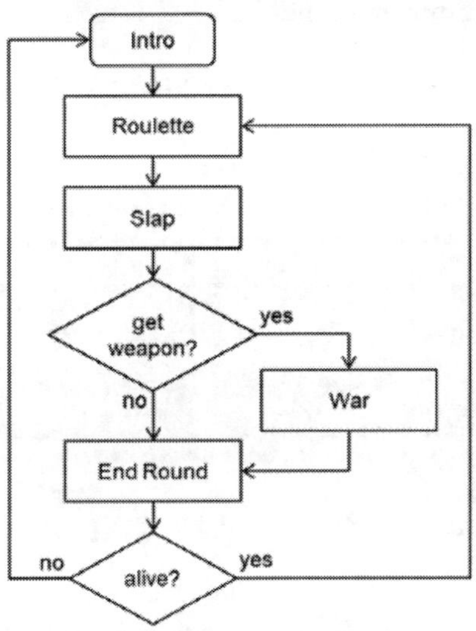

Fig. 2. Flow chart design for "Mafia Wars" gaming mechanism.

2.4.2 Finite-state machine of gaming mechanism

The above flow chart is then transformed to finite-state machine (FSM)[4] diagram for further coding as shown in Fig. 3. Intro, Rolette, Slap, War and End Round are recognized as states in the FSM diagram, in which the transition conditions are also clarified.

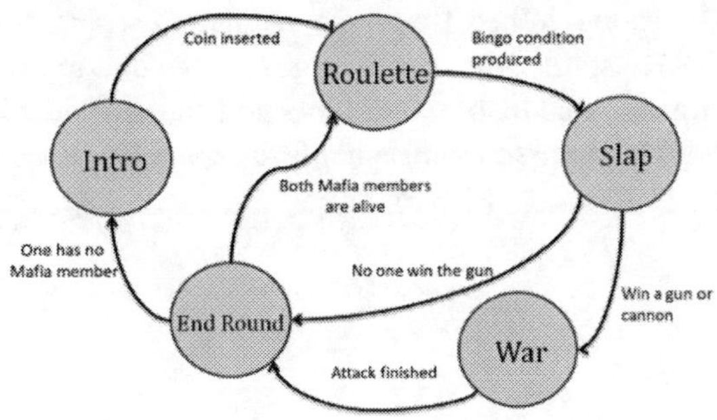

Fig. 3. Finite-state machine diagram for "Mafia Wars" gaming mechanism.

1. "Intro" state:
   - Show title screen
   - Coin LED indicator turns on
   - Transition condition: When a coin is inserted, the state changes to "Roulette" state.
   - Display and music: The display content and music for "Intro" state is shown in Fig. 4.

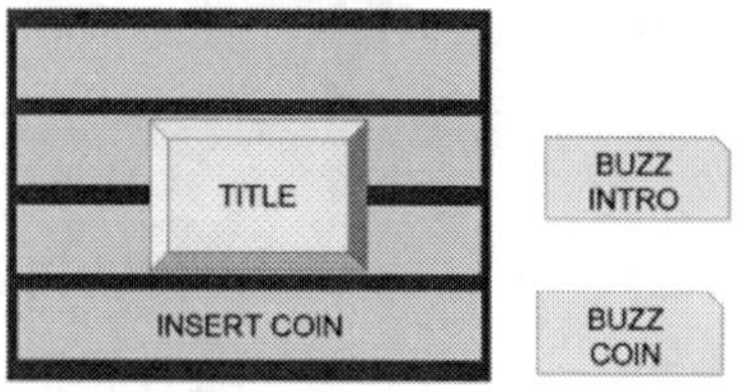

Fig. 4. Display and music contents for "Intro" state.

2. "Roulette" state:
   - A game round starts
   - Flashing player current information
   - Roulette LEDs start to blink.
   - Transition condition: When Roulette Machine has randomly created a "bingo", the game changes to "Slap" state.
   - Display and music: The display content and music for "Roulette" state are shown in Figure 5. BUZZ ROULETTE music maintains during the state.

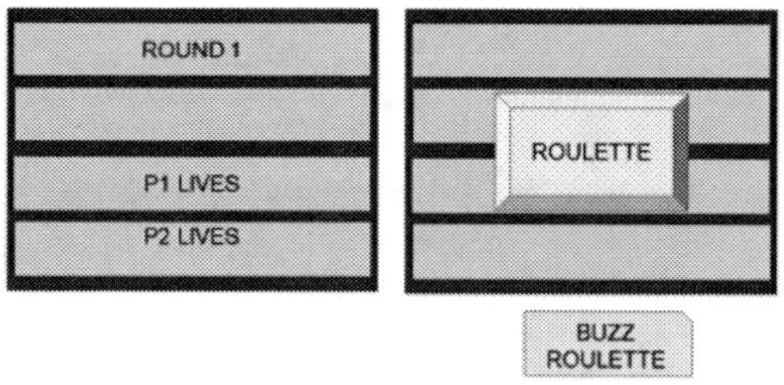

Fig. 5. Display and music contents for "Roulette" state.

3. "Slap" state:
   - The players must react correctly and quickly to win the gun. Fig. 6 illustrates the "Slap" state bingo conditions for players to react.
   - The "Slap" state bingo conditions and the required actions are defined in Fig. 7.
   - Game controller response definition for "Slap" state is shown in Fig. 8.
   - The game will deal with the response from the players and then pick out the correct and the fastest one as the winner. If it is a draw, the game will proceed with a balancing algorithm to pick out a player as the winner.
   - The player who win the gun or cannon will be marked as the "ATTACKER"
   - Transition conditions: When there is a winner (got a gun or cannon), "Slap" state will change to "WAR" state, otherwise the game will change to "End Round" state.
   - Display and music: The display contents and music for "Slap"state are shown in Fig. 9. BUZZ RESULT music is played when the competition results in a winner.

Fig. 6. "Slap" state bingo conditions for players to react.

Fig. 7. Bingo conditions of "Slap" state and the corresponding required responses.

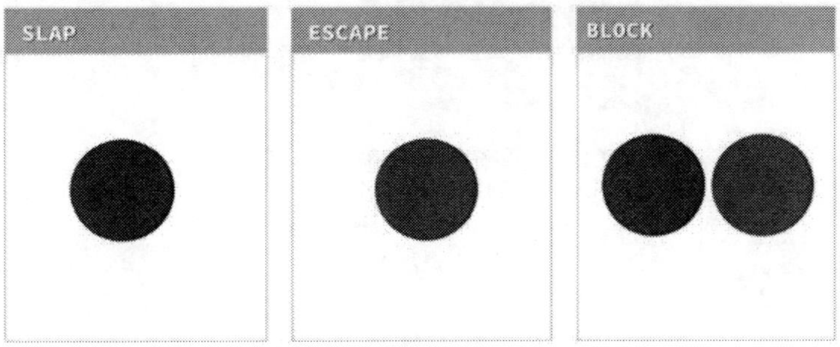

Fig. 8. Game controller response definition for "Slap" state.

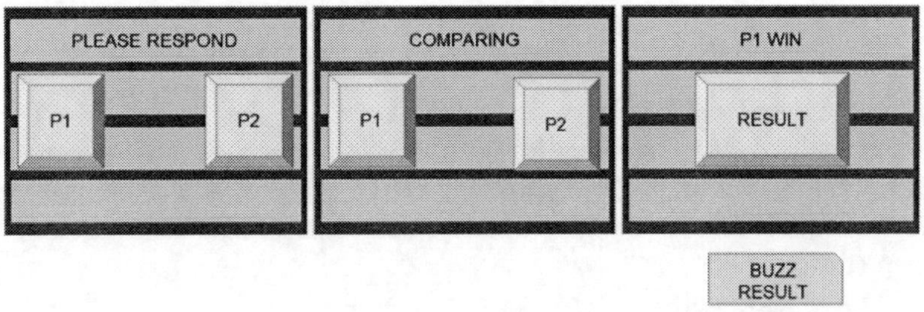

Fig. 9. Display contents of three phases and BUZZ RESULT music setting for "Slap" state.

4. "War" state:
   - The Attacker will choose which side he will attack depending on what weapon he carries.

- Gun can only attack LEFT or RIGHT.
- Cannon can attack LEFT, RIGHT or CENTER.
- The other player must guess well to dodge the attack.
- Transition condition: If the game goes to ATTACK SUCCESS phase, it will change to End Round state.
- Display and music: There are three phases of display content for "War" state as shown in Fig. 10. BUZZ RESULT music is played for "ATTACK SUCCESS" phase.

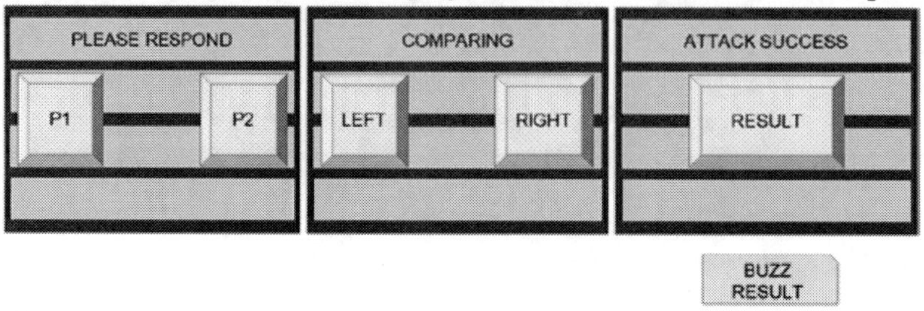

Fig. 10. Display contents of three phases and BUZZ RESULT music setting for "War" state.

5. "End Round" state:
   - Checking the situation to decide whether the game should create the next round or not.
   - Round will always restart as long as both of the players have mafia member alive.
   - Transition conditions: When both of the players have any mafia member alive, the state will change to "Roulette" state, otherwise, it change to "Intro" state.
   - Display and music: The LCD screen would show "Game Over" for a while as illustrated in Fig. 11 when the game goes to end. In the meanwhile, BUZZ GAME OVER music is played.

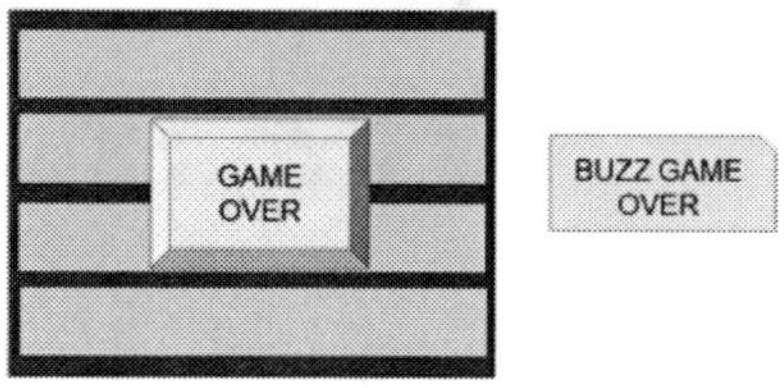

Fig. 11. Display contents of "End Round" state and BUZZ GAME OVER music setting.

3 Implementation

3.1 Bill of material

- Arduino UNO [5, 6]
- 7* LEDS
- 1* LDR Resistor
- 8*100Ω, 6*10kΩ and 2*30KΩ Resistor
- 1* Laser KY008 Module x Laser Detector [7]
- 1* Piezo Buzzer
- 4* 35mm Buttons
- 20x4 LCD I2C Module[8]
- Handmade wood box
- Toy coins (plastic coins)

3.2 Mechanic with Arduino UNO

The Input handler subroutine, for example, is developed with analog1 for player 1 and analog2 for player 2, respectively. The simulation circuit layout and code is illustrated in Fig. 12.

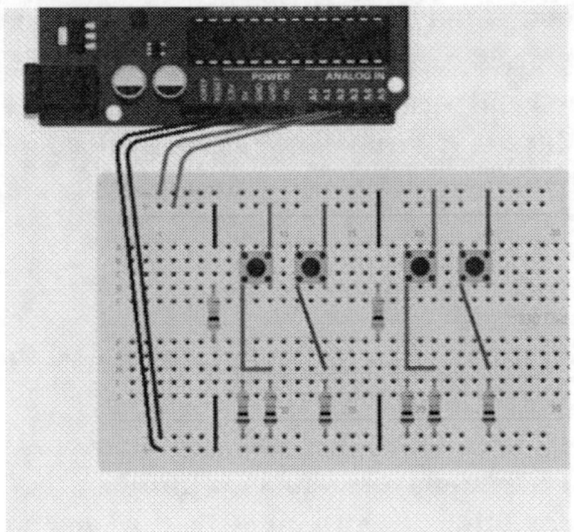

some part of the code for Input handler to judge which button pushed

```
if (readPin > 1000)
        button = NONE;
else if (readPin > 210 && readPin < 250)
        button = RED; // red button pressed
else if (readPin > 120 && readPin < 160)
        button = BLUE; // blue button pressed
else if (readPin > 80 && readPin < 110)
        button = BOTH; // both buttons pressed
```

Fig. 12. Design circuit simulation layout and code for input handler.

3.3 Construction of gaming machine

The gaming machine is constructed by the hardware components within a handmade wood box. The bottom view photograph as shown in Fig. 13 reveals the Arduino UNO, 20x4 LCD I2C module, laser KY008 module x laser detector, electrical continuity, toy coin and the toy coin track, et al. The resultant gaming machine is shown in Fig. 14.

Fig. 13. Bottom view reveals the hardware components and the construction.

Fig. 14. Photograph of the "Mafia Wars" gaming machine.

4 Play testing and discussions

4.1 Play testing

The resultant gaming machine is tested by actual playing. The test result demonstrates the desired functions work very well. The properties of playability are always considered as affected by the different elements of video game architecture [9, 10]. In this study, the intrinsic playability and mechanical playability of "Mafia Wars" gaming machine seem to be evaluated a high mark, however, interactive playability with common value, and artistic playability with low grade.

4.2 Discussions

4.2.1 Discussion about interactive playability

The interactive playability design in the "Mafia Wars" gaming machine contains three stages from Roulette, Slap to War in a round. The players reflect a problem that the LCD screen is too small to recognize the bingo condition. It is thought that a lager LCD panel would provide a wide vision for recognizing.

4.2.1 Discussion about artistic playability

It is well known that the game art setting does influence the playability, which is called artistic playability. The players reflect the poor artistic features about the handmade box. It is thought that a delicate cover paste could beautify the appearance of the gaming machine. 3D printing would be another way to obtain a good-looking appearance for the gaming machine.

5 Conclusions

The interactive gaming machine of "Mafia Wars" is designed and implemented. It provides a novel interactive method to players. A state diagram with FSM is helpful to coding development. The playing test result shows the functions woks well and the gameplay is benefit to all ages. But, it also shows the drawbacks in interactive playability and artistic playability. It is thought that a larger LCD screen, delicate cover and 3D printing could improve effectively.

References

1. A. Järvinen, *Proceedings of the 2009 ACM SIGGRAPH Symposium on Video Games*, pp.95-102 (2009)
2. G. Samuel, *Communications of the ACM*, **54**, 4 (2011)
3. K. Salen, E. Zimmerman, *Rules of Play: Game Design Fundamentals*, Cambridge, Massachusetts: The MIT Press (2004)
4. F. Wagner et al., *Modeling software with finite state machines: a practical approach*, Auerbach Publications (2006)
5. J. Blum, *Exploring Arduino: Tools and Techniques for Engineering Wizardry*, Indianapolis, IN: Wiley (2013)
6. S. Monk, *Programming Arduino: Getting Started with Sketches, Second Edition (Tab) 2nd Edition*, McGraw-Hill Education TAB (2016)

7. S. Bambey, E. Lim, C.J Kai, P. Taheri, *2016 IEEE 7th Annual Information Technology, Electronics and Mobile Communication Conference (IEMCON)*, 1-5 (2016)
8. G. Scott, J. Chin, *Computer Science and Electronic Engineering Conference (CEEC)*, 57-60 (2013)
9. J.L.G. Sánchez, N.P. Zea, F.L. Gutiérrez, *Human-Computer Interaction - INTERACT 2009*, Part I, LNCS **5726**, 356-359 (2009)
10. A. Rollings, D. Morris, *Game Architecture and Design: A New Edition*, New Riders (2003)

MATEC Web of Conferences **185**, 00035 (2018) https://doi.org/10.1051/matecconf/201818500035
ICPMMT 2018

On the design and prototype manufacturing of multi-speed transmission device of bicycle

Tzu-Hsia Chen [1] and *Long-Chang* Hsieh [2,*]

[1]Department of Mechanical Engineering, Minghsin University of Science and Technology, ROC.
[2]Department of Power Mechanical Engineering, National Formosa University, R.O.C.

Abstract. The transmission systems of bicycles are multi-speed systems to satisfy different kinematic requirements. The transmission system of a bicycle can be the external changing-speed system or internal multi-speed system. A complete multi-speed transmission device of a bicycle must contain multi-speed gear unit, control device, and operating device. This paper focused on the systematic design and prototype manufacturing of multi-speed transmission device of a bicycle. Based on the results of kinematic design and meshing efficiency analysis, the systematic design of six-speed transmission device was carried out and prototype was manufactured to verify the design theorem.

1 Introduction

Because of the advantages of light weight, cheap, and pollution-free, bicycles become the more and more popular. They can be used as exercising machines and traffic vehicles. A complete multi-speed transmission device of a bicycle, shown in figure 1, must contain multi-speed gear unit, control device, and operating device.

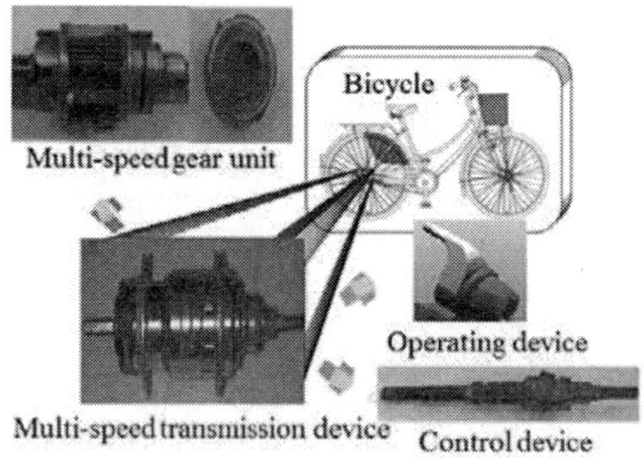

Fig. 1. Multi-speed transmission device of a bicycle.

Planetary gear trains are commonly used in various transmissions due to the reason of compact size, light weight, and multi-degrees of freedom. One of its applications is used as

*Corresponding author: lochsieh@nfu.edu.tw

multi-speed transmission device of the bicycle. This paper focused on the systematic design and prototype manufacturing of multi-speed transmission device of a bicycle. The research subjects of planetary gear trains can be kinematic analysis, kinematic design, and meshing efficiency analysis. Traditionally, the kinematic analysis of planetary gear trains can be accomplished by tabular method or vector method. However, the procedures of such methods are complicated and tedious, often induce human errors. Fundamental circuit method [1-2] is a logical method for the analysis the angular velocities of all links of planetary gear trains. However, train value method [3-4] is the simplest one that is used for the kinematic analysis of planetary gear trains. According to the kinematic analysis, many studies focused on the kinematic design [5-8] of planetary gear trains for specified engineering purpose. The meshing efficiency of planetary gear train is an important factor of planetary gear train. There are some studies focused on the meshing efficiency analysis of planetary gear train. Based on the latent power method [9-14], the meshing efficiency of planetary gear train can be analyzed.

By referring to the train value equation of planetary gear train, the kinematic design algorithm of the multi-speed planetary gear unit was proposed. Based on the latent power theorem, the meshing efficiency equation of planetary gear unit was derived. Then, according to the meshing efficiency of ordinary gear pairs [15-17], the meshing efficiency of six-speed planetary-gear unit was analyzed. Finally, the engineering design of six-speed gear unit (including control device) was accomplished and the corresponding prototype was manufactured to verify the design theorem. In this paper, the control device and operating device were also designed.

2 Train value equation

For a planetary gear train, shown in figure 2, let us denote the first sun gear as i, the last sun gear as j, and the carrier (arm) as k, respectively, the train circuit can be denoted as (i, j; k). The relationship among ω_i, ω_j, and ω_k can be expressed as:

$$\omega_i - \xi_{ji}\omega_j + (\xi_{ji} - 1)\omega_k = 0 \tag{1}$$

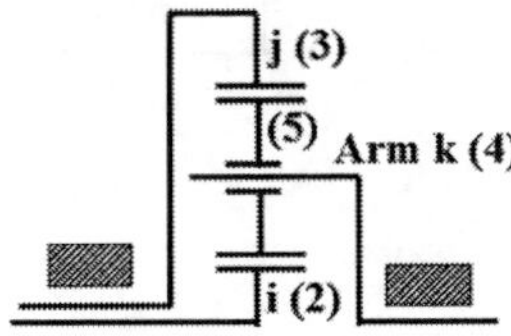

Fig. 2. Planetary gear train.

where ξ_{ji} is the train value of sun gear j to sun gear i. The sum of the coefficients of equation (1) is equal to 0, i.e., $1-\xi_{ji}+(\xi_{ji}-1)=0$. For a planetary gear train with one train circuit, there are three members (i, j, and k) can be adjacent to the input shaft, output shaft, and frame.

[Case I]: If carrier k is adjacent to input shaft (ω_k=ω_{in}), and sun gear j is adjacent to the output shaft (ω_j=ω_{out}), according to equation (1), its reduction ratio (R_r) can be written as:

$$R_r = \frac{\omega_{in}}{\omega_{out}} = \frac{\omega_k}{\omega_j} = -\frac{-\xi_{ji}}{(\xi_{ji}-1)} = \frac{\xi_{ji}}{(\xi_{ji}-1)} \tag{2}$$

For a planetary gear train with ξ_{ji}<0, the reduction ratio 1>R_r>0.

[Case II]: If sun gear j is adjacent to input shaft ($\omega_j=\omega_{in}$), and carrier k is adjacent to the output shaft ($\omega_k=\omega_{out}$), according to equation (1), its reduction ratio (R_r) can be written as:

$$R_r=\frac{\omega_{in}}{\omega_{out}}=\frac{\omega_j}{\omega_k}=-\frac{(\xi_{ji}-1)}{-\xi_{ji}}=\frac{(\xi_{ji}-1)}{\xi_{ji}} \tag{3}$$

For a planetary gear train with $\xi_{ji}>1$, the reduction ratio $1>R_r>0$.

3 Existing design

There are many patents [18-20] of multi-speed gear hubs for bicycles. Figure 3 shows a 5-speed planetary-gear unit which is designed by Bergles in 1987 [18]. Table 1 shows its corresponding clutch sequence. Based on section 2, the corresponding reduction ratios of this 5-speed planetary-gear unit are also shown in table 1.

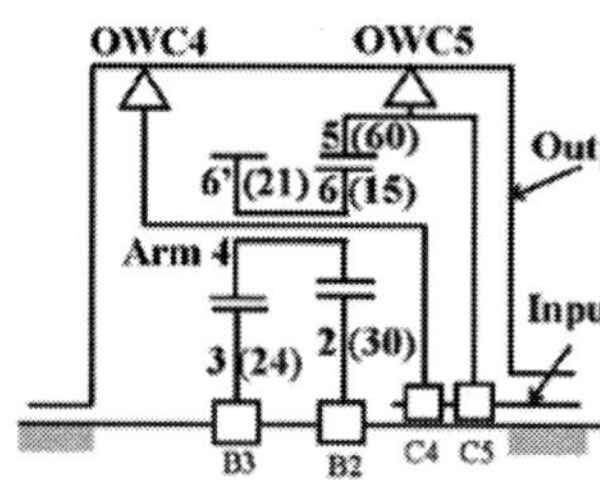

Fig. 3. Kinematic skeleton of five-speed gear unit [18].

Table 1. Clutch sequence of 5-speed planetary-gear unit.

Speed	C4	C5	B2	B3	OWC4	OWC5	*Rr*	Spacing of *Rr*
1		※	※		※		1.5	
2		※		※	※		1.2875	1.1651
3		※		※		※	1	1.2875
4	※			※		※	0.7778	1.1651
5	※		※			※	0.6667	1.2875

※: Engaged

4 Kinematic design

For a 6-speed planetary-gear unit to have the reduction ratios from 1.0 to 0.328 and the spacing of reduction ratio is constant. Table 2 shows the reduction ratio requirements of this 6-speed gear unit.

Table 2. The requirements of the 6-speed gear hub.

Speed	1	2	3	4	5	6
Rr	1	0.8	0.64	0.512	0.410	0.328
Spacing of *Rr*		1.25	1.25	1.25	1.25	1.25

4.1 Design concept

First, the design concept, shown in figure 4, is proposed for the kinematic design of 6-speed planetary-gear unit. According to figure 4, the 6-speed planetary-gear unit is coupled by two planetary gear trains (I and II) in which PGT I has two independent train values ξ_{aI}

https://doi.org/10.1051/matecconf/201818500035

and $\xi_{\alpha 2}$ ($\xi_{\alpha 1}$<0 and $\xi_{\alpha 2}$<0) and PGT II has one independent train value ξ_β (ξ_β>1). Figure 5 shows the corresponding kinematic skeleton. The reduction ratio of the planetary-gear unit is $R_r[\text{I}]\times R_r[\text{II}]$.

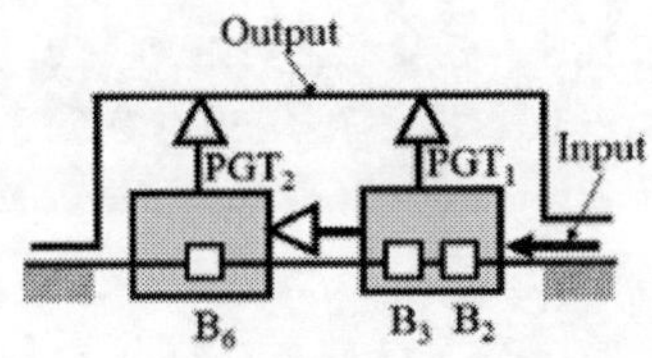

Fig. 4. Design concept.

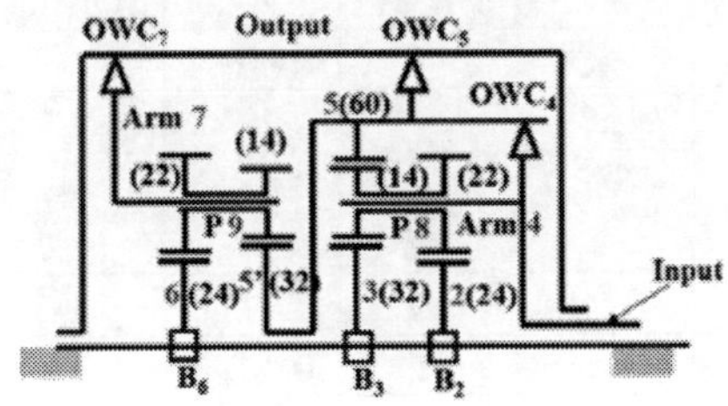

Fig. 5. Kinematic skeleton.

4.2 One way clutch and brake

Generally speaking, the multi-speed planetary-gear unit of bicycle includes the shifting control device (Blakes), planetary gear train, and one way clutches. One-way clutch, shown in figure 6, consists of the outer ring, rollers, inner ring, and spring (not shown). For the structure shown in figure 6(a), the rotating speed of outer ring is larger than the rotating speed of the inner ring, it is in the state of separating and represented as figure 6(c). For the structure shown in figure 6(b), the rotating speed of inner ring is larger than the rotating speed of the outer ring, it is in the state of engaging and represented as figure 6(d).

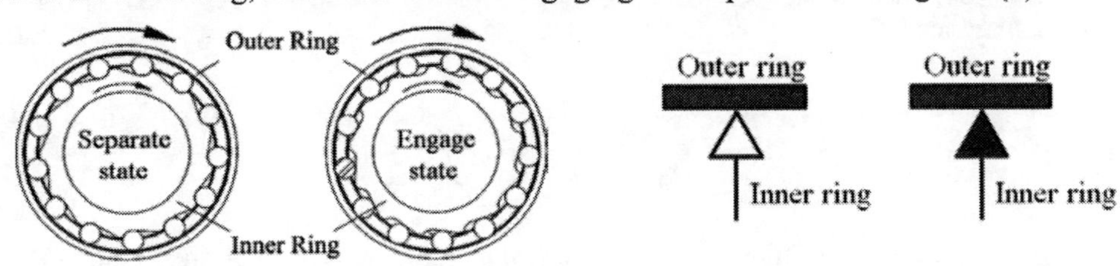

(a)Separating structure (b) engaging structure (c) separating (d) engaging

Fig. 6. One-way clutch.

4.3 Kinematic design of 6-speed planetary-gear unit

Based on the equations. (2)-(3), the reduction ratios of the 6 speeds are expressed as:

$$R_{r1} = 1.0 \tag{4}$$

$$R_{r2} = \xi_{\alpha 1}/(\xi_{\alpha 1} - 1) = 0.8 \tag{5}$$

$$R_{r3} = \xi_{\alpha 2}/(\xi_{\alpha 2} - 1) = 0.64 \tag{6}$$

$$R_{r4} = (\xi_\beta - 1)/\xi_\beta = 0.512 \tag{7}$$

$$R_{r5} = [\xi_{\alpha 1}/(\xi_{\alpha 1} - 1)]\times[(\xi_\beta - 1)/\xi_\beta] = 0.41 \tag{8}$$

$$R_{r6} = [\xi_{\alpha 2}/(\xi_{\alpha 2} - 1)]\times[(\xi_\beta - 1)/\xi_\beta] = 0.328 \tag{9}$$

According to equations. (4)-(9), if $\xi_{\alpha 1}$=-4.026, $\xi_{\alpha 2}$=-1.914, and ξ_{β}=2.103, the reduction ratios are 1, 0.801, 0.657, 0.524, 0.420, and 0.344. For the planetary-gear unit shown in figure 5, if Z_2=24, Z_3=32, Z_5=60, $Z_{5'}$=32, Z_6=24, Z_{8a}=22, Z_{8b}=14, Z_{9a}=22, and Z_{9b}=22, then the train values $\xi_{\alpha 1}=\xi_{52}$=-3.9286, $\xi_{\alpha 2}=\xi_{53}$=-1.875, and $\xi_{\beta}=\xi_{56}$=2.0952. And, the corresponding reduction ratios are 1.0, 0.797, 0.652, 0.523, 0.417, and 0.341. The clutch sequence, reduction ratios, and spacing of reduction ratios are shown in table 3.

Table 3. Clutch sequence, reduction ratios, and meshing efficiency of 6-speed planetary gear unit.

Speed	B2	B3	B6	OWC4	OWC5	OWC7	*Rr*	Spacing of *Rr*	Meshing Efficiency
1				※	※		1.0		100%
2	※				※		0.797	1.255	99.84%
3		※			※		0.652	1.222	99.72%
4			※	※		※	0.523	1.248	98.89%
5	※		※			※	0.417	1.255	98.73%
6		※	※			※	0.341	1.222	98.61%

※: Engaged

5 Latent power theorem and meshing efficiency

For a planetary gear train with sun gear *i*, ring gear *j*, and carrier *k*, let T_i, T_j, and T_k be the torques of members *i*, *j*, and *k*, respectively. Since, sun gear *i*, ring gear *j*, and carrier *k* rotate about the same axis, the torque balance equation can be expressed as:

$$T_i + T_j + T_k = 0 \qquad (10)$$

Furthermore, the latent powers of sun gear *i* and sun gear *j* relative to carrier *k* (P_i^k and P_j^k) are defined as:

$$P_i^k = T_i \times (\omega_i - \omega_k) \qquad (11)$$

$$P_j^k = T_j \times (\omega_j - \omega_k) \qquad (12)$$

Let η_{ij}^k (η_{ji}^k) be the meshing efficiency of planetary gear train if the latent power is transmitted from sun gear *i* (sun gear *j*) to sun gear *j* (sun gear *i*) when carrier *k* is relatively fixed. Based on the concept of latent power [9-14], the relationship among P_i^k and P_j^k can be expressed as:

$$-P_j^k = P_i^k \times \eta_{ij}^k \quad for\ P_i^k \geq 0\ and\ P_j^k \leq 0 \qquad (12)$$

$$-P_i^k = P_j^k \times \eta_{ji}^k \quad for\ P_i^k \leq 0\ and\ P_j^k \geq 0 \qquad (13)$$

The meshing efficiency of a planetary gear train is defined as the ratio of output power to input power, i.e.,

$$\eta_{m(PGT)} = -\frac{P_{out}}{P_{in}} \qquad (14)$$

According to equations. (10)-(14), the meshing efficiencies of case I and case II for a planetary gear train can be expressed as:

$$\eta_{m(PGT)}=\frac{\eta_{ij}^{k}(1-\xi_{ji})}{(1-\xi_{ji}\eta_{ij}^{k})}\quad \text{for } \xi_{ji}<0,\ \omega_k=\omega_{in},\ \omega_i=0,\ \text{and } \omega_j=\omega_{out} \tag{15}$$

$$\eta_{m(PGT)}=\frac{(\xi_{ji}\eta_{ij}^{k}-1)}{\eta_{ij}^{k}(\xi_{ji}-1)}\quad \text{for } \xi_{ji}>1,\ \omega_j=\omega_{in},\ \omega_i=0,\ \text{and } \omega_k=\omega_{out} \tag{16}$$

where η_{ij}^{k} (η_{ji}^{k}) is the meshing efficiency of planetary gear train if the latent power is transmitted from sun gear *i* (sun gear *j*) to sun gear *j* (sun gear *i*) when carrier *k* is relatively fixed. The meshing efficiency of ordinary external gear pair ($\eta_{m(ex)}$) and internal gear pair ($\eta_{m(in)}$) can be expressed as [15-17]:

$$\eta_{m(ex)}=1-\frac{1+1/R_r}{\beta_{A1}+\beta_{R1}}\left[\frac{f_a}{2}\beta_{A1}^2+\frac{f_r}{2}\beta_{R1}^2\right] \tag{17}$$

$$\eta_{m(in)}=1-\frac{1-1/R_r}{\beta_{A1}+\beta_{R1}}\left[\frac{f_a}{2}\beta_{A1}^2+\frac{f_r}{2}\beta_{R1}^2\right] \tag{18}$$

where Rr is the reduction ratio of gear pair, β_{A1} and β_{A2} are the angles of approach of driving and driven gears, f_a and f_r are the average friction coefficient of approach and recess. For external (internal) gear pair, the meshing efficiencies are between 99.2%~99.6% (99.7%~99.9%). For the 6-speed planetary-gear unit, shown in figure 5, if the meshing efficiencies of external and internal gear pairs are identified as 0.994 and 0.998, then the meshing efficiency $\eta_{ij}^{k}=0.994\times0.998=0.992$ for $\xi_{ji}<0$ and $\eta_{ij}^{k}=0.994\times0.994=0.988$ for $\xi_{ji}>1$. Then, according to equations. (15)-(16), the meshing efficiencies of the 6 speeds are 100%, 99.84%, 99.72%, 98.89%, 98.73%, and 98.61% and shown in table 3. Based on the above results, the meshing efficiencies of planetary gear devices are very good.

6 Engineering design and prototype manufacture

6.1 The design of control device

According to the planetary gear unit, shown in figure 5, and clutch sequence, shown in table 3, the corresponding clutch engaging diagrams of the 6 speeds are shown in figures 7(a)-7(f). Based on figures 7(a)-7(f), the control device can be designed and shown in figure 8.

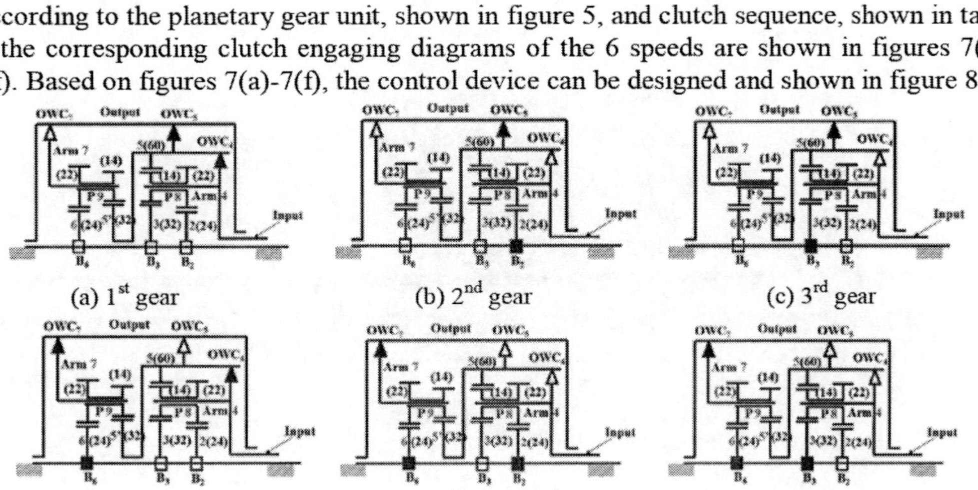

(a) 1st gear (b) 2nd gear (c) 3rd gear
(d) 4th gear (e) 5th gear (f) 6th gear

Fig.7. Clutch engaging diagrams of the 6 speeds.

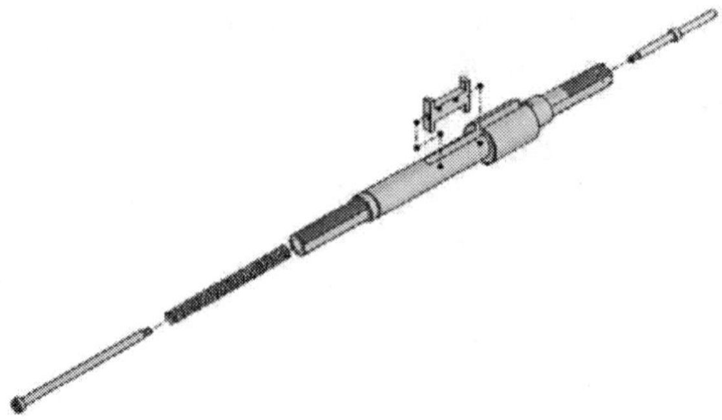

Fig. 8. Exploded view of control device.

6.2 The design of multi-speed planetary gear unit

According to the figure 7, the module is chosen to be 1.0, the engineering design drawings, shown in figures 9(a), 9(b), 9(c), and 9(d), are exploded view, front planetary gear train, rear planetary gear train, and one-way clutch, respectively. Figure 10 shows the dimensions of this 6-speed planetary gear hub. Figures 11(a)~11(f) show the engineering drawing of engaging diagrams of the 6-speed planetary gear hub.

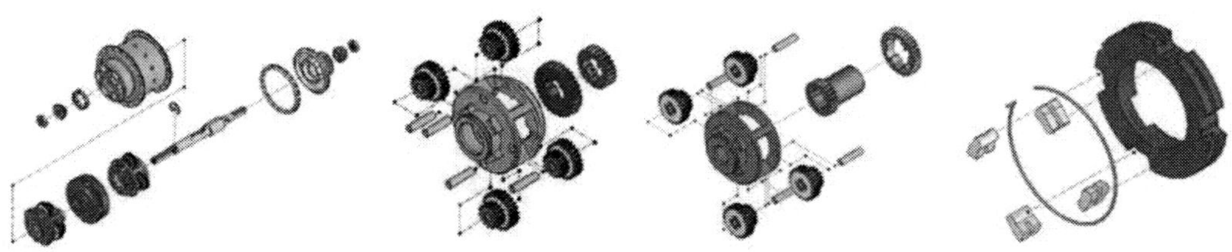

(a) Exploded view (b) Front PGT (c) Rear PGT (d) One-way clutch

Fig. 9. Engineering drawings of six-speed planetary gear unit.

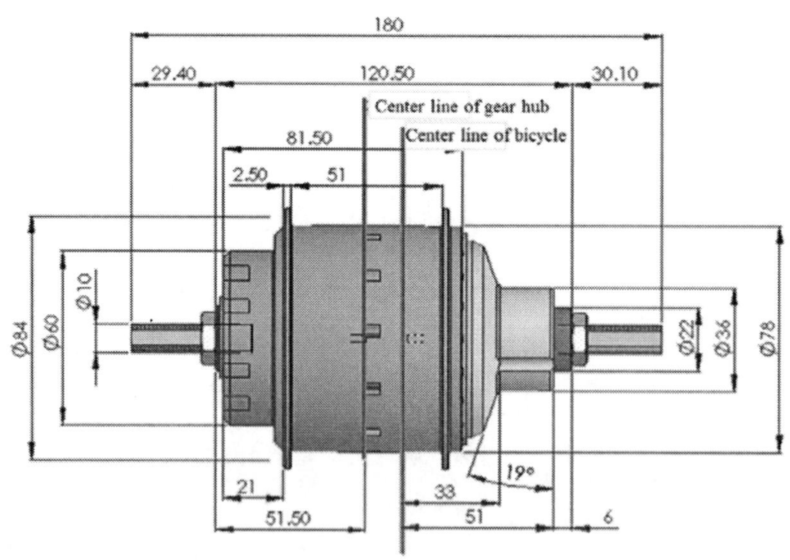

Fig. 10. Dimensions of 6-speed planetary gear unit.

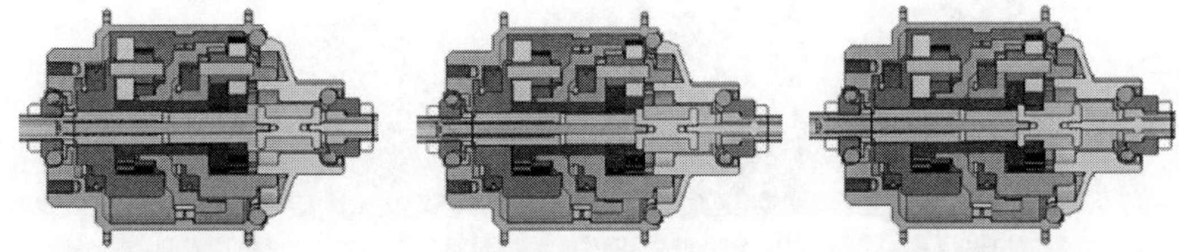

(a) 1st speed (b) 2nd speed (c) 3rd speed

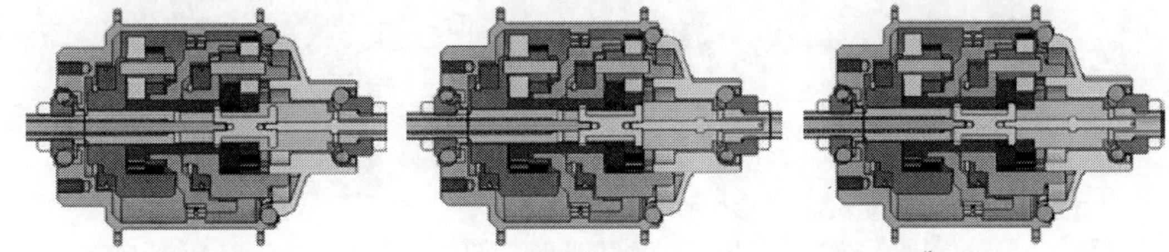

(d) 4th speed (e) 5th speed (f) 6th speed

Fig. 11. Engineering drawing of engaging diagrams of the 6-speed planetary gear unit.

6.3 The design of operating device

The 6-speed transmission device includes planetary-gear unit, control device, and operating device. They are sold together. Therefore, we must design operating device at the same time. The operating device of multi-speed transmission device must contain a fixing portion, manual operating portion, positioning portion, and cable guiding portion. Also, the operating device must have the following features: simple structure, fewer parts, and low production costs. In this paper, the positioning portion and spring are designed to be one part for simplifying the structure and reducing the number of parts. Figures 13(a) and 13(b) show the outer view and the sectional view of the operating device.

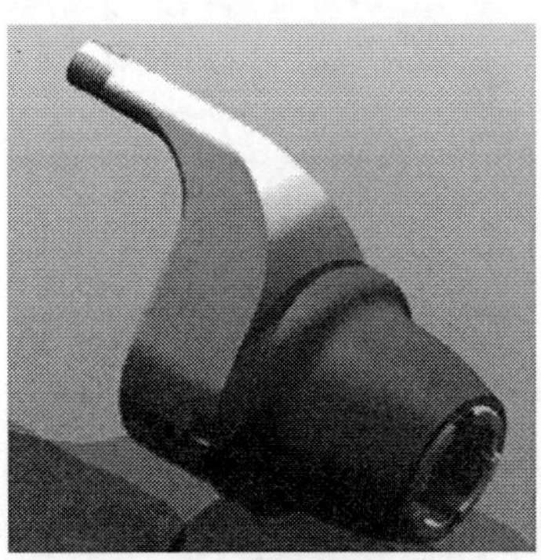

(a) Outer view

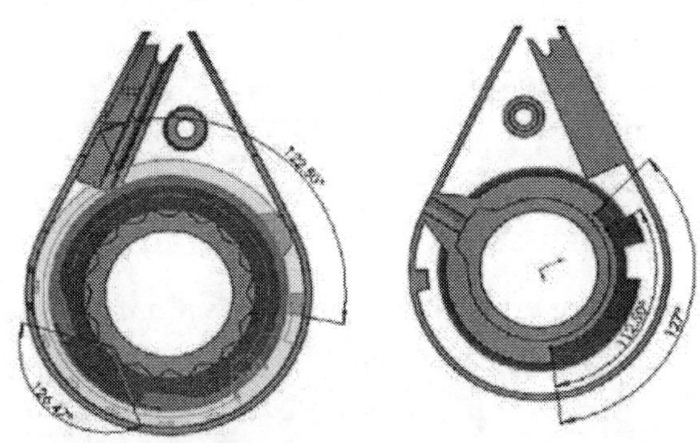

(b) Sectional view

Fig.12. Operating device of 6-speed transmission device.

6.3 Prototype manufacture

According the results of engineering design, the prototype and its parts of this 6-speed transmission device are manufactured and shown in figues 13(a)-13(f).

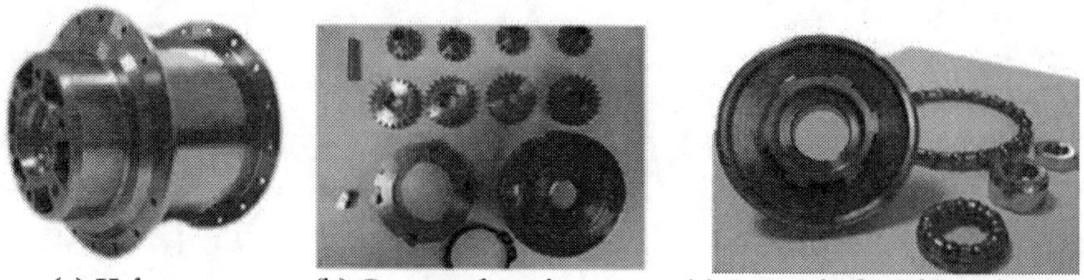

(a) Hub (b) Gears and carriers (c) Input shaft and Bearing

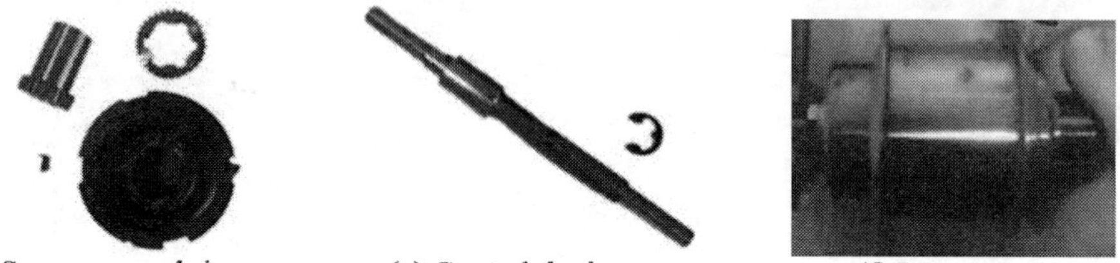

(d) Sun gears and ring gear (e) Control device (f) Layout

Fig.14. Prototype of the six-speed planetary gear unit and control device.

7 Conclusions

This paper focused on the systematic design and prototype manufacturing of multi-speed transmission device of a bicycle. According to the kinematic design, a 6-speed planetary gear unit was synthesized to have reduction ratios 1, 0.797, 0.652, 0.523, 0.417, 0.341 which are good spacing. Based on the proposed design methodology, all multi-speed planetary gear units can be synthesized. According to the latent power theorem, its corresponding meshing efficiencies were calculated to be 99.84% ~ 98.61% for 2nd speed ~ 6th speed. Finally, the engineering design of six-speed planetary gear unit was accomplished and its corresponding prototype was manufactured to verify the design theorem. In this paper, the control device and operating device were also designed.

Acknowledgements

The authors are grateful to the National Science of the Republic of China for the support of this research under MOST 104-2221-E-150-055.

References

1. L.W. Tsai, Proc. 9th Applied Mech. Conf., Kansas City, Missouri, 1, 1-5 (1985)
2. L.C. Hsieh, H.S. Yan and L.I. Wu, J. of the CSME, 10(2), 153-158 (1989)
3. C.H. Hsu, ASME J. Mech. Des, 124(3), 574-576 (2002)
4. M.C. Tsai and C.C. Huang, ASME J. Mech. Des, 132, 065001 (2010)
5. L.C. Hsieh, J.Y. Liu and M.H. Hsu, J. of TCSME, 20(4), 421-435 (1996)
6. W.M. Hwang and Y.L. Huang, J. of TCSME, 29(1), 41-55 (2005)
7. L.C. Hsieh, H.S. Lee and T.H. Chen, Materials Science Forum, 505-507, 1003-1008 (2006)
8. L.C. Hsieh and H.C. Tang, J. of AMM, 232, 955-960 (2012)
9. G. White, Mech. and Machine Theory, 38, 149-159 (2003)
10. C. Chen and J. Angeles, J. of Mechanical Design, 129, 107-113 (2007)
11. L.C. Hsieh and T.H. Chen, J. of Advanced Science Letters, 12, 34-39 (2012)

12. L.C. Hsieh and T.H. Chen, Information-An International Interdisciplinary J. 16(9B), 7025-7032 (2013)
13. L.C. Hsieh and H.C. Tang, J. of AIME, Article ID 686187 (2013)
14. Y.C. Wu, L.A. Chen and C.W, Chang, J. of Mechanical Engineering, 2(8), 266-273 (2014)
15. T.H. Chen, J. of TCSME, 39(3), 467-477 (2015)
16. L.C. Hsieh, T.H. Chen and H.C. Tang, J. of TCSME, 39(3), pp.455-465 (2015)
17. L.C. Hsieh and T.H. Chen, J. of TCSME, 40(5), pp.981-993 (2016)
18. E. Bergles, U.S. Patent, No.4651853 (1987)
19. G. Nurberger, U.S. Patent, No. 5443279 (1995)
20. T. Hino, US Patent No.7,967,718 (2011)

Study on design of temperature - dependence of sublimation transfer roller

Ming-Chu Hsieh[1, *] and *Jung-Sen* Lin[2]

[1]Department of Mechanical Engineering, National Chin-Yi University of Technology, Taiwan

[2] Department of Mechanical Engineering, National Chin-Yi University of Technology, Taiwan

Abstract. Sublimation trasnfer machine's hot roller, a compression type electronic heater, is made of ceramic core which is wrapped with nickel-chromium alloy heater. This roller is insulated with magnesium oxide. When the power is on, the heat is generated. Heat transfers are from high temperature area to low temperature area, and it causes the non-uniform temperature surface and the roller surface distorsion. Such effects always increase with larger diameter of the roller and longer the roller. In order to predict the sublimation transfer machine's roller temperature distribution and validate it, we construct the roller's 3D temperature model and use ANSYS steady state heat transfer analysis module to calculate the power dissipation and simulate it. Final data shall help us to design sublimation transfer machines' roller.

1 Introduction

The phenomenon whereby that the solid materials are directly transferred into the gaseous phase by bypassing the liquid phase at a certain temperature and pressure are called sublimation. Sublimation transfer printing is a technology used to enable the direct transfer of dyes in sublimation inks from the solid into the gaseous phases by heating and pressurizing, then through the condensation in the solid phase to form graphics and text. Correspondingly, this technology is also known as gas transfer printing [1]. During the process of sublimation transfer printing, only the color of the dye will be transferred to produce graphics and text, while other resin substances in the sublimation ink stay attached to the film or paper of the carrier. The "functional intermediate carrier" only contains the printed graphics and text from the sublimate inks, and does not contain release and adhesive layers. Therefore, there are no detachable protective and bonding layers when the printed materials are transfer-printed [2]. This is the difference between sublimation transfer printing and other forms of thermal transfer printing. Since only the dye will be condensed on the surface of printed materials, the printed material always maintains its original characteristics after sublimation. For example, clothes can still hold their original soft nature after sublimation transfer printing. In addition to fabric, the transfer printing can also be applied to ceramics, metals, and other products. However, the printed materials have to possess heat resistance and the gap layer to accommodate dye molecules. During the sublimation transfer printing process, the printing layer and the receiving and accommodation layers of the printed materials must be kept under the compact attaching pressure to ensure good transferred printed graphics and text [3].

This study aims to the electrically heated hot roller's design and the calculation about power to keep well the hot roller's shell mechanism, then have a good allowable thermal

* Corresponding author: smj@ncut.edu.tw

deformation of the roller. Therefore, for effectively controlling the heat transfer process, we need to analyze and calculate its heat resistance during heat conduction process. And then, we use finite element analysis software to determine the appropriate electric heaters' equipped position through uniform temperature analyses and simulations. During heat process, we first apply power. Heaters' temperature rise up gradually, and heat transfers from them to others where temperature are lower. Before reaching constant temperature, it's said to be transient heat conduction. When reaching constant temperature, i.e. reaching a complete thermal equilibrium, that is the heat in equals the heat out, it enters steady-state heat conduction [4].

Heater rollers' surface temperature and working time do a significant impact on the fabric printing quality [5]. While power on, the heaters' temperature reaches their maximum value immediately and rollers' other portion is still on low temperature, it causes thermal deformation longitudinally and radially and has a significant impact on the fabric printing quality. How to improve it? We'll provide practical data to support all of this report's objective.

2 Basic theory

2.1 Heat sublimation transfer printing rollers

2.1.1 Heat transfer principle [6]

All heat transfer modes require temperature differences, since heat transfers from high to low temperature area.
Conduction is effectively an energy transfer process from a matter particle with a higher energy to an adjacent matter particle with a lower energy, which rises the interaction between particles.
Fourier heat conduction equation:

$$q = \frac{Q\ cond}{A} = -k\frac{dT}{dx} \tag{1}$$

$$q_1 = \frac{Q\ cond}{A} = -k\nabla T \tag{2}$$

In equations (1) and (2), q and q_1 are the heat values derived within a unit time period per unit area known as heat fluxes, Q_{cond} is the thermal conductivity rate, A is the surface area, k is the coefficient of thermal conductivity for the material known as the thermal conductivity.
where

mark	description
q , q_1	heat values derived per unit time, unit area and known as heat fluxes.
$Qcond$	thermal conductivity rate
A	surface area
k	coefficient of thermal conductivity for the material known as thermal conductivity
dT/dx and ∇T	temperature increments per unit length along the heat transfer direction, known as the temperature gradient

From equation (1), thermal resistance can be represented as:

$$Q_{cond} = \frac{T_1 - T_2}{R_{cond}} \quad (3)$$

and

$$R_{cond} = \frac{L}{kA} \quad (4)$$

where T_1-T_2 is the temperature difference, R_{cond} is the conduction resistance, and L is the disatnce between two points where temperature difference occurs.

Convection refers to the energy transfer between the solid surface and the adjacent moving liquid or gas. It is only transferred between the solid surface and adjacent liquid via simple thermal conduction when the fluid flow is insufficient. Thus, the convection simultaneously combines fluid flow and heat conduction.

Newton's law of cooling:

$$q = \frac{Q_{cond}}{A} = h(h_s - T_f) \quad (5)$$

$$h = \frac{Q_{conv}}{A(T_s - T_f)} \quad (6)$$

In equation (5), q is the heat flux, Q_{conv} is the convective heat transfer rate, Ts-T_f is the temperature difference between the solid surface and fluid, and h is the convective heat transfer coefficient defined as the heat transfer rate per unit area between the solid surface and the fluid per unit of temperature, which is affected by various factors, such as fluid properties, fluid velocity, wall shape, and roughness.

From equation (5), the concept of thermal resistance can be represented as,

$$Q_{conv} = \frac{T_S - T_f}{R_{conv}} \quad (7)$$

and

$$R_{cond} = \frac{1}{hA} \quad (8)$$

where R_{conv} is the convection thermal resistance, l is the temperature difference between the two ends.

2.1.2 Heat transfer equation

Energy equation:Energy input - Energy output + Energy generated within the unit element = Energy change per unit element.

$$(q_{x1} + q_{y1} + q_{z1}) - (q_{x2} + q_{y2} + q_{z2}) + \rho'(\Delta_x \Delta_y \Delta_z) = \rho(\Delta_x \Delta_y \Delta_z) C_P \frac{\Delta T}{\Delta t} \quad (9)$$

where, q_{x1}, q_{y1}, q_{z1} are heat input transfer rate, q_{x2}, q_{y2}, and q_{z2}, are output transfer rates, and q' is the energy produced per unit volume per unit time ; ρ is the density ; C_p is isobaric specific heat.

Solid heat transfer equation:

$$\rho C_P \frac{\partial T}{\partial t} = Q + \frac{\partial}{\partial x}(k\frac{\partial T}{\partial x}) + \frac{\partial}{\partial y}(k\frac{\partial T}{\partial y}) + \frac{\partial}{\partial z}(k\frac{\partial T}{\partial z}) \tag{10}$$

$$\rho C_P \frac{\partial T}{\partial t} = Q + \nabla \cdot (k\nabla T) \tag{11}$$

Equation (11) is the heat transfer equation derived from the energy conservation law and Fourier's law. The left side of the equation is the energy change. The second item in the right side of equation is the difference of heat transfer rates between input and output. Correspondingly, Q is the heat source, k is the thermal conductivity, T is the temperature, and t is time.

Fluid heat transfer equation:

$$\rho C_P \frac{\partial T}{\partial t} + \rho C_P (v_x \frac{\partial T}{\partial t} + v_y \frac{\partial T}{\partial y} + v_z \frac{\partial T}{\partial z}) = \frac{\partial}{\partial x}(k\frac{\partial T}{\partial x}) + \frac{\partial}{\partial y}(k\frac{\partial T}{\partial y}) + \frac{\partial}{\partial z}(k\frac{\partial T}{\partial z}) + Q \tag{12}$$

$$\rho C_P \frac{\partial T}{\partial t} + \nabla \cdot (-k\nabla T + \rho C_P T_v) = Q \tag{13}$$

where v_x, v_y, and v_z, are the speed values along the directions of the x, y, and z-axes, respectively, and Q is the heat source. In equation (13), the symbols within the brackets denote the heat flow vector that includes the items of diffusion and convection.

Within the research context of this study, as the fluid was assumed to be in a steady flow, $\frac{\partial T}{\partial t} = 0$ and the solid radioactive heat source $Q = 0$. These were substituted into the solid heat transfer equation (11) that yields equation (14),

$$\nabla \cdot (k\nabla T) = 0 \tag{14}$$

Since $\frac{\partial T}{\partial t} = 0$, and the fluid convection heat source $Q = 0$, equation (15) can be obtained from the fluid heat transfer equation (13) as follows:

$$\nabla \cdot (-k\nabla T + \rho C_P T_v) = 0 \tag{15}$$

2.2 Electric heater principle

2.2.1 The working principle of electric heater [7]

Electric heater supplies heat by electric energy conversion. In the actual work, low-temperature fluid flows into the input port through the pipeline at a certain pressure, and heated by electric energy supplied within the pipelines. All the high-temperature thermal energy produced from the electric heat source through this path is required in accordance to the thermodynamics principles for fluids. Therefore, the temperature of the heated medium should increase. According to the signals of the temperature sensor at the output port, the internal control system of the electric heating tubes automatically adjusts the heat output power of the electric heating tubes to keep uniform medium temperature at the output port.

When the heating element is overheated, the independent overheating protection device of the heating element would immediately cut off the heating power to prevent the coking, deterioration, and carbonization of heating materials, as well as the serious burnout of heating elements caused by overheating. This would effectively prolong the service life of the electric heater.

2.2.2 Electric heater design and calculation

Generally, the design and calculation of electric heating are divided into the following three parts:
1. Calculate the power consumption for heating from the initial state to attain a specific temperature within the required time a specific temperature at required time

2. Calculate the thermal energy and power required to keep the medium temperature constant.

3. The total power assigned to is the sum of above two maximum vales times 1.2 for safe operation.

(1) The power consumption required to maintain the heating at the initial stages is,

$$K_w = (C1M1\Delta T + C2M2\Delta T)/863/H{+}P/2 \tag{16}$$

where P is the heat dissipation of the container at the final temperature (unit, kW), H is the time required for heating from the initial temperature to the final required temperature (unit, h), M_1 and M_2 are the masses of the container and medium, respectively, (units, kg), C_1 and C_2, the specific heat capacities of the container and medium, respectively, (units, kcal/(kg·°C), and ΔT is the temperature difference between the final and initial temperatures (unit °C).

(2) The required power consumption for maintaining the medium temperature is defined as,

$$K_w = \mathrm{C2M3\Delta T}/863 + P \tag{17}$$

where M_3 is the mass flow per hour (unit, kcal/h).

3 Methods

3.1 Experimental steps

3.1.1 The mechanism of heat sublimation transfer printing rollers

All heat printing rollers should include the workbench control box, transport equipment, electrically heated hot roller mechanism, detection devices, support tripod, and other parts, as shown in Figure 1.

1. Workbench :The workbench is used to hold fixtures and printing stocks
2. Control system: The control system uses a microcomputer to control the length of the processing roll, and achieve the purpose of accurate material feeding to the machine
3. Conveyor system: Feeding and receiving equipments are also used
4. Electrically heated hot roller: Hot pressing device is a heated roller appliance that uses arc surface rolling to produce hot forming

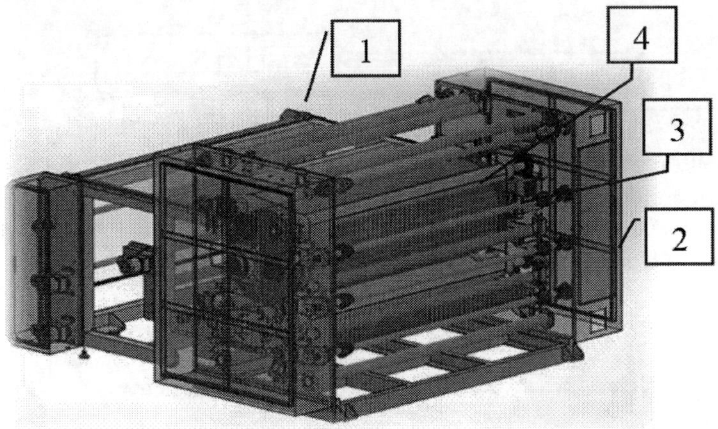

Fig. 1. Heat Sublimation Transfer Printing Rollers basic structure.

3.1.2 Hot rollers' design for sublimation heat printing

For the design of the hot roller for sublimation heat printing, drawing software was used to illustrate the external dimensions of the hot roller structure and the arrangement mode of the electric heaters, is shown in Fig. 2 shows the hot rollers' structure and the arrangement of heaters.

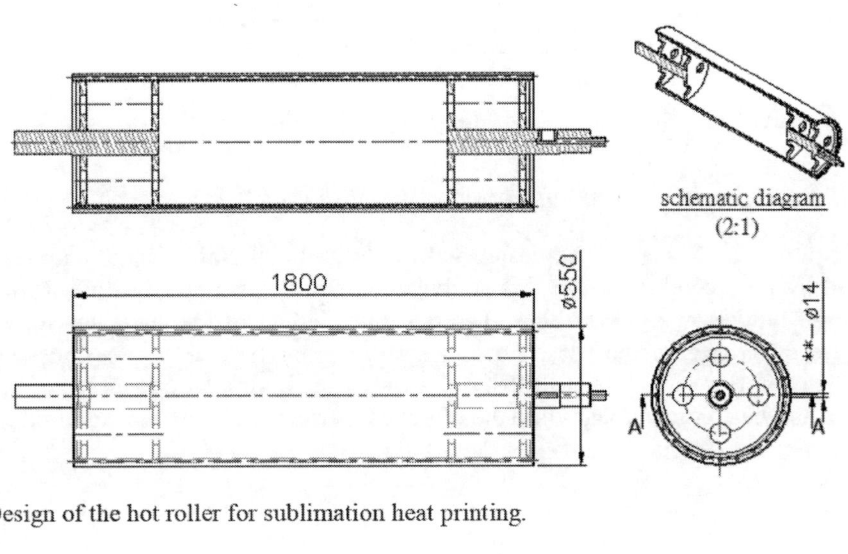

Fig. 2. Design of the hot roller for sublimation heat printing.

3.1.3 Model establishment

The drawing software, *SolidWorks*, was used to draw the geometric schematic of the building structure. It was then imported into the software *ANSYS* for the numerical analysis of the heat flux field of the hot roller. The solution process designed by the study is shown in Figure 3.

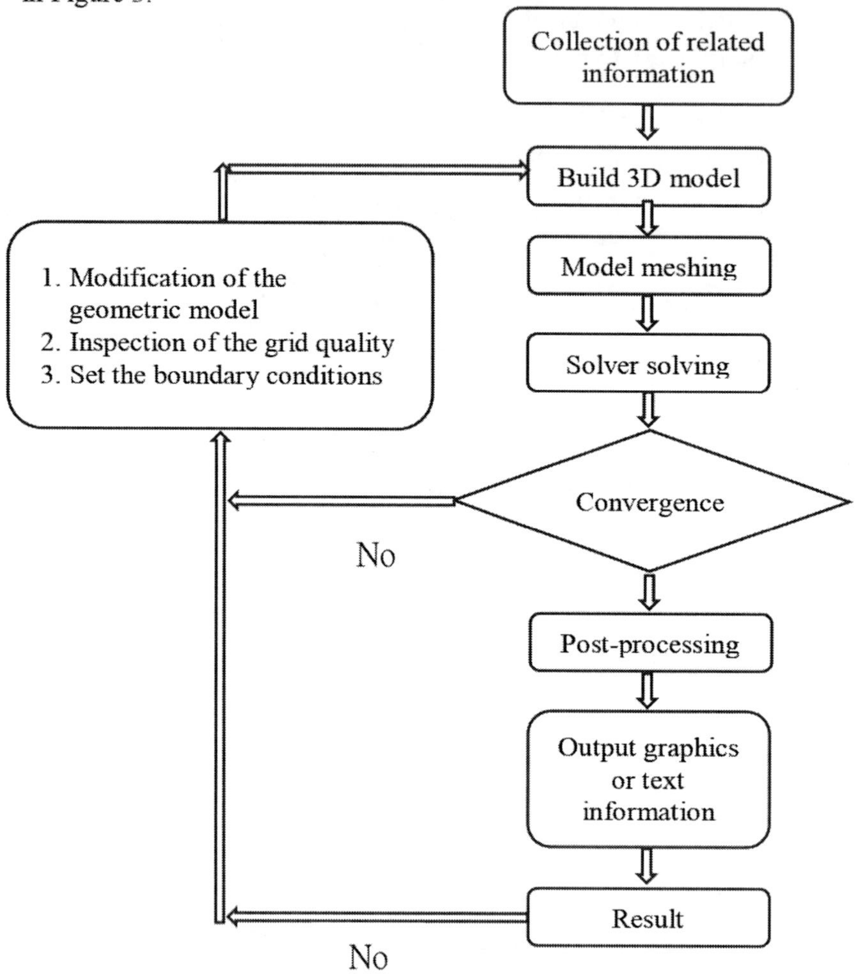

Fig. 3. Flow chart of simulation analysis and design.

3.1.4 Assumptions, equations and boundary conditions

As shown in Figure 4, the electric heater and the heated roller are joined, and heat will be transferred by heat conduction. In this study, we assume it's a steady-state, laminar flow field and the fluid is incompressible. . The buoyancy effect of the air layer caused by the temperature difference in the presence of a gravitational field was also considered. The ambient temperature was set to 25°C. Its numerical simulation included a natural convection and heat conduction, while the effect of thermal radiation was ignored [8].

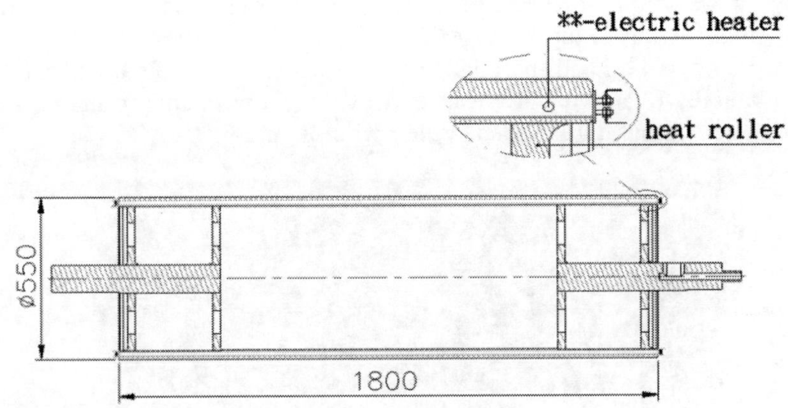

Fig. 4. Module diagram of hot roller and electric heater.

The specific details pertaining to the boundary conditions are as follows:
(1) The thickness of the air layer: The gap thickness between the electric heater and the hot roller was only 0.15 mm, and was integrated with the atmospheric layer. With the specific thickness, if the number of the resolution grid we used is too large, there may be no sufficient computer memory to perform this simulation. If the model grid is not fine enough, the calculation results would be affected. In addition, this gap acts as of thermal resistance. Moreover, the gap would be smaller in the case of thermal expansion. So, we can have a good reasonable simplification.
(2) The thermal conductivity coefficient of air on the outer surface of the hot roller was $15W/m^2 \cdot °C$
(3) The initial temperature on the hot roller was set to 25°C
(4) The heat flux of the electrical heaters on the hot roller were calculated by two different calorific value calculations and analyses, in accordance to the products of 24×800 W and 12×1600 W
(5) Properties of steel
Name: Hot Roller
Material: SS400
Density: 7850 kg/m^3
Specific heat: 480 J/kg·K
Thermal conductivity: 48 W/m·K
(6) Solution: In accordance to the steady-state heat conduction, the system would achieve equilibrium temperature after some time.

4 Results and discussions

4.1 Simulation results and discussions

ANSYS provides a cyclic process that includes a series of analyses, evaluations, and modifications. Correspondingly, we analyzed the initial design, and evaluated the analyzed results according to the design requirements, followed by a subsequent modification of the design. This cyclic process would execute repeatedly until all the requirements were met.

4.1.1 Effect of piping arrangement of electric heater

Fig. 5 shows the temperature distribution of the hot roller, temperatures on both ends are the lowest because of the air convection occurs mainly on both ends. Thus, the temperature

https://doi.org/10.1051/matecconf/201818500036

of the middle part was higher than the temperatures of the left and right sides of the inlet and outlet. This was owing to the larger thermal resistance required for conduction in the middle area of the hot roller, and the relatively long distance to the edge of the hot roller. This distance elicited a relatively slow heat transfer from the middle to the two sides, and thus yielded a higher temperature in the middle part in Figure 5.

Fig. 5. Temperature distribution of the hot roller.

4.1.2. Effect of pipe spacing

The processing pipeline spacing of the hot roller electric heater is shown Figure 6. The pipe spacing between the inlet and outlet was 132.5 mm. As shown in the temperature and thermal conduction distribution of the hot roller, the solid heat transfer rate was reduced. This effect was attributed to the oversized pipe spacing, thus giving rise to a modest uniformity in terms of the temperature distribution. When the pipe spacing for the inlet and outlet pipes was adjusted to 66.5 mm, its heat conduction and temperature distributions are modified in accordance to those shown in Figure 7, whereby the uniformity of the temperature distribution for the hot roller was significantly improved given the same heat flux.

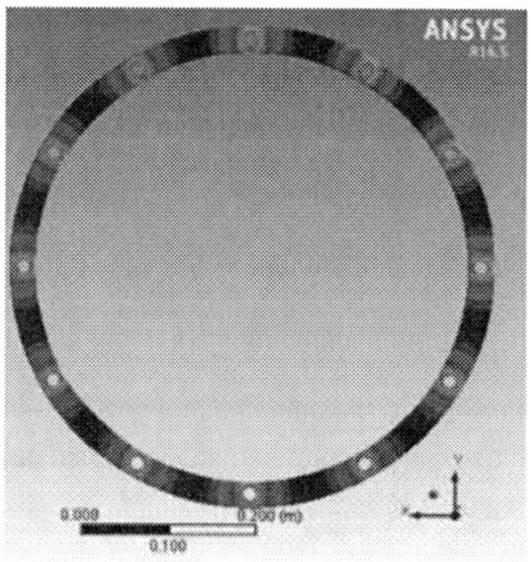

Fig. 6. Temperature profile of left and right sides of the electrically heated hot roller (12 holes).

 https://doi.org/10.1051/matecconf/201818500036

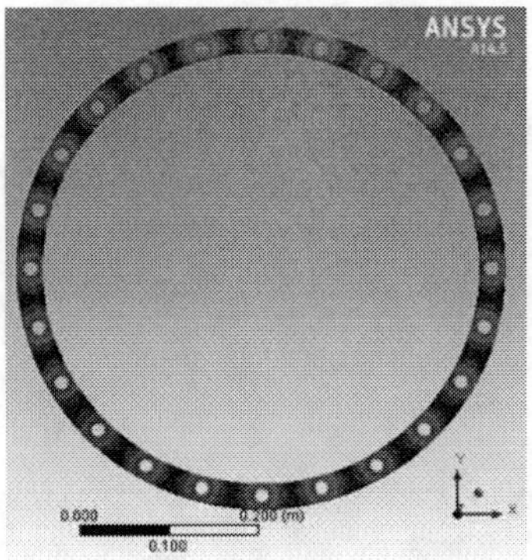

Fig. 7. Temperature profile of left and right sides of the electrically heated hot roller (24 holes).

4.1.3. Performance effect *of temperature uniformity*

From the beginning of heating to the final working temperature, the electrically heated hot roller has yielded optimum temperature uniformity along the axial direction. At increased temperatures, the effects of the uniform temperatures became more pronounced. Therefore is shown Figure 8-9 , the locations and numbers of the electric heaters have an important effect on the temperature distribution of the heat rollers.

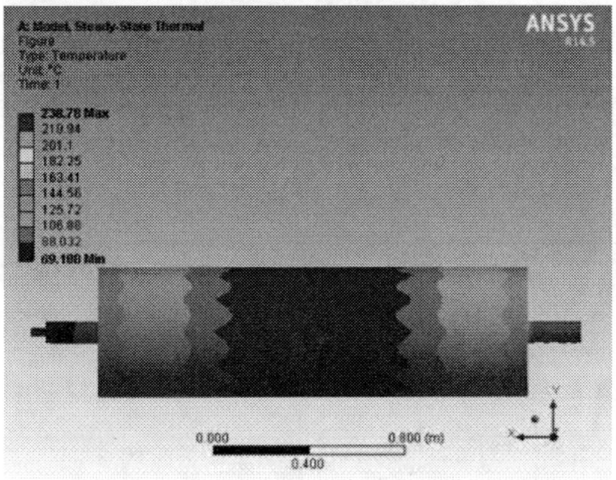

Fig. 8. Temperature profile of the electrically heated hot roller (12 holes with 12 heaters).

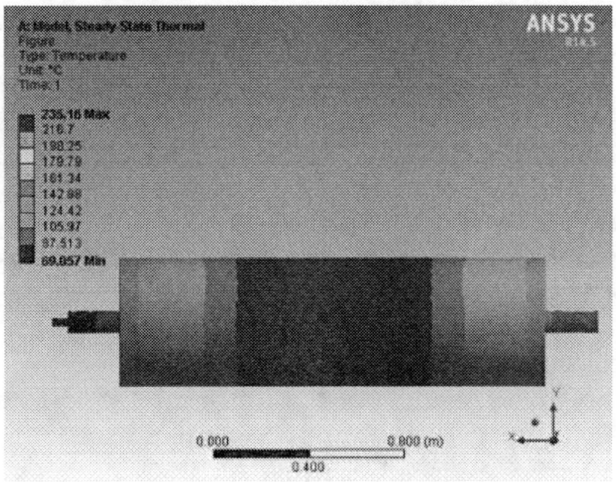

Fig. 9. Temperature profile of the electrically heated hot roller (24 holes with 24 heaters).

4.2 Temperature simulation results for different pipe spacing arrangements on the hot roller

4.2.1 Roller with 12 heaters

This section illustrates the simulation of temperature data at different positions of the electrically heated hot rollers. The positions of the electrically heated hot rollers (12 holes with 12 heaters) are shown in Figure 10. The dimensions of the hot roller model were 1800 mm (L)×550 mm (D). The locations are sequentially numbered from 1 to 28 from the right part of the schematic to the left. The simulation results of the temperature distribution are shown in Table 1.

The temperature difference at section A of the hot roller=235.99-230.67=5.32°C
The temperature difference at section B of the hot roller=220.05-219.63=0.43°C
The temperature difference at section C of the hot roller=202.26-200.53=1.73°C
The temperature difference at section D of the hot roller=182.24-181.01=1.23°C

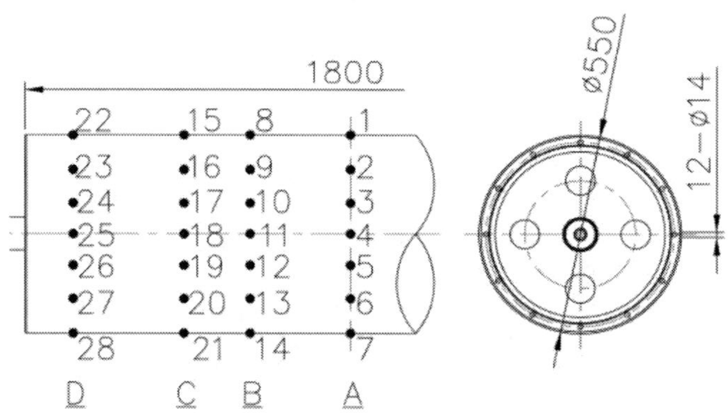

Fig. 10. Position distribution for temperature profile of electrically heated hot roller (12 holes).

Table 1. Temperature profile of the front model (12 holes).

Section A	**1**	**2**	**3**	**4**	**5**	**6**	**7**
Temperature	231.02	231.43	230.98	235.99	230.67	233.55	231.5
Section B	**8**	**9**	**10**	**11**	**12**	**13**	**14**
Temperature	219.84	220.05	219.77	219.85	219.66	219.63	219.85
Section C	**15**	**16**	**17**	**18**	**19**	**20**	**21**
Temperature	202.26	200.53	200.99	200.5	201.25	200.86	201.01
Section D	**22**	**23**	**24**	**25**	**26**	**27**	**28**
Temperature	181.94	182.21	182.24	181.01	181.93	181.79	181.3

4.2.2 Roller with 24 heaters

This section illustrates the simulation data of the temperature on different positions of the electrically heated hot roller. The positions used for the evaluation the temperature distribution of the electrically heated hot roller (24 holes) are shown in Figure 11. The dimensions of the hot roller model were 1800 mm (L)×550 mm (D). The locations are sequentially numbered from 1 to 28 from the right part to the left. The simulation results of the temperature distribution are listed in Table 2.

The temperature difference at section A1 of the hot roller=233.81-233.07=0.74°C

The temperature difference at section B1 of the hot roller=217.03-216.20=0.83°C
The temperature difference at section C1 of the hot roller=198.45-197.29=1.16 ℃
The temperature difference at section D1 of the hot roller=179.87-179.30=0.57 ℃

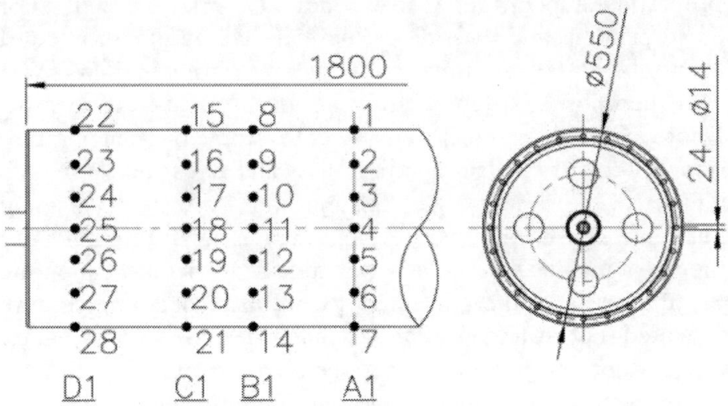

Fig. 11. Position distribution for temperature profile of electrically heated hot roller (24 holes).

Table 2. Temperature profile of the front model (24 holes).

Section A1	**1**	**2**	**3**	**4**	**5**	**6**	**7**
Temperature	233.81	233.07	233.73	233.6	233.48	233.45	233.69
Section B1	**8**	**9**	**10**	**11**	**12**	**13**	**14**
Temperature	216.86	216.26	216.2	216.65	216.7	216.2	217.03
Section C1	**15**	**16**	**17**	**18**	**19**	**20**	**21**
Temperature	198.22	197.36	197.61	198.39	197.29	198.08	198.45
Section D1	**22**	**23**	**24**	**25**	**26**	**27**	**28**
Temperature	179.77	179.3	179.69	178.81	178.58	178.37	179.87

4.2.3 Simulation data for temperature difference with 12 and 24 heaters

For the same heat flux, the effects of pipe spacing which changes the temperature uniformity of the hot roller are shown in Table 3. Elicited results show that the temperature difference became smaller as the spacing changed.

Table 3. Temperature difference for different pipe spacings of hot roller.

Position for the temperature on hot roller(°C)	Temperature profile of the electrically heated(12 holes)	Temperature profile of the electrically heated(24 holes)
Maximum temperature difference between A and A1 (°C)	5.32	0.74
Maximum temperature difference between B and B1 (°C)	0.43	0.83
Maximum temperature difference between C and C1 (°C)	1.73	1.16
Maximum temperature difference between D and D1 (°C)	1.23	0.57
Total temperature difference (°C)	4.89	0.59

(unit: ℃)

5 Conclusions

Sublimation transfer printing technology is currently one of the most important fabric printing technologies in the fabric market. In manufacturing, it is impossible to obtain the best color quality at an inappropriate temperature. Correspondingly, printing with poor quality indirectly results in the bad printing and wasting of the materials in the absence of optimized data. The optimal benefits of the sublimation transfer printing can only be achieved with the combination of the optimized temperature uniformity and heating time. Therefore, in this study, the finite element method was used to study the temperature uniformities of under the various designs of electric heating pipes and hot rollers for with different number of heaters on roller at the same power consumption. As the pipe spacing on the hot roller increased, the heat transfer became poorer, and the temperature uniformity worsened. Smaller hot roller pipe spacing resulted in a better transfer efficiency, pipeline configuration, and power consumption, and therefore led to an increased temperature uniformity of the electrically heated hot roller. Using auxiliary design tool software, computer aided engineering analysis became practical. Engineers can use CAD and CAE software for predesign and development work, which could significantly reduce the time and cost of the development process, and provide an important design reference basis for design and development.

References

1. Y.M. Tsai, The Image Quality Assessment of Digital Transfer Printing, National Taiwan University of Arts/Graduate School of Applied Media Arts, New Taipei City (2006)
2. A.T. Hsiao, A Study of Color Quality Characteristics on Textile Using Different Dye Sublimation Processes, Chinese Culture University, Department of Information Communication, Taipei City (2014)
3. T.R. Chou and K. L. Hung, General Description of Thermal Printing Process and Development, **26**, 1-115 (2010)
4. C.Y. Hsieh, K. M. Tsai, S. R. Juang, L. G. Chen, ANSYS Advanced Examples of Computer Aided Engineer Analysis, QUARK studio, Publisher Easy Book
5. J.H. Chien, New Ink Printing Technology: Beijing: China Light Industry Press, 2000 (in Chinese)
6. Y.A. Cengel, N. Yao, Heat Transfer: A Practical Approach, Liu He Publishing House (2000)
7. M.C. Cheng, The Study of Electric Heater Used in the Thermal Oil Boiler Operation, National Changhua Normal University, Department of Electrical Engineering, Changhua City (2012)
8. J.W. Lee, H.M. Hwang, and I.J. Chung, Finite-element Modeling and Analysis of Time-dependent Thermomechanical Distortion of Optical Sheets in an LCD module, Proceedings of the 6th International Meeting on Information Display and the 5th International Display Manufacturing Conference, pp. 1436-1441 (2006)

Material analysis and characterization of working electrodes of dye-sensitized solar cells

Yeong-Lin Lai[1,*], *Hung-Ru* Hsu[2] , *Yeong-Kang* Lai[3], *Yung-Hua* Chou[1], *Nai-Kun* Hsu[1], and *Chun-Yi* Zheng[1]

[1]Department of Mechatronics Engineering, National Changhua University of Education, Changhua 50007, Taiwan
[2]Green Energy and Environment Research Laboratories, Industrial Technology Research Institute, Hsinchu 31040, Taiwan
[3]Department of Electrical Engineering, National Chung Hsing University, Taichung 40227, Taiwan

Abstract. The aim of this paper is to demonstrate the influence of the TiO_2 with different sintered time regarding the performance of photovoltaic characteristics of DSSCs. A layer of TiO_2 with a thickness of ~8-10 μm and an area of 0.25 cm^2 was prepared by depositing TiO_2 nanoparticles paste onto a fluorine-doped tin oxide (FTO) sbustrate by doctor blade technique, followed by sinterring at 450°C with 4 differennt sintering times: 10 min, 20 min, 30 min, and 40 min. The Pt solution was dripped on FTO substrate works as counter electrode. SEM, XRD and I-V curve were conducted for the material analysis. The photovoltaic characteristics were measured under AM 1.5 sunlight simulator. The results reveal that the different sintered time of TiO_2 working electrode did affect the photovoltaic conversion efficiency. In conclusion, TiO_2 sintered for 30 min yields the highest power conversion efficiency of 6.273%.

1 Introduction

The needs for alternative energy has become a great concern due to the growing energy demand. Solar energy resource is a rich and clean renewable energy and has attracted great attention of researchers in the world. According to the BP Statistical Review of World Energey, it revealed that the reserves/production ratio for oil, natural gas and coal was 50.6 years, 52.5 years, and 153 years, respectively in 2017. [1-2]. Among all of these materials, dye-sensitized solar cells (DSSCs) is one of the most prospective materials for photovoltaic solar cell devices [1,3]. The advantages of DSSCs including easy fabrication, cost-effectiveness, excellent low light performance, highly flexible, and lightweight[1,3].
A conventional DSSCs consists three main components: a mesoporous nanocrystalline TiO_2 anode which sensitized with light-absorbing dye molecules, a counter electrode (CE) which collects electronsand catalyze the redox couple regeneration, and a liquid electrolyte which comprising an iodide/triiodide (I^-/I_3^-) redox couple between two electrodes[3-4].

A layer of TiO_2 with a thickness of ~8–10 μm and area of 0.25 cm^2 was usually prepared by depositing TiO_2 nanoparticles paste onto FTO by doctor blade technique,

*Corresponding author: yllai@cc.ncue.edu.tw

followed by sinterring at 450°C for 30 min. The sintered TiO_2 FTO was then immersing into N719 dye solutionfor dye absorption[5-6].

This paper discusses the effects on differennt annealing times of TiO_2 nanoparticles, in order to find out the best quality of working electrode.

2 Experimental

The main components of DSSCs are fluorine-doped tin oxide (FTO) conductive glass substrate, a TiO_2 film as photoanode, and a dye electrolyte and a counter electrode. The three components were stacked like a sandwich which is shown in Figure 1.

2.1 Fabrication of photoanode

A layer of TiO_2 with a thickness of ~8-10 μm and area of 0.25 cm^2 was prepared by depositing TiO_2 nanoparticles paste onto FTO sbustrate by doctor blade technique, shown in Figure 2.

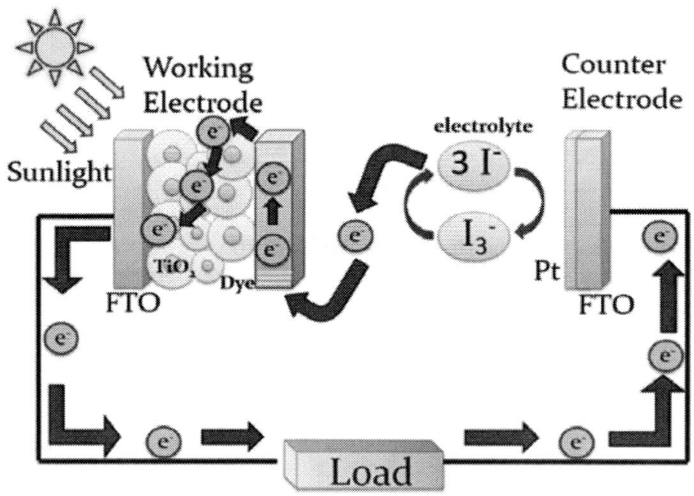

Fig. 1. The configuration of DSSCs.

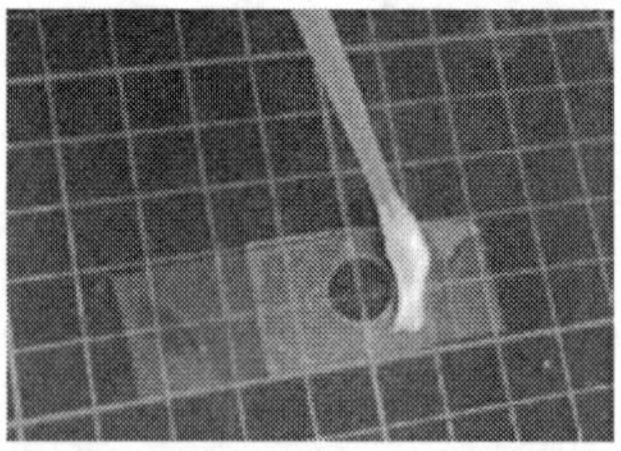

Fig. 2. Depositing TiO_2 nanoparticles paste onto FTO sbustrate.

A piece of transparent tape with 1 cm of adiameter area was cut off and pasted on the FTO substrate. The TiO_2 film was then depositing on the FTO substrate by doctor blade technique, followed by sinterring at 450°C. These four different sintering time samples were indicated by A, B, C, and D, which corresepond to four differennt sintering time conditions, 10 min, 20 min, 30 min, and 40 min, repectively. The fabricated TiO_2 working electrode is shown in Figure 3. It was then soaked in the dye solution for 24 hours at room temperature and subsequently carefully rinsed with ethanol to remove unadsorbed dye.

2.2 Fabrication of counter electrode

The main material for the fabrication of counter electrode (CE) was platium (Pt) by using spin coating method.

The Pt solution was dripped on FTO substrate and then placed on a spin coater machine with a spin speed of 500 rpm for 90 s in order to be evenly coated on FTO stubstrate. The fabricated Pt counter electrode is demonstrated in Figure 4.

2.3 Preparation of dyes

Acetonitrile, *tert*-Butyl alcohol (TBA) and N719 dye powder were the materials to prepare for the dye. Acetonitrile and TBA were mixed with the porpotion of 1:1 and the dye concentration was 5×10^{-4} mol/L. The prepared TiO_2 photoanode was then soaked in the dye solution for 24 hours at room temperature.

2.4 Preparation of electrolyte

The chemicals used to prepare the electrolyte including 0.5 M tert-butylpyridine, 0.005 M iodine (I_2), 0.6 M 1,2-dimethyl-3-propylimidazolium iodide, and 0.1 M lithium iodide (LiI). The above chemicals were all mixed with acetonitrile.

Fig. 3. Fabricated TiO_2 film working electrode.

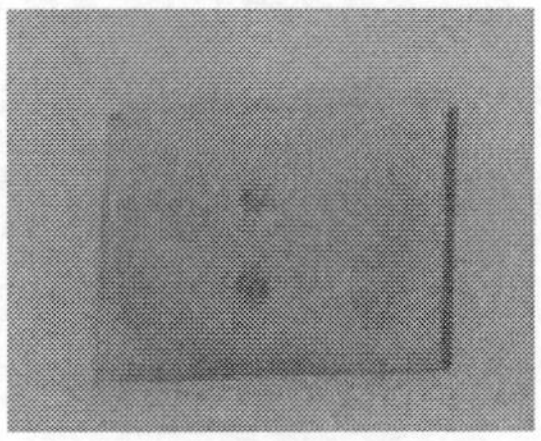

Fig. 4. Fabricated Pt counter electrode.

2.5 Assembly of DSSCs

The DSSCs components were assembled started with applying a membrance (Surlyn) with an area of 1.5 cm × 1.5 cm and a thickness of 60 μm on TiO_2 working electrode. A area with idameter of 1cm was cut off in order to expose the TiO_2 layer. The counter electrode was aligned and assembled with the membrance. The components were sealed together by heat press sealing maching with temperature of 145°C and 1.5 kg/mm^2 presure. The electrolyte was injected to the drilled hole of counter electrode and sealed with a 3M transparent tape. An assembled DSSCs device was shown in Figure 5.

Fig. 5. Fabricated DSSCs.

3 Results and Discussion

3.1 SEM analysis

Figure 6 shows the SEM (Scanning Electron Microscopy) images of surface morphology of TiO_2 sintered at 450°C with different time (a) 10 min (b) 20 min (c) 30 min (d) 40 min. The uniformity, thickness, porosity, and grain size of TiO_2 were observed through SEM. The result indicates that the adhesion of TiO_2 on FTO does not change with different sintered times. In addition, the surfaces of TiO_2 were coral shape and without serious cracking which helps dye absorption.

3.2 XRD analysis

XRD analysis was conducted and shown in Figure 7. It was found that TiO_2 appears to be anatase phase crystals with different sintered time. However, the diffraction peak did not increase significantly as sintered time increased. It results in the best anatase phase crystals and diffraction peak under 20 and 30 min of sintered time.

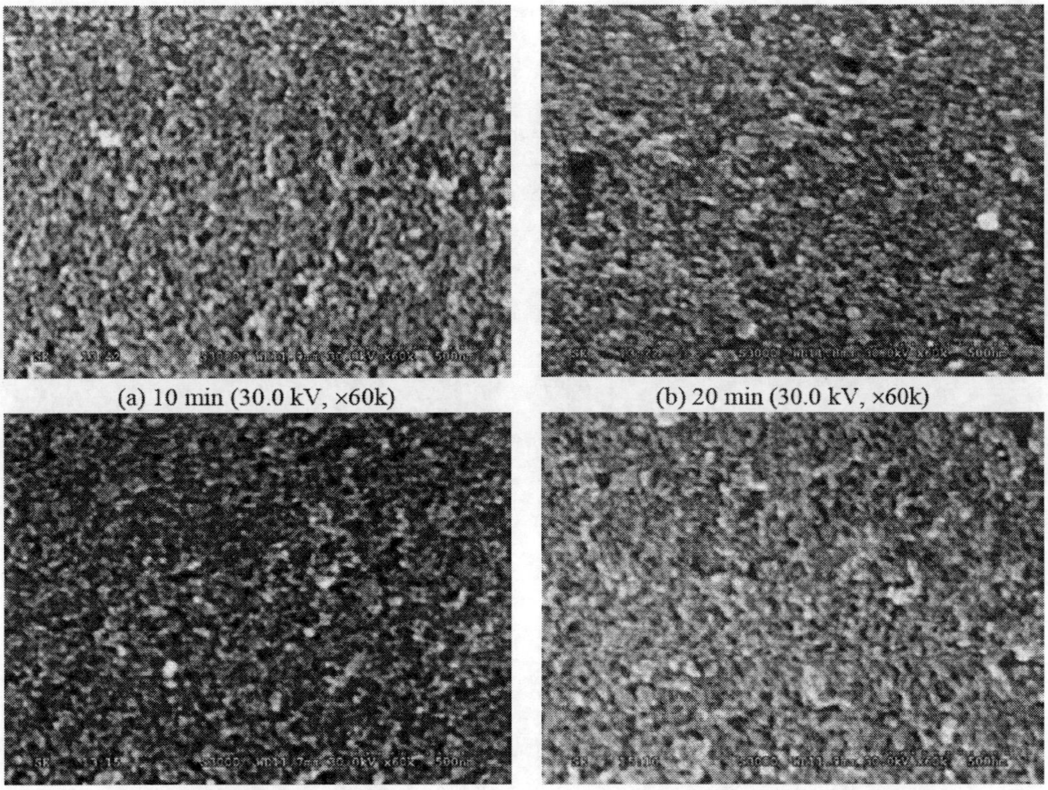

(a) 10 min (30.0 kV, ×60k) (b) 20 min (30.0 kV, ×60k)

(c) 30 min (30.0 kV, ×60k) (d) 40 min (30.0 kV, ×60k)

Fig. 6. Top SEM images of TiO_2 sintered at 450°C with different time (a) 10 min (b) 20 min (c) 30 min (d) 40 min.

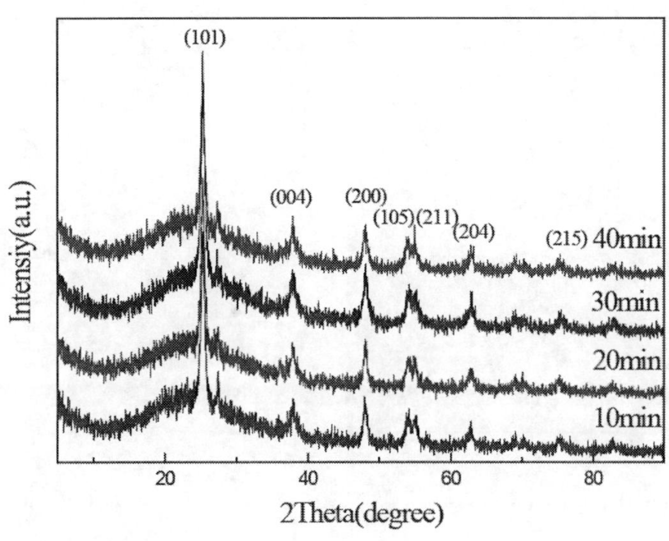

Fig. 7. XRD pattern of TiO_2 with different sintered times.

3.3 Measurement of photovoltaic characteristics of the DSSCs

The photovoltaic characteristics of the DSSCs were measured and shown in Figure 8. The characteristics including open-circuited voltage (V_{oc}), short-circuited current density (J_{sc}), fill factor (FF), and photovoltaic conversion efficiency. The results with different parameters are listed in Table 1. The J_{sc} of the DSSCs with the TiO_2 working electrode fabricated under condition A, B, C, and D is 16.182, 16.284, 16.734, 16.423 mA/cm^2, respectively. The photoelectric conversion efficiency is 5.885%, 6.010%, 6.273%, and 5.634%, respectively.

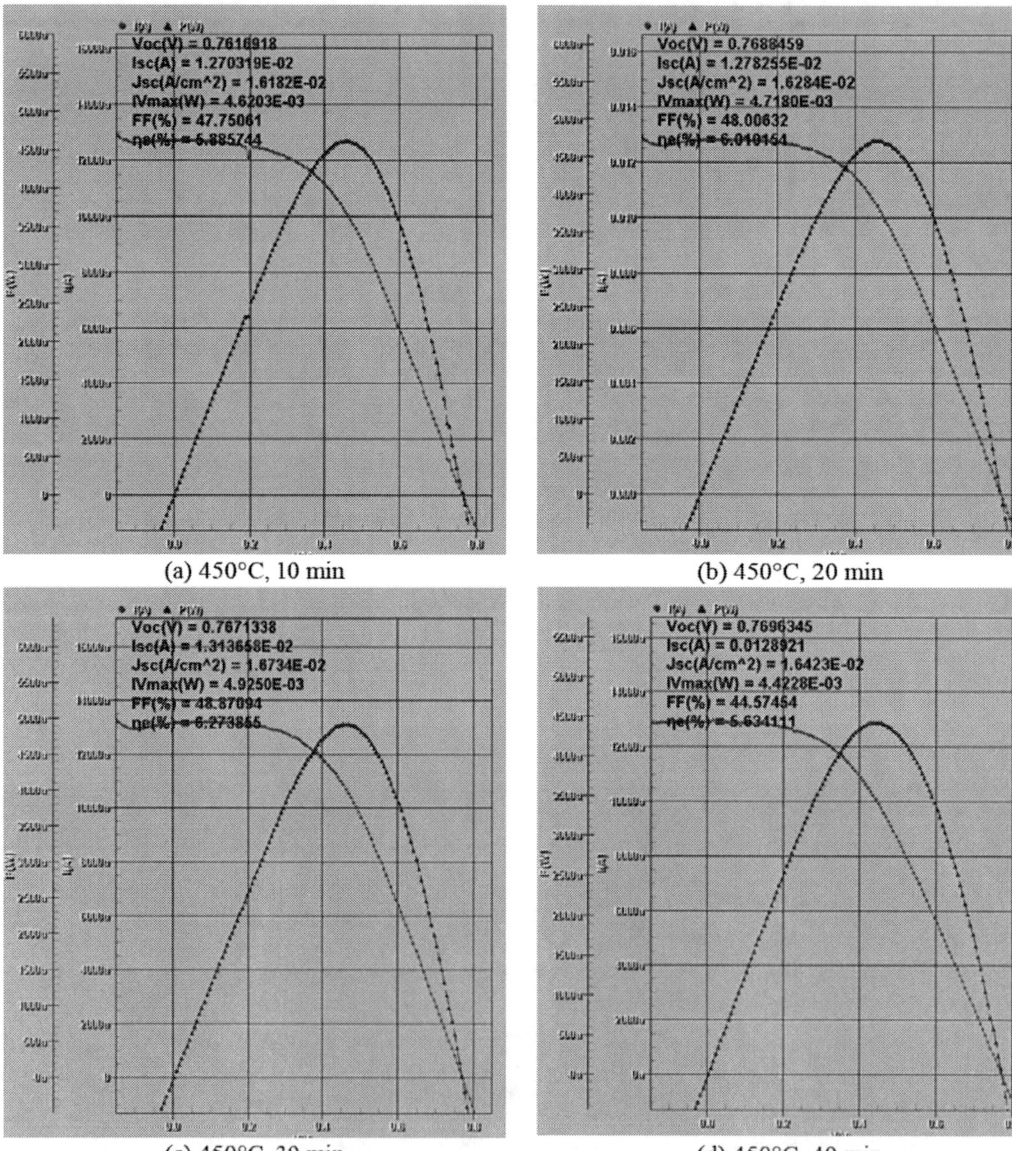

(a) 450°C, 10 min (b) 450°C, 20 min

(c) 450°C, 30 min (d) 450°C, 40 min

Fig. 8. The measurement of photovoltaic characteristics of the DSSCs with the TiO_2 working electrodes. (a) 450°C, 10 min (b) 450°C, 20 min (c) 450°C, 30 min (d) 450°C, 40 min.

Table 1. The photovoltaic characteristics of the DSSCs with the TiO_2 working electrodes.

Time	V_{oc} (V)	J_{sc}(mA/cm^2)	FF (%)	Efficiency (%)
10min	0.761	16.182	47.750	5.885
20min	0.768	16.284	48.006	6.010
30min	0.767	16.734	48.870	6.273
40min	0.769	16.423	44.574	5.634

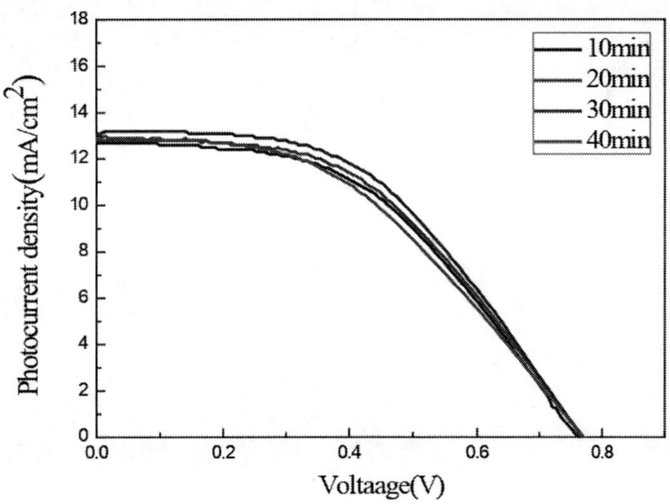

Fig. 9. The I-V curves of the DSSCs with the TiO_2 working electrodes under different sintered times.

As we input all of the characteristics results into Origin software, the I-V curves figure was yield as Figure 9. The I-V curves indicates that different parameter has an effect on the photovoltaic characteristics of DSSCs. It also reveals that the photocurrent increases as the TiO_2 sinter time increases. Based on Table 1, we also found that as the TiO_2 sintered time increases, it improves J_{sc} and FF. It explains the result of XRD in Figure 7. When TiO_2 sintered at 450°C with enough time, anatase phase crystals and diffraction peak will be increased. In addition, TiO_2 sintered for 30 min will yield the highest power conversion efficiency of 6.273%.

4 Conclusions

In summary, we have demonstrated the influence of the TiO_2 with different sintered time (10 min, 20 min, 30 min, 40 min) of the working electrode on the DSSCs performance. It includes V_{oc}, J_{sc}, fill factor, and power conversion efficiency. SEM, XRD and I-V curve were conducted for the material analysis of the TiO_2 photoanode. The photovoltaic characteristics were measured under AM 1.5 sunlight simulator. Different sintered time of TiO_2 working electrode has an influence on the photovoltaic conversion efficiency. TiO_2 sintered for 30 min will yield the highest power conversion efficiency of 6.273%.

References

1. Y.L. Lai, H.R. Hsu, Y. K. Lai, C.Y. Zeng, Y. H. Chou, N.K. Hsu, G.Y. Lung. ICPMMT (2017)
2. BP Statistical Review of World Energy June 2017
3. X. Yuan, X. Li, X. Zhang, Y. Li, L. Liu. Journal of Alloys and Compounds. 731(2018) 685-692
4. N. Huang, G. Li, Z. Xia, F. Zheng, H. Huang, W. L, C. Xiang, Y. Sun, P. Sun, X. Sun. Electrochimica Acta. 235 (2017) 182-190
5. J. Liu, Q. Tang, B. He. Journal of Power Sources. 268(2014) 52-62
6. M. Motlak, N. A. M. Barakat, A. G. El-Deen, A. M. Hamza, M. Obaid, O.B, Yang, M. S. Akhtar, K. A. Khalil. Applied Catalysis A : General 501 (2015) 41-47

Automatic waste water treatment, aeration, and retrieval for a hybrid aquarium/plant system

An-Bang Cheng[1], *Min-Chie* Chiu[2,*], and *Che-Min* Chiu[3]

[1]Department of Industrial Education and Technology, National Changhua University of Education, Changhua City, Taiwan, R.O.C.
[2]Department of Mechanical and Automation Engineering, Chung Chou University of Science and Technology, Taiwan, R.O.C.
[3]Institute of Biomedical Engineering, National Tsing Hua University, Taiwan, R.O.C.

Abstract. Because of the greenhouse effect, global climate changes will adversely influence the water supply in the future. As can be seen in traditional reservoirs, the crisis will be exacerbated due to the reservoir's sedimentation. From a global view, 70 % of freshwater resources are used for agricultural purposes. Thus, water management and infrastructure changes are needed to improve the situation if current climate conditions continue into the future. In order to save water resources, an attempt to reuse a pond's wastewater for agricultural irrigation is proposed. Utilizing waste water treatment via nitrification (with nitrifying bacteria) and aeration processes, the plant will receive enough fertilizer extracted from fish dung. Additionally, to assure a safe environment, the temperature, dissolved oxygen, and pH (potential of hydrogen) value of the aquarium will be monitored online and shown on the interface of the server device. Various automatic functions such as an aquarium filtering system, a waste water treatment system, a plant irrigation system, a waste water aeration system, and a water retrieval system are also designed in the hybrid aquarium/plant system. Consequently, a prototype of a PC-based automatic plant/aquarium hybrid system is established.

1 Introduction

Rosegrant's research indicates that the effects of climate change on hydrology and water resources are tremendous [1, 2]. This change will lead to decreasing precipitation rates and increasing drought events [3]. Based on FAO's report, 70 % of the freshwater resources are used for agricultural purposes [4]. Ostrom's investigation also indicates that there has been a threefold increase in the total area of irrigated agriculture during the past century [5]. Research from Perry's study [6] reveals that 20 % of the total cropland (irrigated agriculture) produced more than 40 % of the world's total agricultural output. Therefore, water is closely related to the crop's production. Thus, water management and recycling are needed to improve the situation if the current climate conditions continue into the future [7].

*Corresponding author:minchie.chiu@msa.hinet.net

For aquaculture farmers, the over-load pumping of underground water on an island results in sinkholes which may destroy the aquaculture farm [8]. Temperatures also tremendously affect aquarium animals [9, 10, 11, 12]. To overcome this drawback, Chiu [13] proposed an alternative method using an inner water circulation system in conjunction with an automatic water treatment process and a thermal control for a pool.

Regarding feeding and water irrigation/recycling of an aquarium and plants, a hybrid system having a filter, waste water treatment, waste water aeration, water irrigation, and a water retrieving function is presented. Because a plant cannot absorb the components of NH4+ and NH_3 from fish dung, the waste water pumped from the aquarium will be treated in the waste water treatment tank via nitrifying bacteria. Here, the nitrifying bacteria will convert NH_4+ and NH_3 into NO_3−. With this, the plant will receive fertilizer from the hybrid system. An advanced waste water treatment using an aeration process is also adopted. Furthermore, to assure a safe aquarium environment, several environmental factors such as temperature, dissolved oxygen, and pH values will be monitored online.

Consequently, to facilitate an operational process, a PC-based interface connected to the PLC is established.

2 Automatic waste water treatment, aeration, and retrieval system for a hybrid aquarium/plant system

The structure of an automatic waste water treatment, aeration, and retrieval monitoring/controlling system for a hybrid aquarium/plant system is depicted in Fig. 1. As indicated in Fig. 1, three kinds of sensors (a pH detecting sensor, a dissolved oxygen detecting sensor, and a temperature detecting sensor) used in the aquarium is adopted and fed back to the server device. To automatically manipulate a waste water treatment and a water recycling process, five sub-functions (an aquarium filter, waste water treatment, waste water aeration, plant irrigation, and water retrieval) connected to the PLC are designed into the system. In order to improve the pH value of aquarium water, a waste water treatment system is designed. Here, waste water treatment will be initiated when the solenoid valve#1 is actuated. The water of the aquarium will be conveyed to the waste water tank via the solenoid valve#1. A water level gauge installed in the tank will be actuated and send back a signal to the PLC to close solenoid valve#1 when the water reaches the top of the water's level gauge. Because plants cannot directly absorb the NH_4+ and NH_3 from fish dung, the process of nitrification from NH_4+ and NH_3 into NO_3− via nitrifying bacteria within the tank is required. Furthermore, because aeration can help prevent low dissolved oxygen levels from negatively impacting pond fish populations [14], an aeration motor used for the waste water aeration process is designed. In addition, the nitrified water can be conveyed to the plants for irrigation via solenoid valve#2. Subsequently, the water filtered through the soil will be conducted into the recycling tank. The filtered water can also be conveyed to the aquarium via the water pump.

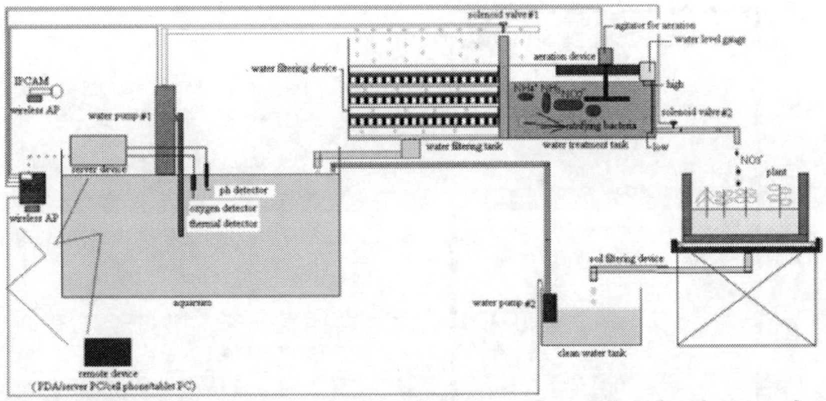

Fig. 1. The structure of the automatic waste water treatment, aeration, and retrieval system for a hybrid aquarium/plant system.

The related input/output diagram of the PLC connection is shown in Fig. 2. In order to remotely manipulate waste water treatment, aeration, and the retrieval monitoring/controlling system via the server PC, the auxiliary relays defined in Table 1 have been added into the PLC's ladder program shown in Fig. 3. As indicated in Fig. 3, the auxiliary relays are marked in red in place of the button functions. The server pc can then submit the command to actuate the related function for the waste water treatment, aeration, and retrieval monitoring/controlling system. Moreover, the interface of the server PC programmed by the VB(Visual Basic) is also illustrated in Fig. 4.

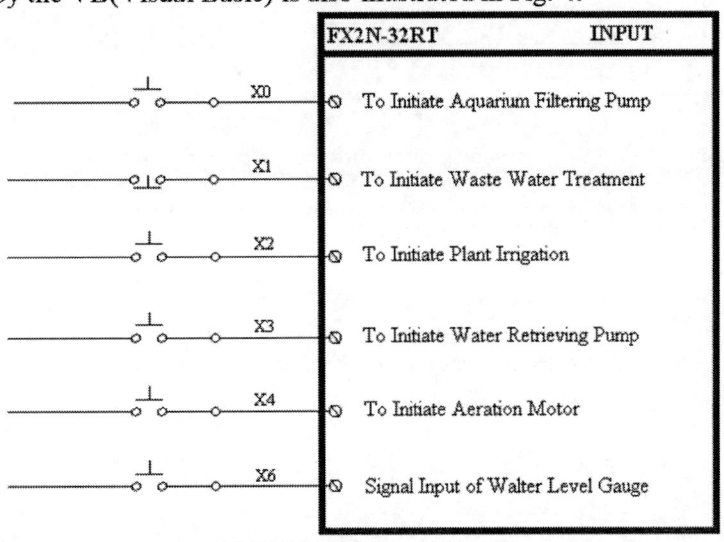

(a) PLC input from X0 to X6

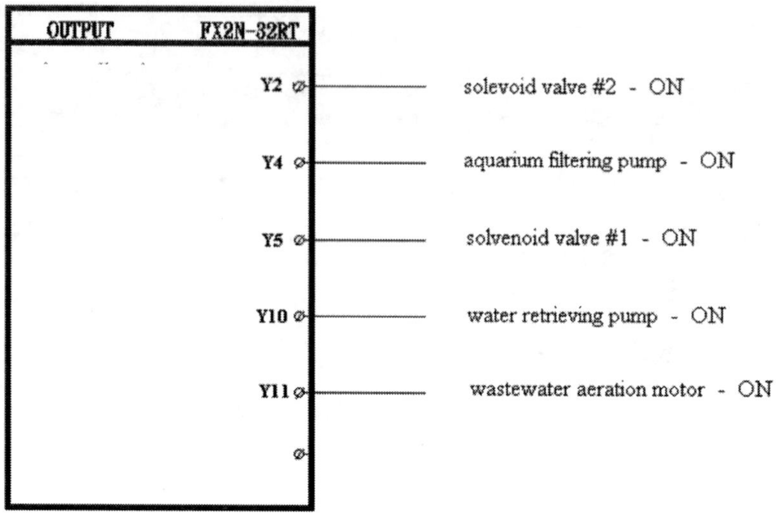

(b) PLC output from Y2 ~Y11

Fig. 2. The related wiring diagram of the PLC connection.

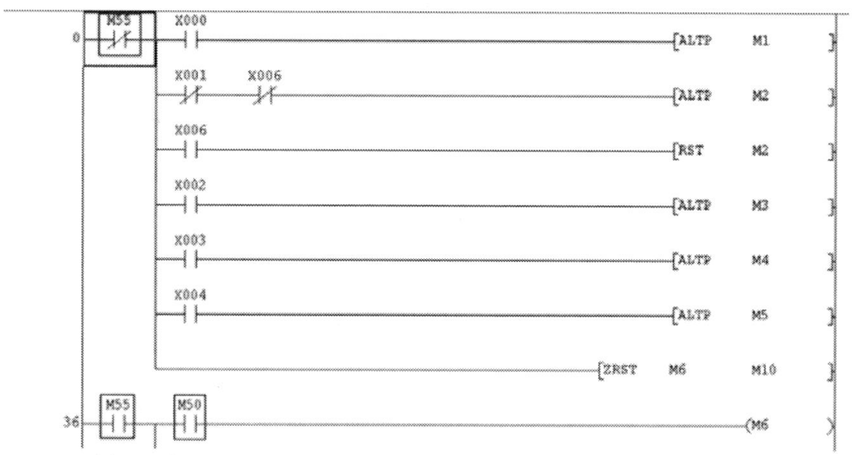

(a) Abstracted ladder diagram #1 using auxiliary relay marked in red.

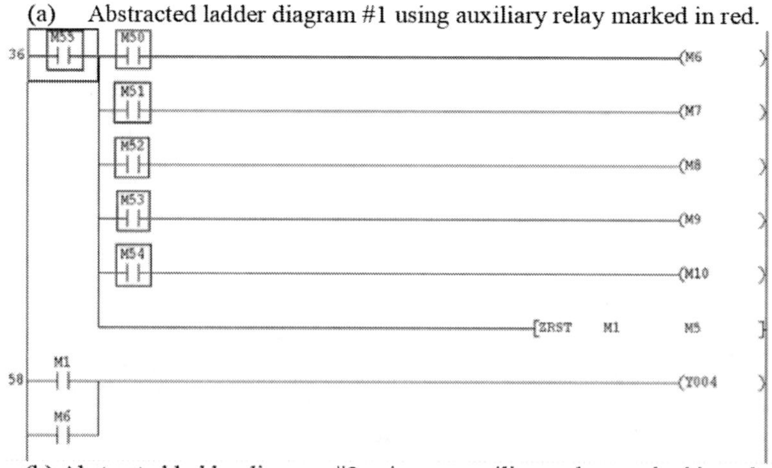

(b) Abstracted ladder diagram #2 using an auxiliary relay marked in red.

Fig. 3. The related ladder program built into the PLC.

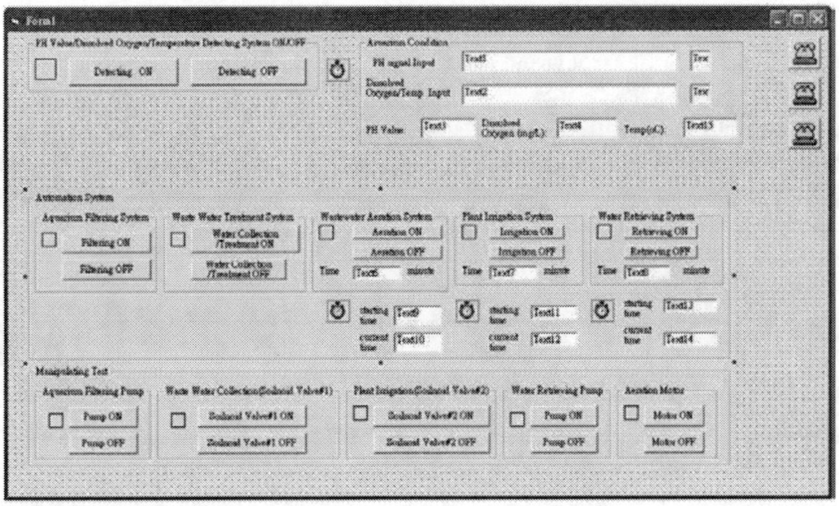

Fig. 4. The interface in the server PC.

3 System established and communication formated automatic

As indicated in Fig. 1 and Table 1, a level gauge installed inside a waste water treatment tank and connected to a PLC is used to detect the status of the water level. The PLC is used as a controller to actuate the actuators (two water pumps, two solenoid valves, and an aerating motor) of the automatic waste water treatment, aeration, and retrieval monitoring/controlling system. Also, three kinds of aquarium sensors used to detect the pH value, the dissolved oxygen, and the temperature of the aquarium that are installed and connected to the server PC will feedback the pH value, the dissolved oxygen value, and the thermal data to the server PC. The related wiring diagram of the system is shown in Fig. 5. An interface at the server PC is programmed by the VB and is used to implement the system by sending a command to the PLC controller. The abstract of the related interface program built into the server PC is depicted in Fig. 6. To facilitate communication between the PLC and the PC via a RS232/RS422, the picket format for the MISTISUE FX2n-PLC's primary port communication is shown in Table 2. Also, the definition of the picket format is shown in Table 3. The MISTISUE FX2n-PLC's extended port communication is shown in Table 4. In addition, the communication parameter used in the extended port is illustrated in Table 5. Consequently, the command description of the MISTISUE FX2n-PLC's extended port communication protocol type 1 is indicated in Table 6.

Table 1. The auxiliary relays used to remotely control the action of the PLC program.

Auxiliary Relay Number	Action	PLC Program
M50	Aquarium Filtering Pump On/Off	SET，RST M50
M51	Waster Water Treatment/Collection On/Off	SET，RST M51
M52	Plant Irrigation/Collection On/Off	SET，RST M52
M53	Water Retrieving Pump On/Off	SET，RST M53
M54	Waste Water Aeration Motor On/Off	SET，RST M54
M55	M50-M54 Off	RST M50-M54

Table 2. The communication packet format of MISUBISHI FX2n-PLC's primary port.

Header code	Command code	Address and data	End code	Inspection code
STX(02)	CMD	DATA	ETX(03)	Checksum

Table 3. Definition of the communication packet for the MISUBISHI FX2n-PLC's primary port.

Command category	Command code	Object	Function description
Read	0	X,Y,M,S,T,C,D	Read the object value
Write	1	Y,M,S,T,C,D	Write data to the object
Forcing on	7	Y,M,S,T,C	Forcing node to be on
Forcing off	8	Y,M,S,T,C	Forcing node to be off

Table 4. Communication packet format of the MISUBISHI FX2n-PLC's extended port.

Header code	Station number	Computer number	command	Object number/data	End code	Inspection code

Table 5. Parameters of the extended port's communications protocol for the MISUBISHI FX2n-PLC.

D8120Bit	Name	Content
15	Protocol type	0: type 1, 1: type 4
14	protocol	0: No, 1: Yes
13	Inspection code	0: manipulate, 1: auto
12	Type of control line(II)	none
11	DTR inspection mode	none
10	Type of control line(I)	none
9	End code	none
8	Header code	none
7	Transmission speed	0011:300 0222:4800
6		0100:600 1000:9600
5		0101:1200 1001:19200
4		0110:2400
3	Stop bit	0: 1 bit, 1: 2 bits
2	Parity check	00: none 02: even
1		01: odd
0	Data bit	0: 7 bits, 1: 8 bits

Table 6. Command of the extended port's communications protocol for the MISUBISHI FX2n-PLC.

Command	Description	Available object
BR	Read PLC's bit data	X, Y, M, S
WR	Read PLC's bit_set data	D, TN, CN
BW	Write data to PLC's bit	Y, M, S
WW	Write data to PLC's bit_set	D, TN, CN
BT	Hybrid type of the writing for bit	Y, M, S
WT	Hybrid type of the writing for bit_set	D, TN, CN
RR	Actuate PLC's running	
RS	Actuate PLC's stop	
PC	Read the PLC's category	

 https://doi.org/10.1051/matecconf/201818500038

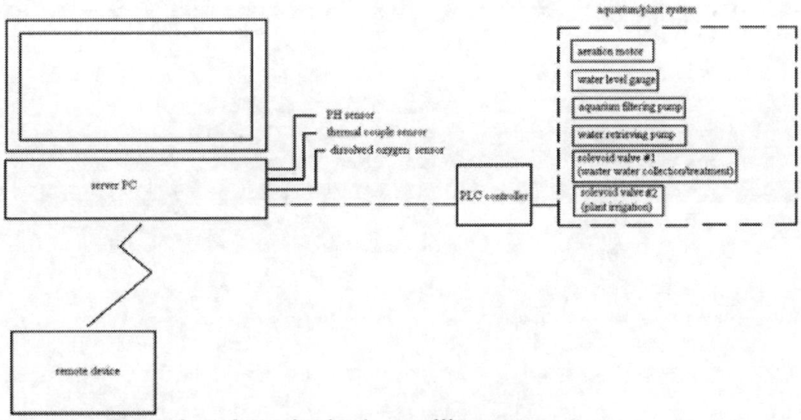

Fig. 5. The structure of a PC-based monitoring/controlling system.

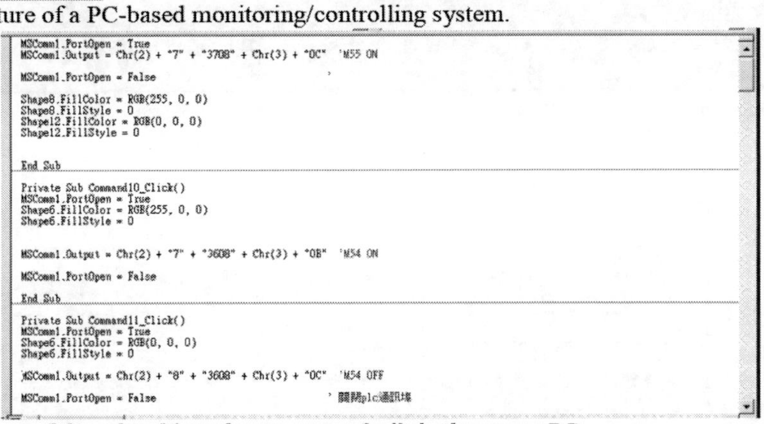

```
MSComm1.PortOpen = True
MSComm1.Output = Chr(2) + "7" + "3708" + Chr(3) + "0C"  'M55 ON

MSComm1.PortOpen = False

Shape8.FillColor = RGB(255, 0, 0)
Shape8.FillStyle = 0
Shape12.FillColor = RGB(0, 0, 0)
Shape12.FillStyle = 0


End Sub

Private Sub Command10_Click()
MSComm1.PortOpen = True
Shape6.FillColor = RGB(255, 0, 0)
Shape6.FillStyle = 0


MSComm1.Output = Chr(2) + "7" + "3608" + Chr(3) + "0B"  'M54 ON

MSComm1.PortOpen = False

End Sub

Private Sub Command11_Click()
MSComm1.PortOpen = True
Shape6.FillColor = RGB(0, 0, 0)
Shape6.FillStyle = 0

MSComm1.Output = Chr(2) + "8" + "3608" + Chr(3) + "0C"  'M54 OFF

MSComm1.PortOpen = False
```

Fig. 6. The abstract of the related interface program built in the server PC.

4 The program design for serial communication between the server PC and the client PC

The TCP/IP communication between the server pc and the client PC is performed using the WINSOCK object [15, 16, 17, 18, 19]. The host (server PC) can also be connected simultaneously by the client's PCs. The interface of the client PC is depicted in Fig. 7. The related communication diagram is shown in Fig. 8. The communication process for the server port and the client port is described below.

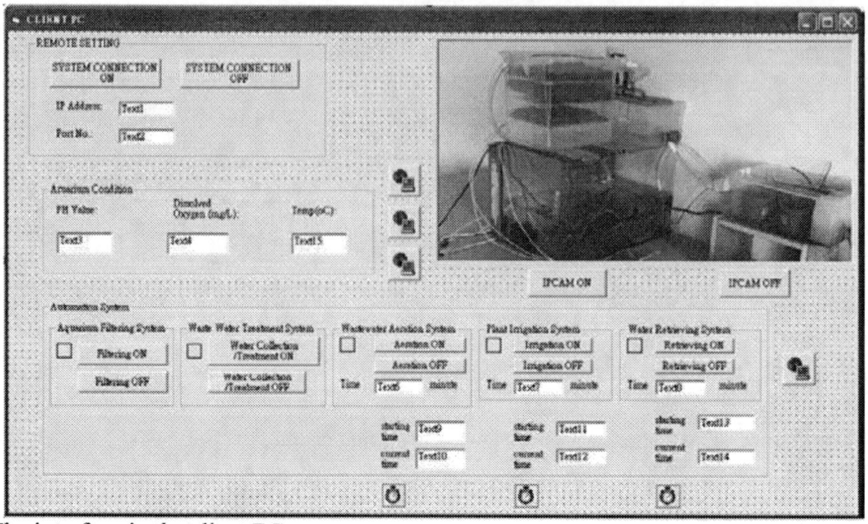

Fig. 7. The interface in the client PC.

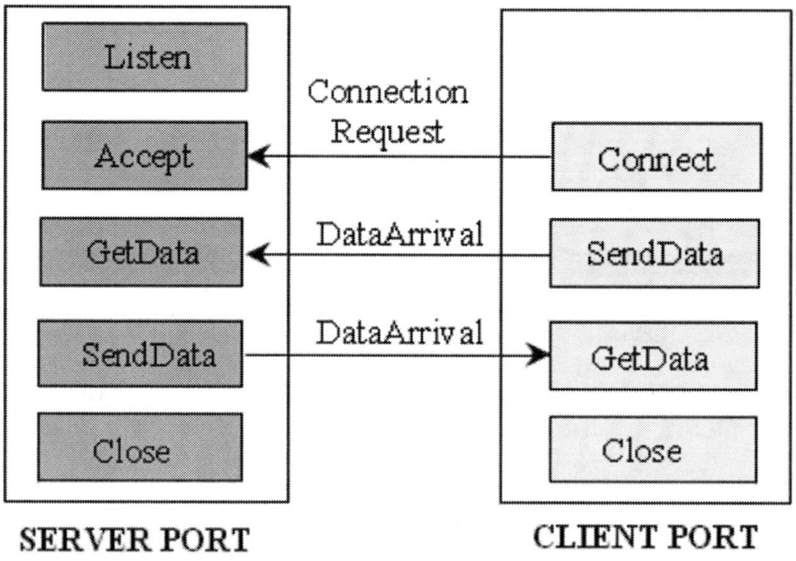

Fig. 8. The communication process between the client port and the server port.

4.1 The communication process (server port)

Four steps performed in the communication process are

(1) Listen:
Because the server port will establish a connection platform for the client port, a condition of "Listen" for the server PC is required. The server port waits for the connection from the client port when "Listen" is ready.

(2) Accept:
The server port will receive an event ConnectionRequest from the client port when the client port submits a "Connect" for an internet connection. The server port also accepts the connection request from client port using "Accept" via another WINSOCK.

(3) GetData:

The server port receives an event "DataArrival" from the client port when the client port sends data back to the server port. The server port will also receive data by using "GetData."

(4) SendData:
The server port sends data to the client port by using "SendData."

4.2 The communication process (client port)

Four steps performed in the communication process are

(1) Connect:
A connection request will be submitted from the WINSOCK of the client port by using "Connect." The WINSOCK will assign the server port's "Port Number" (at "Listen") to be the value of its "Remote Port."

(2) SendData:
The client port will send a data set to the server port using "SendData" when the internet connection between the server port and the client port is established.

(3) GetData:
An event "DataArrival" will be generated from the client port when the server port sends a data set to the client port. The client port can receive the data using "GetData."

(4) Close:
The end connection using "Close" is required when the communication is finished.

5 Results and discussions

5.1 Results

As indicated in Fig. 1, a prototype of an automatic waste water treatment, aeration, and retrieval monitoring/controlling system is established and shown in Fig. 9. A PLC serves as a controller to control various actuators (two water pumps, two solenoid valves, and one aerating motor). Water pump#1 is used to filter the aquarium water. To avoid the overflow of the waste water tank, a water level gauge installed inside the waste water tank will detect the water level when the aquarium water is conveyed to the tank via the opened solenoid valve #1. In addition, in order to nitrify the waste water for further plant irrigation, nitrifying bacteria will be added inside the waste water tank. Also, to purify the waste water, an aeration process using an aerating motor inside the tank is adopted. To monitor the status of the aquarium online, three kinds of sensors (a pH value detector, a dissolved oxygen detector, and a thermal detector) connected to the server device are used. The nitrified water will be conveyed to the plant by opening solenoid valve #2. After the irrigation process, the water filtered by the soil will be conducted to the water recycling tank and re-pumped up to the aquarium via water pump #2. To realize the current status of the system, a remote distributing, monitoring, and controlling network in conjunction with an IP CAM via the TCP/IP network is built and shown in Fig. 7. To assure the operation of the automatic waste water treatment, aeration, and retrieval monitoring/controlling system, a visual monitoring system using the IP CAM is adopted.

Fig. 9. A prototype of the automatic waste water treatment, aeration, and retrieval monitoring/controlling system.

5.2 Discussions

The main issues concerning a PC-based automatic waste water treatment, aeration, and retrieval monitoring/controlling system follow:

(1) Aquarium filtering, waste water treatment, waste water aeration, plant irrigation, and water retrieval.

(2) Automation for processing the aquarium water to be used for plants and to be reused in the aquarium.

(3) Online monitoring of the status of the aquarium's pH value, the dissolved oxygen, and the temperature.

(4) Establishing the client-server control system between the server port and the client port.

(5) Saving manpower on site by using an interface between the server port and the system.

(6) Establishing system's hardware and setting up a communication format between the PLC and the server PC.

(7) Assuring the operation of the automatic waste water treatment, aeration, and retrieval monitoring/controlling system using the visual monitoring system of IP CAM.

6 Conclusions

It has been shown that to reduce manpower and to nitrify/aerate and retrieve aquarium water, a PC-based automatic waste water treatment, aeration, and retrieval monitoring/controlling system is necessary. This network is utilized using an inquiry between the server PC and the client PC via the TCP/IP network. To monitor the quality of aquarium water online, three kinds of sensors (a pH detector, a dissolved oxygen detector, and a thermal detector) are adopted and connected to the server device. Moreover, to realize the current status of the system, a remote distributing, monitoring, and controlling network in conjunction with an IP CAM via the TCP/IP network is built. Consequently, to assure on-site operation of an automatic waste water treatment, aeration, and retrieval monitoring/controlling system, a visual monitoring system of an IP CAM is established.

References

1. M. Rosegrant, X. Cai, S.A. Cline, *World water and food for 2025: dealing with scarcity* (Int. Food Policy Research Inst., 2002)
2. M. Rosegrant, S. Msangi, C. Ringler, T.B. Sulser, T. Zhu, S.A. Cline, *International model for policy analysis of agricultural commodities and trade (IMPACT): model description* (Int. Food Policy Research Inst., 2008)
3. IPCC, *Climate change 2013: the physical science basis. working group I contribution to the fifth assessment report of the intergovernmental panel on climate change* (Cambridge University Press, 2013)
4. FAO, *Aquastat – the FAO's global water information system* (Food and Agri. Organization of the United Nations, FAO, 2010)
5. E. Ostrom, *The rudiments of a theory of the origins, survival, and performance of common property institutions* (In: Bromley D (ed) Making the Commons Work: Theory, Practice, and Policy. ICS Press, 1992)
6. C. Perry, Irrigation Drain, **56** (2007)
7. T. Zhu, C. Ringler, *Climate change implications for water resources in the limpopo river basin* (Int. Food Policy Research Inst., 2010)
8. M. Addington, D.L. Schodek, *Smart Materials and Technologies for the Architecture and Design Professions* (Elsevier, 2004)
9. L.W. Chang, *The mechanism design of ankle-joint lower limps for a apoplexy patient* (Eng. Sci. & Tech. Bulletin, NSC Report, 1999)
10. G.W. Chen, C.H. Chang, C.M. Chiu, Int. Symp. on biomed. Eng. (2004)
11. K.C. Wang, K.C., *A study on design and control of oscillating actuator fabricated by shape memory alloy* (Master thesis, Tatung University, 2004)
12. Y. Saito, K. Kikuchi, H. Negoto, T. Oshima, T. Haneyoshi, T., *Development of externally powered lower limb orthosis with bilateral-servo actuator*. (IEEE 9th Int. Conf. on Rehabilitation Rob., 2005)
13. M.C. Chiu, Inf. Tech. J., **9**(7) (2010)
14. American Fisheries Society, *Texas farm ponds: stocking, assessment, and management recommendations* (the American Fisheries Soc., 2005)
15. H.C. Cheng, M.C. Chiu, C.L. Huang, P.C. Chuo, Inf. Tech. J., **11**(6) (2012)
16. H.C. Cheng, C.B. Chiou, M.C. Chiu, J. Appl. Eng. Tech., **2**(1) (2014)
17. H.C. Cheng, M.C. Chiu, Appl. Mech. and Mat., **336-338** (2013)
18. H.C. Cheng, M.C. Chiu, J. Inf. & Opt. Sciences, **34**(6) (2013)
19. M.C. Chiu, H.C. Cheng, J. Interdisciplinary Math., **18**(5) (2015)

A wireless rainwater collecting, filtering, and storing system using a cell-phone

An-Bang Cheng[1], *Min-Chie* Chiu[2,*], and *Che-Min* Chiu[3]

[1]Department of Industrial Education and Technology, National Changhua University of Education, Changhua City, Taiwan, R.O.C.
[2]Department of Mechanical and Automation Engineering, Chung Chou University of Science and Technology, Taiwan, R.O.C.
[3]Institute of Biomedical Engineering, National Tsing Hua University, Taiwan, R.O.C.

Abstract. Because of the greenhouse effect, global climate considerably changes and will worsen the water supply situations in future. Also, as can be seen in traditional reservoir, the crisis of water lacking will exacerbate due to the reservoir sedimentation. In order to widen the storage of water resources, a design of portable rain-collecting system used for irritation and daily use for people is presented. To facilitate the monitoring and manipulating, a rainwater collecting and storing system with green energy/automation /monitoring/control function is established. Considering the electricity shortage in remote island/mountain regions, two kinds of green energy generators (wind electrical generator and solar-board electrical device) are built and linked with the system. Here, a rain detector installed onto the roof can actuate the rainwater collecting board to raise and to collect water when raining event occurs. For the purpose of water purity, water filter is also set at the entrance of the water tank. In order to efficiently store/use the rainwater, several water storage tanks in conjunction with ecological pool are preset and adjusted via the water level gauge sensors and the pipe's solenoid valve. To meet the weather-proof ability, PLC is adopted as the controller. Moreover, a human-machine-interface is login on the internet and remotely communicated with the cell phone.Consequently, a prototype of the cell-phone base automatic rainwater collecting and storing system is established.

1 Introduction

Water is essential to the human life. The supply of potable water (from natural sources, desalination or importation) needs to provide at least 100 m^3 /year per person [1]. Previous studies from Rosegrant [2, 3] indicated that the effects of climate change on hydrology and water resources are tremendous. It will lead to decreasing precipitation rates and increase drought events [4]. Thus, water management and infrastructure changes have been expected to improve the situation if current climate conditions continue into the future [5]. On a global view, 70 % of the freshwater resources are used for agricultural purposes, and only

[*]Corresponding author: minchie.chiu@msa.hinet.net

19 % for industrial and 11 % for domestic purposes (including drinking, washing, food preparation and sanitation) [6]. According to Ostrom report [7], there is an almost threefold increase in the total area of irrigated agriculture during the past century. Perry's study [8] also indicated that 20 % of total cropland (irrigated agriculture) produces more than 40 % of the world's total agricultural output. Therefore, expansion of irrigated agricultural areas is essential in nation-building processes [9]. Attempts to reuse domestic wastewater for diverse purposes, such as for agricultural irrigation, have proven to be successful in many parts of the world already [10]. Under current recycling technology, each m^3 allocated for domestic use provides between 0.6 m^3 and 0.65 m^3 of treated (recycled) water [11].

In addition, desalinated water has been regarded as one of the main sources of domestic and industrial water [12, 13, 14]. Israel reacted to the water crisis with an enforced building of seawater desalination facilities and wastewater treatment plants for the reuse of water and with extensive structural and price reforms. The challenges and benefits of those reforms are discussed in Bismuth *et al.* [15]. However, because desalinated water has an impact on the quality of water system, how to reduce salts to very low levels becomes a crucial issue [16, 17]. Beside, it has been seen that salt water intrusion to the coastal aquifers will considerably decrease the groundwater levels and cause the rising of salinity levels [18]. Also, a continuously lowered water level will lead to rapid salinity due to the penetration of saline water from underground sources into the aquifer [19].

Moreover, groundwater resources have been regarded as one of the essential water resources [20]. But, for a region that has a high evaporation rate and surface water runoff in mountains areas, only 10 to 14 % of the total precipitation percolates to recharge the shallow groundwater aquifers [21]. Furthermore, for aquaculture farmers on an island, the over-load pumping of underground water that is widely used in large outdoor pools [22] results in sinkholes which may ruin the aquaculture farm. For example, ten percent of Taiwan's level land is sinking, some seriously, particularly in Ilan, Changhua, Yunlin, Chiayi and Pingtung Counties. As water resources are decreasing, water may be priced according to its cost of supply in future [23]. Pricing water in this way may influence farmers' choices of crops and will encourage them to switch irrigation water away from expensive natural water into marginal (recycled and saline) water.

Nevertheless, in order to overcome this drawback, a portable rain-collecting system used in irritation and daily use for people is presented. Considering the electricity lack at remote mountain regions and island, two kinds of green energy supply using wind electrical generation and solar board electrical generation are planned and linked to the system. Here, the system's rain detector attached onto the roof can actuate the rainwater collecting board to raise and efficiently collect water. To assure the rainwater cleanness, a filtering device installed at the entrance of first water tank is applied. In addition, to efficiently collect rainwater, several water storage tanks in conjunction with an ecological pool will be set up in advance and flexibly adjusted via the signal of tanks' level gauge sensors and the pipe's solenoid valve. Because of weather-proof characteristic, PLC is adopted as the controller. Moreover, a human-machine-interface used to login the internet and to remotely communicate with the cell phone is established. Consequently, the system having automation function and remote controlling ability can provide manpower saving and promote the rainwater storage efficiency.

2 Automatic rainwater collecting and storing system

The structure of automatic rainwater collecting and storing system is depicted in Fig. 1. Also, the related wiring diagram of the PLC connection is illustrated in Fig. 2. In order to remotely manipulate the rainwater collecting and storing system via the human/machine interface (server), the auxiliary relays defined in Table 1 have been added into the PLC's ladder program shown in Fig. 3. Fig. 3 indicated that the auxiliary relays remarked in red can be in place of the button functions in PLC. Therefore, the human/machine interface (server) connected to the PLC can submit the command to actuate the related operation functions of the automatic rainwater collecting and storing system via the touching panel.

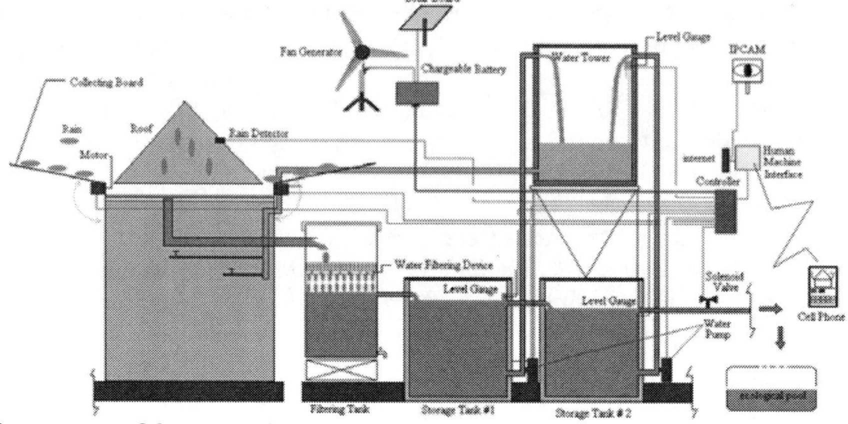

Fig. 1. The structure of the automatic waste water treatment, aeration, and retrieval system for a hybrid aquarium/plant system.

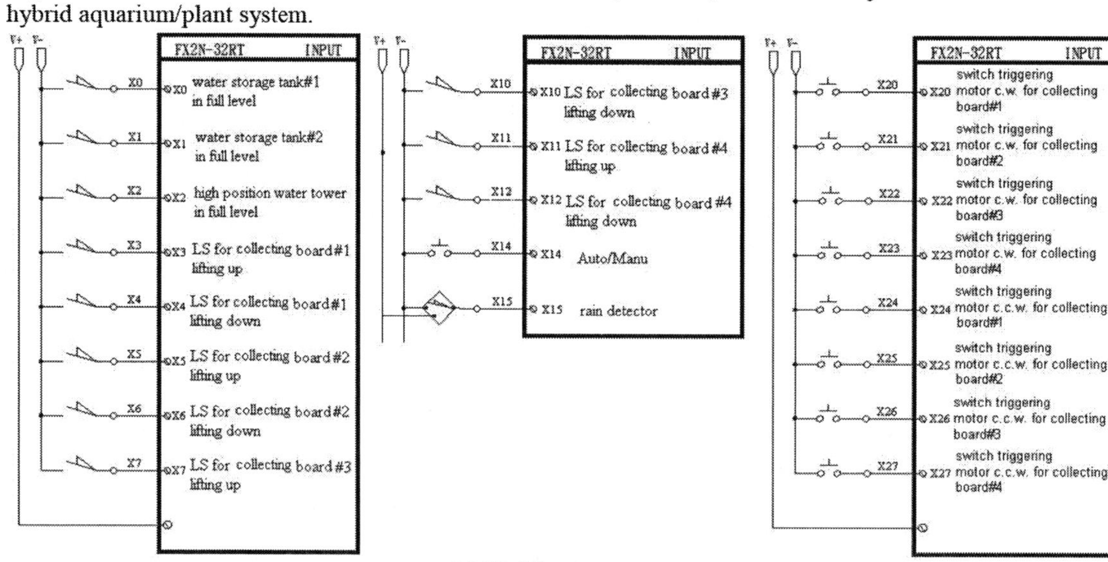

(a) PLC input.

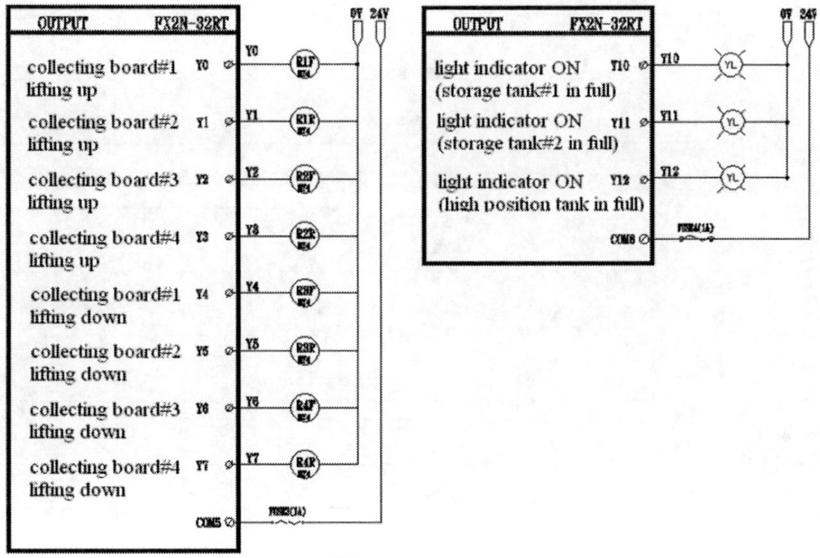

(b) PLC output.

Fig. 2. The related wiring diagram of the PLC connection.

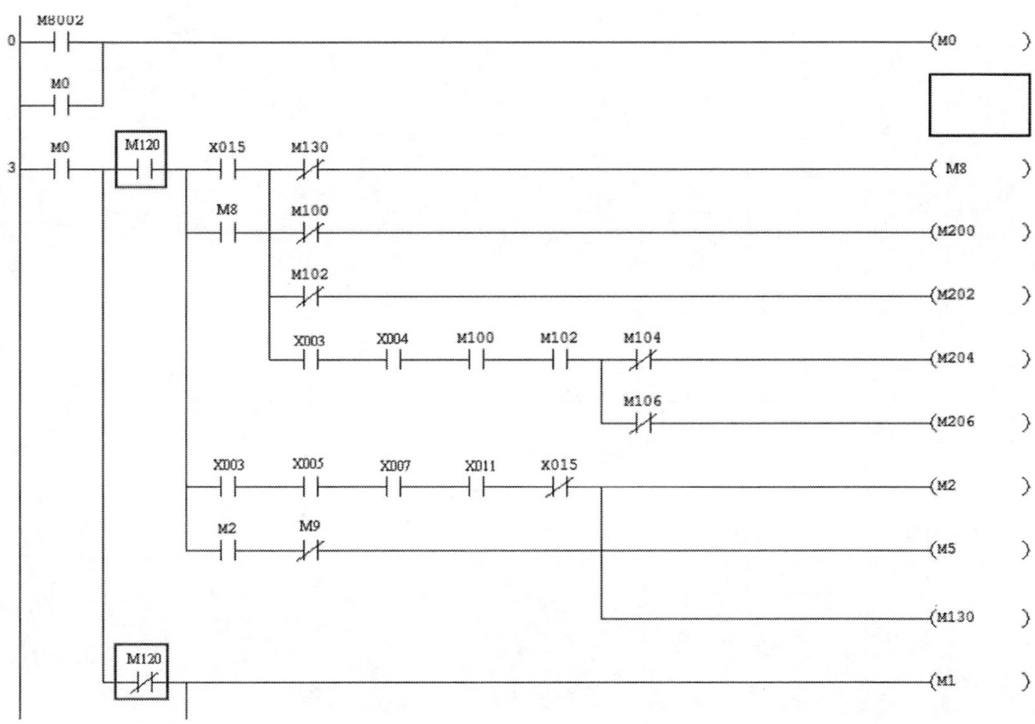

https://doi.org/10.1051/matecconf/201818500039

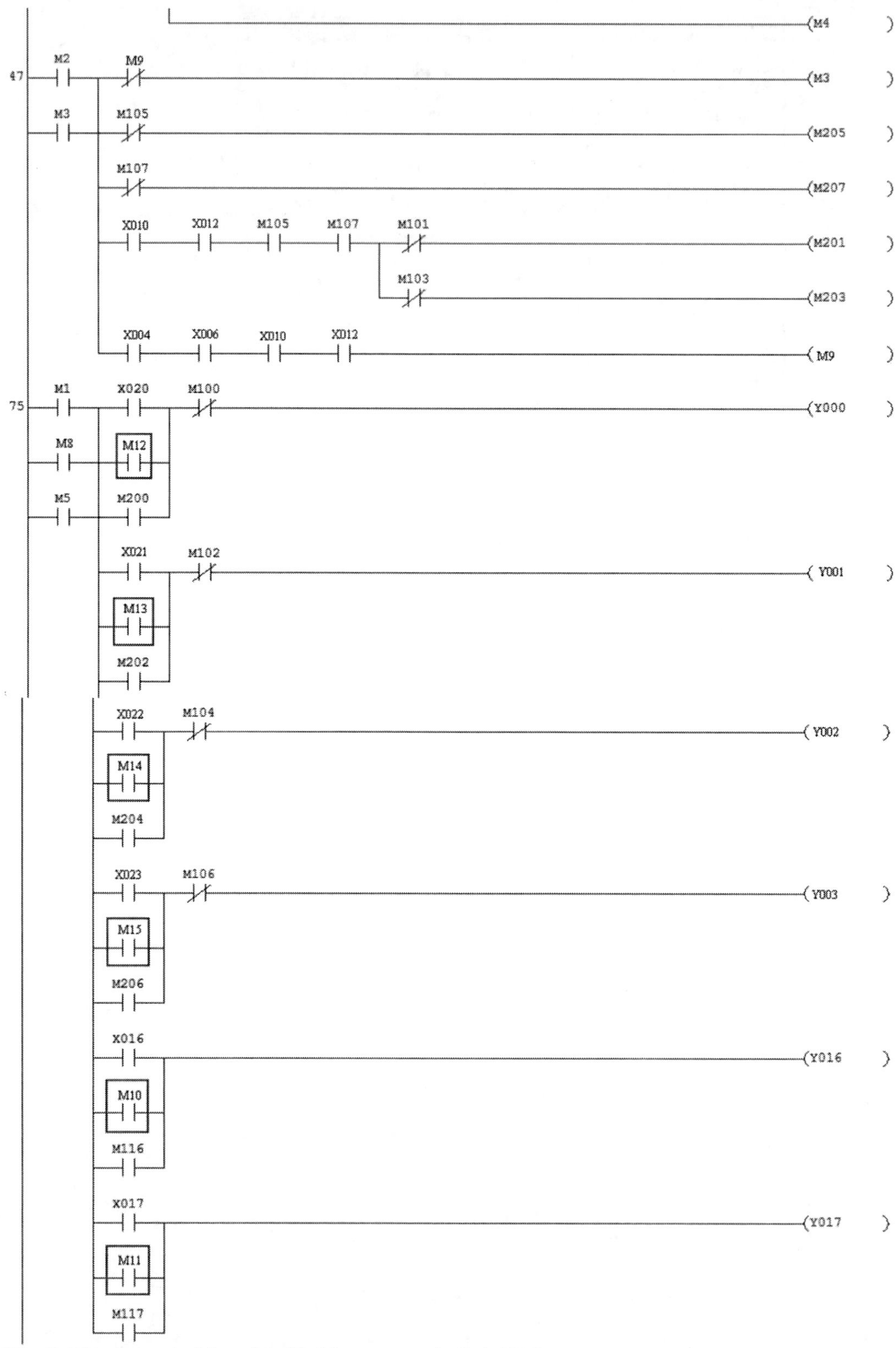

Fig. 3. The abstract of the related ladder program built in PLC.

Table 1. The auxiliary relays used to remotely control the action of the PLC program.

Auxiliary Relay Number	Action	PLC Program
M10	Water pump#1 On/Off	SET，RST M10
M11	Water pump#2 On/Off	SET，RST M11
M12	collecting board#1 Up	SET，RST M12
M13	collecting board#2 Up	SET，RST M13
M14	collecting board#3 Up	SET，RST M14
M15	collecting board#4 Up	SET，RST M15
M16	collecting board#1 Down	SET，RST M16
M17	collecting board#2 Down	SET，RST M17
M18	collecting board#3 Down	SET，RST M18
M19	collecting board#4 Down	SET，RST M19
M120	System Start Auto/Manu	SET，RST M120

As indicated in Fig. 1 and Table 1, a PLC used as a controller will actuate the actuators including four motors (for collecting board#1~#4), two water storage pumps (for water storage tank #1~#2), and one solenoid valve (for inducing superfluous rainwater to the ecological pool). The related wiring diagram of motors is depicted in Fig.4.

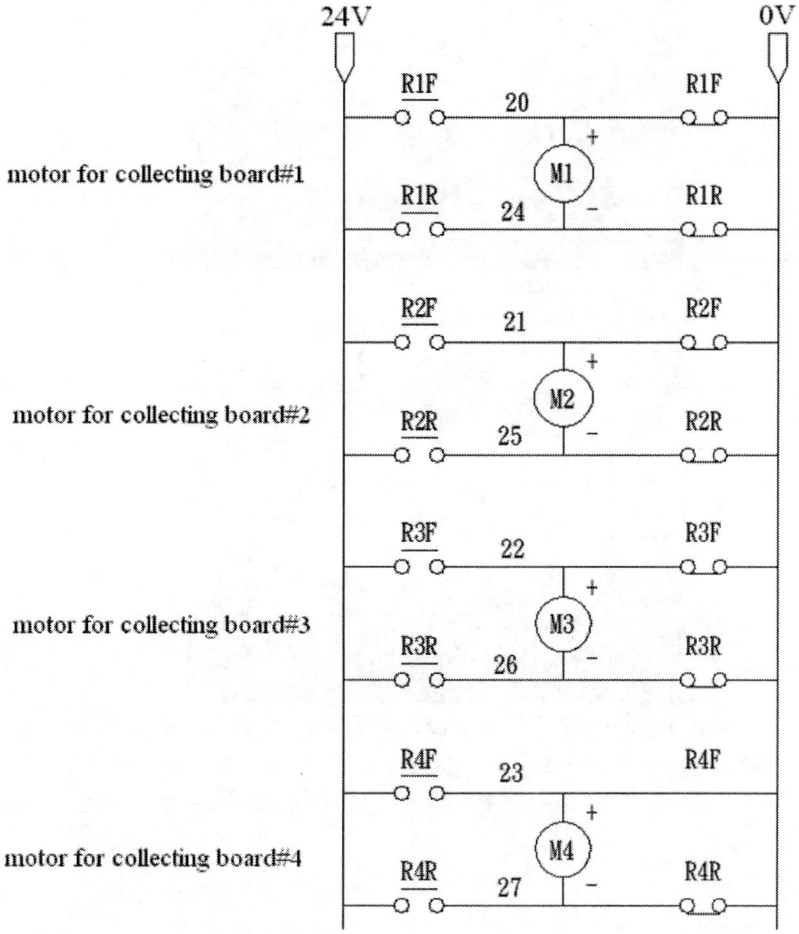

Fig. 4. The related wiring diagram of the motors.

In order to overcome the electricity lack at remote island/mountain regions, two kinds of green energy (wind electrical generator and solar board) that can provide the necessary electrical power for people and system are applied and linked to the system. A rain detector attached onto the roof is also used for detecting the raining event. The collecting boards driven by their motors can be rotated clockwise/counter clockwise depending on the raining situation. In addition, in order to allocate the upper and lower bound for each rain-collecting board during board-lifting process, one pair of limit switches (LS) is installed at both the top and bottom sides. To realize the water levels inside the storage tank's water, three level gauges (two for water storage and one for high position water tower) are equipped. With the signals emitted from the sensors, the rainwater can be efficiently stored. In order to remotely monitor/control the rainwater collecting and storing system, a human/machine interface shown in Fig.5 is login an internet. Also, an IPCAM used to capture the site vision is set up in the system. As indicated in Fig. 6, the internet of the human/machine interface is preset. And, a wireless communication between the human/machine interface (server) and the cell phone (client) via the internet protocol is also established. Here, the human/machine interface (server) shown in Fig. 7 is used to implement the system by sending a command to the PLC controller via a touching panel's buttons.

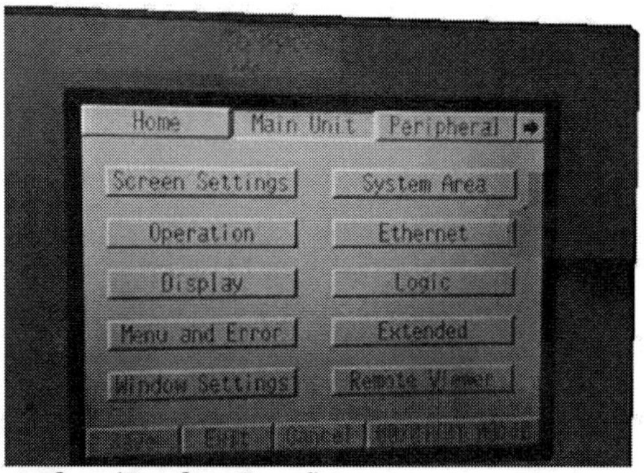

Fig. 5. A human/machine interface (Pro-face Brand).

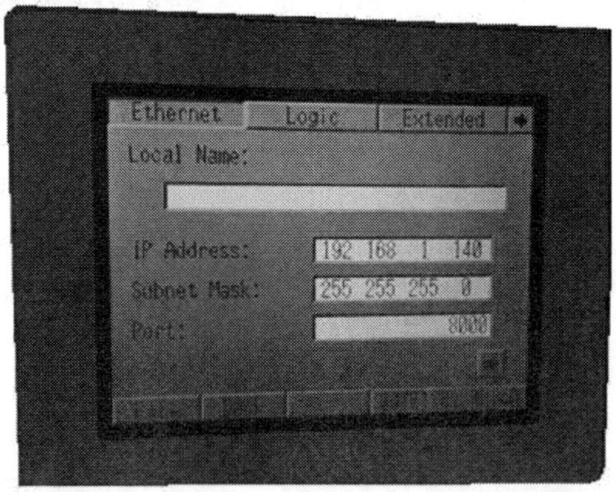

Fig. 6. The internet setting of a human/machine interface (Proface Brand).

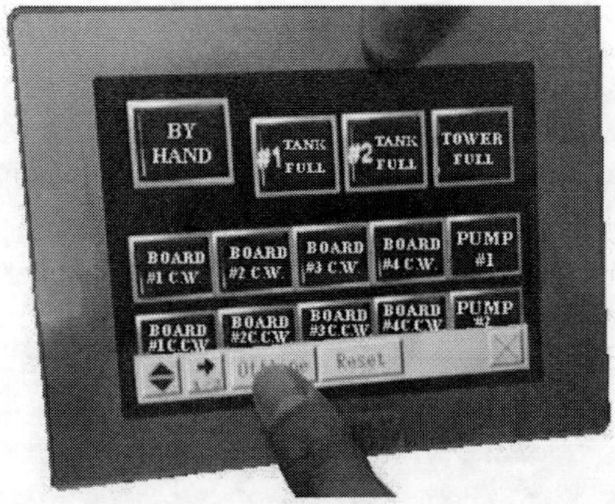

Fig. 7. The completed function setting of a human/machine interface (Pro-face Brand).

To facilitate communication between the PLC and the human/machine interface, a RS232/RS422 protocol is used. The picket format for the MISTISUE FX2n-PLC primary port's communication is illustrated in Table 2. And, the definition of the picket format is depicted in Table 3. Consequently, the command description of the MISTISUE FX2n-PLC extended port's communication protocol type 1 is shown in Table 4.

Table 2. The communication packet format of MISUBISHI FX2n-PLC's primary port.

Header code	Command code	Address and data	End code	Inspection code
STX(02)	CMD	DATA	ETX(03)	Checksum

Table 3. The definition of communication packet for MISUBISHI FX2n-PLC's primary port.

Command category	Object	Command code	Function description
Read	X,Y,M,S,T,C,D	0	Read the object value
Write	Y,M,S,T,C,D	1	Write data to the object
Forcing on	Y,M,S,T,C	7	Forcing node to be on
Forcing off	Y,M,S,T,C	8	Forcing node to be off

Table 4. The parameter of the extended port's communications protocol for the MISUBISHI FX2n-PLC.

D8120Bit	Name	Content
15	Protocol type	0: type 1, 1: type 4
14	protocol	0: No, 1: Yes
13	Inspection code	0: manipulate, 1: auto
12	Type of control line (II)	none
11	DTR inspection mode	none
10	Type of control line(I)	none
9	End code	none
8	Header code	none
7	Transmission speed	0011:300 0222:4800
6		0100:600 1000:9600
5		0101:1200 1001:19200
4		0110:2400
3	Stop bit	0: 1 bit, 1: 2 bits
2	Parity check	00: none 02: even
1		01: odd
0	Data bit	0: 7 bits, 1: 8 bits

3 The communication between the server and the client

In previous studies, the TCP/IP communication between server (human/machine interface) and client (cell phone) has been established [24, 25, 26, 27]. The communication process for server (human/machine interface) and client (cell phone) is also depicted in Fig. 8 and described as below.

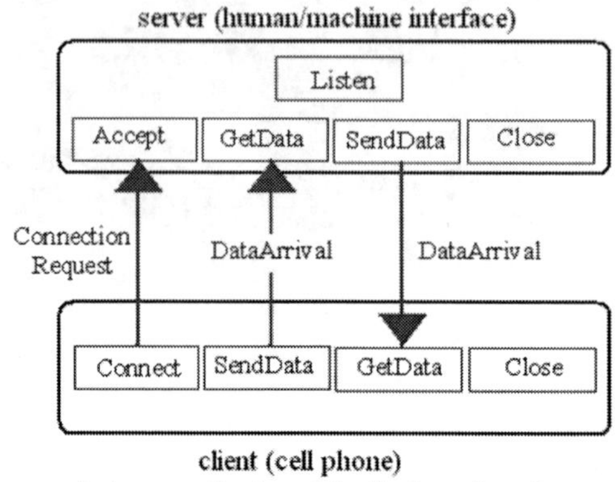

Fig. 8. The communication process between client port (cell phone) and server port (human/machine interface).

3.1 The communication process (server viewport) [28, 29]

Four steps used in the server's communication process are

(1) Listen:
Because the server port will establish a connection platform for the client port, the condition of "Listen" for the server port is needed. The server port waits for the connection from the client port when "Listen" is ready.

(2) Accept:
The server port will receive an event ConnectionRequest from the client port when the client port submits a "Connect" for a further internet connection. The server port also accepts the client's connection request using "Accept" via another WINSOCK.

(3) GetData:
The server port receives an event "DataArrival" from the client port when the client port sends data back to the server port. At the present time, the server port will also receive data by using "GetData."

(4) SendData:
The server port sends data to the client port by using "SendData."

3.2 The communication process (client viewport) [28, 29]

Four steps used in the client's communication process are

(1) Connect:
The WINSOCK in the client port submits a connection request using "Connect." At this time, the WINSOCK will assign the server port's "Port Number" (at "Listen") to be the value of its "Remote Port."

(2) SendData:

The client port sends the data set to the server port using "SendData" when the internet connection between the server port and the client port is built.

(3) GetData:

The client port will generate an event "DataArrival" when the server port sends a data set to the client port. For the moment, the client port can receive the data using "GetData."

(4) Close:

The end connection using "Close" is needed when the communication is finished.

4 Results and discussions

4.1 Results

As indicated in Fig. 1, a cell-phone base automatic rainwater collecting and storing system is established. The interface of the cell phone is shown in Fig. 9. The prototype of the automatic rainwater collecting and storing system is also depicted in Fig. 10. A PLC is used as a controller to control various actuators, including four panel-lifting motors, two water pumps, and a solenoid valve. One rain detector is equipped to detect the raining event. Four pairs of limit switches installed around four pieces of the rain-collecting boards are applied to detect/confine the board position. Three sets of level gauges installed inside the water tanks and high position water tank are used to online detect the water levels. According to the water levels, water pumping process and excessive rain discharging process (to the ecological pool) can be automatically initialized. A communication between the human/machine interface (server) and the rainwater collecting/storage system is made via the RS232/RS422 protocol to the PLC. The human/machine interface (server) can implement the system function by sending a command to actuate the PLC via triggering the auxiliary relays programmed in the ladder program. Moreover, to realize current rainwater collecting and storage status, a cell phone having remote distributing, monitoring, and controlling function is built and shown in Fig. 9.

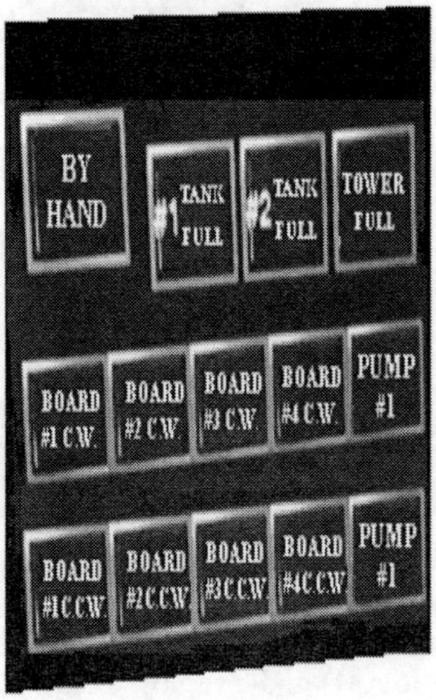

Fig. 9. The complete function setting of a cell phone.

Fig. 10. A prototype of the automatic rainwater collecting and storing system.

4.2 Discussions

The main issues concerning a cell-phone base automatic rainwater collecting and storing system have

(1) Water and green electricity supplement.
(2) Automation in rainwater collection and storage
(3) Manpower saving.
(4) Efficient rainwater collection and storage using level gauge sensors and solenoid valve.
(5) Establishing a client-server control system between the human/machine interface and the cell phone.
(6) Assuring the operation of rainwater collecting and storage using a visual monitoring system (IP CAM).

5 Conclusions

It has been seen that to overcome the electricity lack at remote island/mountain, two kinds of green energy generators (wind generator and solar board generator) have been built and linked to the water collecting/storage system. In order to facilitate the water collecting/storage operation and improve the rainwater collecting and storage efficiency, a remote distributing, controlling and monitoring network that analyzing the online status (such as the raining event, the upper/lower limit switches for the lifting boards, and the water levels) of system and triggering related actuators (motors for lifting board, motors for water pumps, and solenoid valve for pipe outlet) is established. Here, this network is also utilized using an inquiry between the human/machine interface (server) and the cell phone (client) via the TCP/IP network.

Consequently, an automatic rainwater collecting and storage system equipped with a visual monitoring device (IP CAM) and a remote cell-phone control is established.

References

1. P. H. Gleick, Water Int., **21** (1996)
2. M. Rosegrant, X. Cai, S.A. Cline, *World water and food for 2025: Dealing with Scarcity* (Int. Food Policy Research Inst., 2002)
3. M. Rosegrant, S. Msangi, C. Ringler, T. B. Sulser, T. Zhu, S. A. Cline, *International model for policy analysis of agricultural commodities and trade (IMPACT): model description* (Int. Food Policy Research Inst., 2008)
4. IPCC, *Climate change 2013: the physical science basis. working group I contribution to the fifth assessment report of the intergovernmental panel on climate change* (Cambridge University Press, 2013)
5. T. Zhu, C. Ringler, *Climate change implications for water resources in the limpopo river basin* (Int. Food Policy Research Inst., 2010)
6. FAO, *Aquastat – the FAO's global water information system* (Food and Agriculture Organization of the United Nations, FAO, 2010)
7. E. Ostrom, *The rudiments of a theory of the origins, survival, and performance of common property institutions* (In: Bromley D (ed) Making the Commons Work: Theory, Practice, and Policy. ICS Press, 1992)
8. C. Perry, Irrigation Drain, **56** (2007)
9 H. Kreutzmann, *From upscaling to rescaling – the fergana basin's transformation from tsarist irrigation to water management for an independent uzbekistan* (In : Huettl RF, Bens O, Bismuth C, Hoechstetter S (eds) Society – water – technology: a Critical Appraisal of Major Water Engineering Projects. Springer, 2015)
10. G. Oron, C. Campos, L. Gillerman, Agriculture Water Management, **38** (1999)
11. A. Cohen, D. Feiman, M. Ladal, *Sewage collection and treatment for irrigation: national survey 2006/2007/2008* (http://siteresources.worldbank.org/EXTREDSEADEADSEA/Resources/5174616-1416839444345/RSDS-Summary_of_Final_FS_Report.pdf)
12. A. Mangoosh, *Integrated vision for water resources environment for UAE. office of H.H. the President* (Department of Water Resources Studies, Abu Dhabi, UAE, 2004)
13. M. Mohamed, A. Almualla, Water Resources Management, **24**(14) (2010)
14. M. N. S. Mohamed, M. Sherif, Bulletin of Environmental Contamination and Toxicology, **84**(4) (2010)
15. C. Bismuth, B. Hansjurgens, I. Yaari, Technologies, *Incentives and cost recovery: is there an Israeli role model?* (In: Huettl RF, Bens O, Bismuth C, Hoechstetter S (eds) Society – water – technology: a Critical Appraisal of Major Water Engineering Projects. Springer, 2015)
16. N. Kedmi, N., *Integrated water resources management in Israel* (2005) (http://www.unece.org/fileadmin/DAM/env/water/meetings/payment_ecosystems/Discpapers/Israel.pdf)
17. R. F. Hüttl, O. B. C. Bismuth, S. Hoechstetter, *Water resources development and management* (Springer, 2016)
18. D. Markel, U. Shamir, P. Green, *Operational management of lake kinneret and its watershed* (In: Zohary T, Sukenik A, Nishry A (eds) Lake Kinneret. Ecology and Management. Springer, 2014)
19. M. Haran, R.. Samuels, S. Gabbay, Israel J. Chem., **42** (2008)
20. M. M. Mohamed, N. Al-Suweidi, A. Ebraheem, M. Al Mulla, J. of Water Resource and Hydraulic Eng., **4**(4) (2015)
21. ESCWA, *Water scarcity in the Arab world* (Population and Development Report. Beirut, Lebanon, 2003)

22. M. Addington, D. L. Schodek, *Smart materials and technologies for the architecture and design professions* (Elsevier, 2004).
23. Y. Tsur, Annual Rev. Resour. Econ., **1** (2009)
24. H. C. Cheng, M.C. Chiu, C.L. Huang, P.C. Chuo, Inf. Tech. J., **11**(6) (2012)
25. H. C. Cheng, M.C. Chiu, Appl. Mech. and Mat., **336-338** (2013).
26. H. C. Cheng, M.C. Chiu, J. Inf. & Opt. Sciences, **34**(6) (2013).
27. M. C. Chiu, H.C. Cheng, J. Interdisciplinary Math., **18**(5) (2015)
28. H. C. Cheng, C.B. Chiou, M.C. Chiu, J. Appl. Eng. Tech., **2**(1) (2014)
29. H. C. Cheng, M.C. Chiu, I. J. on Smart Sens. and Intel. Sys., 7(3) (2014)

Investigation on rotating compression of porous cylinder under constant shear friction using FEM simulation

Gow-Yi Tzou[1], *Yeong-Maw* Hwang[2], and *Sai-Chih* Pan[2,*]

[1]Department of Mechanical and Automation Engineering, Chung Chou University of Science and Technology, Yuanlin, Changhua County 510, Taiwan, R.O.C.
[2]Department of Mechanical and Electro-Mechanical Engineering, National Sun Yat-Sen University, Kaohsiung 804, Taiwan, R.O.C.

Abstract. In this paper, a series of systematic studies on the rotating compression of porous cylinders by the FEM using DEFORM-3D analysis software are carried out. For a series of simulations, the forming conditions such as (1) the angular velocities from small to large are 0, 0.1, 0.25, 0.5, and 1 (rad/sec), respectively;(2) aspect ratio of porous cylinder (height-diameter ratio) are 0.75, 1, and 1.25, respectively;(3) the frictional factors are 0.3 and 0.7. Through different combinations of the above-mentioned variation factors, the characteristics of the rotating compression of porous cylinder can be explored. These rotating forming characteristics such as the compression force, the bulging ratio, the change in density, the rotating torque and so on can be investigated. The rotating compression characteristics of both porous and sound cylinders reveal lots of similarities and some differences. The main conclusions of this study are as follows: (1) as the rotating speed increases fast, the compression force of porous cylinder is whereas reduced. This phenomenon is different from the compression of sound cylinder; (2)the compression force under the lower aspect ratio can be reduced by the lubrication; (3)the relative density can be increased by increasing the rotating speed and reduction ratio; (4)the rotating torque is increased with higher rotating speed, higher frictional factor, and higher aspect ratio; (5)the bulging effect can be reduced by increasing the rotating speed, however it is increased by increasing the frictional factor.

1 Introduction

Porous materials can be manufactured by the process of casting or powder metallurgy. For the contracted volume of metal slabs or the constrained air within the metal slabs during the casting, there are tiny voids appeared; especially on or within the thick part easy to generate the voids. It is hard to observe the existed voids without a microscope. These tiny voids will affect the pressure resistance and strength of the products. For the products manufactured by the power metallurgy, if the technology of powder forging technology can

* Corresponding author:scpan@seed.net.tw

be used, more functional and good mechanical properties as well as the alloys in different compositions will be acquired.

In the paper of Yang et al. [1] explored the porous materials on the relationship of bulging deformation, forging force and density change during open die forging. By FEM analysis and other experiment results, they investigated the applicability of FEM simulations. Lee et al. [2] emphasized the importance of the empty hole effect and the density distribution in the study of porous materials. In addition to their guideline on plastic yield, they also compared with other guidelines on yield in various studies. Using the stress-strain curve data in the sintered copper experiment, Shima et al. [3] proposed the plasticity theory for porous metals and applied it to a friction-free closed-dies compression experiment to verify its practicability. Doraivelu et al. [4] reviewed many yield criteria proposed by various scholars. They detailed the compressible powder metallurgy materials, proposed a new yield function, and used aluminum alloy (A7091) uniaxial compression experiment to test the accuracy and practicability. Dias et al. [5] used the plasticity equation to analyze the parameters of the sintered porous materials after forged. He confirmed the experiment results agreeable to the theoretical analysis of the proposed model, which can be applied to the cold forging of sintered preforms. Jha et al. [6] detailed the experiments on forming characteristics and failure mechanisms from the start of spraying iron metal powder to the completion of sintering processing. In the forming properties, they discussed the size of powder particles in detail, and also classified the types of crack to the failure mechanisms.

Nguyen et al. [7] investigated the mechanical behavior of porous functionally graded (FG) beams, they indicated that it is necessary to take the effect of porosities into consideration in the vibration analysis of FG beams traversed by moving forces. Tzou et al. [8] used Taguchi Method to explore the wire rod drawing with rotating die under constant shear friction and further, they used DEFORM-3D commercial FEM software to simulate and obtain the effective stress, the effective strain, the velocity field, and the drawing force. Tzou et al. [9] carries out an FEM simulation on rotating compression forming of double-layer Bounded clad ring by DEFORM-3D software. Effects of various compression conditions such as the clad ring height ratio, the frictional factor ratio, the rotating angular velocity etc on the compression characteristics are explored in this study.

Tzou's 2014 study [10] clearly indicates that the rotating compression forming of double-layer ring can reduce the compressive force and make the deformation uniform. The 2017's research [11] establishes a plastic mechanics analytical model based on slab method to explore the effects of rotating effect on the characteristics of rotating compression of porous cylinder. In addition, the latest study of Tzou et al. [12] also compared the various data of traditional compression forming of porous cylinder between slab method and FEM. This study used Deform-2D to explore the compression forming without rotating die. Therefore this present research uses Deform-3D to explore the rotating compression process of porous cylinders. Besides, the variations of compression characteristics with and without rotation can be investigated.

2 FEM simulation

In this paper, the commercial software, Deform 3D, is used to simulate the rotating compression of porous cylinder. This software can provide the rigid, plastic, elastic-plastic and porous material simulation of the metal forming in large deformation. The FEM software Deform 3D enables to do the tool design and product design. The FEM simulations provide 3D analysis with the tetrahedron elements for meshing the work-piece of bulk forming and the auto re-mesh step is enveloped in the software package if the case is occurred. Using the FEM software, it can reduce the time and cost consumption of die and tool design significantly. The effective stress distribution, flow pattern, effective strain distribution, velocity field of work-piece and load stoke can be simulated by FEM software. These simulation results can be used to obtain the product geometric profile and material properties required. In the pre-process of modelling, the geometric profile of porous cylinder can be constructed easily. The porous material analysis in rotating compression process is chosen and the material properties such as Young's modulus, Poisson's ratio, are needed. For the first stage of exploring the rotating compression of porous cylinder, the isotropic work-hardening rule is assumed in flow rule due to plastic strain hardening. The green yield criterion is used. The initial conditions of work-piece are set up, and the contact along work-piece and dies are defined.

The top die with a downward speed is 1 mm/sec, and the bottom die with the angular velocity are 0, 0.1, 0.5, 1 rad/sec, all rigid contact bodies are assumed as insulation, and the heat conduction phenomenon during the porous cylinder is neglected. In the rotating compression of porous cylinder, the relationship between strain and stress is nonlinear. The DEFORM 3D has been developed to solve the nonlinear problem. The green yield criterion is available in DEFORM-3D commercial software. The contact bodies are regarded as rigid surface. The Deform pre-process are shown in Fig. 1.

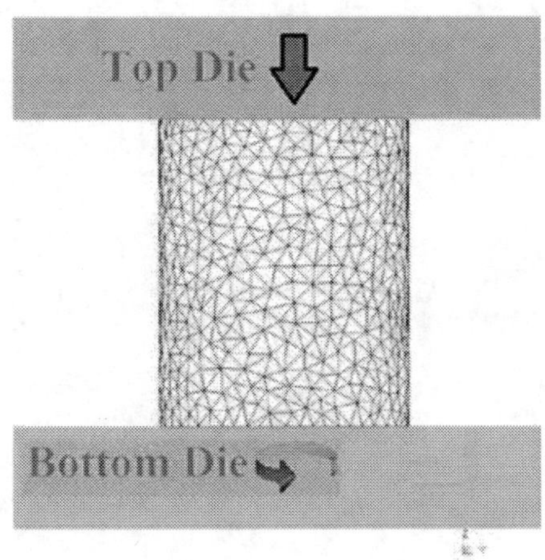

Fig. 1. The pre-process of rotating compression for porous cylinder.

3 Results and discussions

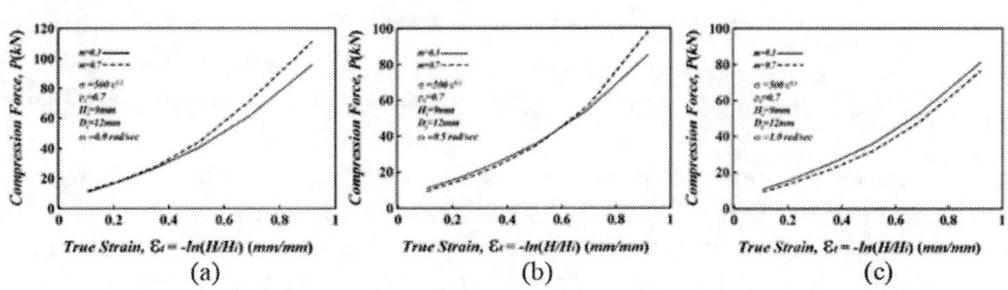

Fig. 2. Variations of the compression force with true strain for different rotating speeds (Hi=9mm, Hi/Di=0.75).

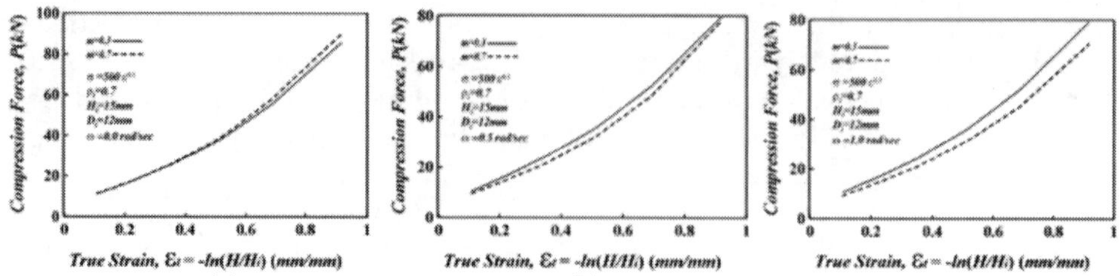

(a) (b) (c)

Fig. 3. Variations of the compression force with true strain for different rotating speeds (Hi=15mm, Hi/Di=1.25).

Figure 2 shows variations of the compression force with true strain for different rotating speeds (H_i=9mm, H_i/D_i=0.75). For the compression of sound cylinder with and without rotation, as the frictional factor increases the compression force increases. However, for the porous cylinder compression, as the rotating angular speed increases gradually, whereas the compression force with higher frictional factor is smaller than that with smaller frictional factor. Figure 3 shows variations of the compression force with true strain for different rotating speeds (H_i=15mm, H_i/D_i=1.25). From this figure, the smaller compression force for higher frictional factor under Hi/Di=1.25 due to the increase of rotating angular speed is more obvious than that for Hi/Di=0.75.

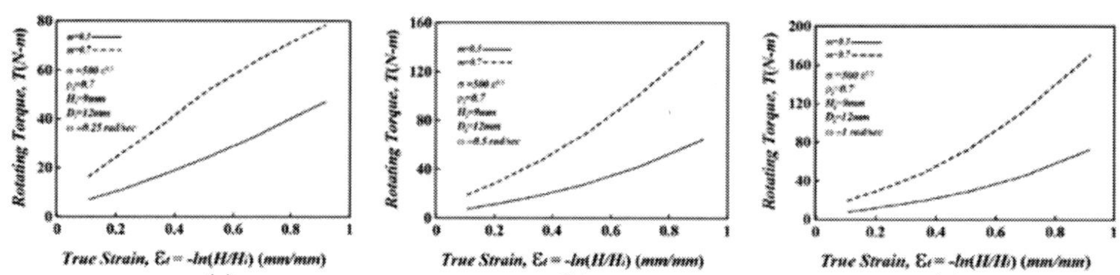

(a) (b) (c)

Fig. 4. Variations of the rotating torque with true strain for different rotating speeds (Hi=9mm, Hi/Di=0.75)

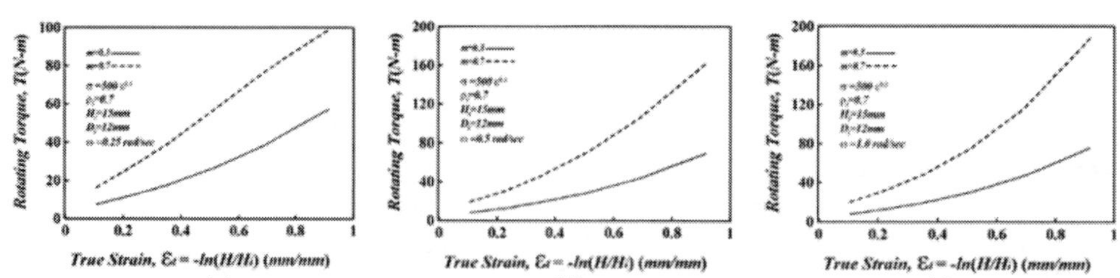

(a) (b) (c)

Fig. 5. Variations of the rotating torque with true strain for different rotating speeds (Hi=15mm, Hi/Di=1.25).

Figure 4 shows variations of the rotating torque with true strain for different rotating speeds (H_i=9mm, H_i/D_i=0.75). Figure 5 shows variations of the rotating torque with true strain for different rotating speeds (H_i=15mm, H_i/D_i=1.25). Regardless of sound cylinder or porous cylinder with and without rotation, as the frictional factor and rotating speed increase the rotating torque increases. Under the higher aspect ratio (H_i/D_i=1.25), the rotating torque is higher.

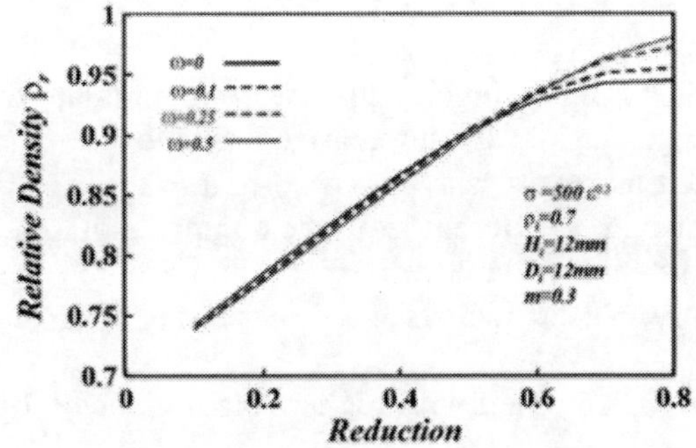

Fig. 6. The effect of different rotating speeds on relative density.

Figure 6 shows the effect of different rotating speeds on relative density. The relative density increases with the increase of reduction and rotating speed. This result can be seen in Figure 7. In Figure 7, the bulging effect can be reduced with the increase of rotating speed. And higher and uniform density can be seen under the higher rotating speed.

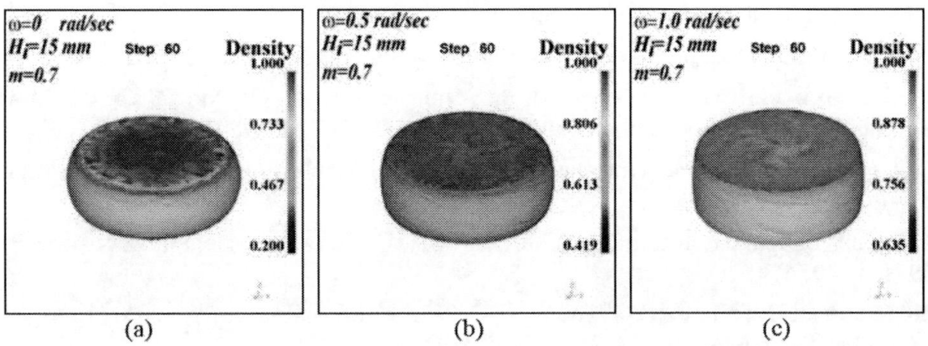

Fig. 7. Variations of density and bulging with different rotating speeds for r=60%.

Figure 8 shows bulging ratio for various frictional factors. Definition of bulging ratio is that subtraction between the maximum bulging diameter and top diameter is divided by the top diameter. As the bulging ratio is increased, that indicates the bulging effect is obvious. The bulging ratio is increased with the increase of reduction and frictional factor.

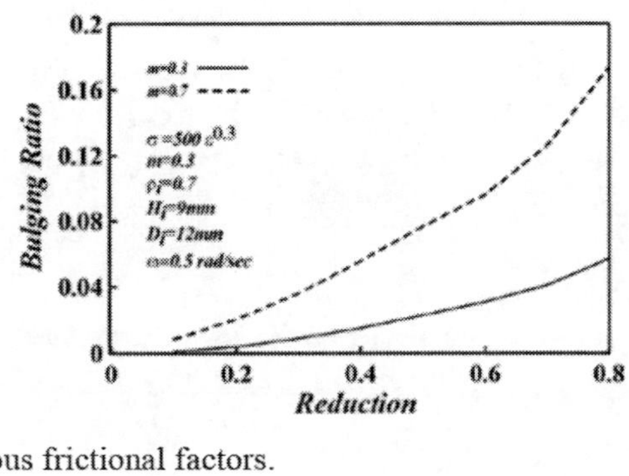

Fig. 8. Bulging ratio for various frictional factors.

4 Conclusions

This study uses Deform-3D commercial software to simulate the rotating compression of porous cylinder, the major results are summarized as below:
(1)As the rotating speed increases fast, the compression force of porous cylinder is whereas reduced. This phenomenon is different from the compression of sound cylinder.
(2)The compression force under the lower aspect ratio can be reduced by the lubrication.
(3) The relative density can be increased by increasing the rotating speed and reduction ratio.
(4) The rotating torque is increased with higher rotating speed, higher frictional factor, and higher aspect ratio.
(5)The bulging effect can be reduced by increasing the rotating speed, however it is increased by increasing the frictional factor.

References

1. T. S. Yang, Y. C. Hsu, Journal of Materials Processing Technology, **177**, 154-158 (2006)
2. D. N. Lee, H. S. Kim, Powder Metallurgy, **35**, 275-279 (1992)
3. S. Shima, M. Oyane, International Journal of Mechanical Sciences, **18**, 285-291 (1976)
4. S. M .Doraivelu, H. L. Gegel, J. S. Gunasekera, J. C. Malas, J. T. Morgan, International Journal of Mechanical Sciences, **26**, 527-535 (1984)
5. A. D. M. Dias, L. Schaeer, International Journal of Machining Tools and Manufacture, **33**, 577-586 (1993)
6. A. K. Jha, S. Kumar, International Journal of Machine Tool Design & Research, **26**, 369-384 (1986)
7. D.K. Nguyen, T.T. Tran, S.G. Buntara, V.T. Bui, International Journal of Engineering and Technology Innovation **6**, 3, 173-189 (2016)
8. G.Y. Tzou, Y.M. Ling, S.Y. Hsia, Journal of Precision Machinery and Manufacturing Technology **5**, 2, 8-16 (2015)
9. G.Y. Tzou, T.L. Chiang, C.Y. Chiu, Journal of Precision Machinery and Manufacturing Technology **2**, 2, 36-46 (2012)
10. G.Y. Tzou, Y.M. Hwang, H.Y. Teng, International Journal of Precision Engineering and Manufacturing-Green Technology **1**, 3, 191-199 (2014)
11. G.Y. Tzou, Y.M. Hwang, S.C. Pan, Transactions of the Canadian Society for Mechanical Engineering, **41**, 571-580 (2017)
12. G.Y. Tzou, Y.M. Hwang, S.C.Pan, MATEC Web of Conferences, **123**, 1-7 (2017), DOI: 10.1051/matecconf/201712300002

AUTHOR INDEX

AUTHOR INDEX

EDP Sciences – Web of Conferences
17, Avenue du Hoggar
Parc d'Activité de Courtabouf
BP 112
F-91944 Les Ulis Cedex A
France

ISBN 978-1-5108-6859-5

9 781510 868595